ROBERT I. TILLING

Volcanoes of North America

United States and Canada

Volcanoes of North America

United States and Canada

Compiled and edited by

CHARLES A. WOOD
Johnson Space Center

and

JÜRGEN KIENLE
University of Alaska

Cambridge University Press
Cambridge
New York Port Chester
Melbourne Sydney

Published by the Press Syndicate of the University of Cambridge
The Pitt Building, Trumpington Street, Cambridge CB2 1RP
40 West 20th Street, New York, NY 10011, USA
10 Stamford Road, Oakleigh, Melbourne 3166, Australia

First published 1990

Library of Congress Cataloging-in-Publication Data

Volcanoes of North America : United States and Canada /
edited by Charles A. Wood and Jürgen Kienle.

p. cm.

Includes bibliographical references.

ISBN 0-521-36469-8

1. Volcanoes – United States. 2. Volcanoes – Canada
I. Wood, Charles Arthur, 1942– . II. Kienle, Jürgen.
QE524.V66 1990
551.2′1′0973–dc20 90–1516

British Library Cataloguing-in-Publication Data

Volcanoes of North America.

1. North America. Volcanoes
I. Wood, Charles A. II. Kienle, Jürgen
551.21097

ISBN 0-521-36469-8 hardback

Contents

List of contributors *page vii*

Introduction
Charles A. Wood 1

Alaska

Volcano tectonics of Alaska
J. Kienle and C. J. Nye 8

Volcanoes of Alaska 17
Buldir; Kiska; Segula; Davidof and Khvostof; Little Sitkin; Semisopochnoi; Gareloi; Tanaga; Bobrof; Kanaga; Moffett and Adagdak; Great Sitkin; Kasatochi; Koniuji; Atka; Seguam; Amukta; Chagulak; Yunaska; Herbert; Carlisle; Cleveland and Chuginadak; Uliaga; Kagamil; Vsevidof; Recheschnoi; Okmok; Bogoslof; Makushin; Akutan; Gilbert; Pogromni; Westdahl; Fisher; Shishaldin; Isanotski; Roundtop; Cold Bay; Amak; Dutton; Emmons and Hague; Pavlof and Pavlof Sister; Dana; Kupreanof (Stepovak Bay); Veniaminof; Black Peak; Aniakchak; Yantarni; Chiginagak; Kialagvik; Ugashik and Peulik; Ukinrek; Kejulik; Martin; Mageik; Trident; Novarupta, Falling Mountain, and Cerberus; Katmai; Griggs; Snowy; Denison, Steller, and Kukak; Devils Desk; Kaguyak; Fourpeaked; Douglas; Augustine; Iliamna; Redoubt; Spurr; Hayes; Buzzard Creek; Drum; Sanford; Wrangell; Capital; Tanada; Eastern Wrangell; White River Ash; Old Crow Tephra; Edgecumbe; Duncan Canal; Tlevak Strait and Suemez Island; Behm Canal and Rudyerd Bay; St. Paul; St. George; Nunivak; Yukon Delta (Ungulungwak Hill–Ingrichuak Hill, Ingakslugwat Hills, Kochilagok Hill, Nushkolik Mountain, Ingrisarak Mountain, Nelson Island); Togiak; St. Michael; Kookooligit; Teller; Espenberg; Imuruk; Koyuk-Buckland; Porcupine; Prindle

Canada

Volcano tectonics of Canada
J. G. Souther 111

Canadian Cordillera: Volcano vent map and table
C. J. Hickson 116

Volcanoes of Canada 118
Volcano Mountain; Tuya Butte; Level Mountain; Edziza; Maitland; Hoodoo; Iskut-Unuk River Cones; The Thumb; Milbanke Sound Cones; Rainbow Range; Ilgachuz Range; Itcha; Nazko; Chilcotin Basalt; Wells Gray-Clearwater; Silverthrone; Bridge River Cones; Meager Mountain; Cayley; Garibaldi

Western United States

Volcano tectonics of the Western United States
Charles A. Wood and Scott Baldridge 147

Volcanoes of Washington 155
Baker; Glacier Peak; Rainier; Goat Rocks; St. Helens; Adams; Simcoe; Indian Heaven; Marble Mountain–Trout Creek

Volcanoes of Oregon 169
Western Cascades; Boring Lava; High Cascades (Columbia River to Mount Hood); Hood; High Cascades (East of Mount Hood to Clear Lake, Clear Lake to Olallie Butte); Jefferson; High Cascades (South of Mount Jefferson to Santiam Pass); Sand Mountain; Washington; Belknap; Broken Top; Three Sister; Bachelor; High Cascades (South of Three Sisters to Willamette Pass, Willamette Pass to Windigo Pass, Windingo Pass to Diamond Lake); Thielsen; Crater Lake; Pelican Butte; Sprague River Valley; McLoughlin; Brown Mountain; Newberry; Devils Garden; Fort Rock basin; Bald Mountain; Silicic Domes of South-Central Oregon; Iron Mountain; Silver Creek; Diamond Craters; Saddle Butte; Jordan Craters; Jackies Butte

Volcanoes of California 212
Medicine Lake; Shasta; Lassen; Pre-Lassen Centers (Yana, Snow Mountain, Dittmar, Maidu); Lassen: Other cones; Eagle Lake; Tuscan Formation; Sutter Buttes; Clear Lake; Sonoma; Aurora–Bodie; Adobe Hills; Mono Craters; Inyo Craters; Long Valley; Big Pine; Ubehebe; Saline Range; Coso; Cima; Black Mountain; Pisgah; Amboy; Salton Buttes

Volcanoes of Idaho 246
Snake River Plain; Craters of the Moon; King's Bowl; Menan Buttes; Split Butte; Wapi

Volcanoes of Nevada 256
Sheldon–Antelope; Buffalo Valley; Steamboat Springs; Lunar Crater; Reveille Range; Clayton Valley; Timber Mountain

Volcanoes of Wyoming 263
Yellowstone Plateau; Leucite Hills

Volcanoes of Colorado 268
Dotsero

Volcanoes of Utah 269
Honeycomb Hills; Fumarole Butte and Smelter Knolls; Black Rock Desert; Twin Peaks; Mineral Mountains and Cove Fort; Kolob and Loa

Volcanoes of Arizona 277
Uinkaret; San Francisco; Sunset Crater; Morman; Hopi Buttes; Springerville; Carlos; Sentinel Plain; Geronimo

Volcanoes of New Mexico 290
Taos; Raton–Clayton; Ocate; Jemez; Cerros del Rio; San Felipe; Taylor; Albuquerque; Cat Hills; Zuni–Bandera; Lucero; Red Hill; Carrizozo; Jornada del Muerto; Kilbourne Hole; Potrillo

Volcano tectonics of the Hawaiian Islands
Charles A. Wood 315

Volcanoes of Hawaii 319
Kaula; Niihau; Kauai; Waianae; Koolau; Honolulu Vents; West Molokai; Wailau (East Molokai); Kalaupapa and Mokuhooniki; Lanai; Kahoolawe; West Maui; Haleakala; Kohala; Hualalai; Mauna Kea; Mauna Loa; Kilauea

Contributor index 349

Volcano index 351

CONTRIBUTORS

Carl Allen
Westinghouse NW
Environmental Center
Richland, WA

John E. Allen
Portland State University
Portland, OR

Jayne C. Aubele
Brown University
Providence, RI

Charles R. Bacon
US Geological Survey
Menlo Park, CA

W. Scott Baldridge
Los Alamos National
Laboratory
Los Alamos, NM

Jim Beget
University of Alaska
Fairbanks, AK

Mary Lou Bevier
Geological Survey of Canada
Ottawa, Canada

David A. Brew
US Geological Survey
Menlo Park, CA

James G. Brophy
Indiana University
Bloomington, IN

Anne Charland
McGill University
Montreal, Canada

Lawrence A. Chitwood
US Forest Service
Bend, OR

Robert L. Christiansen
US Geological Survey
Menlo Park, CA

David A. Clague
US Geological Survey
Menlo Park, CA

Michael A. Clynne
US Geological Survey
Menlo Park, CA

C. D. Condit
University of Massachusetts
Amherst, MA

Richard M. Conrey
Washington State University
Pullman, WA

Bruce M. Crowe
Los Alamos National
Laboratory
Los Alamos, NM

L. S. Crumpler
Brown University
Providence, RI

John C. Dohrenwend
US Geological Survey
Menlo Park, CA

Julie M. Donnelly-Nolan
US Geological Survey
Menlo Park, CA

Wendell A. Duffield
US Geological Survey
Flagstaff, AZ

Helen L. Foster
US Geological Survey
Menlo Park, CA

John Fournelle
Smithsonian Institution
Washington, DC

Don Francis
McGill University
Montreal, Canada

Michael Garcia
University of Hawaii
Honolulu, HI

Alan R. Gillespie
University of Washington
Seattle, WA

Ronald Greeley
Arizona State University
Tempe, AZ

Nathan L. Green
University of Alabama
Tuscaloosa, AL

T. L. T. Grose
Colorado School of Mines
Golden, CO

David Gust
University of New Hampshire
Durham, NH

T. S. Hamilton
Pacific Geoscience Centre
Sidney, Canada

Paul E. Hammond
Portland State University
Portland, OR

William K. Hart
Miami University
Oxford, OH

David S. Harwood
US Geological Survey
Menlo Park, CA

Catherine J. Hickson
Geological Survey of Canada
Vancouver, Canada

Wes Hildreth
US Geological Survey
Menlo Park, CA

Jerry Hoffer
University of Texas
El Paso, TX

Robin Holcomb
US Geological Survey
Seattle, WA

Robert W. Kay
Cornell University
Ithica, NY

Jürgen Kienle
University of Alaska
Fairbanks, AK

John S. King
University of New York
Buffalo, NY

A. M. Kudo
University of New Mexico
Albuquerque, NM

Florence Lee-Wong
US Geological Survey
Palo Alto, CA

Peter W. Lipman
US Geological Survey
Denver, CO

Jack P. Lockwood
US Geological Survey
Hawaii Volcano Observatory, HI

Daniel Lynch
Franklin College
Boston, MA

Bruce D. Marsh
Johns Hopkins University
Baltimore, MD

William H. Mathews
University of British Columbia
Vancouver, Canada

Thomas P. Miller
US Geological Survey
Anchorage, AK

Betsy Moll-Stalcup
US Geological Survey
Menlo Park, CA

L. J. Patrick Muffler
US Geological Survey
Menlo Park, CA

James D. Myers
University of Wyoming
Laramie, WY

William Nash
University of Utah
Salt Lake City, UT

Christopher J. Nye
University of Alaska
Fairbanks, AK

Patrick Pringle
US Geological Survey
Vancouver, WA

Donald H. Richter
US Geological Survey
Anchorage, AK

James R. Riehle
US Geological Survey
Anchorage, AK

William E. Scott
US Geological Survey
Vancouver, WA

Stephen Self
University of Texas
Arlington, TX

David R. Sherrod
US Geological Survey
Menlo Park, CA

John Sinton
University of Hawaii
Honolulu, HI

James G. Smith
US Geological Survey
Menlo Park, CA

Jack G. Souther
Geological Survey of Canada
Vancouver, Canada

Mark Stasiuk
University of British Columbia
Vancouver, Canada

Donald A. Swanson
US Geological Survey
Vancouver, WA

Samuel E. Swanson
University of Alaska
Fairbanks, AK

Edward M. Taylor
Oregon State University
Corvallis, OR

Eileen Theilig
Jet Propulsion Laboratory
Pasadena, CA

Tracy Vallier
US Geological Survey
Menlo Park, CA

George P. L. Walker
University of Hawaii
Honolulu, HI

George W. Walker
US Geological Survey
Menlo Park, CA

Karen J. Wenrich
US Geological Survey
Denver, CO

Howard B. West
University of Hawaii
Honolulu, HI

John Westgate
University of Toronto
Scarborough, Canada

Donald E. White
US Geological Survey
Menlo Park, CA

Stanley N. Williams
Louisiana State University
Baton Rouge, LA

Frederic H. Wilson
US Geological Survey
Anchorage, AK

William S. Wise
University of California
Santa Barbara, CA

Edward W. Wolfe
US Geological Survey
Hawaii Volcano Observatory, HI

Charles A. Wood
Johnson Space Center
Houston, TX

M. Elizabeth Yount
US Geological Survey
Anchorage, AK

INTRODUCTION

Charles A. Wood

In 1897 Israel Russell authored the first and only detailed publication on volcanoes in North America. That book was necessarily incomplete, being largely based on the first scientific explorations of the American West by US Geological Survey pioneers and other early travelers (King, 1959). Russell described some of the stratovolcanoes of the Cascades, and various other smaller volcanoes, but was unaware of the largest and most dangerous calderas such as Yellowstone, Long Valley, and Valles. St. Helens and Lassen, the only active Cascade volcanoes of this century, were mentioned only briefly, and many other conspicuous volcanic centers were overlooked altogether.

During the 1980s volcanoes were much in the news due to dramatic and tragic eruptions that have killed more people and have done more damage than any others in our lifetimes. The fascination and fear stemming from this volcanic activity is in the awareness of society's increased vulnerabilities; future eruptions can only have worse effects. This is because moderate eruptions, such as those that occurred during the 19th century at Baker, Rainier, St. Helens, Hood, and perhaps elsewhere in the western states, will send their ashes, lavas, and mudflows toward densely populated and intensely used lands.

Volcanic eruptions have also been recognized as apparently significant perturbations of climate. Eruptions of Tambora and Krakatau in Indonesia during the last century led to measurably lower temperatures in various parts of the world (Stommel and Stommel, 1983; Self and Rampino, 1988). An eruption of the Laki volcano in Iceland in 1783 caused famines which devastated that island, and may have triggered a remarkably cold American winter that made George Washington and Thomas Jefferson snowbound, and caused ice floes to clog the Mississippi River at New Orleans.

Volcanoes have also assumed a more cosmic significance to geologists and astronomers as space exploration has demonstrated that volcanism is a fundamental process which occurs on all of the solid planets in our solar system. Indeed, volcanism is known with certainty to occur today on Jupiter's moon Io, and there is good circumstantial evidence that Venus and Mars may still experience volcanic activity (see, e.g., Wood and Francis, 1988).

We have acquired a vastly greater knowledge of volcanoes than was available to Russell more than 90 years ago. Although various volcanoes in the Aleutians, the Cascades, and the Canadian Cordillera have not yet received even reconnaissance mapping, a large number of volcanoes throughout the USA and Canada have been studied in detail, and photographs and topographic maps are available for most volcanic areas in North America. Additionally, much information to guide volcano researchers is readily available in two major technical works. The Smithsonian Institution's *Volcanoes of the World*, compiled by Simkin and colleagues (1981), tabulates all documented eruptions of the last 10,000 years for North America and the rest of the world. Luedke and Smith (1978–1986) of the US Geological Survey have compiled a series of maps showing locations, ages, and chemical compositions of all known volcanic rocks in the United States for the last 15 million years. These two essential resources make it possible to identify which volcanoes actually exist in North America, and are the foundation for this book.

Volcanoes of North America: United States and Canada (*VNA*) capitalizes on the the tremendous body of volcano literature now available to provide brief descriptions of more than 250 volcanoes. In this book, "North America" includes Alaska, Canada, the continental USA, and Hawaii. The many volcanoes of Mexico, Central America, and the Caribbean, undeniably in North America, are omitted for two reasons: Their tectonic settings are distinct from those of the areas described here, and their history of exploration and understanding is separate from volcano studies in the United States and Canada. Perhaps a companion volume on *Volcanoes of Latin America* is required.

VNA is restricted to volcanoes younger than 5 million years old that are morphologically distinct. Young volcanic flows and cones buried by sediments or later volcanic material from other vents are omitted. Such volcanic rocks often provide much insight into magmatic chemistry and internal structure of volcanoes, but they are not the subject of this book. Five million years is a physically significant boundary because most volcanic systems have lifetimes of a few million years or less. Any volcanic region which has erupted within the last five million years may possibly be the site of future activity (Smith and Luedke, 1984).

No single person has the detailed knowledge of volcanoes in the Great Basin, the Pacific Northwest, Alaska, Canada, and Hawaii necessary to compile *VNA*. Thus, in this book nearly every volcano is described by a volcanologist who is an expert on that volcano. These 81 contributors are among the leading volcanologists in North America today, and their participation made *VNA* a credible reality. Inevitably, to fill what would otherwise be gaps, the editors wrote a few descriptions, based on scanty information. Nonetheless, a few small volcanic fields which should be included are not, because of a nearly total lack of information. Book reviewers and students will undoubtedly discover these omissions.

Compilation of *VNA* has been an enlightening experience. There are more volcanoes in North America than anyone would have imagined; no one can look through the book without discovering volcanoes beyond their knowledge. Additionally, the diversity of processes that built the volcanoes include virtually all those described in textbooks. There are plateau or trap-style volcanic plains, water and ice interaction volcanoes, shield volcanoes (both mafic and silicic), submarine volcanoes, stratovolcanoes, domes, cinder and ash cones, and ashflow calderas. Jürgen Kienle and I hope this volume will stimulate and guide future research and interest in volcanism so that a *revised* version will be necessary soon.

How the Book Works

VNA contains descriptions of 262 volcanoes and volcanic fields. Some of the volcanoes (e.g., St. Helens and Kilauea) are so intensely studied that selection of material for the limited space is the main challenge. Other volcanoes are so little known that this book contains their first published descriptions, which are, regrettably, often meager. Thus, the volcano descriptions are highly variable.

To try to standardize basic volcano characteristics, each description begins with a block of statistical information (in italicized type) on volcano location and physical features. A basic core of data is thus presented for all volcanoes; additional information on volcano and caldera size is presented where appropriate. The following information is included in the statistics blocks and in the main text:

Volcano Name and Geographic Location: Volcanoes in North America are described from north to south, in general, starting with Alaska, passing through Canada and the western United States, and ending with Hawaii. The volcanoes along the Aleutian arc are listed from east to west and followed by the more widespread volcanics on the Alaskan mainland.

Volcano names are taken from *Volcanoes of the World* (Simkin et al., 1978) and Smith and Luedke (1984; Table 4.1) whenever possible. In some cases new names were created, based on nearby geographic entities, simply to have a means to refer to previously unstudied volcanic fields. As with all of the data in the statistics blocks, even names cause some problems. There is ambiguity when the same name has been applied to two or more volcanoes (e.g., Sugarloaf, Black Mountain, and the unfortunate use of Mauna Loa on two Hawaiian islands). Additionally, a few volcanoes have multiple names (Pitt = McLoughlin). A third problem with volcano names is that most are too long. Nearly every volcano is "Mount" something or other, or such and such "Volcanic Field," or "Volcanic Complex." These nominal appendages have not been used in the statistics block nor in the index because almost every volcano is a "Mount" (except for human-scale maars, which often have wonderfully descriptive names like "Hole-in-the Ground"). Furthermore, almost all volcanoes have more than one vent and, once mapped, all are found to be complex.

The spine of the Cascades in Oregon includes many small shield volcanoes and cinder cones that are considered to be monotonously similar; most remain unstudied. Contrary to the format of the remainder of *VNA*, these volcanoes are not described separately; rather their features are briefly presented in a series of regional descriptions.

Below each volcano designation is the name of the state or province containing the volcano. Alaska has so many volcanoes, and the state is so large, that the the state name alone gives little clue to the volcano's location. Thus, Alaskan volcanoes are further listed by their geographic regions, such as island groups (e.g., Rat Islands), Alaska Peninsula, East Alaska, Bering Sea, etc. Similarly, for Hawaiian volcanoes the name of the island that contains the volcano is given.

Type: One of the frustrating aspects of most past listings of volcanoes has been the very small selection of terms used to characterize volcanic landforms. Apparently, whenever there was doubt about the nature of a volcano it was labeled a "stratovolcano" or a "composite cone." In fact, there are many varieties of volcano morphology that have not been recognized by existing moribund schemes of volcano classification. To name three, there are dome clusters, lines of overlapping stratovolcanoes, and, possibly the most common type of volcanic assemblage, monogenetic volcanic fields (e.g., Wood and Shoan, 1984). Further complication is added by the obvious observation that nearly all volcanic centers include a variety of different types: Stratovolcanoes may have summit calderas and often cinder cones and maars on their flanks. In *VNA*, the "type" listed is that of the major volcanic landform; the text usually includes related volcanic features in the described area.

The following volcano types are listed in *VNA:* tuff ring, maar, cinder cone, monogenetic volcano field, lava field, dome, dome cluster, stratovolcano, stratovolcano cluster, caldera, shield, shield caldera, linear ridge, ash-flow caldera, and volcanic plain. Some volcanoes are described as combinations of these types, and sometimes, as in ice-buried cones, with uncertainty (i.e., "?") as to type. Most of these types are familiar or self explanatory, a few are not. Ashflow calderas (e.g., Yellowstone) are the largest calderas on Earth, and are characterized by their very low relief, broad calderas, and immense ash deposits (Wood, 1984). There has never been an adequate name for such features, other than *resurgent calderas* (Smith and Bailey, 1968), but not all ash-flow calderas have resurgent centers.

Monogenetic volcanic fields are collections of cinder cones and/or maar vents and associated lava flows and pyroclastic deposits. Sometimes a stratovolcano is at the center of the field, as at the San Francisco Volcanic Field in Arizona. Monogenetic volcano fields have systematic growth patterns that suggest they represent single magmatic systems in the same way that stratovolcanoes do, but monogenetic volcanoes grow latcrally rather than vertically (Wood and Shoan, 1984). Detailed mapping indicates that some cinder cones in monogenetic fields (e.g., Cima, California, and Timber Mountain Volcanic Field, Nevada) may have had multiple eruptions separated by tens to hundreds of thousands of years (Wells et al., 1989). Thus, some monogenetic cones may actually be polygenetic. This possibility differs from observations of historic eruptions (Wood, 1979), and is physically difficult to understand because cinder cone conduits are narrow and must solidify within a few years of eruption. Nonetheless, such multiple eruptions appear to have happened at least twice: Other cinder cones need to be closely reexamined.

Lat/Long: The geographic coordinates listed are for the center of the volcano or volcanic field. Usually, the values are given to a hundredth of a degree, but sometimes only to a tenth, depending upon how well the coordinates are known.

Elevation: The height (in meters, or kilometers for large cones) of the tallest peak, or the range of elevations for broad volcanic groups, above sea level.

Relief: This is the height of a volcanic landform above its base level. For cinder cones, maars, and many shields and stratovolcanoes this quantity can be established reasonably accurately from topographic maps. For monogenetic volcano fields and closely spaced complexes of stratovolcanoes, meaningful measurements are often unobtainable.

Volcano Diameter: The basal diameter (or dimensions for non-circular features) in kilometers of relatively well-defined volcanoes or volcanic fields. Volcano and caldera diameters are as given by contributors; an excellent, relatively homogeneous source of independently derived diameters and height is Pike (1978).

Caldera Diameter: The diameter in kilometers of summit (or rarely, flank) calderas. Definitions of calderas (Wood, 1984) are generally unsatisfying, but in reality there is usually very little question whether a volcanic depression is a crater or caldera. Typically, any volcanic hole

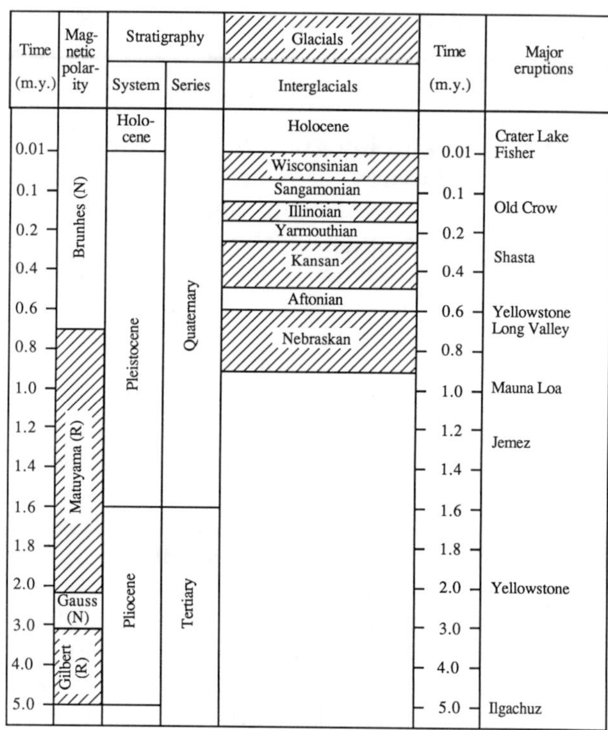

INTRO FIGURE1: Methods of determining ages of volcanic eruptions that have occurred during the last 5 Ma. Dates of a few larger eruptions are also indicated.

wider than 2 kilometers is a caldera by any definition.

Eruptive History: Critical information for every volcano is the chronology of its activity. Since *VNA* attempts to include all morphologically significant volcanoes younger than 5 million years, there are great differences in the known eruptive histories. For volcanoes like Kilauea, a listing of dates of historic eruptions would take much space without providing much information other than that it is often active. For other volcanoes with only a few historic eruptions, those dates are given. Thus, whereas *VNA* tries to indicate periods of major activity, it should not be considered as complete. *Volcanoes of the World* (Simkin et al., 1981) remains the best source for complete eruption listings for the last 10,000 years. *VNA* is the best (only?) single source for general eruption information for North American volcanic activity during the last 5 million years.

Outside of Hawaii and Alaska, most North American volcanoes have had little or no historic activity (although it is quite surprising how many eruptions have occurred in the last 1,000 years). In general, the information given summarizes a volcano's date of origin and main episodes of activity as determined by various dating methods, including K-Ar, ^{14}C, paleomagnetism, glacial advances, and oxygen isotopes. The divisions of time for the last 5 million years, as indicated by these various methods, are shown in Figure 1, which is based on the Elsevier Geologic Time Chart.

Composition: The dominant rock type for a volcanic center, and often the range of types, is tabulated, using terminology supplied by each contributor. There is an uneven level of rock characterizations, with petrologists tending to provide more modifiers to basic terms such as basalt or andesite.

Volcano Description: The narrative description of each volcano includes information on the morphology and structure of the volcanic landform, a brief mention of notable eruptions or stages in the volcano's evolution, characteristic and unusual rock types, and other notable features. Because the descriptions were provided by many different volcanologists, emphases vary considerably, even after editorial homogenization. In general, however, details of geochemistry, isotopes, tephra layers, flow morphologies, and eruption histories have been eliminated. Each description is intended to be understood by general geologists, not just specialists.

How to Get There: The original desire for a book like *VNA* derived from frustration in trying to find out about volcanoes in areas planned for visits. Commonly the only information available was petrological, with little indication of the nature of the volcano or any of its morphological features of interest. To aid other first time visitors *VNA* includes capsule directions providing basic access information and warnings. For many volcanoes in the continental USA, paved roads lead to scenic points, but especially in Canada and Alaska access requires substantial planning and money, and can be dangerous. Perhaps the single most useful mode of transportation is the helicopter (which – except for tourist areas – typically only national geologic agencies can afford), but snow-ski and pontoon-equipped bush planes are often called upon, as well as boats, jeeps, dog sleds, and hiking boots. Dangers include suddenly inclement weather (in Hawaii as well as Alaska), rattlesnakes, ice crevasses, stormy seas, bears, and suspicious drug growers.

The most detailed information on the access to and things to see at volcanoes is usually in regional guidebooks. Fortunately, three comprehensive guidebook series are newly available. The Geological Society of America has recently published a series of *Centennial Field Guides* for all regions of North America, and additional, exhaustive guide volumes were published in 1989 in association with international meetings held in the USA by both the International Association of Volcanology and Chemistry of the Earth's Interior (IAVCEI) and the International Geological Congress. Finally, a fine guide to parts of the Cascades (Johnston and Donnelly-Nolan, 1981) is available from the US Government Printing Office.

References: The editors scrupulously limited references to only two entries, much to the chagrin of many contributors who wanted to give full credit for each piece of information used, as is common in scientific publications. The purpose of the two references is not to document each fact, but to introduce the reader to the available scientific literature for that volcano. Usually, the most recent publication with a good bibliography is listed. Alternatively, classic papers are given. For many volcanoes there are few or no published references, and doctoral dissertations or masters theses are listed, even though such documents are often difficult to acquire. In general, the contributor of each volcano description is the best living authority on that volcano and is the person to contact when field research for that volcano is being planned.

Photographs: Most scientific articles about volcanoes are published by petrologists fascinated by geochemistry and the origin and evolution of the magmas that ultimately erupt to produce the source of their samples. Such articles rarely include photographs of the actual volcano from which the marvelous rocks come, even though there is considerable evidence that the style of an eruption and the morphology of the resulting volcano are related to various petrological variables (e.g., Jakobsson et al., 1978). Similarly, all previous catalogs of volcanoes have been sparingly illustrated, if at all. *VNA* provides pictorial antidotes to previous omissions by striving to include one or more photographs of every volcano described. Some volcanoes in Alaska and Canada are very seldom visited and no pictures seem to exist. In these and similar cases, radar images and/or spacecraft photos have been used where available. Monotonously non-spectacular, low volcanoes in the Cascades also seem to have escaped photographic documentation. Because of the variety of types of photographs (and the immense range in volcano sizes) included in *VNA*, there can be no consistent viewing perspective or scale. Photograph sources are listed in each caption *only* when someone other than the contributor took the picture.

Maps: To complement the photograph a small scale map is given for every volcano where possible. The maps too come from a variety of sources and hence have no consistent format or scale. Most are cut from US Geological Survey topographic maps with scales from 1:250,000 to 1:20,000. Others are simple sketch maps showing the locations of principal volcanic features. Because of the small space available, some of the details are not legible, but the maps still present a feel for the volcanic topography. For low-relief volcanic fields, maps often show nothing of volcanological interest, and vertical aerial photographs are used instead. In some cases, no map or photograph is available.

Acknowledgments

VNA was prepared with the aid of many contributors who provided volcano descriptions and photographs. Without these people this book would be only a dream. Jürgen Kienle and I especially wish to thank a few individuals who not only provided their own volcano descriptions, but helped organize entire chapters, suggesting other contributors, finding photos, and helping to write tectonic overviews. Chris Nye helped with Alaska, Jack Souther and Cathie Hickson with Canada, Scott Baldridge with New Mexico, and Dave Clague with Hawaii. Dave Sherrod helped similarly with Oregon, and also took me on a delightful field trip through some of the multitude of small, little known cones overshadowed by the majestic Cascade stratovolcanoes. Had I known how many small volcanoes exist in the Oregon Cascades I would have been too daunted to attempt this book. Robert Kay, David Griggs, and Lyn Topinka are thanked for providing photographs. Last, Jürgen and I thank Don Swanson and Don Mullineaux for meticulous reviews of the entire book.

References

Jakobsson, S. P., Jonsson, J., and Shido, F., 1978, Petrology of the western Reykjanes Peninsula, Iceland: *J. Petrol. 19*, 669-705.

Johnston, D. A., and Donnelly-Nolan, J., 1981, Guides to some volcanic terranes in Washington, Idaho, Oregon, and Northern California: *USGS Circ. 838*, 189 pp.

King, P. B., 1959, *The Evolution of North America*: Princeton Univ. Press, see pp. 93-97.

Luedke, R. G., and Smith, R. L., 1978, Map showing distribution, composition and age of Late Cenozoic volcanic centers in Arizona and New Mexico: *USGS Misc. Invest. Ser. Map I-1091-A.*

Luedke, R. G., and Smith, R. L., 1978, Map showing distribution, composition and age of Late Cenozoic volcanic centers in Colorado, Utah and southwestern Wyoming: *USGS Misc. Invest. Ser. Map I-1091-B.*

Luedke, R. G., and Smith, R. L., 1981, Map showing distribution, composition and age of Late Cenozoic volcanic centers in California and Nevada: *USGS Misc. Invest. Ser. Map I-1091-C.*

Luedke, R. G., and Smith, R. L., 1982, Map showing distribution, composition and age of Late Cenozoic volcanic centers in Oregon and Washington: *USGS Misc. Invest. Ser. Map I-1091-D.*

Luedke, R. G., and Smith, R. L., 1983, Map showing distribution, composition and age of Late Cenozoic volcanic centers in Idaho, western Montana, western South Dakota, and northwestern Wyoming: *USGS Misc. Invest. Ser. Map I-1091-E.*

Luedke, R. G., and Smith, R. L., 1986, Map showing distribution, composition and age of Late Cenozoic volcanic centers in Alaska: *USGS Misc. Invest. Ser. Map I-1091-F.*

Pike, R., 1978, Volcanoes on the inner planets: Some preliminary comparisons of gross topography: *Proc. Lunar Planet. Sci. Conf. 9th*, 3239-3273. Also see Pike, R. J., and Clow, G. D., 1981, Revised classification of terrestrial volcanoes and catalog of topographic dimensions, with new results on edifice volume: *USGS Open-file Rpt. 81-1038*, 40 p.

Russell, I., 1897, *Volcanoes of North America*: Macmillan, London, 346 pp.

Self, S., and Rampino, M., 1988, The relationship between volcanic eruptions and climate change: still a conundrum? *EOS, Trans. Am. Geophys. Un. 69*, 74-75, 85-86.

Simkin, T., Siebert, L., McClelland, L., Bridge, D., Newhall, C., and Latter, J. H., 1981, *Volcanoes of the World*: Hutchinson Press, Stroudsburg, PA, 233 pp.

Smith, R. L., and Bailey, R. A., 1968, Resurgent cauldrons, in *Studies in Volcanology*, R.R. Coats et al. (eds.): *Geol. Soc. Am. Mem. 116*, 613-662.

Smith, R. L., and Luedke, R. G., 1984, Potentially active volcanic lineaments and loci in western coterminous United States: in *Explosive Volcanism: Inception, Evolution, and Hazards*, Natl. Acad. Press, Washington, DC, 47-66.

Stommel, H., and Stommel, E., 1983, *Volcano Weather: The Story of 1816: The Year Without a Summer*, Seven Seas, Newport, RI, 177 pp.

Wells, S. E., Renault, C. E., and McFadden, L. D., 1989, Geomorphic and pedologic criteria for recognizing polycyclic volcanism at small basaltic centers in the western USA: *N. Mex. Bur. Mines Min. Res. Bull. 131*, 289.

Wood, C. A., 1980, Morphometric evolution of cinder cones: *J. Volc. Geotherm. Res. 7*, 387-413.

Wood, C. A., 1984, Calderas: A planetary perspective: *J. Geophys. Res. 89*, 8391-8406.

Wood, C. A., and Francis, P., 1988, Venus lives! *Proc. Lunar Planet. Sci. Conf. 18th*, 659-664.

Wood, C. A., and Shoan, W., 1981, Growth patterns of monogenetic volcanic fields: *EOS, Trans. Am. Geophys. Un. 62*, 1061.

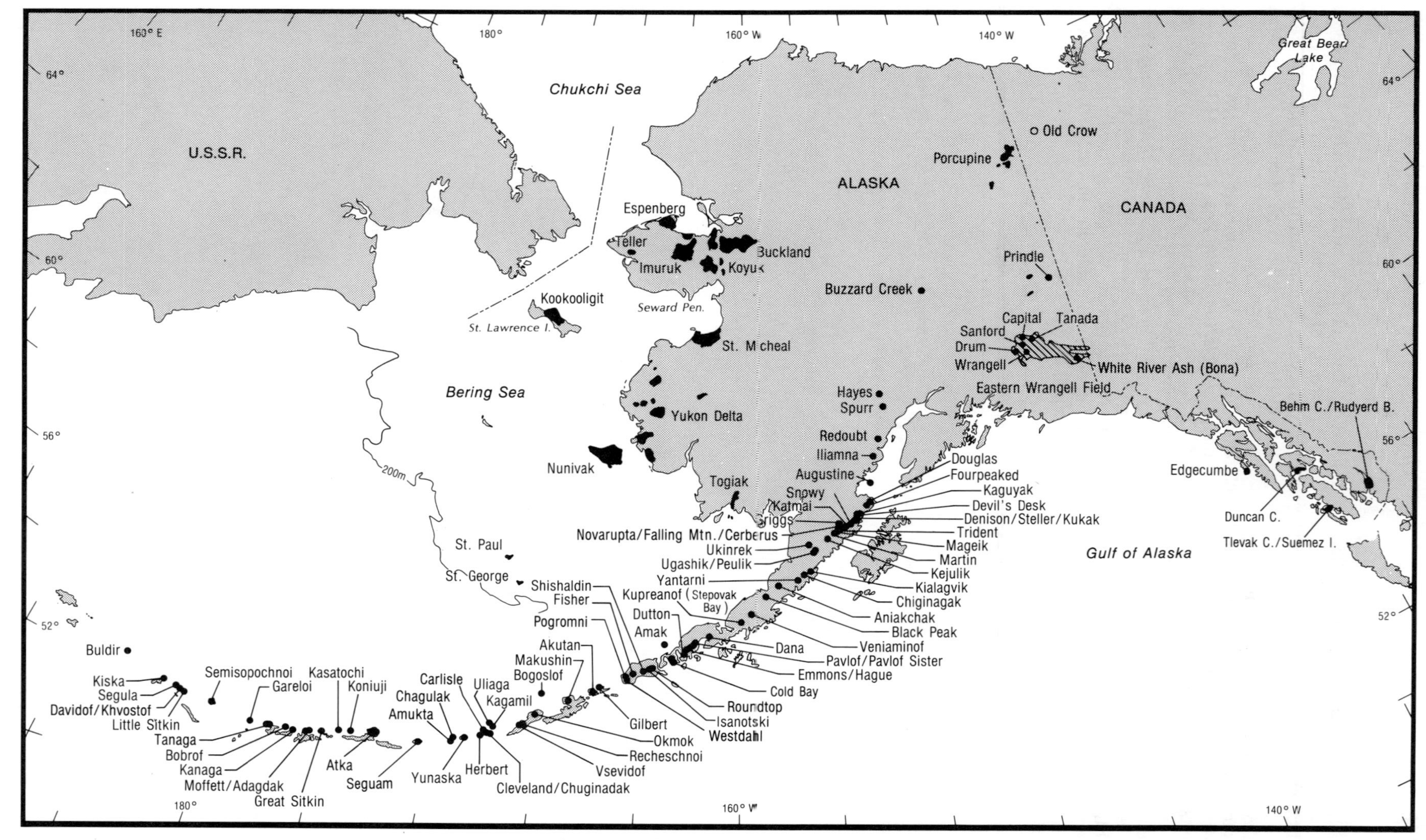

ALASKA FIGURE 1: Late Cenozoic volcanic centers in Alaska. (Modified from Kienle et al., 1983, and Luedke and Smith, 1986.)

VOLCANO TECTONICS OF ALASKA

J. Kienle and C. J. Nye

Most Alaskan volcanoes are in the Aleutian arc which extends ~2,500 km along the southern edge of the Bering Sea and Alaskan mainland (Fig. 1). This classic volcanic arc contains some 80 Quaternary stratovolcanoes and calderas. Aleutian arc volcanism is the result of active subduction of the Pacific Plate beneath the North American Plate. The 3,400-km-long Aleutian trench that extends from the northern end of the Kamchatka trench to the Gulf of Alaska marks the boundary between the two plates.

Compressional tectonics resulting from plate collision also gives rise to orogenic arc volcanism in the Wrangell Mountains. The Wrangell volcanic arc segment is offset 400 km to the east from the northeastern end of the Aleutian arc and strikes at right angles to it (Fig. 1). There are several possible plate geometries that could explain this offset. One is that the subducting Pacific Plate has a tear in the Gulf of Alaska region and that the two segments of the Pacific Plate on either side of the tear are subducting in different directions. Another is that the subducting plate is continuous but buckles through 90°, and that one limb of this buckle underlies the Wrangell volcanoes. This is the model recently proposed by Page et al. (in press) based on seismicity studies in the Gulf of Alaska.

In southeastern Alaska, there are a number of small basaltic lava fields and one major calc-alkaline volcanic center, Mount Edgecumbe, which has erupted basalt through rhyolite. This volcano lies within 15 km of the Fairweather–Queen Charlotte transform fault, but the ultimate cause of volcanism is unknown. Further to the southeast, crustal extension in the vicinity of the transform plate boundary may be the cause of small scale basaltic volcanism represented by the southeastern Alaska fields.

Behind the Aleutian–Wrangell subduction system lies a region of diffuse, exclusively basaltic volcanism that is associated with broad tectonic extension in the backarc region. This basaltic volcanism occurs primarily in Beringia, a region that includes the islands of the Bering Sea shelf, the Seward Peninsula, and the Bering Sea coast of Western Alaska. A few small basalt fields are found in a similar extensional tectonic setting in Interior and Eastern Alaska (Fig. 1).

Descriptions of individual volcanoes and volcanic fields are organized within the tectonic framework just outlined, starting with the volcanoes of the compressional Aleutian–Wrangell plate margin, followed by the basalt fields of the extensional backarc region. There are 92 individual entries (identified on Fig. 1). Some entries group several discrete vents into one single volcanic center.

The first volcano described is tiny Buldir in the westernmost Aleutian arc. The descriptions that follow are organized geographically from west to east, following the volcanic front toward mainland Alaska. The next set of descriptions is for the Wrangell volcanoes. Entries for Mount Edgecumbe and the southeastern Alaska basalt fields conclude this first west to east sweep along the north Pacific Plate margin. We then go to the basalt fields of Beringia and interior Alaska, starting again in the west with the Pribilof Islands, St. Paul, and St. George, followed by descriptions of Nunivak Island and the Yukon Delta and Togiak fields, moving north to the St. Michael field and the Kookooligit field on St. Lawrence Island, then on to the Seward Peninsula and east to the Porcupine field in interior Alaska, ending with Prindle volcano on the Alaska–Yukon border.

Travelers to Alaska who wish to see some of the volcanic centers should realize that access to the volcanoes is difficult. They cannot be reached by car, nor, in most cases, by scheduled aircraft, and trips require careful expeditionary planning. Boat landings in the Aleutians can be outright dangerous and there is usually little or no published material for guidance. This is why special attention has been given to the *How to get there* section for individual volcano descriptions in this book. The three most important volcanic regions in Alaska are discussed below.

Aleutian Arc

The Aleutian volcanic arc contains 44 of the 54 historically active volcanoes in the USA (Simkin et al., 1981). In the last decade, hardly a year has passed without a reported volcanic eruption in the Aleutians. This is probably not because the arc has become more active but rather because eruption detection has improved. In recent years the eruption record has become more complete because there has been increased surveillance, both from space and from commercial air traffic.

There are about 80 named volcanoes in the Aleutian arc (Coats, 1962; Simkin et al., 1981; Luedke and Smith, 1986), of which ~44 have erupted, several repeatedly, since 1741 when the written record begins in Alaska. At least 21 of the Quaternary volcanic centers have calderas (Miller and Barnes, 1976; Miller and Smith, 1987); as many as 19 may have formed in the Holocene.

The following summary of the tectonic setting of the Aleutian arc draws heavily on important review articles, several of which have appeared as chapters in books on the geography, physiography, geology and geophysics, structure, evolution and tectonics, or petrology and geochemistry of the Aleutian arc. The reviews by Coats (1962), Marsh (1982), Kienle et al. (1983), Jacob (1986), Scholl et al. (1987), Stone (1988), Kay and Kay (in press), Marsh (in press), and Miller and Richter (in press) are good introductions to the now rather extensive body of literature on the Aleutian arc for the interested reader who is not familiar with Alaska.

The Quaternary Aleutian arc spans about 2,500 km of the Alaska mainland and the Aleutian Islands, and is built on continental crust in the east and on oceanic crust in the west. The dividing line between the two parts is the Bering Sea continental shelf, which intersects the arc in the vicinity of Unalaska, Akutan, and Unimak Islands. The volcanic front is sharp (Marsh, 1982) and closely aligned with the 100-km depth contour to the Wadati–Benioff zone (Jacob, 1986; Kienle et al., 1983). A weak, more potassic, secondary volcanic front represented by only 2 volcanoes, Bogoslof and Amak, has developed 50 km behind the volcanic front in the eastern Aleutians (Marsh, 1979). On the Alaska Peninsula, the Ukinrek Maars, the nearby Gas Rocks, and three unnamed basaltic scoria cones in Katmai National Park may be part of this secondary front.

The island-forming Quaternary volcanoes of the oceanic part of the arc are perched on top of the Aleutian Ridge, which rises 3,000 m from the Bering Sea floor to the north and more than 6,000 m from the Aleutian trench to the south. The entire structure is more than 200 km wide from the trench to the abyssal floor of the Bering Sea and includes an important accretionary

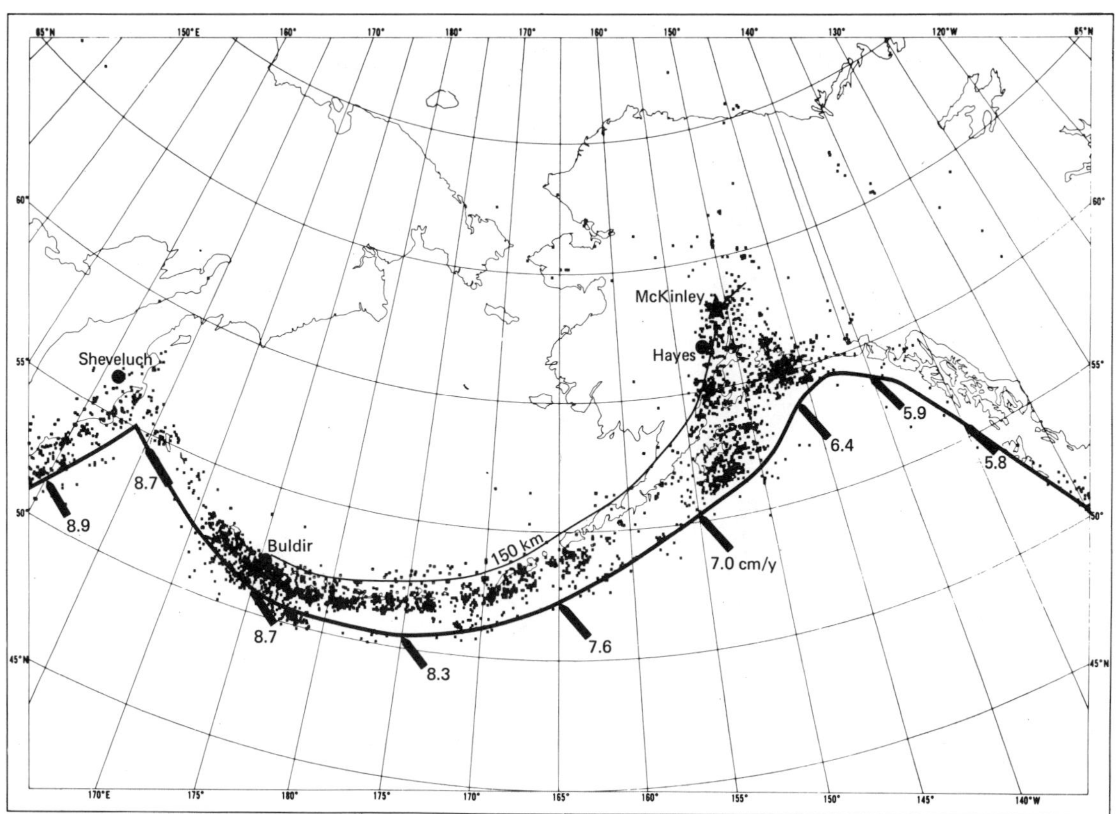

ALASKA FIGURE 2: Relative motion vectors between Pacific and North American plates after Minster and Jordan (1978) model RM2, and seismicity of the Aleutian arc (WWSSN teleseismic locations 1962-1969). Pacific-North American Plate boundary and 150-km depth contour to Wadati–Benioff zone shown by solid lines. (From Kienle et al., 1983; figure modified from Jacob et al., 1977.)

forearc basin. Scholl et al. (1987) describe a three-tiered rock sequence which records the major evolutionary steps of the Aleutian Ridge:

(1) Arc volcanism in the early to middle Eocene (55 to 50 Ma) formed the bulk of the ridge. The igneous and volcaniclastic basement rocks of the ridge from this time are called the lower series.

(2) Oligocene and Miocene rocks are associated with greatly reduced magmatic growth. Erosional processes became more important and the flanks of the ridge became buried by thick marine sedimentary units of the middle series. Plutonism is important from the Oligocene through the early Miocene (35 to 8 Ma). These rocks form the middle series.

(3) Rocks of Pliocene and Quaternary age (younger than 5.3 Ma) form the upper series. These are young, slope mantling, and crestal basin sedimentary sequences that unconformably overlie the older units of the middle and early series. Regional subsidence and block faulting affected the ridge crest in the late Cenozoic and the ridge crest was cut by wave erosion creating a prominent summit platform. The Quaternary volcanic chain developed near the northern edge of the summit platform.

Whereas no terrestrial rocks older than Tertiary have been found in the oceanic part of the arc, Mesozoic and even Paleozoic rocks make up the basement of the continental part of the arc on the Alaska Peninsula and in the Cook Inlet region (Burk, 1965; Beikman, 1980).

Present day subduction under the Aleutian arc changes from normal to oblique, from east to west along the arc (Fig. 2). The subareal arc ends at Buldir Island, where plate motion changes from convergent to transform slip. Arc magmatism resumes where plate motion becomes again convergent in the Kamchatka–Kuril arc; Sheveluch is the northernmost major volcanic center in Kamchatka (see Fig. 2). The 3,400-km-long Aleutian trench is continuous from the Kuril–Kamchatka trench to the Gulf of

Alaska, but changes trend near Buldir Island. West of Buldir it loses curvature and becomes more linear along the arc–arc transform plate boundary.

At the eastern end of the Aleutian arc the relationship between subduction and volcanism is not as clear as at the western end. Although the Wadati–Benioff zone continues for another 300 km beyond the last volcano in Cook Inlet (Hayes Volcano) to a point beyond Mount McKinley (see Fig. 2), that segment of the subduction zone is devoid of volcanoes with the exception of one minor Holocene basalt occurrence at the Buzzard Creek craters. The craters are situated on the Aleutian volcanic trend at the northern end of the Wadati-Benioff zone, but their chemistry suggests a closer affinity with the basalt fields of interior Alaska.

The geometry and attitude of the subducting Pacific Plate is quite well known from seismicity studies. Jacob (1986) gives a good review of subduction seismicity studies carried out in the Gulf of Alaska region with local networks operated by Columbia University of New York, the University of Alaska Fairbanks, and the US Geological Survey. In the oceanic part of the subduction zone, the Wadati–Benioff zone is defined to a depth of 250–300 km and dips at ~45° beneath the Aleutian arc (Engdahl, 1977). The distance between the trench and the volcanic arc, the arc-trench gap, increases systematically from west to east and is about 170 km wide in the central Aleutians (Jacob et al., 1977). The arc-trench gap widens in the continental section of the arc, probably because accretion of material from the oceanic plate to the overriding North American Plate has forced the trench to move several hundred kilometers seaward. On the Alaska Peninsula, the arc-trench gap is 300 km wide; near Hayes volcano it is 570 km wide, increasing northward. The attitude of the Wadati–Benioff zone in the continental part of the arc is illustrated by Figure 3, showing three hypocentral cross sections across the eastern Aleutian arc in the Kodiak, Katmai, and Cook Inlet regions. The eastward widening of the arc trench gap can readily be seen. The Wadati–Benioff zone consists of 2 segments, a shallow-dipping thrust zone with a dip of ~10° and a steeply dipping segment with a dip of ~45° that plunges beneath the volcanic arc to a depth of ~200 km. Subduction along the shallow thrust

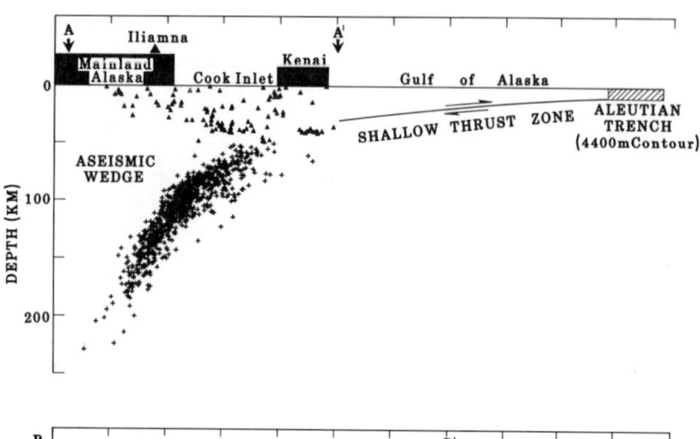

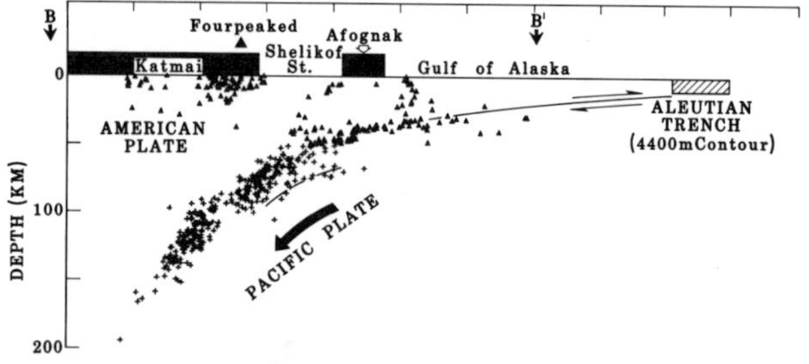

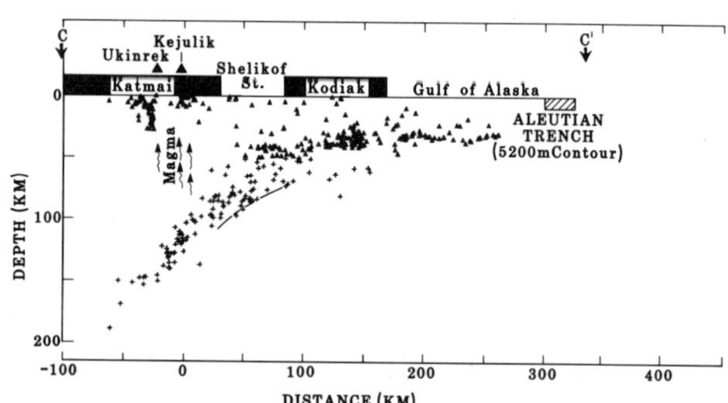

ALASKA FIGURE 3: Cross sections A–A' (Kenai), B–B' (Afognak Island), and C–C' (Kodiak Island) with hypocenters projected onto these sections (refer to Fig. 5 for location of sections). This is no vertical exaggeration. (From Kienle et al., 1983.)

zone results in strong plate coupling and large rupture areas of great earthquakes, such as the great Alaskan earthquake of 1964 (Mw 9.2, Kanamori, 1977).

The volcanic front is strikingly sharp and is broken into linear segments of volcanoes (Marsh, 1979, 1982; Kay et al., 1982; Kienle et al., 1983). In the oceanic part of the arc, segment boundaries appear to correlate with breaks in the downgoing plate, while on the continental part of the arc, structural features of the crust seem to influence volcano positioning (Fisher et al., 1981). In Katmai there is evidence that the close spacing of volcanoes and the alignment of the volcanic front between Mageik and Kaguyak are controlled by a deep-seated crustal fault (Kienle and Swanson, 1983).

The composition, origin, and evolution of Aleutian magmas have been discussed extensively over the last decade or so. The published data base now contains over 1,000 analyses, which have been reported in over 50 papers, reports, and theses. The following paragraphs are broad, first-order observations based on these data.

Aleutian magmas have major and trace element and isotopic compositions broadly typical of relatively mature arcs built on oceanic or thin continental crust. They are dominantly medium-K (Gill, 1981) calc-alkaline and tholeiitic basalts through dacites. The few centers a short distance behind the main arc are more alkaline than volcanic front magmas, as is typical of other arcs. Rocks alkaline enough to belong to the trachytic series of Le Bas et al. (1986) are rare. Rhyolite is volumetrically minor and occurs as glass shards, ash flows (e.g., the Katmai eruption of 1912), and as small pods and domes on the flanks of a few volcanoes.

Aleutian magmas are typically porphyritic, with plagioclase almost always predominant. Mafic lavas usually have olivine and clinopyroxene phenocrysts. Intermediate and silicic lavas usually contain orthopyroxene and clinopyroxene. Iron–titanium oxides are ubiquitous in minor amounts. Complex mineral assemblages and morphologies are common, especially in evolved calc-alkaline lavas, presumably as a result of magma mixing. Amphibole phenocrysts are less common, but do occur throughout the entire compositional range of Aleutian calc-alkaline magmas. Biotite is even rarer, but has been reported from a few centers. Kay and Kay (1985) provide a summary of the composition and occurrence of Aleutian mafic phenocrysts.

Magmas from the eastern portion of the arc are dominantly calc-alkaline andesite; basaltic andesite, dacite, and tholeiitic magmas are rare (Fig. 4). Magmas from the central portion of the arc are dominantly tholeiitic basalt and basaltic andesite. The central portion of the arc also contains the largest volcanoes. The western arc contains both calc-alkaline and tholeiitic basalt, basaltic andesite, and andesite.

The body of petrogenetic literature concerning Aleutian magmas is extensive enough to preclude detailed summarization. Convenient entry points into the literature are review papers by Kay and Kay (in press) and Marsh (in press). There is general consensus that parent magmas are mafic (basalts or low-silica basaltic andesites), and that silicic andesites, andesites, and dacites are produced by a spectrum of processes dominated by fractional crystallization but also including magma mixing and lithospheric contamination. There is also general, although not universal, consensus that primary magmas carry a strong chemical signal generated by the recycling of upper crustal components during the subduction process, either by dewatering of subducted oceanic crust or by subduction of sediments on top of the subducted plate. There is not yet consensus about whether the ultimate source of primary magmas is the subducted slab itself or asthenospheric mantle above the slab.

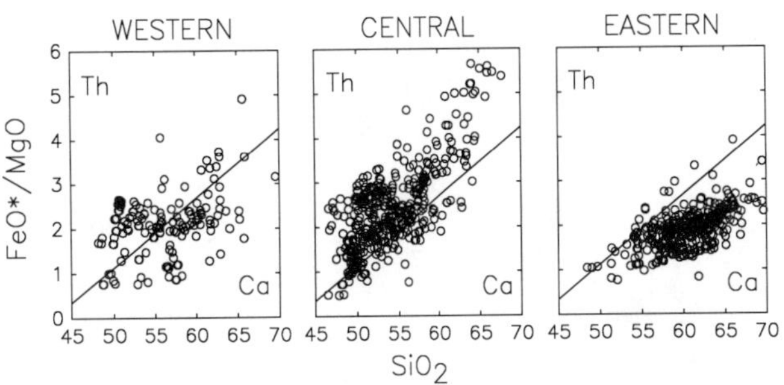

ALASKA FIGURE 4: Summary of available Aleutian geochemical data. The tholeiitic (Th)–calc-alkaline (Ca) dividing line is from Miyashiro (1974). For the purposes of this figure the boundary between the western and central arc is 172°W; that between the central and eastern arc is 159°W. Note that the central arc is built on both oceanic and continental crust.

Wrangell Volcanoes

The Wrangell volcanoes form a cluster of exceptionally large andesitic stratovolcanoes concentrated at the western edge of a Neogene and Quaternary volcanic field. Their geographic and tectonic setting is shown on Figure 5. The Aleutian arc subduction system is separated from the Pacific–North American transform boundary by a wide transition zone. In this zone a small sliver of continental crust, the Yakutat Block, is moving north and impinging on the North American continent. The Yakutat Block has moved to its present position as a passenger

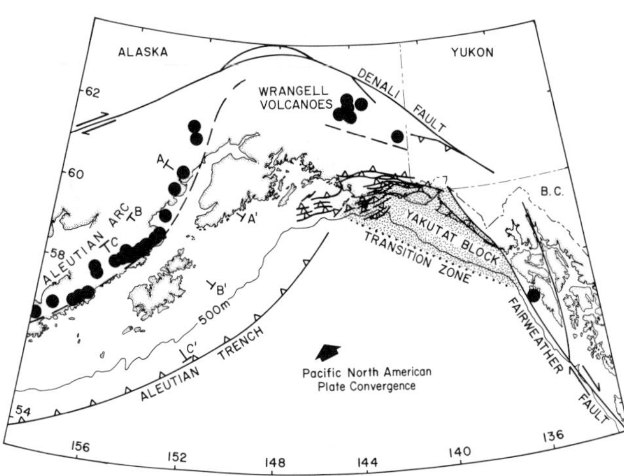

ALASKA FIGURE 5: Volcanoes (black dots) and structural elements of the Gulf of Alaska region. The dashed line is the 85-km depth contour to the Wadati–Benioff zone. Positions of cross sections in Figure 3 are also shown

on the Pacific Plate from the place of its origin far to the south. There is presently limited slip between the Yakutat Block and the Pacific Plate (Lahr and Plafker, 1980), which is presumably generated as the Yakutat Block detaches from the Pacific Plate and is accreted to North America. An actively deforming fold and thrust belt lies at the northern margin of the Yakutat Block.

Subduction ahead of the Yakutat Block generated the volcanic field now dominated by the Wrangell volcanoes. There is a weakly defined Wadati-Benioff zone beneath the Wrangell volcanoes which has only recently been discovered (Stephens et al., 1984). This Wadati–Benioff zone is at right angles to that underlying the easternmost part of the Aleutian arc, as discussed above. The dashed line on Figure 5 is the 85-km depth contour of the subducted plate.

The Wrangell volcanoes represent exceptionally vigorous andesitic volcanism, a fact belied by the indistinct Wadati–Benioff zone which underlies them. Many of the individual volcanoes are huge by arc standards. Several of the volcanoes exceed 600 km^3, and the largest is almost 900 km^3. These volumes are impressive compared to the worldwide mean of convergent margin calc-alkaline volcanoes of under 100 km^3. These volumes are also much greater than the largest of the Aleutian tholeiitic volcanoes (which are generally more voluminous than Aleutian calc-alkaline volcanoes). In rare instances, other convergent margin volcanoes are as big as Wrangell volcanoes, but these are usually scores of kilometers from their nearest neighbors. The large size of the volcanoes seems to reflect rapid growth, rather than growth at normal rates over a long time period, of individual centers. The detailed eruptive histories of many volcanoes are still little known but Mount Wrangell itself has grown to be the largest of the volcanoes in ~750,000 yr. Other volcanoes seem to have had total lifetimes of around 1 million years, similar to the lifetimes of other Aleutian centers. The Wrangells are tightly clumped and of about the same age, which requires cumulative growth rates that are equal to or greater than the maximum reported for other arcs and that exceed those of some other arcs by as much as a factor of 10. The exceptional volume and growth rate have yet to be explained.

Several of the Wrangell volcanoes are shield-shaped, despite the fact that they are medium-K silicic andesites. Wrangell magmas are compositionally similar to those which produce steep-sided stratocones elsewhere in the world. The shield shapes have yet to be clearly explained, but may be related to high effusion rates. Despite the unusual size and morphology, the composition and mineralogy of Wrangell magmas are markedly similar to those of other Aleutian calc-alkaline volcanoes.

Beringia Basalts

Basalt of late Tertiary and Quaternary age is widespread in the Beringia region of Western Alaska (Fig. 1). Two individuals have done most of the fieldwork on the Beringia basalts, David Hopkins and the late Joseph Hoare. Betsy Moll-Stalcup (in press) has recently compiled an important geochronologic and petrologic review of all available data on the Beringia basalts.

The Beringia basalt fields form a diffuse southwest–northeast trending volcanic belt about 500 km wide that stretches some 1,000 km from the Pribilof Islands (St. Paul and St. George) through Nunivak Island and the Yukon Delta to St. Lawrence Island and the Seward Peninsula (Fig. 1). The belt is crudely subparallel to the eastern part of the Aleutian arc but lies at a great distance behind it. Some of the basalt fields form very extensive blankets of predominantly tholeiitic basalt. The Buckland, Imuruk, St. Michael, combined Yukon Delta and Nunivak Island fields all exceed 2,000 km^2 in area, Buckland being the largest with over 5,000 km^2 (Swanson and others, personal communication). Other basalt occurrences are just individual small cinder cones or lava flows. Tholeiitic basalt fields are sometimes associated with relatively minor volumes of highly alkalic basalts (basanites and nephelinites). Tholeiitic basalt typically forms low shields and cinder cones that sometimes feed very long lava flows (up to 35 km). Alkalic basalts often form steep necks and cones, and short flows. As in other areas of the world, some of the alkali basalt vents contain mantle xenolith suites (for a recent review of the western Alaska xenoliths, see Swanson et al., 1987). Maar volcanoes are common in the Espenberg field (Hopkins, 1988) on the Seward Peninsula, and also on Nunivak Island. Geologic field data on the Beringia basalt fields are quite variable and largely unpublished. A few of the Beringia basalt fields have been well studied while others remain virtually untouched. Nunivak Island is probably the best studied Beringia basalt field.

Late Cenozoic basaltic volcanism was more or less synchronous throughout Beringia ranging from 5.9 Ma to the present (Moll-Stalcup, in press). The much older Kugruk basalts from the Imuruk field on the Seward Peninsula (26–29 Ma) are isolated in time and probably represent a different magmatic event (Hopkins, 1963; Swanson, 1981). Recent lavas, perhaps best represented by the Lost Jim and Camille flows in the Imuruk lava field, are found at most of the Beringia localities.

Several of the volcanic fields of Beringia are associated with young basins bounded by normal faults which suggests an extensional tectonic environment. Turner et al. (1981) speculated that basaltic volcanism and hot spring activity on the Seward Peninsula might be associated with incipient continental rifting. Stone and Wallace (1987) think that basaltic volcanism occurs along fault-bounded crustal blocks that are locally jostled in response to major regional transcurrent faulting or terrain accretion. Nakamura et al. (1977, 1980) have proposed that the entire Beringian region is under broad tensional stress. This conclusion is based on regional orientation of dikes and other volcanic features. It should be emphasized, however, that widespread normal faulting is not well developed in the region and that basaltic volcanism is the only indication of regional deep-seated extension.

References

Beikman, H. M., 1980, Geologic Map of Alaska: *USGS Map*, 1:250,000.

Burk, C. A., 1965, Geology of the Alaska Peninsula island arc and continental margin: *Geol. Soc. Am. Mem. 99.*

Coats, R., 1962, Magma type and crustal structure in the Aleutian arc: *in* Macdonald, G., and Kuno, H., (eds.), *The Crust of the Pacific Basin*, Am. Geophys. Un. Monograph 6, 92-109.

Engdahl, E. R., 1977, Seismicity and plate subduction in the central Aleutians: *in* Talwani, M., and Pitman III, W. C. (eds.), *Island Arcs, Deep Sea Trenches, and Back-Arc Basins*, Maurice Ewing Series 1, Am. Geophys. Un., Washington, DC, 259-272.

Fisher, M. A., Bruns, T. R., and Von Huene, R., 1981, Transverse tectonic boundaries near Kodiak Island, Alaska: *Geol. Soc. Am. Bull. 92*, 10-18.

Gill, J., 1981, *Orogenic Andesites and Plate Tectonics:* Springer Verlag, New York, 390 pp.

Hoare, J. M., Condon, W. H., Cox, A., and Dalrymple, G. B., 1968, Geology, paleomagnetism, and potassium-argon ages of basalts from Nunivak Island, Alaska: *in* Coats, R. R., Hay, R. L., and Anderson, C. A. (eds.), *Studies in Volcanology*, Geol. Soc. Am. Mem. 116, 377-413.

Hoare, J. M., and Coonrad, W. L., 1980, The Togiak Basalt, a new formation in southwestern Alaska: *USGS Bull. 1482-C*, C1-C11.

Hopkins, D. M., 1963, Geology of the Imuruk Lake area, Seward Peninsula, Alaska: *USGS Bull. 1141-C*, 1-101.

Hopkins, D. M., 1988, The Espenberg Maars: a record of explosive volcanic activity in the Devil Mountain-Cape Espenberg area, Seward Peninsula, Alaska: *in* Schaaf, J. (ed.), *The Bering Sea Land Bridge: an Archeological Survey*, US National Park Service, in press.

Jacob, K. H., 1986, Seismicity, tectonics, and geohazards of the Gulf of Alaska regions: *in* Hood, D. W., and Zimmerman, S. T. (eds.), *The Gulf of Alaska, Physical Environment and Biological Resources*, US Dept. of Commerce, NOAA, and other organizations, US Dept. of the Interior, Minerals Management Service, 145-184.

Jacob, K. H., Nakamura K., and Davies J. N., 1977, Trench-volcano gap along the Alaskan–Aleutian arc: Facts, and speculations on the role of terrigenous sediments: *in* Talwani, M., and Pitman III, W. C. (eds.), *Island Arcs, Deep Sea Trenches, and Back-Arc Basins*, Maurice Ewing Series, 1, Am. Geophys. Un., Washington, DC, 243-258.

Kanamori, H., 1977, The energy release in great earthquakes: *J. Geophys. Res. 82*, 2981-2988.

Kay, S. M., and Kay, R. W., 1982, Tectonic controls on tholeiitic and calcalkaline magmatism in the Aleutian Arc: *J. Geophys. Res. 87*, 4051-4072.

Kay, S. M., and Kay, R. W., 1985, Aleutian tholeiitic and calcalkaline magma series I: the mafic phenocrysts: *Cont. Min. Petr., 90*, 276-290.

Kay, S. M., and Kay, R. W., in press, Aleutian magmas in space and time: *in* Plafker, G., Jones, D.L.., and Berg, H. C. (eds.), *Geology of Alaska*, Geol. Soc. Am., The Geology of North America series.

Kienle, J., and Swanson, S. E., 1983, Volcanism in the eastern Aleutian arc: Late Quaternary and Holocene centers, tectonic setting and petrology:*J. Volc. Geotherm. Res. 17*, 393-432.

Kienle, J., Swanson, S. E., and Pulpan, H., 1983, Magmatism and subduction in the eastern Aleutian Arc: *in* Shimozuru, D., and Yokoyama, I. (eds.), *Arc Volcanism: Physics and Tectonics*, Terra Sci. Pub. Co., Tokyo, 191-224.

Lahr, J. C., and Plafker, G., 1980, Holocene Pacific-North American plate interaction in southern Alaska: implications for the Yakataga seismic gap: *Geology 8*, 483-486.

Le Bas, M. J., Le Maitre, R. W., Streckeisen, A., and Zanetin, B., 1986, A chemical classification of volcanic rocks based on the total alkali-silica diagram: *J. Petrol. 27*, 745-750.

Luedke, R. G., and Smith, R. L., 1986, Map showing distribution, composition, and age of Late Cenozoic volcanic centers in Alaska: *USGS Misc. Geol. Invest. Map I-1091-F.*

Marsh, B. D., 1979, Island-arc volcanism: *Am. Sci. 67*, 161-172.

Marsh, B. D., 1982, The Aleutians: *in* Thorpe, R.S. (ed.), *Andesites*, John Wiley & Sons, New York, 99-114.

Marsh, B. D., in press, Age, character, and significance of Aleutian arc volcanic rocks: *in* Plafker, G., Jones, D. L., and Berg, H. C. (eds.), *Geology of Alaska*, Geol. Soc. Am., The Geology of North America series.

Miller, T. P., and Barnes, I., 1976, Potential for geothermal energy development in Alaska: *in* Halbouty, M. T., Maher, J. C., and Lian, H. M. (eds.), *Circum-Pacific Energy and Mineral Resources*, Am. Assoc. Pet. Geol. Mem. 25, 149-153.

Miller, T. P., and Smith, R. L., 1987, Late Quaternary caldera-forming eruptions in the eastern Aleutian arc, Alaska: *Geology 15*, 434-438.

Miller, T. P., and Richter, D. H., in press, Quaternary volcanism in the Alaska Peninsula and Wrangell Mountains, Alaska: *in* Plafker, G., Jones, D. L., and Berg, H. C. (eds.), *Geology of Alaska*, Geol. Soc. Am., The Geology of North America series.

Minster, J. B., and Jordan, T. H., 1978, Present-day plate motions: *J. Geophys. Res. 83*, 5331-5354.

Miyashiro, A., 1974, Volcanic rock series in island arcs and active continental margins: *Am. J. Sci. 274*, 321-355.

Moll-Stalcup, E., in press, Cenozoic magmatism in mainland Alaska: *in* Plafker, G., Jones, D. L., and Berg, H. C. (eds.), *Geology of Alaska*, Geol. Soc. Am., The Geology of North America series.

Nakamura, K., Jacob, K. H., and Davies, J. N., 1977, Volcanoes as possible indicators of tectonic stress orientation – Aleutians and Alaska: *Pure Applied Geophys. 115*, 86-112.

Nakamura, K., Plafker, G., Jacob, K. H., and Davies, J. N., 1980, A tectonic stress trajectory map of Alaska using information from volcanoes and faults: *Bull. Earthquake Res. Inst. 55*, 89-100.

Page, R. A., Stephens, C. D., and Lahr, J. C., in press, Seismicity of the Wrangell and Aleutian WadBen zones and the North American Plate along the Trans-Alaska Crustal Transect, Chugach Mountains and Copper River basin, southern Alaska: *J. Geophys. Res.*

Scholl, D. W., Vallier, T. L., and Stevenson, A. J., 1987, Geologic evolution and petroleum geology of the Aleutian Ridge: *in* Scholl, D. W., Grantz, A., and Vedder, J. G. (eds.), *Circum-Pacific Energy and Mineral Resources*, Earth Science Series 6, Geology and Resource Potential of the Continental Margin of Western North American and adjacent Ocean basins-Beaufort Sea to Baja California, 123-189.

Simkin, T., Siebert, L., McClelland, L., Bridge, D., Newhall, C., and Latter, J. H., 1981, *Volcanoes of the World*: Smithsonian Inst., Hutchinson Ross Pub. Co., Stroudsburg, Pennsylvania, 232 pp.

Stephens, C. D., Fogleman, K. A., Lahr, J. C., and Page, R. A., 1984, The Wrangell WadBen zone, southern Alaska: *Geology 12*, 373-376.

Stone, D. B., 1988, Bering Sea–Aleutian Arc, Alaska: *in* Nairn, A. E. M., Stehli, F. G., and Uyeda, S. (eds.), *The Ocean Basins and Margins, 7B*, Plenum Publishing Co., 1-84.

Stone, D. B., and Wallace, W. K., 1987, A geological framework of Alaska: *Episodes 10*, 283-289.

Swanson, S. E., Kay, S. M., Brearley, M., and Scarfe, C.M., 1987, Arc and back-arc xenoliths in Kurile–Kamchatka and western Alaska: *in* Nixon, P. H. (ed.), *Mantle Xenoliths*, John Wiley and Sons, 303-318.

Swanson, S. E., Turner, D. L., Forbes, R. B., and Hopkins, D. M., 1981, Petrology and geochronology of Tertiary and Quaternary basalts from the Seward peninsula, western Alaska: *Geol. Soc. Am. Abst. with Program.13*, 563.

Turner, D. L., Swanson, S. E., and Wescott, E., 1981, Continental rifting – a new tectonic model for geothermal exploration of the central Seward Peninsula, Alaska: *Geotherm. Res. Council Trans. 5*, 213-216.

VOLCANOES OF ALASKA

Buldir, 18
Kiska, 18
Segula, 19
Davidof and Khvostof, 20
Little Sitkin, 20
Semisopochnoi, 21
Gareloi, 22
Tanaga, 23
Bobrof, 24
Kanaga, 24
Moffett and Adagdak, 25
Great Sitkin, 27
Kasatochi, 28
Koniuji, 28
Atka, 29
Seguam, 31
Amukta, 32
Chagulak, 33
Yunaska, 33
Herbert, 34
Carlisle, 34
Cleveland and Chuginadak, 35
Uliaga, 36
Kagamil, 36
Vsevidof, 37
Recheschnoi, 37
Okmok, 38
Bogoslof, 40
Makushin, 41
Akutan, 43
Gilbert, 44
Pogromni, 44
Westdahl, 45
Fisher, 46
Shishaldin, 48
Isanotski, 49
Roundtop, 49
Cold Bay, 50
Amak, 51
Dutton, 51
Emmons and Hague, 52
Pavlof and Pavlof Sister, 53
Dana, 54
Kupreanof (Stepovak Bay), 55
Veniaminof, 56
Black Peak, 58
Aniakchak, 59
Yantarni, 60
Chiginagak, 61
Kialagvik, 62
Ugashik and Peulik, 63
Ukinrek, 65
Kejulik, 66
Martin, 67
Mageik, 67
Trident, 68
Novarupta, Falling Mountain, and Cerberus, 70
Katmai, 71
Griggs, 72
Snowy, 73
Denison, Steller, and Kukak, 73
Devils Desk, 75
Kaguyak, 75
Fourpeaked, 77
Douglas, 78
Augustine, 79
Iliamna, 80
Redoubt, 81
Spurr, 83
Hayes, 84
Buzzard Creek, 85
Drum, 86
Sanford, 87
Wrangell, 88
Capital, 89
Tanada, 90
Eastern Wrangell, 91
White River Ash, 92
Old Crow Tephra, 92
Edgecumbe, 93
Duncan Canal, 94
Tlevak Strait and Suemez Island, 95
Behm Canal and Rudyerd Bay, 95
St. Paul, 96
St. George, 97
Nunivak, 98
Yukon Delta, 99
- *Ungulungwak Hill–Ingrichuak Hill, 100*
- *Ingakslugwat Hills, 100*
- *Kochilagok Hill, 101*
- *Nushkolik Mountain, 101*
- *Ingrisarak Mountain, 102*
- *Nelson Island, 102*

Togiak, 102
St. Michael, 103
Kookooligit, 104
Teller, 105
Espenberg, 106
Imuruk, 106
Koyuk–Buckland, 107
Porcupine, 108
Prindle, 108

BULDIR
Western Aleutian Islands

Type: *Stratovolcano cluster*
Lat/Long: *52.35°N, 175.0°E*
Elevation: *656 m*
Eruptive History: *No historic activity*
Composition: *High alumina basalt to andesite*

Buldir Island is the westernmost volcanic center of the present Pleistocene to Recent Aleutian volcanic front. The next westward subaerial volcanism is in Kamchatka. Buldir is a small (~2 km^3), isolated, and mountainous island consisting of two volcanoes, the older of which is Buldir volcano and the younger **East Cape** volcano. Although broadly of similar age, a significant lapse of time between their formation allowed considerable marine and subaerial erosion, the products of which fill the lowlands. Buldir volcano, which once had a parasitic cone, consists of a few thin (3-m), olivine-bearing, high alumina basalt flows and much volcaniclastic debris. East Cape volcano has two vents: the principal vent forms an eruptive cone cored by a late stage plug, whereas the secondary vent is a large flank dome of hornblende andesite.

Buldir Island is unusual in its restricted flora relative to neighboring islands, suggesting that it is comparatively young and not a fragment of a much older, larger subaerial island. The once nearly extinct Aleutian goose (a lesser Canada goose) was rekindled from relict nestings on Buldir.

How to get there: *Buldir is remote and inaccessible except by special marine transportation. The only geologic visit has been by R. R. Coats in 1947. Various personnel of the US Fish and Game Department (from which permission to visit must be obtained) have visited and may continue to visit Buldir for bird population studies. It may be possible to arrange regional transportation from Adak Island with the US Coast Guard.*

Reference

Coats, R. R., 1951, Geology of Buldir Island, Aleutian Islands, Alaska. *USGS Bull. 989-A*, 26 pp.

Bruce D. Marsh

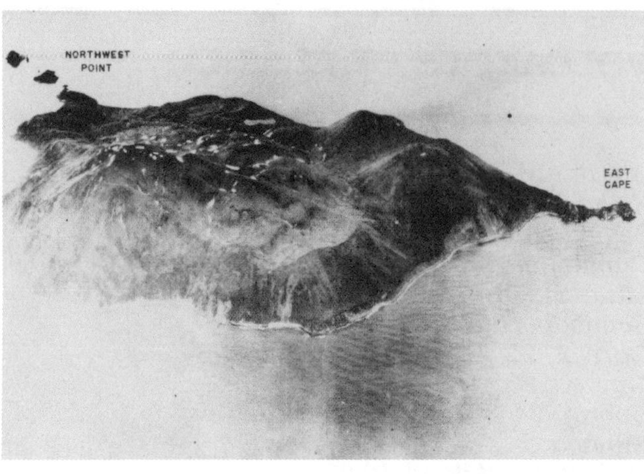

BULDIR: Aerial view, looking north (US Coast and Geodetic Survey).

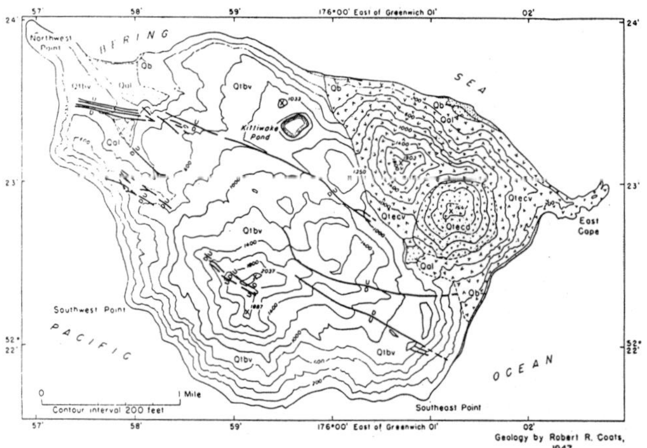

BULDIR: Simplified geologic map showing Quaternary basalt units of Buldir volcano on the left and East Cape volcano on the right. (From Coats, 1951.)

KISKA
Western Aleutian Islands

Type: *Stratovolcano*
Lat/Long: *52.10°N, 177.60°E*
Elevation: *1,220 m*
Dimensions: *8.5 × 6.4 km at base*
Eruptive History: *1907?, 1927?, 1962, 1964, 1969*
Composition: *Olivine basalt, andesite, dacite*

Kiska is a classic conical volcano composed of blocky lava flows and pyroclastic material. The active volcano sits on the remnants of an older volcano making up the Kiska Harbor Formation. The recent lava flows have fronts as

KISKA: Radar image (X band) from USGS Radar Image Mosaic Kiska sheet, 1982.

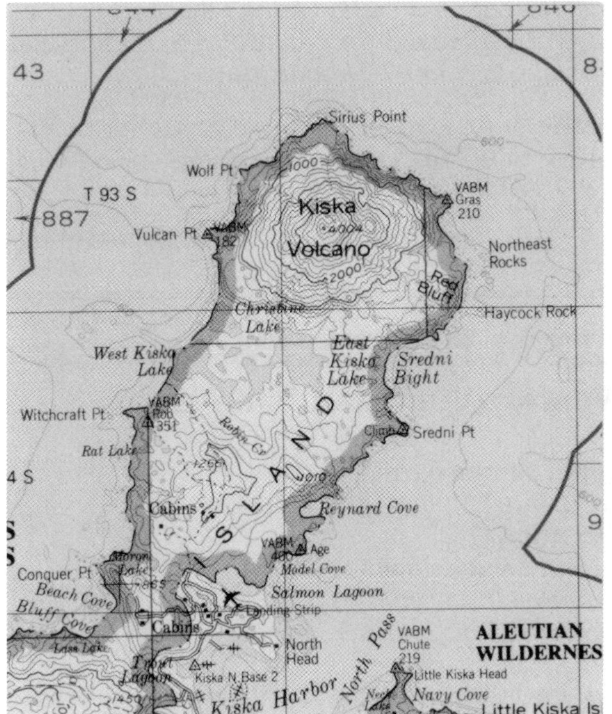

KISKA: Topographic map, USGS 1:250,000 series, Kiska sheet, contour interval 200 ft.

steep as 30 m, and erupted from both the summit and flanks. Evidence for the nature of the 1907 and 1927 activity is scanty. A moderate to large eruption in 1962 resulted in a 30-m-high cinder cone and accompanying lavas on the north flank, 3 km from the summit of the main cone. Lava flows were emplaced in 1964, and an ash column observed in 1969 was part of a moderate eruption. The surface of Kiska volcano is not glaciated and thus was formed after the Pleistocene glacial period.

Lavas are two-pyroxene and olivine basalts and two-pyroxene andesites. Hornblende andesite pyroclastics are common; crystal vitric ash – some from neighboring volcanoes – is largely dacitic.

How to get there: *Kiska is uninhabited and is ~150 km northwest of Constantine Harbor, Amchitka. Adak, the nearest permanent settlement from which a charter is possible, is ~365 km distant.*

References

Coats, R. R., Nelson, W. H., Lewis, R. Q., and Powers, H. A., 1961. Geologic Reconnaissance of Kiska Island, Aleutian Islands, Alaska: *USGS Bull. 1028-R.*

Panuska, B. C., 1980. Stratigraphy and sedimentary petrology of the Kiska Harbor Formation and its relationship to the Near Island-Amchitka lineament, Aleutian Islands: M.S. Thesis, University of Alaska.

Robert W. Kay

SEGULA

Western Aleutian Islands

Type: *Stratovolcano*
Lat/Long: *52.02°N, 178.13°E*
Elevation: *1,159 m*
Dimensions: *7.5 × 6.7 km*
Eruptive History: *No historic activity*
Composition: *Olivine basalt, andesite*

Segula volcano is built upon a 100-m-deep marine platform by the accumulation of flows and pyroclastic material emanating from a single center. Although there are no dated eruptions, some flows and pyroclastic deposits are so little weathered that ages of only a few hundred years are estimated. Pyroclastic material covers much of the cone's surface and a cinder cone resides in the middle of the summit crater.

How to get there: *Segula is uninhabited and is ~110 km from Constantine Harbor, Amchitka.*

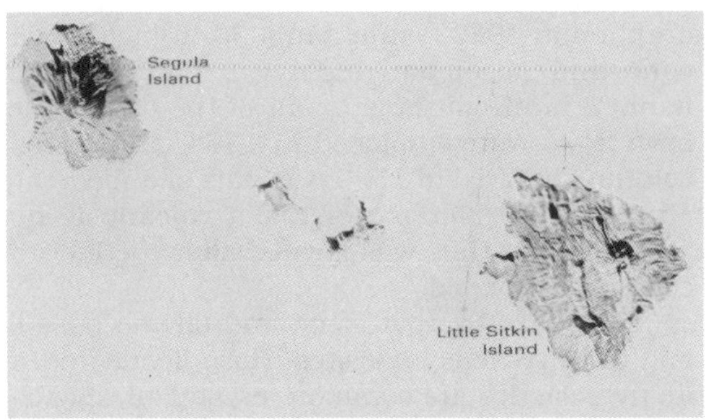

SEGULA, DAVIDOF AND KHVOSTOF, and LITTLE SITKIN: Radar image (X band) from USGS Radar Image Mosaic Rat Islands sheet, 1982.

Adak, the nearest permanent settlement from which a boat charter is possible, is >330 km distant. Landing on the island is difficult; sea cliffs greatly restrict access from the beach inland.

Reference

Nelson, W. H., 1959. Geology of Segula, Davidof and Khvostof Islands, Aleutian Islands, Alaska: *USGS Bull. 1028-K.*

Robert W. Kay

DAVIDOF AND KHVOSTOF

Western Aleutian Islands

Type: *Caldera remnants*
Lat/Long: *51.95°N, 178.30°E (Davidof); 51.96°N, 178.28°E (Khvostof)*
Elevation: *320 m (Davidof); 259 m (Khvostof)*
Caldera Diameter/Depth: *2.7 km/ 350 m*
Eruptive History: *No historical activity*
Composition: *Unknown*

Davidof, Khvostof, and nearby small islands Pyramid and Lopy rise 100 m above a submarine platform as the remnants of a collapsed caldera. This "Aleutian Krakatoa" is thought to have formed during the Late Tertiary, but the volcano is essentially unstudied. The islands are covered by vegetation; however lava flows can be recognized on aerial photographs. Lavas and pyroclastic layers form the islands, and rocks on the northern part of Davidof and Lopy Island are intensely hydrothermally altered.

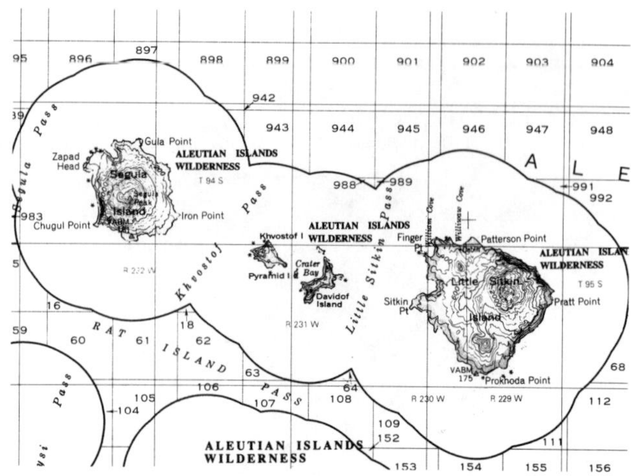

SEGULA, DAVIDOF AND KHVOSTOF, and LITTLE SITKIN: Topographic map, USGS 1:250,000 series, Rat Islands sheet, contour interval 200 ft.

Boulder Beach on the east coast of Davidof appears from the topographic map to be a breached maar crater.

How to get there: *Davidof and Khvostof are uninhabited and are 95 km northwest of Constantine Harbor, Amchitka. Adak, the nearest permanent settlement from which a boat charter is possible, is over 320 km distant.*

Reference

Nelson, W.H., 1959. Geology of Segula, Davidof and Khvostof Islands, Aleutian Islands, Alaska: *USGS Bull. 1028-K.*

Robert W. Kay

LITTLE SITKIN

Western Aleutian Islands

Type: *Stratovolcano*
Lat/Long: *51.95°N, 178.53°E*
Elevation: *1,202 m*
Island Dimensions: *11 × 9 km*
Eruptive History: *1776, 1828*
Composition: *Andesite, basalt, dacite*

Little Sitkin island is the site of repeated central volcano constructions with two caldera collapses separating three periods of cone growth. The oldest caldera is ~4.8 km in diameter; the youngest, 2.7 × 4.2 km. A small summit crater caps the youngest, active cone which has a basal diameter of 4.5 km. A small explosive eruption occurred in 1776. In 1828 two aa flows were extruded, one from the central crater, and the other from a fissure in one of the older calderas. The two oldest volcanic series are dominated by andesite and less abundant basalt; the youngest flows, domes, and pyroclastic units have dacitic compositions.

How to get there: *Little Sitkin is uninhabited and is 80 km from Constantine Harbor, Amchitka. Adak, the nearest permanent settlement from which a boat charter is possible, is over 300 km distant.*

References

Snyder, G. L., 1959. Geology of Little Sitkin Island, Alaska: *USGS Bull. 1028-H.*

Wolf, D. A., 1987, Identification of endmembers for magma mixing in Little Sitkin volcano, Alaska: M.S. thesis, SUNY, Albany, 201 pp.

Robert W. Kay

SEMISOPOCHNOI
Western Aleutian Islands

Type: *Shield caldera with subsequent stratovolcanoes*
Lat/Long: *51.93°N, 179.60°E*
Elevation: *855 m*
Island Dimensions: *21 × 18 km*
Eruptive History: *1873*
Composition: *Basalt, andesite, dacite*

Semisopochnoi is the largest young volcanic island in the western Aleutians and is composed of a variety of volcanic landforms. Basaltic pyroclastic material built a shield of ~20 km wide (at sea level) which culminated in a post-glacial pumice and ash eruption of dacite and andesite, producing an 8-km-wide caldera. Smaller composite cones are both pre- and post-glacial. Mount **Cerberus** is the most active of the three younger cones within the caldera. These young cones are dominantly two-pyroxene, high-alumina basalt, and andesite. One young composite cone (**Sugarloaf**) has olivine basalt. Dacite and andesite are found among the eruptive products of the pre caldera shield. Much of the island is covered by basaltic to andesitic ash derived from the younger cones. Semisopochnoi's tholeiitic differentiation trend (iron is enriched as silica increases) and relatively large volume are common in volcanoes near segment boundaries. Semisopochnoi is also on a small submarine ridge that extends northward as a part of the scorpion-tail-shaped Bowers Ridge; it is unclear if this setting influences its volcanism.

An historic eruption of Semisopochnoi was reported in 1873, and at least four others may have occurred in the previous hundred years, but documentation is scanty. These eruptions apparently emanated from the flanks of Mount Cerberus; the most recent flow appears to be less than a century old.

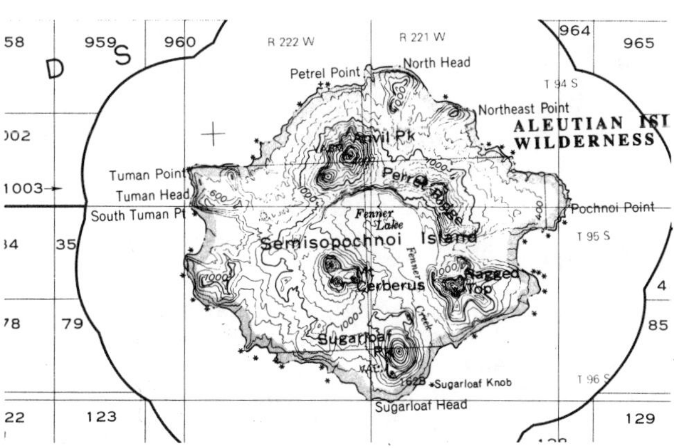

SEMISOPOCHNOI: Topographic map, USGS 1:250,000 series, Rat Islands sheet, contour interval 200 ft.

SEMISOPOCHNOI: Radar image (X band) from USGS Radar Image Mosaic Rat Islands sheet, 1982. This and other radar images in this chapter are often easier to interpret if turned upside down (i.e., south up).

How to get there: *Semisopochnoi is uninhabited. It is 60 km northeast of Constantine Harbor, Amchitka, and readily accessible when Amchitka is inhabited, as it is intermittently. Otherwise boat charters from Adak, nearly 360 km distant, may be available.*

References

Coats, R. R., 1959. Geologic reconnaissance of Semisopochnoi Island, Western Aleutian Islands: *USGS Bull. 1028-O.*

Delong, S. E., Perfit, M. T., McCulloch, M. T., and Ach, J., 1985, Magmatic evolution of Semisopochnoi Island, Alaska: *J. Geol. 93*, 609-618.

Robert W. Kay

GARELOI: Seen from the southwest. The site of the fissure eruption of 1929 is nearly along the flank horizon on the right (photo: US Coast and Geodetic Survey).

GARELOI

Western Aleutian Islands

Type: *Stratovolcano*
Lat/Long: *51.80°N, 178.80°W*
Elevation: *1,573 m*
Eruptive History: *1760?, 1790, 1791, 1792, 1828?, 1873, 1922, 1929-30, 1950-51, 1951, 1952, 1980, 1982*
Composition: *High alumina basaltic andesite*

Gareloi is a small (~10-15 km^3), simple conical volcano consisting of thin (~3 m) lavas interbedded with scoria and volcaniclastic materials. Two distinct periods of activity and cone building are apparent: an older stage followed by a significant eruptive hiatus with prolonged erosion of sea cliffs, culminating in renewed activity at essentially the same vent, but slightly northward. The composition of the lava and the style of volcanism have remained similar throughout the history of the volcano, except perhaps for the striking fissure eruption of 1929. In this eruption a series of 12 craters formed along a 4-km fissure striking southeast from the summit to the sea. Phreatic, pyroclastic, and lava eruptions occurred, producing deposits of accidental, pumiceous, and heterogeneous basaltic lava flows.

How to get there: *Gareloi is remote and inaccessible except by special marine transportation. It may be possible to arrange regional transportation from Adak Island with the US Coast Guard; permission must be gained from the US Fish*

GARELOI: Radar image (X band) from USGS Radar Image Mosaic Gareloi Island sheet, 1982.

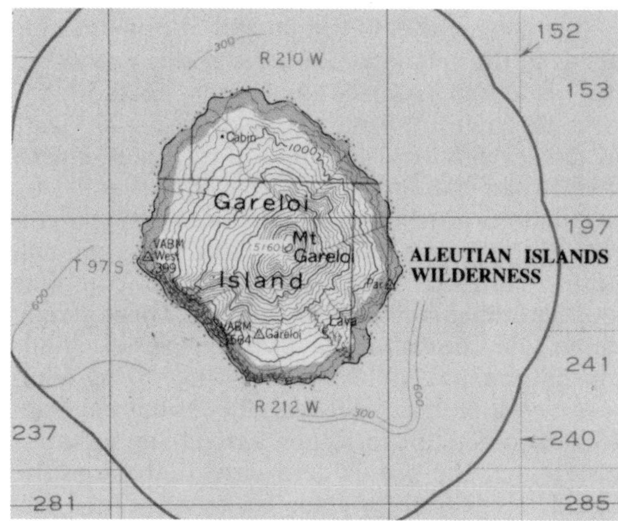

GARELOI: Topographic map, USGS 1:250,000 series, Gareloi Island sheet, contour interval 200 ft.

and Wildlife Service. The first (and evidently only) geologic visit was by R. R. Coats in 1946.

Reference

Coats, R.R., 1959, Geologic reconnaissance of Gareloi Island, Aleutian Islands, Alaska: *USGS Bull. 1028-J*,. 249-256.

Bruce D. Marsh

TANAGA
Central Aleutian Islands

Type: *Stratovolcano cluster*
Lat/Long: *51.88°N, 178.15°W*
Elevation: *1,806 m*
Eruptive History: *1763, 1791?, 1829, 1914*
Composition: *High alumina basalt*

Tanaga Island, one of the larger islands in the central Aleutians, is the second largest volcanic center (~75 km^3); Atka is the largest. Tanaga is typical of all large Aleutian Islands in consisting of a broad southern, mountainous expanse of Tertiary rocks and a Pleistocene to Holocene volcanic center on its northern extremity. The volcanic center itself is somewhat unusual in its approximately east–west alignment of volcanoes rather than the more common central clustering. The principal volcanoes are, from west to east, **Sajaka** (1,316 m), the two cones of Tanaga (1,784 m), and the truncated cone of **Takawangha** (1,462 m); each of these is an active volcano. Adjacent to Takawangha on the east are older, deeply eroded volcanoes, presumably of the next earlier pulse of volcanism, of which little is known.

Most Holocene activity has been from Tanaga volcano, whose two major cones have replaced an earlier large cone destroyed in a caldera-forming event; the sharp, ~400-m scarp of this caldera is still clear southeast of Tanaga volcano. Takawangha has an ice-filled caldera with minor late tephra cones on the flanks, on the rim, and within the caldera. Some of the caldera cones have had the most recent activity. Sajaka is probably the youngest of these volcanoes. The flows are generally thin ($\lesssim$3 m) and interbedded with volcaniclastic debris, which increases in abundance upward in the volcanic sequence.

TANAGA: Volcanoes (near to far, looking east-southeast): Sajaka, Tanaga, Takawangha. (Photo: US Coast and Geodetic Survey.)

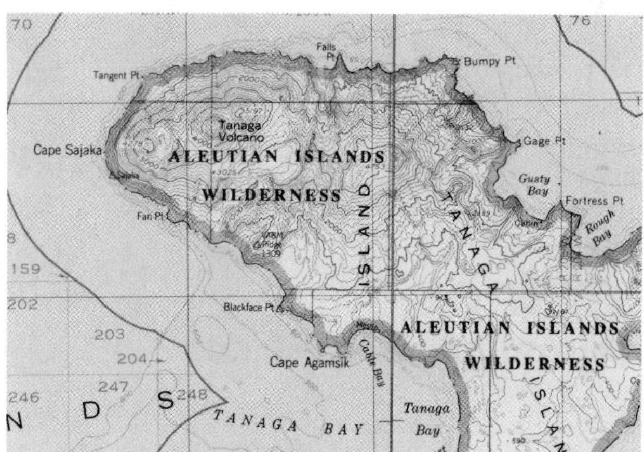

TANAGA: Topographic map from USGS 1:250,000 series.

The lavas themselves are predominantly crystal-rich, high alumina basalt, with phenocrysts of plagioclase, olivine, magnetite, and clinopyroxene. The most siliceous lavas are glassy dacite (~64% SiO_2) with phenocrysts of plagioclase, orthopyroxene, magnetite, and augite. The overall suite is low in MgO (7–2%) and shows a hint of both high and low potash trends, similar to Atka.

How to get there: *Tanaga is a remote, uninhabited, and inaccessible island. Access by sea might be possible from the US Coast Guard through Adak Island, and permission must be gained from the US Fish and Wildlife Service.*

Reference

Coats, R. R., and Marsh, B. D., 1984, Reconnaissance geology of northern Tanaga, Aleutian Islands, Alaska.: *Geol. Soc. Amer. Abst., with Program, 16*, 474.

Bruce D. Marsh

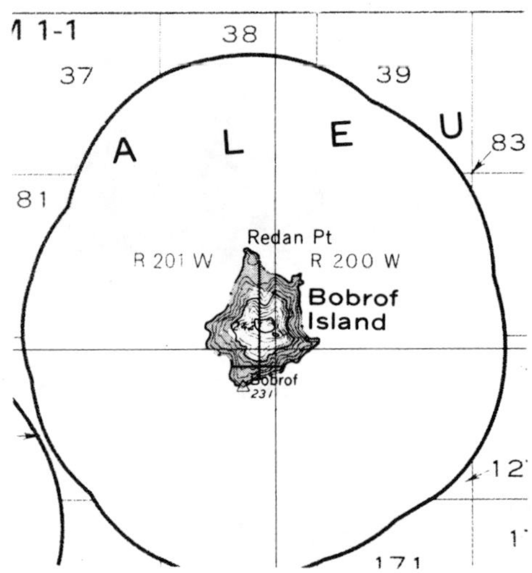

BOBROF: Topographic map, USGS 1:250,000 series, Adak sheet, contour interval 200 ft.

BOBROF
Central Aleutian Islands

Type: *Stratovolcano*
Lat/Long: *51.90°N, 177.43°W*
Elevation: *738 m*
Island Dimensions: *3 × 4 km*
Eruptive History: *No historic activity*
Composition: *High-alumina basalt and andesite*

Andesitic pyroclastic flows sampled at a seismic station (Adak network) are indicative of explosive eruptions; lavas at the same site indicate that more quiescent extrusive activity also occurred. Texturally, the rocks resemble those of Moffett volcano 50 km to the east. The geology of this small island and the timing of its volcanism are unknown.
How to get there: *Bobrof is uninhabited and is accessible by boat charter from Adak, 50 km to the east.*

Reference

Kay, S. M., and Kay, R. W., 1982, Tectonic controls on tholeiitic and calc-alkaline magmatism in the Aleutian arc: *J. Geophys. Res. 87*, 4051-4072.

Robert W. Kay

KANAGA: Seen from the southeast, looking across an unnamed lake. On the map: Tb = oldest lavas, Tkb = Ancient Mount Kanaton, QTb = Round Head lavas, Qkv = Kanaga Volcano, Qkb = recent lava flows.

KANAGA
Central Aleutian Islands

Type: *Stratovolcano*
Lat/Long: *51.90°N, 177.12°W*
Elevation and Height: *1,312 m*
Volcano Diameter: *10.3 km*
Eruptive History:
Ancient Mount Kanaton: Late Pliocene to early Pleistocene;
Kanaga: Quaternary to present; 1768, 1783, 1904, 1906, 1933, 1942
Composition: *Andesite*

Kanaga Island, one of the most southerly of the Aleutian Islands, is dominated by the presently active Kanaga volcano, a nearly perfect cone-shaped stratovolcano. The cone has been constructed within the caldera of an older late Pliocene to early Pleistocene volcano called **Ancient Mount Kanaton** which, in turn, has engulfed a still older small eruptive center. Their remains form a 760-m-high arcuate ridge (Kanaton Ridge) to the south and east of Kanaga volcano. All three volcanic sequences are dominated by hypersthene-augite andesite. Forming the eastern end of the island is **Round Head**, the remnants of a prominent basaltic satellitic cone formed on the lower slopes of Mount Kanaton. Kanaga Island is perhaps best known for a collection of ultramafic inclusions located in an isolated outcrop of alkaline basalt several

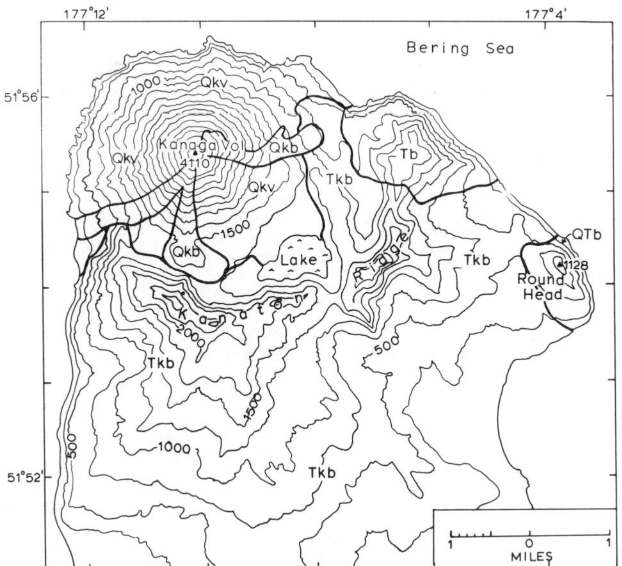

KANAGA: Topographic map, simplified from Coats (1956), showing historic and late-prehistoric basalt flows (Qkb), older volcanics (Qkv), and Tertiary rocks of Ancient Mount Kanaton (Tkb) and older (Tb).

KANAGA: Radar image (X band) from USGS Radar Image Mosaic.

kilometers to the southwest of the active volcanic center.

How to get there: *Kanaga is ~24 km west of Adak Island, site of the largest US military facility in the Aleutians. Access to Kanaga is possible only by boat or seaplane. Commercial charter services are not available in Adak, though arrangements might be made with military or civilian personnel who own small private boats. Enquiries should be directed towards the Small Boat Owners Association of Adak. Kanaga lies within the Aleutian Island Unit of the Alaska Maritime National Wildlife Refuge, administered by the US Fish and Wildlife Service. Applications for entry permits may be obtained from the Refuge Manager, Adak.*

References

Coats, R.R., 1952, Magmatic differentiation in Tertiary and Quaternary volcanic rocks from Adak and Kanaga Islands, Aleutian Islands, Alaska: *Geol. Soc. Am. Bull. 63*, 485-514.

Coats, R.R., 1956, Geology of Northern Kanaga Island, Alaska: *USGS Bull. 1028-D*, 69-81.

James G. Brophy

MOFFETT AND ADAGDAK

Central Aleutian Islands

Type: *Stratovolcano cluster*
Lat/Long: *51.55°N, 176.40°W*
Elevation: *1,200 m (Mount Moffett)*
Eruptive History: *No historic activity*
Composition: *High alumina basalt and andesite*

Adak is a large Tertiary island in the central Aleutians with a small (~40 km^3) volcanic center at its northern extremity. Kanaga lies to the west and Great Sitkin to the east. Because of its easy access, Adak is certainly the most frequently visited and sampled island in the Aleutians. **Andrew Bay** (~350 m, oldest), Mount Moffett (1,200 m), and Mount Adagdak (650 m) volcanoes have produced essentially all the Recent volcanic material. Only the erosional vestiges of Andrew Bay volcano remain (now filled by Andrew Bay and Lake); it was apparently obliterated by caldera formation, encroachment of the sea, and glaciation. Heavily glaciated, Mount Moffett consists principally of thick andesite flows, flank domes, and a substantial parasitic cone of many thin basalt flows. The scoriaceous, blocky dome on the outward south flank of Mount Moffett may be one of the youngest volcanic features of this center. Mount Adagdak is a model composite cone with a distinct lower shield of one or two basalt flows and interbedded scoria. At ~350 m the small stratovol-

MOFFETT AND ADAGDAK: Parasitic cone of Mount Moffett (foreground), Andrew Bay (lake and cliff fragment in middle ground), and Adagdak (background). Far in the distance on the right is Great Sitkin volcano.

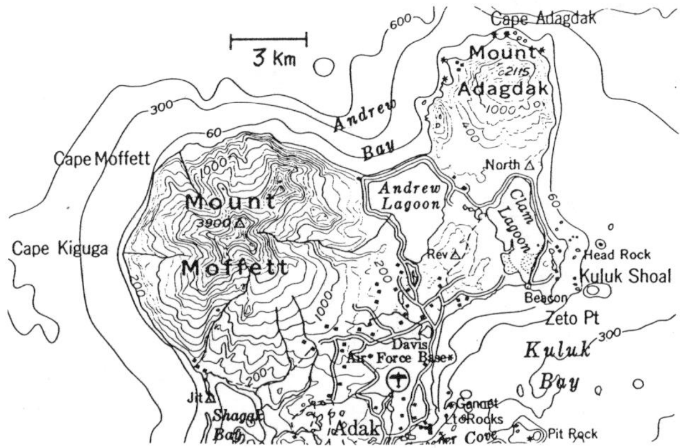

MOFFETT AND ADAGDAK: Topographic map from USGS 1:250,000 map series.

MOFFETT AND ADAGDAK: Radar image (X band) from USGS Radar Image Mosaic.

cano begins, consisting mainly of volcaniclastic debris and an occasionally thin, fragmentary andesitic flow. The summit crater, with a well-defined south rim, is occupied by a hornblende andesite plug; in places it has vertical, smooth walls peppered with indigenous xenoliths. At one time this plug may have been partly a Pelean spine, large blocks of which are scattered across the shield. A northwesterly directed explosion and ash flow may have strongly breached the summit crater prior to the vent-filling event. A late stage basaltic dome lies on the southeast flank.

Although Moffett and Adagdak are certainly volcanoes, when compared to large Aleutian volcanoes, both vents represent small, almost futile outpourings. This is perhaps reflected in the heterogeneous nature and composition of the erupted materials. Mafic, olivine-rich xenolithic material is common in a thick andesite flow on the north shore of Mount Moffett, and also in an apparently phreatic vent on the west shield of Mount Adagdak; gabbroic and dioritic xenoliths are found along the bouldery beach north of Mount Adagdak.

How to get there: *Adak is perhaps the most readily accessible Aleutian island. The US Navy maintains a Naval Station and ancillary support on Adak with a population of around 5,000. The island is serviced daily by commercial aircraft (Reeve Aleutian Airways), but permission to enter Adak must be obtained from the Station Captain well in advance of arrival. Transportation on the island itself is most easily gained by renting private vehicles from Naval personnel.*

References

Coats, R. R., 1956, Geology of northern Adak, Alaska: *USGS Bull. 1028-C*, 47-67.

Marsh, B. D., 1976, Some Aleutian andesites: their nature and source. *J. Geol. 84*, 27-42.

Bruce D. Marsh

GREAT SITKIN: Photograph by R. W. Kay.

GREAT SITKIN
Central Aleutian Islands

Type: *Stratovolcano*
Lat/Long: *52.07°N, 176.13°W*
Elevation: *1,740 m*
Volcano Diameter: *8 × 10 km*
Eruptive History: *1792, 1933, 1945, 1949, 1953, 1974*
Composition: *Basaltic andesite to dacite*

Great Sitkin is an eroded, but active stratovolcano with a summit caldera, the whole rising from the caldera of an older shield volcano. Lavas and pyroclastic products (**Sand Bay** Volcanics) of this earlier shield are exposed south of Great Sitkin. The summit caldera (~1.5 × 2.5 km) of Great Sitkin has been the site of frequent eruptive activity this century, building a broad dome of olivine-bearing, two pyroxene andesite. An eruption of ash in 1974 was monitored from nearby Adak.

Three plugs of andesite occur in a northwest-trending line ending at the summit dome; two additional plugs formed north of the line. A thick dacitic ignimbrite is exposed on the western side of the volcano, and the entire island is blanketed with light brown to black pumice. Some pumice blocks contain compositionally distinct (basaltic andesite and high silica andesite) contiguous bands indicative of magma mixing. Another noteworthy constituent of the pumice deposits are cumulative gabbro xenoliths (up to 10 cm across) composed of plagioclase, clinopyroxene, olivine, and magnetite.

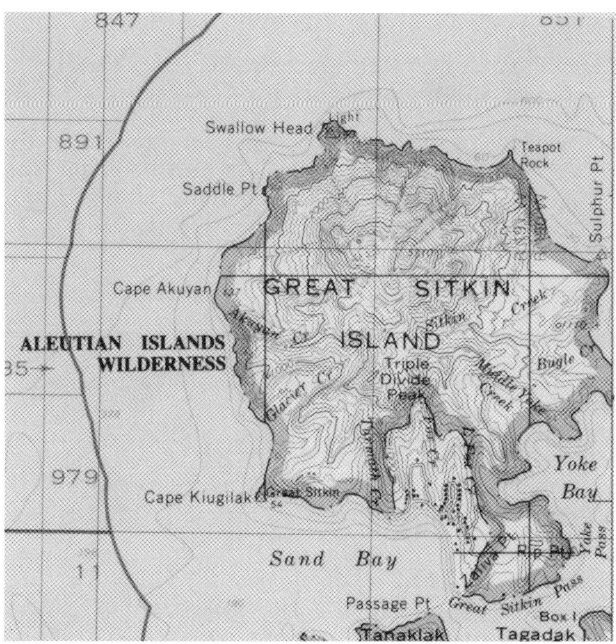

GREAT SITKIN: Topographic map, USGS 1:250,000 series, Adak sheet, contour interval 200 ft.

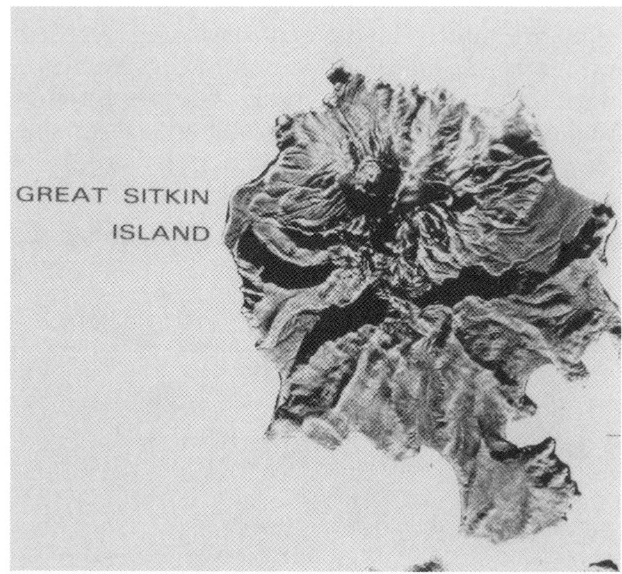

GREAT SITKIN: Radar image (X band) from USGS Radar Image Mosaic, Adak sheet.

How to get there: *Boat charters to this uninhabited island from Adak naval base (40 km to the west) can be arranged. Landing is easiest at Sand Bay (10 km south of Great Sitkin volcano), but is possible in many other places depending on sea conditions. Great Sitkin is in the Aleutian Island Wildlife Refuge.*

References

Simons, F. S., and Mathewson, D. E., 1955, Geology of Great Sitkin Island, Alaska: *USGS Bull. 1028-B*, 21-43.

Neuweld, M. A., 1987, The petrology and geochemistry of the Great Sitkin suite: Implications for the genesis of calc-alkaline magmas: M.S. Thesis, Cornell University, Ithaca, NY, 174 pp.

Robert W. Kay

KASATOCHI
Central Aleutian Islands

Type: *Stratovolcano*
Lat/Long: *52.18°N, 175.50°W*
Elevation: *301 m*
Island Dimensions: *2.7 × 3.3 km*
Eruptive History: *1760, 1827-8?*
Composition: *Basalt, andesite, dacite*

Kasatochi is a small island volcano with a broad crater. On the northwest side of the volcano are interbedded basalt flows and pyroclastic material, including xenolithic fragments of hornblende-plagioclase rock. A dome intrudes the northwest side of the volcano at an elevation of ~150 m. The 1760 activity emplaced a lava flow. Lava flows and scoria are olivine basalt; pyroclastics and dome lavas are hornblende andesite and dacite.

KASATOCHI: Topographic map, USGS 1:250,000 series, Atka sheet, contour interval 200 ft.

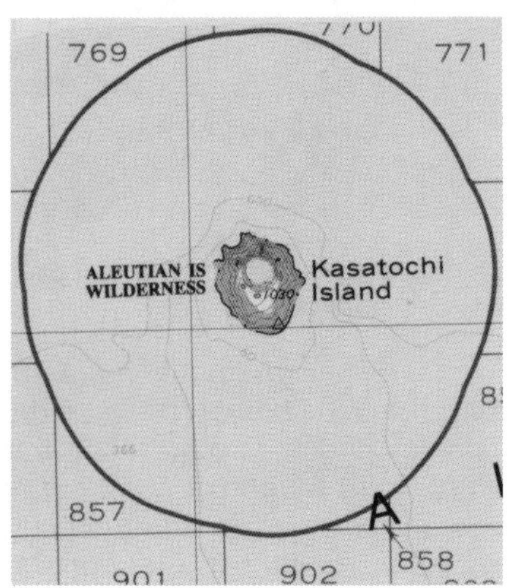

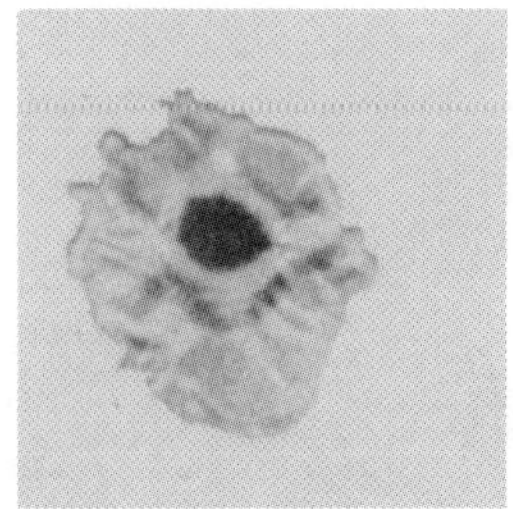

KASATOCHI: Radar image (X band) from USGS Radar Image Mosaic Atka sheet, 1982.

How to get there: *Kasatochi is uninhabited, and accessible from Adak (85 km to the west) or Atka only in favorable sea conditions, as there are no sheltered parts of the coastline.*

Reference

Kay, S. M., and Kay, R. W., 1985, Aleutian tholeiitic and calc-alkaline magma series, I: The mafic phenocrysts: *Cont. Min. Petr. 90*, 276-290.

Robert W. Kay

KONIUJI
Central Aleutian Islands

Type: *Small cone*
Lat/Long: *52.22°N, 175.12°W*
Elevation: *272 m*
Volcano Diameter: *5 km at seafloor*
Island Dimensions: *1 × 1.5 km*
Eruptive History: *No historic activity*
Composition: *High alumina basalt, andesite*

Koniuji is the top of a mostly submerged dormant stratovolcano. It lies between volcanoes of Atka Island on the east and Kasatochi Volcano on the west. The volcano has not been mapped geologically and the only known geologic study was completed during a short two-hour stop along the north shoreline in 1981 when geologists from the USGS research vessel S. P. LEE landed in a small boat to collect samples. The collected rocks are from andesite flows and plugs, thick basalt flows, and pyroclastic deposits. A northeast-trending fault cuts the island on the southeast. A possible vent occurs just north of the highest point on the island.

Hornblende-rich andesite flows contain diorite and andesite porphyry xenoliths that have approximately the same composition as the

KONIUJI: Photograph looking southwest.

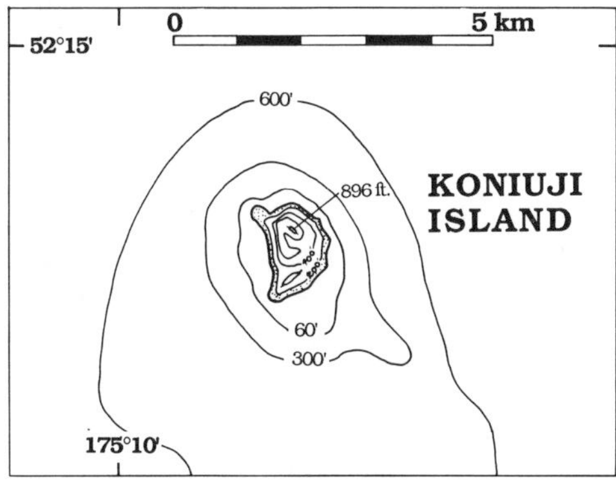

KONIUJI: Topography from USGS Alaska Topographic Series, Atka, Alaska, 1:250,000

flows. Basalt flows are >10 m thick in places. Red and gray tuff beds form bold outcrops along the north shore.

How to get there: *The best access to the volcano is by means of a ship and inflatable boat. A landing strip near the village of Atka on the southeast side of Atka Island makes the region accessible by plane from both Adak and Dutch Harbor (Unalaska Island). Once on Atka, it might be possible to hire someone to ferry geologists and equipment to Koniuji in a small boat. The distance and the amount of open water between Atka and Koniuji islands, however, would make the trip both expensive and dangerous.*

References: None

Tracy Vallier

ATKA
Central Aleutian Islands

Type: *Cluster of stratovolcanoes and caldera*
Long/Lat: *52.33°N, 174.15°W*
Elevation: *1,479 m (Korovin), 1,464 m (Kliuchef), 1,135 m (Konia), 1,066 m (Sarichef)*
Eruptive History: *Korovin: 1829?, 1830?, 1844?, 1907, 1951, 1974, 1987; Kliuchef: 1812; Konia, none recorded, but probable historic; Sarichef: 1812 as recorded, but was most likely Kliuchef*
Composition: *High alumina basalt with some andesite and dacite.*

Atka is the largest (~200 km^3) volcanic center in the central Aleutians. There are no larger centers westward, and the closest larger center is Umnak, some 300 km east. Seguam lies directly to the east and Great Sitkin to the west. The overall structure of the center is that of a broad central shield which once supported a large (~2,200 m) central cone (**Atka** volcano) ringed by as many as 7 or 8 smaller satellite volcanoes. The central cone was lost to caldera formation, shutting down the whole system, and the satellite vents still remain at various stages of erosional decay. **Sarichef** is perhaps the youngest satellite vent and has survived erosion largely unscathed. More often, summit ice buildup has breached the crater walls, forming active cirques, which have deeply incised the satellite vents. Tangential to these vents are U-shaped valleys, formed by moving ice.

Kliuchef volcano grew on the north rim of the now ice-filled Atka caldera and formed a series of five vents striking northeast. The two main summit vents and the easternmost vent are fresh; the latter is most likely the source of the 1812 eruption attributed to Sarichef. Double-coned **Korovin** volcano next appeared, although overlapping considerably in time with Kliuchef. Six km north of Kliuchef, **Konia** volcano occupies the middle ground between Korovin and Kliuchef and is as old as much of Korovin itself. Korovin has been and remains the principal active volcano on Atka. It is unusual in that its summit crater marks an open, cylindrical vent reaching nearly to sea level; it has been observed by pilots flying over to sometimes

ATKA: Aerial photograph looking northeast, showing, from left to right: Korovin, Konia, Kliuchef. US Coast and Geodetic Survey photograph.

ATKA: Radar image (X band) from USGS Radar Image Mosaic Atka sheet, 1982.

contain a crater lake and at other times to be brimming with magma. This vent is the source of most recent eruptions.

All the volcanoes consist principally of crystal-rich, thin (<~3 m) basaltic lavas with interbedded scoria yielding increasing amounts of pyroclastic debris, autobreccias, and lahars. The summit of Kliuchef is mostly glassy dacite as are some late flows of Korovin and Konia. A thick (~400 m) pink dacite with pumiceous and glassy cooling units was erupted upon formation of Atka caldera, but no ash flows have been found. Thick, expansive lahar aprons fill many early Pleistocene glacial valleys. These have

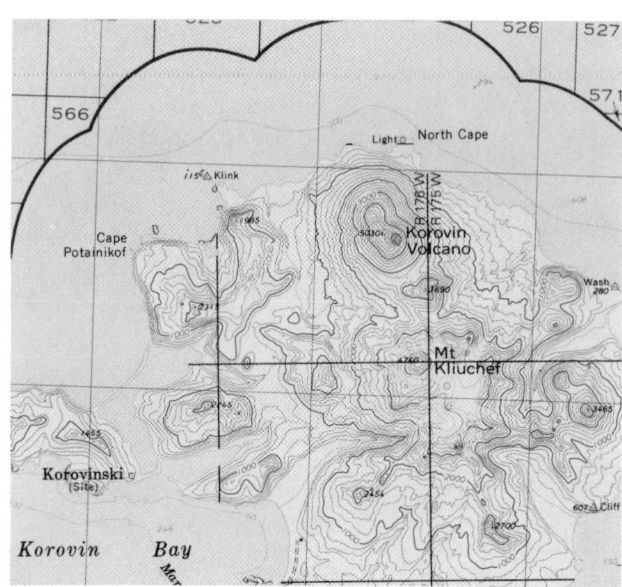

ATKA: Topographic map from USGS 1:250,000 Atka sheet; contour interval 200 ft.

been cut by parse dikes spanning the volcanic center, whose emplacement apparently attended caldera collapse. The lavas themselves are overwhelmingly (≥90% by volume) high-alumina basalt, strikingly free of xenoliths. The basalt contains plagioclase, orthopyroxene, magnetite, and clinopyroxene. The andesite and dacite contain orthopyroxene at the expense of olivine; trace amounts of biotite are also found in the dacite. No hydrous phases are found in any basalt or andesite lava. Three hot spring areas are found apparently associated with Kliuchef and Atka caldera, and a fourth spring occurs some 7.5 km west of Kliuchef.

How to get there: *Atka has a long-standing Aleut village on Nazan Bay which is serviced at least monthly by either ship or aircraft. Details of this transportation can be obtained from either the Aleut Corporation or the Bureau of Indian Affairs (Anchorage, Alaska). Permission to visit Atka must also be obtained from the village itself. Transportation in and around Atka is by hired open boat and by foot.*

References

Marsh, B. D., 1982, The Aleutians, *in*: Thorpe, R. S. (ed.): *Andesites*, J. Wiley and Sons, N.Y., 99-114.

Myers, J. D., Marsh, B. D., and Sinha, A. K., 1985, Strontium isotope and selected trace element variations between two Aleutian volcanic centers (Adak and Atka): implications for the development of arc plumbing systems. *Cont. Min. Petr.* *91*, 221-234.

Bruce D. Marsh

SEGUAM
Central Aleutian Islands

Type: *Two calderas*
Lat/Long: *52.32°N, 172.48°W*
Elevation: *1,054 m*
Island Diameter: *24 × 11.5 km*
Eruptive History: *1786-90, 1827, 1891, 1892, 1902, 1927?, 1977*
Composition: *Basalt to rhyodacite*

Seguam is an elliptical island with two calderas, each with an associated central volcano. The western caldera is 1.7 km in diameter, and its walls reach ~685 m above sea level. A central cone, **Pyre Peak**, rises 1,054 m above the caldera floor, and is topped by a 300-m-wide and 75-m-deep crater. Most historic eruptions are probably from the vicinity of Pyre Peak. In 1977 a 2.5-km-long fissure eruption occurred 2.5 km southeast of the peak, and in 1987 a thermally active region was observed just south of Pyre Peak. On the southwest end of the island a 717-m peak may represent another major volcanic vent.

A second unnamed caldera and central volcano occur on the eastern end of the island. The caldera wall is preserved only along the eastern margin and rises steeply from the caldera floor at 532 m to over 880 m. The central volcano of this caldera has a basal diameter of 1.9 km and is 313 m high. Immediately west of this complex are three large silicic lava flows. East of the caldera is a 589-m-high volcanic edifice. This end of the island does not appear to have been volcanically active recently.

Very recent flows have erupted from the central part of the island and flowed south to the shore. In the west the lavas are aa basalt that appears to have originated at Pyre Peak. Near Lava Cove (slightly east of the center of the island) the flows are rhyodacite up to 30 m thick. Geographically, these siliceous flows are spatially close to the chemically similar flows from the eastern caldera. Thus far, only abstracts describing the southern side of the is-

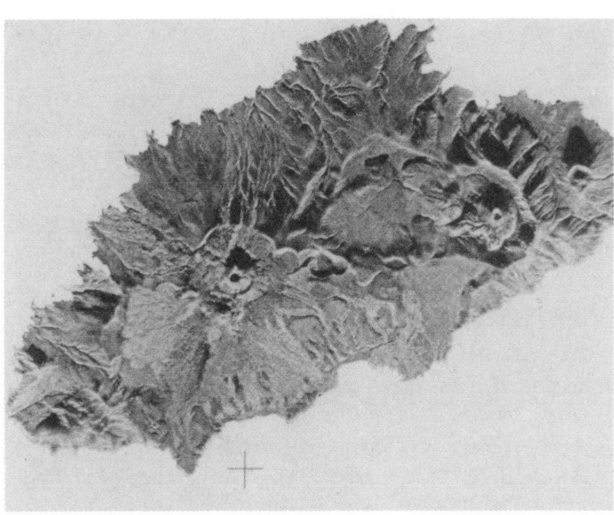

SEGUAM: Radar image (X band) from USGS Radar Image Mosaic Seguam sheet, 1982.

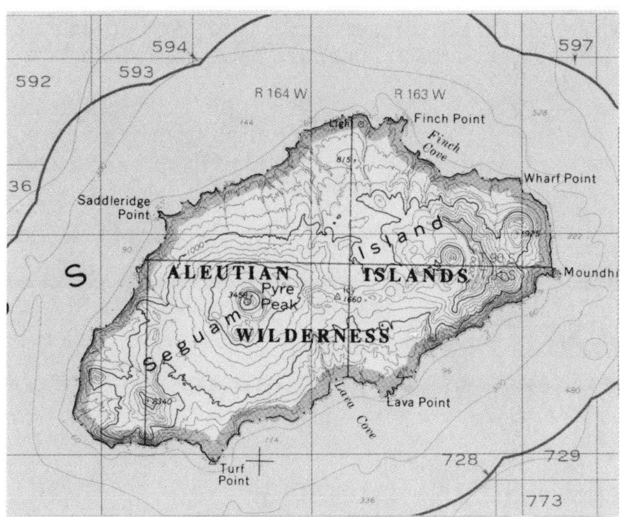

SEGUAM: Topographic map, USGS 1:250,000 series, Seguam sheet, contour interval 200 ft.

SEGUAM: Photograph of Seguam rising above cloud bank; photograph by R. W. Kay.

land are available; additional work is in progress.

How to get there: *The islands of the central Aleutians arc (Seguam to Chuginadak) are uninhabited and exceedingly difficult to reach. Although the Islands of Four Mountains region may be reachable by floatplane from Dutch Harbor, Alaska, the islands to the west (Seguam to Yunaska) are accessible only by boat. Useful information about possible landing and camp sites can be obtained from the US Coast Pilot. Because the islands are commonly subject to extremely severe storms, called williwaws, fieldwork can be very dangerous as well as difficult. The islands are part of the Aleutian Maritime Wildlife Refuge, and permission must be obtained from the US Fish and Wildlife Service to conduct fieldwork on the islands.*

References

Myers, J. D., and Singer, B. S., 1987, Seguam Island, Central Aleutian Islands: I. Geologic field relations: *EOS 68*, 1525.

Singer, B. S., and Myers, J. D., 1988, Major and trace element characteristics of lavas from Seguam Island, central Aleutian Islands, Alaska: *Geol. Soc. Am. Abst. with Program, 20*, 196.

James D. Myers

AMUKTA

Central Aleutian Islands

Type: *Stratovolcano*
Lat/Long: *52.50°N, 171.25°W*
Elevation: *1,064 m*
Island Diameter: *8 km*
Eruptive History: *1786-91, 1876?, 1963, 1984, 1987*
Composition: *Basaltic andesite*

AMUKTA and CHAGULAK: Radar image (X-band, NW look).

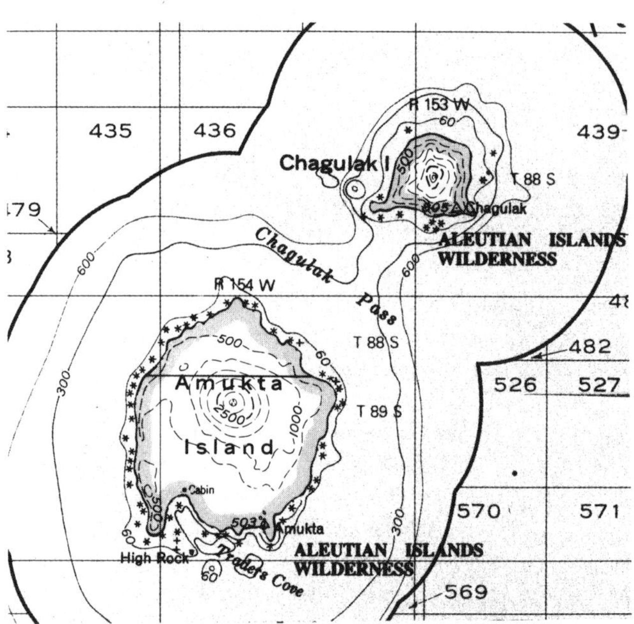

AMUKTA and CHAGULAK: USGS topographic map.

Amukta is an unstudied volcanic island, with a young stratovolcano resting on a broad basal shield. The volcanic cone, which is displaced to the north from the center of the island, has a summit crater, and a small cinder cone occurs on its northeast flank. An arcuate ridge on the southwest side of the island may be a remnant of a caldera wall from an earlier volcano. The 1963 activity included explosions from both the central crater and parasitic vents, producing ash and lava flows. Activity in the 1980s consisted of summit eruptions only. A single grab sample from the north-northeast shore of Amukta is a low potassium basaltic andesite (54% SiO_2, 0.60% K_2O).

How to get there: *Amukta is an uninhabited island in the Aleutian Island Wildlife Refuge, with no good harbors. Access is by charter boat from Nikolski, 118 km to the east; landing by small boat in favorable sea conditions is possible.*

Reference

Kay, R.W., 1988, unpublished chemical data.

James D. Myers

CHAGULAK
Central Aleutian Islands

Type: *Stratovolcano*
Lat/Long: *52.57°N, 171.13°W*
Elevation: *1,142 m*
Island Diameter: *3 km*
Eruptive History: *No historic activity*
Composition: *Basaltic andesite*

Chagulak is a small stratovolcano separated from Amukta by ~7 km of ocean, but the two coalesce at depth. Chagulak is unstudied; the only geologic information available is a single chemical analysis. A low-potassium, high-alumina basaltic andesite (51% SiO_2, 0.59% K_2O) was recovered from the north-northwest shore of the island, but the sample's geologic setting is unknown.

How to get there: *Chagulak is an uninhabited island in the Aleutian Island Wildlife Refuge with no good harbors. Access is by charter boat from Nikolski, 109 km to the east; landing by small boat in favorable sea conditions is possible.*

Reference

Kay, R.W., 1988, unpublished chemical data.

Robert W. Kay

YUNASKA
Central Aleutian Islands

Type: *Caldera and linear ridge*
Lat/Long: *52.63°N, 170.70°W*
Elevation: *914 m*
Island Dimensions: *22.9 ×8.9 km*
Eruptive History: *1817?, 1824, 1830, 1873, 1937*
Composition: *Andesite*

Yunaska Island is elongated in a northeast-southwest direction and is composed of two major, but nameless, volcanic centers. The northeast end of the island consists of a large strato-

YUNASKA: Radar image of Yunaska Island (X-band, northwest look).

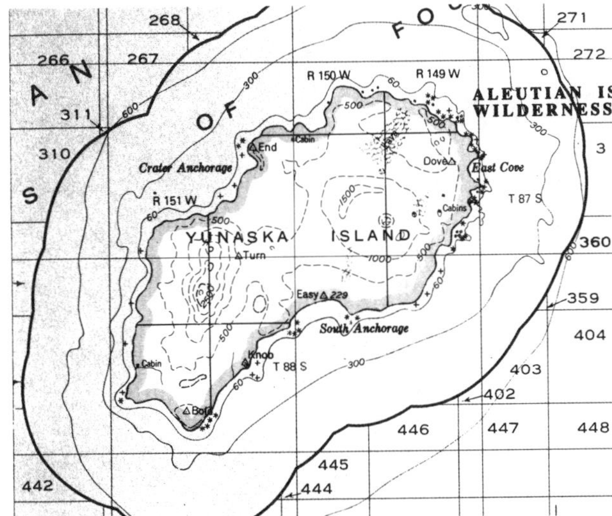

YUNASKA: USGS topographic map.

volcano with a central caldera. The volcano rests on an 11kmwide basal shield with elevations between 150 and 300 m. Small domes or cinder cones are evident on the northwest part of the shield. The 3.5kmwide caldera is located slightly to the southeast of the shield center, and has well-preserved walls on all but the southwest side, where it appears to be breached by post-caldera eruptions. The caldera walls are ~150 m high on the west and north, and have a maximum height of 457 m. A small dome (or perhaps a cone) occurs on the eastern floor of the caldera.

The western half of the island, which is markedly different topographically, is elongate in a north–south direction,with a length of 13.5 km and a width of 6.8 km. A linear ridge (fissure?) nearly 915 m high runs through the center of this part of the island, paralleling the trend of the elongation. No published geologic studies of this volcanic center are currently available. A sample from the north-northwest shore of the island is an andesite (63% SiO_2, with 1.13 wt. % K_2O); the geologic setting of the sample is unknown.

How to get there: *As with Seguam, access to Yunaska is difficult. The island is part of the Aleutian Wildlife Refuge.*

Reference

Kay, R. W., 1988, unpublished chemical data.

James D. Myers

HERBERT

Central Aleutian Islands

Type: *Caldera and linear ridge*
Lat/Long: *52.75N, 170.12°W*
Elevation: *1,290 m*
Volcano Diameter: *9.8 km*
Eruptive History: *No historic eruptions*
Composition: *Andesite*

Herbert is a classic stratovolcano with a 2.1-km-wide caldera, which the topographic map indicates is breached to the northwest. Radar images do not show the breach. No geologic studies have been published on this volcano. A grab sample from the tip of the island's southwest point is an andesite (59% SiO_2, with 0.82% K_2O).

How to get there: *See Seguam.*

Reference

Kay, R. W., 1988, unpublished chemical data.

James D. Myers

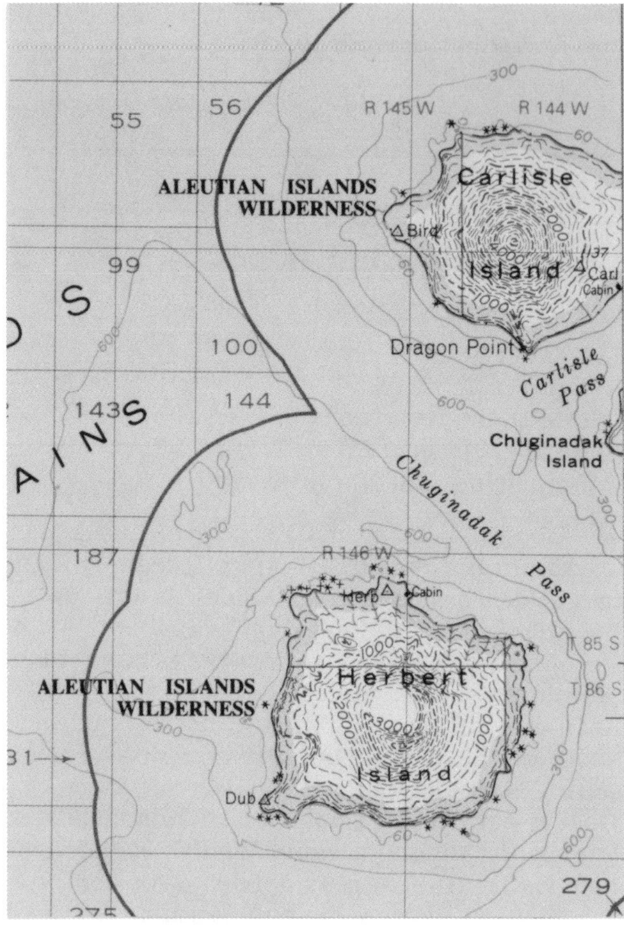

HERBERT and CARLISLE: USGS topographic map.

CARLISLE

Central Aleutian Islands

Type: *Stratovolcano*
Lat/Long: *52.90°N, 170.05°W*
Elevation: *1,620 m*
Volcano Diameter: *7 km*
Eruptive History: *1774, 1828, 1838?, 1987*
Composition: *Andesite*

Carlisle is a steep-sided volcano, lacking a caldera. Radar images suggest there may be two closely spaced volcanic cones. No geologic studies have been published on this volcano; however, a sample recovered from the eastern shore has 59.4% SiO_2 and 1.09 wt.% K_2O. The geologic field relations of this sample are unknown.

How to get there: *See Seguam.*

HERBERT and CARLISLE: USGS radar image (X-band, northwest look).

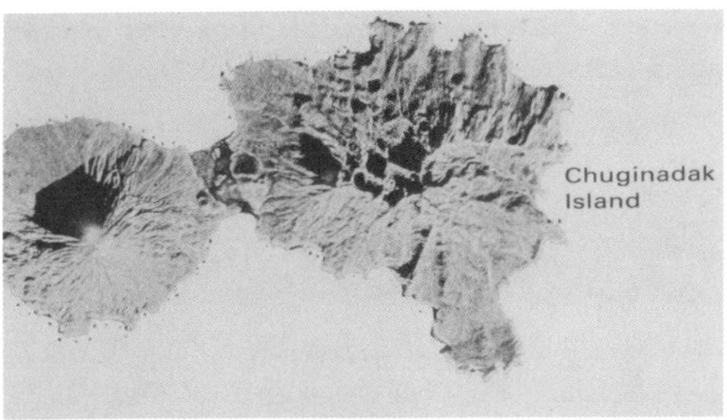

CLEVELAND and CHUGINADAK: USGS radar image (X-band, northwest look).

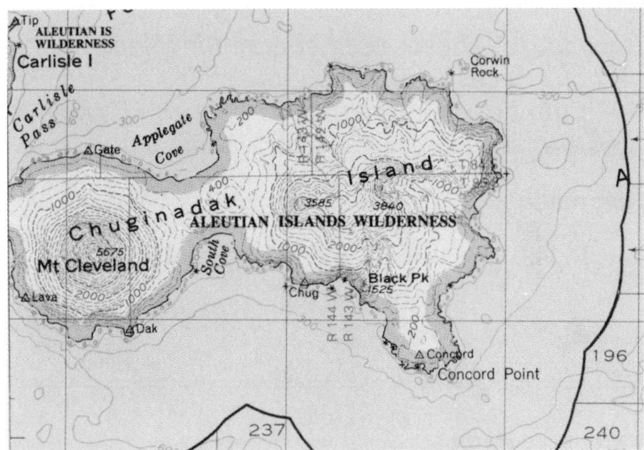

CLEVELAND and CHUGINADAK: USGS topographic map.

Reference

Kay, R. W., 1988, unpublished chemical data.

James D. Myers

CLEVELAND AND CHUGINADAK
Central Aleutian Islands

Type: *Stratovolcanoes*
Lat/Long: *52.82°N, 169.95°W*
Elevation: *1,730 m*
Island Dimensions: *23 × 14 km*
Eruptive History: *(Mount Cleveland): 1893, 1897, 1929?, 1932, 1938, 1944, 1951?, 1975?, 1984, 1985, 1986, 1987*
Composition: *Unknown*

Chuginadak Island is dumbbell-shaped, with its long axis oriented almost east–west. The eastern part of the island is irregular in outline with a diameter of ~14 km, and is composed of two prominent, but unnamed, volcanoes (1,170 m and 1,093 m in elevation). This part of the island has been extensively eroded and does not appear to have been volcanically active for some time. A low-silica rhyolite has been recovered from the east side of Concord Point.

The western half of the island is formed by Mount Cleveland, a 1,730-m-high and 8-km-wide stratovolcano. Numerous large lava flows are evident on the flanks of this almost perfectly symmetrical volcano, but no caldera is present. Mount Cleveland is one of the most active volcanoes in the Aleutian chain. In 1944, a US Army serviceman was killed by an eruption of Mount Cleveland; the volcano has been especially active since 1984.

The two parts of the island are joined by a narrow (2.7 km) and low-lying (<121 m) isthmus.

CLEVELAND and CHUGINADAK: Low oblique aerial view, looking north. Photograph by R. W. Kay.

Two small, broad volcanoes occur on the isthmus; one is on the flank of Mount Cleveland, and the larger is at the eastern end of the isthmus.

How to get there: *See Seguam for general information. Access by boat charter from Nikolski, 50 km to the east, with landing by small boat, is possible in favorable sea conditions.*

Reference

Kay, R. W., 1988, unpublished chemical data.

James D. Myers

ULIAGA

Central Aleutian Islands

Type: *Stratovolcano*
Lat/Long: *53.07°N, 169.77°W*
Elevation: *887 m*
Island Dimensions: *3.3 km*
Eruptive History: *No historic activity*
Composition: *Unknown*

Uliaga is a triangular-shaped island composed of an eroded stratovolcano. No evidence for a caldera is seen, but little is known geologically about this volcano.

How to get there: *See Seguam.*

References: None

James D. Myers

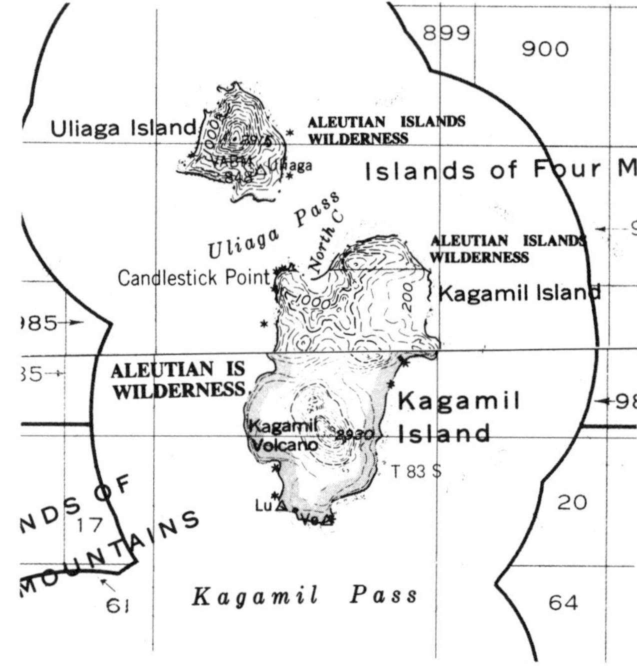

ULIAGA and KAGAMIL: USGS topographic map.

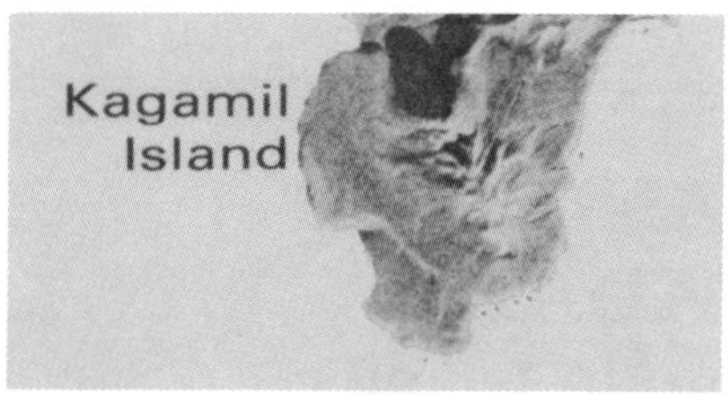

KAGAMIL: Radar image, northwest look.

KAGAMIL

Central Aleutian Islands

Type: *Small stratovolcano*
Lat/Long: *52.98°N, 169.72°W*
Elevation: *893 m*
Island Dimensions: *10 × 5 km*
Eruptive History: *1929*
Composition: *Unknown*

Kagamil Island has a crudely rectangular shape with its long axis aligned almost north-south. Kagamil volcano occupies part of the southern end of the island and is ~893 m high. The northern part of the island is flat and low (much of it below 300 m). Short ridge arcs both in the north and south of the island suggest a possible earlier caldera. The early explorer Veniaminov (1840) reported that Kagamil formerly flamed and smoked. No geologic studies have been published describing this island.

How to get there: *See Seguam.*

References: None

James D. Myers

VSEVIDOF

Eastern Aleutian Islands

Type: *Stratovolcano*
Lat/Long: *53.15°N, 168.69°W*
Elevation: *2,109 m*
Volcano Diameter: *12 km*
Eruptive History: *1817, 1830, 1878, 1957*
Composition: *Basaltic andesite through dacite*

Vsevidof is a moderate-size stratovolcano on central Umnak Island 10 km west of the glaciated stratovolcano Mount Recheschnoi. The western and southern flanks of Vsevidof reach sea level, and its western flank overlies Mount Recheschnoi. Vsevidof is conical and steep and has a surface only slightly modified by glacial erosion. Surface slopes near the summit approach 30° while the lower flanks are ~15°. The 1.2-km-diameter summit crater is ice-filled and feeds two small glaciers which flow to the north and west. Upper parts of the cone are covered by dacitic pyroclastic rocks, while lower parts are composed dominantly of basaltic andesite flows. The three flows which are freshest morphologically issue from the upper parts of the north, west, and south sides of the cone and flowed to, or near to, sea level. The western flow may be historic (1878) and is dacite. There is a 3-km-long, east–west trending zone of post-glacial scoria cones on the western flank between altitudes of 390 and 1,200 m. This flank fissure has been the locus of many historic eruptions. Vsevidof is probably mostly Holocene in age.

VSEVIDOF: From the south. Irregularities on the left skyline are scoria cones on the western flank.

How to get there: *There is commercial air service to Dutch Harbor on Unalaska Island, 150 km to the east. Access from Dutch Harbor is by chartered boat or aircraft. The nearest maintained airstrip is in the village of Nikolski, 25 km to the southwest.*

Reference

Byers, F. M. Jr., 1959, Geology of Umnak and Bogoslof Islands, Aleutian Islands, Alaska. *USGS Bull. 1028-L*, 267-369.

Christopher J. Nye

RECHESCHNOI

Eastern Aleutian Islands

Type: *Stratovolcano*
Lat/Long: *53.15°N, 168.54°W*
Elevation: *1,984 m*
Height: *1,920 m*
Volcano Diameter: *22 km*
Eruptive History: *No historic activity; Holocene flank eruptions*
Composition: *Basalt(?) through rhyolite; dominantly andesite*

Mount Recheschnoi is a large, heavily glaciated stratovolcano on central Umnak Island, 10 km east of Mount Vsevidof, its nearest volcanic neighbor. The central part of the cone is built on a flat erosional surface ~60 m above sea level cut on Tertiary plutonic and sedimentary rocks. The east and northeast flanks overlie ~300 m of lava flows from older volcanoes. The central 40-

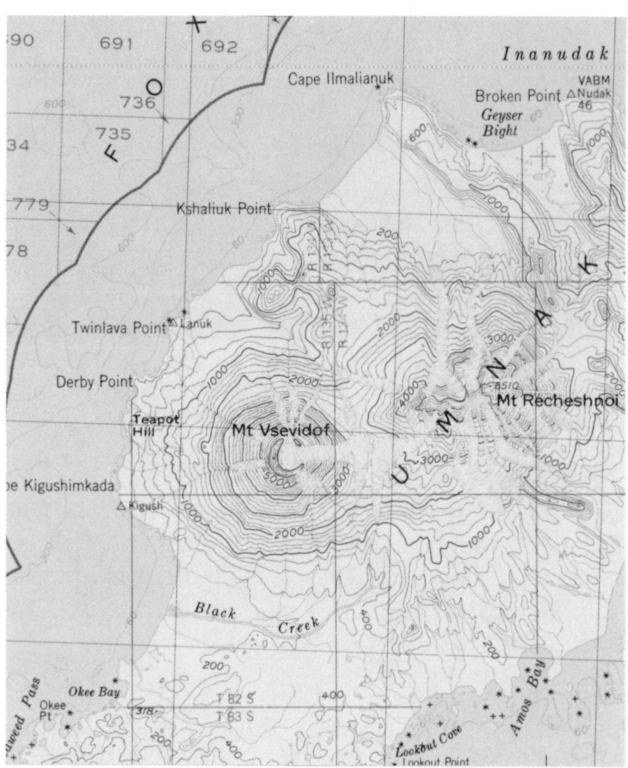

VSEVIDOF and RECHESCHNOI: Topographic map.

VSEVIDOF and RECHESCHNOI: Radar image.

50 km^2 (above 1,000 m elevation) consists of pyroclastic beds and a vent complex. This area has been heavily eroded by 9 small valley glaciers and retains none of its original constructional volcanic form. The summit area consists of a 4-

RECHESCHNOI : From the southwest. View is oblique to the long, narrow summit crest.

km-long east–west ridge which may reflect construction from an older eastern vent and a younger western vent. Below 1,000 m the volcano consists of andesite flows with minor pyroclastic interbeds. The original constructional surface is preserved in some upland areas between glaciers, especially on the western flank. The central cone is probably mostly Pleistocene in age. A glaciated basaltic parasitic cone exists on the northwest flank. Post-glacial deposits are limited to andesitic cinder cones and flows on the east and west flanks, a series of small rhyolite domes on the west flank and a quartz-olivine andesite dome on the east-northeast flank. There are a large zone of hot springs and a few geysers 8 km northeast of the summit.

How to get there: *Commercial airlines serve Dutch Harbor on Unalaska Island, 150 km to the east. Access from Dutch Harbor is by aircraft or chartered boat. The nearest maintained airstrip is in the village of Nikolski, 30 km southwest of Recheschnoi.*

Reference

Byers, F. M. Jr., 1959, Geology of Umnak and Bogoslof Islands, Aleutian Islands, Alaska. *USGS Bull. 1028-L*, 267-369.

Christopher J. Nye

OKMOK

Eastern Aleutian Islands

Type: *Shield volcano with nested summit calderas*
Lat/Long: *53.43°N, 168.65°W*
Elevation: *1,073 m/335 m (caldera rim/floor)*
Volcano/Caldera Diameter: *35/9.3 km*
Caldera Depth: *600 m*

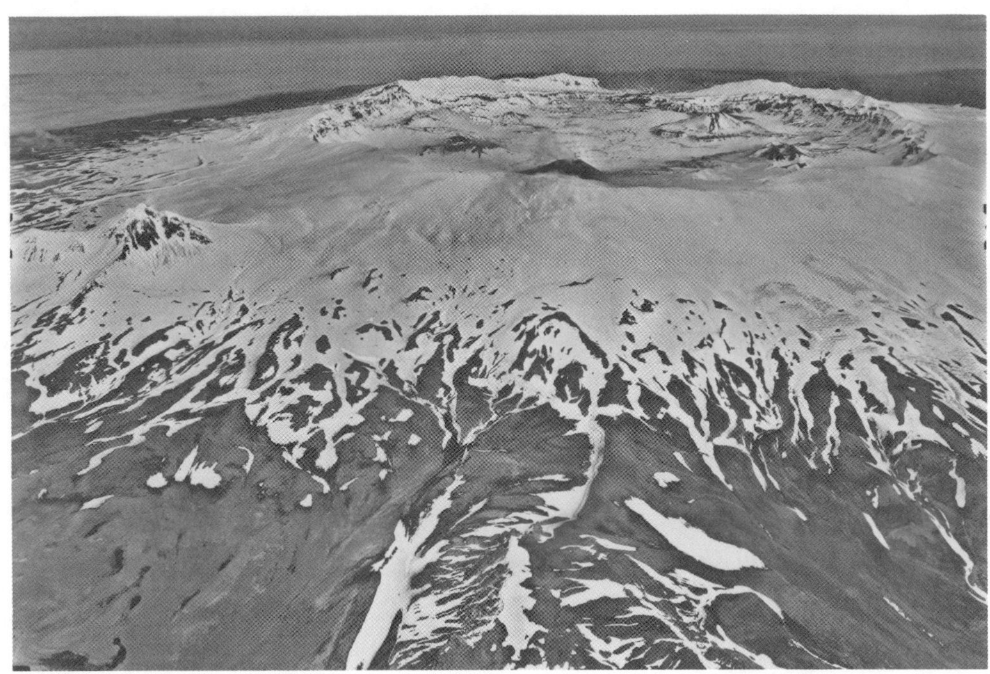

OKMOK: Caldera seen from the southwest. Pillow basalt terrace which erupted into the now-drained crater lake is on the cone in middle distance. The active cone (ash covered) is in the near side of the caldera, and produced the lava flow in the foreground in 1958. Photograph by North Pacific Aerial Surveys, Inc.

Eruptive History: *Pleistocene shield building, caldera collapse at 8,250 and 2,400 yr BP, post-collapse intracaldera flows, two of which are historic (1945, 1958)*

Composition: *Dominantly basalt and basaltic andesite, some dacite during caldera formation, rare rhyolite*

Okmok, a large shield volcano that had two Holocene caldera-forming eruptions, occupies the northeastern half of Umnak Island. It is one of the best described Aleutian volcanoes. The shield is mostly basaltic and was constructed throughout the Pleistocene. The earliest known lavas are ~2 Ma, and the youngest are only lightly glaciated. There are three crude stratigraphic units within the shield. Most early deposits are porphyritic lavas with phenocrysts of mafic minerals. Later deposits are palagonitized pyroclastic rocks which contain xenolithic as well as juvenile material. Continuous thicknesses of these pyroclastic units in excess of 300 m are exposed in the caldera walls. Latest shield lavas are most often aphyric or plagioclase phyric basalt flows.

Six large pre-caldera vents are located on the east, southeast, and southwest flanks of the shield. They rise 300 to 600 m above the adjacent shield, have a maximum diameter of ~3.5 km, and are composed of plagioclase-olivine basalt flows which dip as much as 30° away from their vents. There are also a few minor pre-caldera basaltic vents and a rhyolite dome 1.6 km in maximum dimension which is on the outer rim of the first of two calderas.

Shield growth was terminated by the first of two caldera eruptions ~8,250 yr BP, which produced a caldera ~10 km in diameter and a widespread pyroclastic blanket. In the ensuing 5,500 yr the caldera was partially filled by basalt flows presumably derived from intracaldera cones. The exposed cumulative thickness of these flows is ~150 m. An eruption at ~2,400 yr BP formed the present caldera, which mostly overlaps the previous one. Deposits from caldera-forming eruptions average 100 m in thickness at the caldera rim and 30 m near the coast. Proximal deposits consist of spatter, ash, bombs, and blocks with small amounts of xenolithic material which are loosely consolidated though densely welded. Agglomerate that is also densely welded can be as much as 60 m thick.

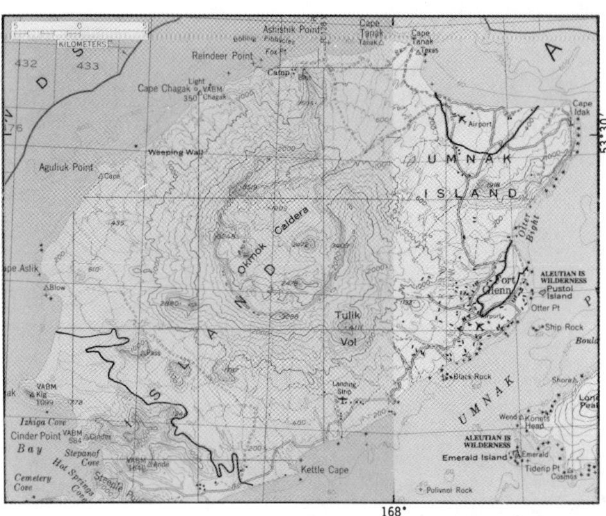

OKMOK: Portions of the USGS 1:250,000-scale Umnak and Unalaska quadrangles showing northern Umnak Island and the extent of Okmok lavas.

Distal deposits are non-welded. The few reported analyses of juvenile material erupted during caldera formation are basaltic andesite and dacite.

The caldera was partially filled by a lake which reached a maximum depth in excess of 200 m before it overtopped the older intracaldera lava flows. Water exited to the north, carving a canyon some 180 m deep which is the only drainage from the caldera. As the lake was forming, basaltic intracaldera cones were growing. These cones have a basal platform of pillow lavas ~100 m thick overlain by a flat, relatively thin platform of subaerial basalt, in turn overlain by basaltic cinder cones. Continued rise of the lake resulted in wavecut benches on the cinder cones. The lake drained and subsequent intracaldera volcanism has been subaerial.

There are six major intracaldera cones and a few minor vents. Additionally, several post-caldera basaltic cones and flows occur on the flanks of the volcano. Historic volcanism has been restricted to the southernmost of the intracaldera cones. This cone frequently emits low plumes of steam and occasional plumes of steam and ash which rise a few thousand meters. Lava flows issued from the base of the cone in 1945 and 1958. The 1958 flow crossed nearly the entire caldera floor.

How to get there: *There is commercial air service to Dutch Harbor on Unalaska Island, 120 km east of Okmok Volcano, and there is a useable airstrip accessible by charter at the old Fort Glenn on the western flank of the volcano. There are no year-round residents on northern Umnak Island.*

OKMOK: USGS radar image, northwest look.

References

Byers, F. M. Jr., 1959, Geology of Umnak and Bogoslof Islands, Aleutian Islands, Alaska: *USGS Bull. 1028-L*, 267-369.

Miller, T. P., Smith, R. L., 1987, Late Quaternary caldera-forming eruptions in the eastern Aleutian arc, Alaska: *Geology 15*, 434-438.

Christopher J. Nye

BOGOSLOF
Eastern Aleutian Islands

Type: *Largely submarine stratovolcano*
Lat/Long: *53.93°N, 168.03°W*
Elevation: *101 m; rises ~1,800 m above seafloor*
Island Dimensions: *1.5 × 0.6 km*
Eruptive History: *1796, 1804, 1806, 1883, 1906, 1909, 1913?, 1926, 1931, 1951?*
Composition: *Andesite to basalt*

BOGOSLOF: Castle Rock, formed in 1796.

Bogoslof has evolved greatly since its first appearance as an unnamed island (later called Shiprock) on a nautical track chart in 1769. Captain Cook showed **Ship Rock** on the chart of his voyage of 1778-9. Eruptions in 1796 resulted in the formation of a new island, south of Ship Rock, named Bogoslof by the natives of Unalaska village. **Castle Rock** is the spine from the 1796 eruption. The 1883 activity, north of Ship Rock, produced a volcanic dome, and culminated in an eruption of ash which reached Unalaska village, 100 km to the east. The resulting island, **New Bogoslof** (also called **Grewingk**), was connected to Ship Rock and Old Bogoslof in 1887, with New Bogoslof the largest, having pinnacles reaching 150 m. Some of the new features (**Metcalf Domes** and **McCulloch Peak**) formed during the eruptions of 1906-7 were recorded by an expedition from the Massachusetts Institute of Technology, but a violent eruption 23 days after the expedition left destroyed many of the landforms. In 1910, **Tahoma Peak** rose on the site of McCulloch Peak, but was itself destroyed by explosion, collapse, and erosion in 1922. During 1926-7 a conical lava dome, 300 m in diameter and 60 m high, formed a new island.

Bogoslof is 50 km northwest (parallel to the seismic slip direction) of the main volcanic front, between and behind the Four Mountains and Cold Bay volcanic segments. All sampled rocks from Bogoslof are potassium-rich. Castle Rock is made of hornblende andesite; the 1927 flows are hornblende basalts.

How to get there: *Bogoslof is an uninhabited island accessible by small landing craft in favorable sea conditions. Charter from Unalaska (Dutch Harbor) is suggested. The island is occupied by a sea lion colony, and is in the Aleutian Island Wildlife Refuge.*

References

Arculus, R., Delong, S., Kay, R. W., Brooks, C., and Sun, S., 1977, The alkalic rock suite of Bogoslof Island, eastern Aleutian Arc, Alaska: *J. Geol. 85*, 177-186.

Byers, F. M., 1959, Geology of Umnak and Bogoslof Islands, Alaska: *USGS Bull. 1028-L.*

Robert W. Kay

MAKUSHIN

Eastern Aleutian Islands

Type: *Stratovolcano*
Lat/Long: *53.89°N, 166.92°W*
Elevation: *2,036 m*
Volcano/Caldera Diameter: *15/2.5 km*
Eruptive History: *Pleistocene cone building, Holocene caldera formation, and construction of large flank cones; historic fumaroles and plumes from summit*
Composition: *Basalt through dacite*

Makushin volcano is a large, ice-covered volcano on northern Unalaska Island, and is the most recent of a series of vents that are probably mostly Pleistocene in age. The oldest known age of lavas is 0.93 Ma. The present volcano is ~2,036 m tall, with a summit caldera ~2.5 km in diameter which contains an active cinder cone. Glaciation has slightly modified the original constructional surface. Erosional ribs of lava rise a few hundred meters or less above the ice. Ice is generally present above 1,000 m today, but reached down to elevations of a few hundred meters in some drainages during Holocene glacial advances.

Makushin volcano is built on late Tertiary pyroclastic rocks which have been intruded by gabbronorite which yields mid-Pleistocene K-Ar ages. Bedrock is exposed as high as 975 m on the southeast flank of the volcano but is completely concealed beneath volcanic rocks to the northwest. Pleistocene lavas are dominantly basalt and basaltic andesite, although some andesite and dacite also erupted. Early Holocene lavas span a wide compositional range and are,

MAKUSHIN: From the east. The flat-topped, snow-free terraces in the left background are composed predominantly of ash flows which are the remnants of fans produced during Holocene caldera formation. The "lava ramp," produced by voluminous outpouring of lava from the central cone, forms terraces in front of and above and to the right of the ash flows. The small cinder cone in the middle distance is Sugarloaf. Photograph by G. D. March.

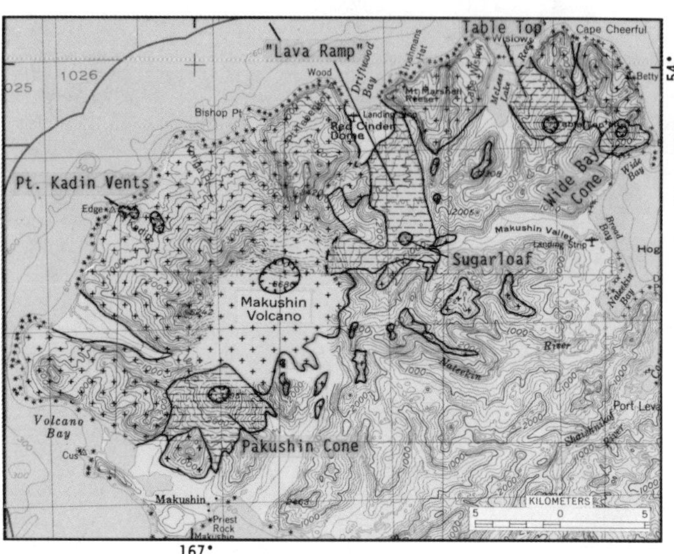

MAKUSHIN: A portion of the USGS 1:250,000-scale Unalaska quadrangle showing northern Unalaska Island. Eruptive products of Makushin volcano and preceding polygenetic volcanoes are shown by small pluses. Products of monogenetic flank vents are shown by horizontal dashes.

on average, more silicic than preceding lavas. Late Holocene lavas are andesite and dacite and are notably more silicic than Pleistocene lavas. Makushin lavas characteristically contain chemical and mineralogical evidence of frequent mixing of small, independent, batches of magma.

The small summit caldera of Makushin formed in the early Holocene. Younger and older limiting dates are 4,280 and 7,950 yr. The eruption resulted in deposition of aprons up to 90 m thick in the heads of nearby valleys. These deposits typically have basal till, mudflows, flood deposits, or slightly welded ash-flow tuffs overlain by slightly welded or sintered ash-flow tuffs and airfall tuffs which are in turn overlain by ~2 m of late Holocene tephra. Ash flows predominate. Juvenile clasts in the ash flows are high-silica andesite and low-silica dacite.

There are several monogenetic satellitic vents which, along with Makushin volcano, form a broad southwest to northeast trending band. Most of these vents have been slightly glaciated but blanket or fill late Pleistocene topography and are therefore probably latest Pleistocene or early Holocene in age. Summit elevations and approximate volumes are, from the southwest; **Pakushin** cone, 1,050 m, 1-2 km^3; **Sugarloaf**, 580 m, <0.5 km^3; **Table Top** Mountain, 800 m, 1 km^3 and **Wide Bay** cone, 640 m, 0.5 km^3. Cone height cannot be reliably determined because of the extreme irregularity of the underlying topography. Pakushin cone and Table Top Mountain are cinder cones surrounded by broad lava aprons; the other vents lack lava flows. These cones were all produced by the eruption of relatively large volumes of fairly homogeneous magma over short periods. Lavas are generally mafic, homogeneous within individual centers and different between centers. Extremely small volume eruptive centers are at Cape Wislow and on a west-northwest trending fracture on the northwestern flank of Makushin. The dozen or so explosion pits and small cinder cones along this fracture are termed the **Point Kadin** vents.

A feature shown as the "lava ramp" on the map is a volcanological oddity. This is a large volume package of co-eruptive basaltic andesite and andesite flows which erupted from Makushin volcano at about the same time as the other flank vents. This package of flows has a volume of at least 5 km^3 and is odd because such large volumes of magma rarely erupt from Aleutian polygenetic cones.

A geothermal resource capable of meeting the power requirements of the Dutch Harbor and Unalaska communities underlies the eastern flank of Makushin volcano.

How to get there: *There is commercial air service to Dutch Harbor, 25 km to the east. Further access is by chartered helicopter or by skiff and foot.*

References

Drewes, H., Fraser, G. D., Snyder, G. L., Barnett, H. F., Jr., 1961, Geology of Unalaska Island and adjacent insular shelf, Aleutian Islands, Alaska: *USGS Bull. 1028-S*, 583-676.

Nye, C. J., Queen, L. D., and Motyka, R. J., 1984. Geologic map of the Makushin geothermal area, Unalaska Island, Alaska: *Alaska Div. Geol. Geophys. Surv. Report of Invest. RI-84-3*, 2 sheets, 1:24,000.

Christopher J. Nye

AKUTAN
Eastern Aleutian Islands

Type: *Stratovolcano with caldera*
Lat/Long: *54.08°N, 165.92°W*
Elevation: *1,300 m*
Volcano Diameter: *22 km*
Eruptive History: *1848, 1852, 1883, 1887, 1907-08, 1929, 1946-47, 1974, 1978, 1983*
Composition: *Basalt to dacite*

Akutan volcano is situated in the west-central part of Akutan Island. The island is composed of an older sequence of lava flows and lahar deposits unconformably overlain by lavas from Akutan Volcano. The summit of the volcano contains a caldera, 2 km in diameter, enclosing a cinder cone and small lake. The caldera is breached on the north; a lava flow issued from the caldera through this gap in 1978 and came within 2 km of the sea.

Glacial erosion has deeply scoured the older volcanic rocks on Akutan Island. U-shaped valleys up to 500 m deep and extending to the coast form a radial pattern around the volcano. Lower valley walls are composed of volcanic breccia, while the ridges between the valleys are capped by lava flows. K-Ar dates of these older volcanic rocks range from 1.5 to 1.1 Ma. The source for these older lavas appears to have been in the present area of Akutan volcano. Lavas of Akutan volcano are superimposed on the glaciated,

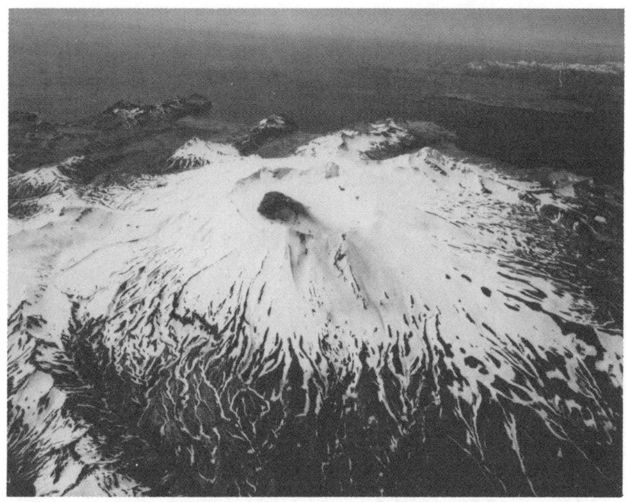

AKUTAN: From the northeast. The summit crater contains the currently active cinder cone, which frequently emits steam and ash plumes reaching altitudes of a few thousand meters. In 1978 a lava flow crossed the breach in the crater (foreground) and nearly reached sea level. Glacially scoured valleys in the background cut Pleistocene lava flows. Photograph by North Pacific Aerial Surveys.

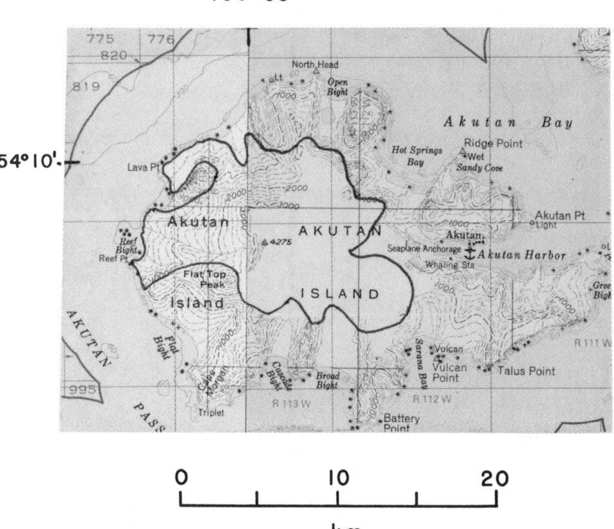

AKUTAN: USGS topographic map (1:250,000).

older volcanic rocks and fill the glacial valleys on the western side of the island, resulting in a smooth slope.

Two volcanic centers occur outside the caldera on the northwest corner of Akutan Island. One of the vents, **Lava Peak**, consists of thin lava flows symmetrically disposed around an eroded vent. A K-Ar age of 1.1 Ma shows that

Lava Peak belongs to the older volcanic units. **Lava Point** is formed by a young lava flow that issued from a cinder cone in 1852. The flow extended the coastline of the island and added ~4 km^2 to its total area. Post-1852 erosion has produced some spectacular arches where the sea has exposed lava tubes.

Hot springs and fumarole fields are found at several sites around Akutan volcano. A geothermal investigation conducted by the State of Alaska suggests the presence of hot (200°C) geothermal fluids at shallow depths around the flanks of the volcano.

How to get there: *Commercial air service is available from Anchorage to Dutch Harbor. A plane or boat can be chartered at Dutch Harbor for a trip to Akutan Island.*

Reference

Romick, J. D., 1982, *The igneous petrology and geochemistry of northern Akutan Island, Alaska:* M.S. Thesis, University of Alaska, Fairbanks, 151 pp.

Swanson, S. E., and Romick, J. D., 1988, Geology of northern Akutan Island, *in* Motyka, R.J., and Nye, C. J. (eds.), A geological, geochemical and geophysical survey of the geothermal resources at Hot Springs Bay Valley, Akutan Island, Alaska: *Alaska Div. Geol. Geophys. Surv. Rpt. of Investigations 88-3*, 11-28.

Samuel E. Swanson

GILBERT
Eastern Aleutian Islands

Type: *Stratovolcano?*
Lat/Long: *54.25°N, 165.66°W*
Elevation: *818 m*
Volcano Diameter: *6 km*
Eruptive History: *No historic or post-glacial eruptions*
Composition: *Unknown*

Mount Gilbert is a small extinct volcano on northernmost Akun Island. Its northern flank is steep and heavily eroded, presumably by the sea, while its southern flank is little dissected. The summit region has been completely eroded and was presumably north of, and higher than, the present summit. A 20,000-m^2 zone of slightly older fumarolically altered rock exists 1.5 km northeast of the current summit. Prior to 1948, active fumaroles in this area produced steam plumes visible from a distance; fumarolic activity has since stopped. The altered ground was the site of an unsuccessful attempt to mine sulfur in the 1920s.

GILBERT: Northern Akun Island from the west. The snowy peak in the foreground is the remains of Gilbert volcano. The southern slope retains its constructional morphology while the northern slope is deeply eroded. Volcanoes on Unimak Island (the large snow-covered island to the east) are Westdahl, the shield in the right foreground; Pogromni, left foreground; Shishaldin, middle distance; and Isanotski, background. Photograph by North Pacific Aerial Surveys Inc.

How to get there: *There is commercial air service to Dutch Harbor on Unalaska Island, 70 km to the west of Gilbert, and periodic air service from Dutch Harbor to the village of Akutan, 16 km southwest of Gilbert. There are no maintained airstrips or permanent residents on Akun Island.*

Reference

Byers, F. M., Jr. and Barth, T. F. W., 1953, Volcanic activity on Akun and Akutan Islands: *Proc Seventh Pacific Sci. Cong. 2*, 382-397, Auckland and Christchurch, New Zealand.

Christopher J. Nye

POGROMNI
Eastern Aleutian Islands

Type: *Stratovolcano*
Lat/Long: *54.57°N, 164.85°W*
Elevation: *2,002 m*
Volcano Diameter: *4 km*
Eruptive History: *1795, 1796, 1826, 1827, 1830*
Composition: *Basalt*

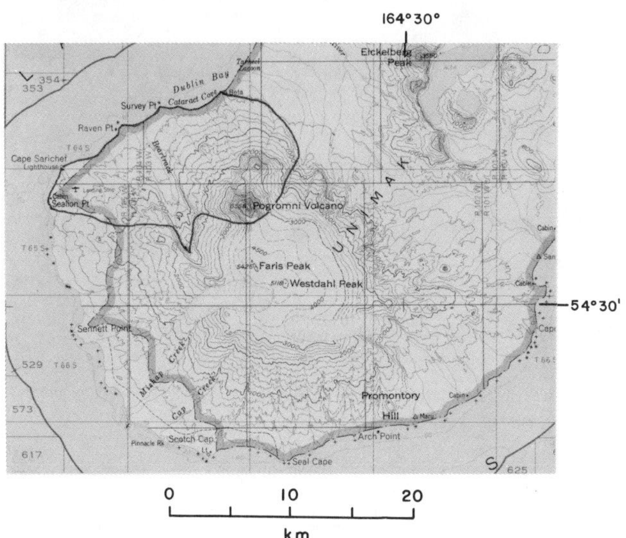

POGROMNI: USGS topographic map (1:250,000), extent of Pogromni deposits outlined.

POGROMNI: Right to Shishaldin (far left).

Pogromni volcano is a steep-sided stratovolcano on the southwest end of Unimak Island. All of the lavas analyzed from Pogromni are tholeiitic basalt. A single glacier is found on Pogromni. Outcrops in the glacially scoured walls of Pogromni suggest a caldera collapse was followed by building of the present cone. Another peak (**Pogromni's Sister**, 1,230 m) may be a remnant of this older vent.

Five cinder cones are aligned with Pogromni volcano along a northwest–southeast trend. The monogenetic cones are 50-100 m high and are strongly oxidized on the surface.

Pogromni volcano is little known. Reconnaissance petrologic studies report a few analyses from the satellitic cinder cones.

How to get there: *Commercial air service is available to Cold Bay on the Alaska Peninsula; aircraft can be chartered there for the trip to Pogromni.*

Reference

Neal, R. J., and Swanson, S. E., 1983, Petrology and geochemistry of Westdahl and Pogromni Volcanoes, Unimak Island, Alaska: *EOS, Trans. Am. Geophys. Un. 64*, 893.

Samuel E. Swanson

WESTDAHL
Eastern Aleutian Islands

Type: *Shield volcano*
Lat/Long: *54.00°N, 164.85°W*
Elevation: *1,560 m*
Volcano Diameter: *25 km*
Eruptive History: *1964-65, 1978*
Composition: *Basalt to andesite*

Westdahl is a broad, glacier-covered shield volcano on the southwest end of Unimak Island. Most of the Westdahl lavas are tholeiitic basalts, whose fluidity controls the form of the volcano. Two peaks, **Faris** and Westdahl, protrude from the glacial ice near the summit of the volcano. A new crater in the ice cap marks the site of the 1978 eruption. Like nearby Pogromni, Westdahl volcano is poorly represented in the published literature.

At least 17 satellite vents are located southeast and northwest of the summit along a N75°W trend. These vents are both monogenetic cinder cones and polygenetic cones composed of cinders and lava flows. The cones show variable amounts of erosion, but all appear to be postglacial (6,700-8,400 yr ago).

Two historic eruptions have been witnessed on Westdahl. A fissure eruption on the east side of Westdahl started in March 1964 and continued at irregular intervals through early 1965. Lava flows produced during this eruption cover ~35 km^2 of the eastern flank of the volcano. A vulcanian eruption in February 1978 blasted a crater 1.5 km in diameter and 0.5 km deep through the glacial ice 1 km south of the sum-

WESTDAHL: Oblique aerial profile of the glacier-clad shield volcano.

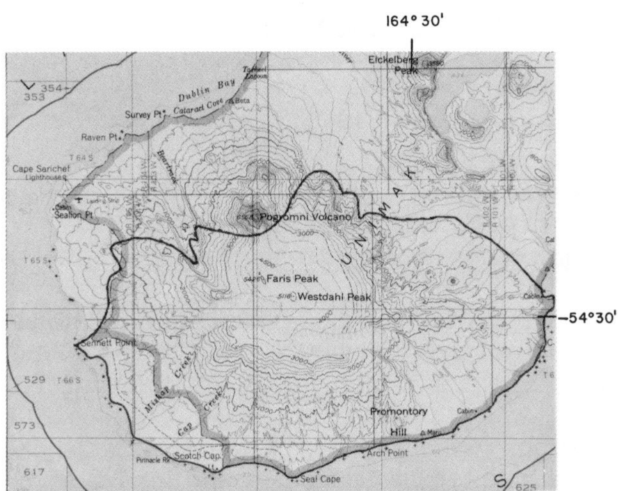

WESTDAHL: USGS topographic map (1:250,000), extent of Westdahl deposits outlined.

mit. The eruption sent ash 11 km into the atmosphere. Ash thickness was 18 cm at a distance of <15 km from the crater. A lahar produced by melting of snow and ice extended 12 km.

A thick sequence of pre-glacial basalt lava flows comprises most of Westdahl volcano. Many of these lavas are too young to date by the whole-rock K-Ar method, but the lowest flow in one section yielded an age of 0.18 Ma.

How to get there: *Commercial air service is available to Cold Bay; aircraft must then be chartered for the trip to Westdahl volcano.*

Reference

Neal, R. J., and Swanson, S. E., 1983, Petrology and geochemistry of Westdahl and Pogromni Volcanoes, Unimak Island, Alaska: *EOS, Trans. Am. Geophys. Un. 64*, 893.

Samuel E. Swanson

FISHER
Eastern Aleutian Islands

Type: *Stratovolcano with caldera*
Lat/Long: *54.63°N, 164.42°W*
Elevation: *1,095 m*
Height: *962 m*
Volcano Diameter: *21 km*
Caldera Diameter: *11 × 18 km*
Eruptive History: *Caldera formation: 9,100 yr BP; 1826*
Composition: *Basalt to dacite*

The Fisher volcanic center consists of a large (~300 km^3), deeply eroded, and glaciated andesitic stratovolcano complex with an oval-shaped summit caldera elongated in a northeast direction. Two large lakes cover ~25 percent of the caldera floor. Little is known about the geology of the volcano or of the area in general, as topographic maps have only recently become available and the only studies of the volcano have been reconnaissance investigations of ash-flow tuffs surrounding the caldera. No mapping has been done on the cone-building and post-caldera volcanic rocks; their descriptions are based chiefly on aerial observations and photo interpretation.

The physiography of the stratovolcano flanks suggests it may have had multiple vents, and the irregular and scalloped caldera rim suggests more than one collapse. Several intra-caldera cinder and spatter cones as large as 2 km in

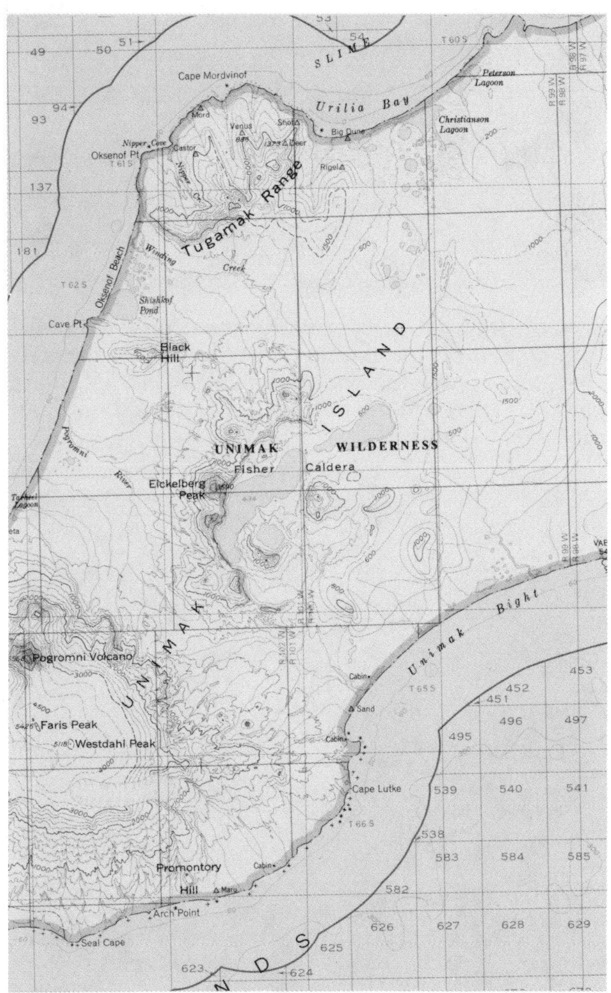

FISHER: USGS topographic map.

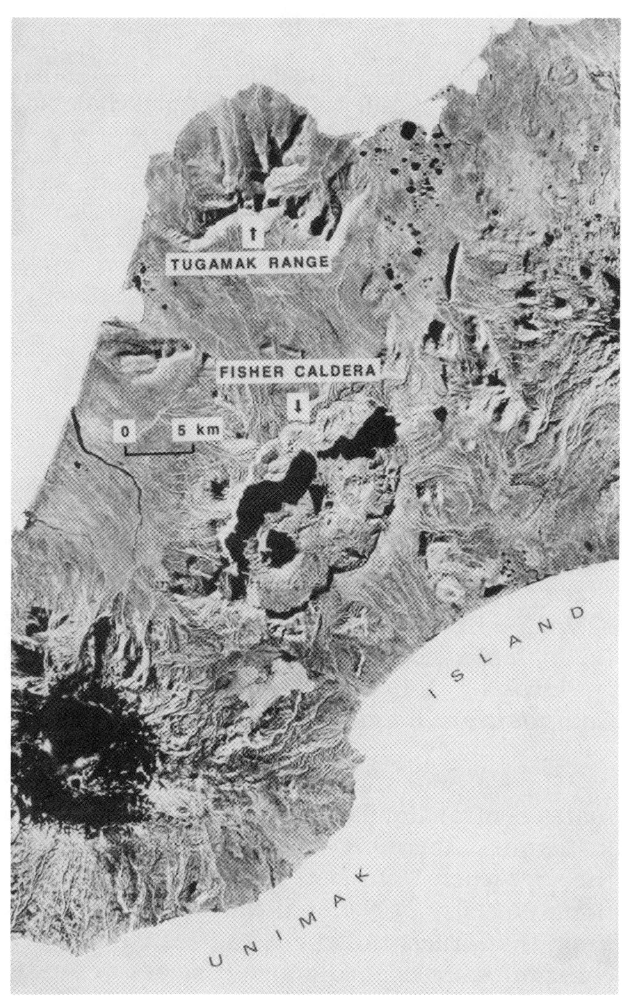

FISHER: USGS radar image, X-band NW look.

diameter and 400 m high are aligned in a northeasterly direction across the caldera. Much of the southern part of the caldera is covered by a thick layer of tephra. The northwestern caldera rim is much higher than the remainder of the rim. The southern half of the caldera is drained by a breach in the southwest caldera wall.

Ash-flow tuffs surrounding the caldera are non-welded and compositionally zoned; dacitic (64.4% SiO_2) basal tuff is overlain by andesitic basalt (52.5% SiO_2) tuff. The pyroclastic flows that deposited these tuffs were extremely mobile; pyroclastic flows moving down the north flank of the volcano, for example, swept 15 km across the adjoining lowland and climbed at least 500 m over the Tugamak Range before descending into the Bering Sea.

A poorly documented ash eruption in late 1826 is the only historic activity reported for Fisher volcano. The source of ash is unknown, although a large area of fumarolic activity occurs on the west side of an intracaldera cone 2.5 km west of the southernmost of the two large intracaldera lakes. Holocene activity consists of a major caldera-forming eruption ~9100 yr BP and the formation of several large intracaldera cones and craters. The stratovolcano is assumed to be chiefly Plio-Pleistocene age, similar to other Aleutian arc volcanoes.

How to get there: *Fisher volcanic center is near the west end of Unimak Island, eastern Aleutian Islands, in the Unimak Wilderness area. No roads occur in the region; access is by floatplane from Cold Bay to one of the lakes in the caldera or by boat to the coast and then by foot to the caldera.*

References

Miller, T. P., and Smith, R. L., 1977, Spectacular mobility of ash-flows around Aniakchak and Fisher calderas, Alaska: *Geology* 5, 173-176.

Miller, T. P., and Smith, R. L., 1987, Late Quaternary caldera-forming eruptions in the eastern Aleutian arc, Alaska: *Geology* 15, 434-438.

Thomas P. Miller

SHISHALDIN
Eastern Aleutian Islands

Type: *Stratovolcano and monogenetic field*
Lat/Long: *54.75°N, 163.97°W*
Elevation: *2,857 m*
Size of Volcanic Center: *~9 × ~6 km; ~300 km^3*
Eruptive History: *Virtually continuous smoking with occasional ash eruption; lahar 1976; lavas 1830, 1932*
Composition: *Basalt with minor dacite*

The ice- and snow-capped Shishaldin dominates central Unimak Island; the Aleuts named it "Sisquk," meaning "mountain which points the way when I am lost." It is listed in the National Registry of National Natural Landmarks. From the earliest reports, it has been in a nearly constant state of mild activity; steam or smoke is commonly seen rising from its summit.

Shishaldin is a post-glacial stratovolcano, built atop an older, glaciated base. There are remnants of an older, ancestral volcano ("Somma") on both the west and northeast sides at 1,500-1,800 m. Spread over the northwest flank, down to <100 m, are massive aa flows; at lower elevations, most are covered with soils and vegetated by grasses and mosses. These flows have erupted from both the main and parasitic vents, with one ~12 km from the main vent (at ~100 m elevation). Over two dozen monogenetic cones are scattered over the northwest flank.

A post-glacial welded pyroclastic flow (or lahar) is present at lower elevations to the west; accidental inclusions suggest it came from Shishaldin. Below it is a sequence of ash, pumice (salmon color), and black cinders, with mixed salmon-black pumice at the pumice-cinder boundary. Exposed at ~1,000 m by melting, ephemeral glacial streams or fissures in ice, are thick sequences of ash, cinders, and pumice interbedded with pebbles from glacial runoff.

SHISHALDIN: View from the northwest. Small plumes such as the one in the photograph are common. Monogenetic peaks on northwest flank. Isanotski Peaks (left) are 15 km east of Shishaldin. Photograph by D. Johnson.

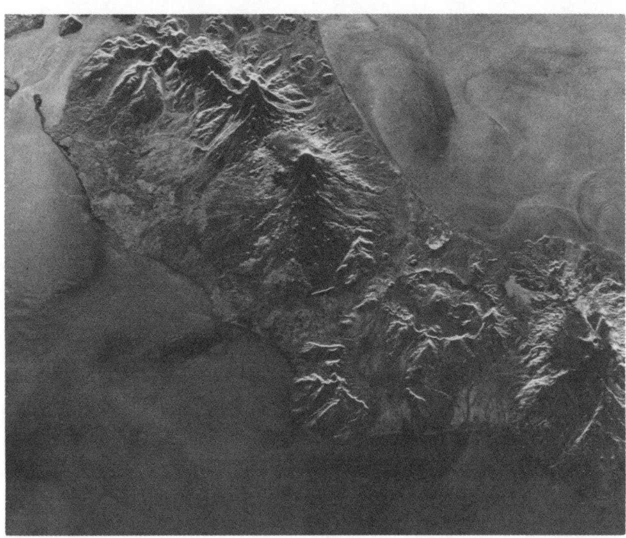

SHISHALDIN: Seasat satellite radar image of Unimak Island showing 6 volcanoes: from south to north: Westdahl, Pogromni, Fisher caldera, Shishaldin, Isanotski, and Roundtop. Image provided by R. Blom, JPL.

Older, preglacial lavas are exposed in only a few locations on the western flank. An older **Whaleback** is located west northwest of Shishaldin. It is composed of several million year old (2-4?) lavas and hypabyssals, cut by dikes, and overlain by steeply dipping pyroclastic flows; all units are glaciated.

Shishaldin's volcanics are distinctive. They are predominantly basaltic, with three different types present: plagioclase-rich high alumina basalt, phenocryst-poor FeTi basalt, and high-magnesian basalt. The later basalt contains a notable set of xenocrysts of Fo-rich olivine (+ chromite) and diopsidic clinopyroxene, suggesting the rising arc magmas entrained material from an underlying layered intrusion (the Border Ranges Complex).

How to get there: *Unimak Island is part of the Aleutian Islands National Wildlife Refuge. Request permit from US Fish and Wildlife Service, Cold Bay; amphibious aircraft may be chartered from Peninsula Airways, Cold Bay (landings permitted only on water bodies or beaches). Be prepared for bears.*

References

Fournelle, J. H., 1988, Geology and Petrology of Shishaldin, Ph.D. thesis: The Johns Hopkins University, 507 pp.

Finch, R. H., 1934, Shishaldin Volcano: *Proc. Fifth Pacific Sci. Conf. Canada 3*, 2369-2376.

John Fournelle

ISANOTSKI

Eastern Aleutian Islands

Type: *Stratovolcano*
Lat/Long: *54.75°N, 153.73°W*
Elevation: *2,446 m*
Eruptive History: *1825, 1840*
Composition: *Unknown*

Isanotski, known locally as **"Ragged Jack,"** is a rugged, ice- and snow-capped volcano that has been heavily incised. It has not erupted since the 1840s; a large eruption in 1825 blasted away a large part of it. A short dark lava (?) flow in a valley on its southeastern margin was sighted during a fly-by in 1985.

An unnamed volcano appears to be between, and slightly north of, Roundtop and Isanotski.

How to get there: *Unimak Island is part of the Aleutian Islands National Wildlife Refuge. Request permit from US Fish and Wildlife Service, Cold Bay; amphibious aircraft may be chartered from Peninsula Airways, Cold Bay (landings permitted only on water bodies or beaches). Be prepared for bears.*

References: None

John Fournelle

ISANOTSKI: Aerial photo (8/22/42). The highly glaciated and ragged summit of Isanotski. View from the south.

ROUNDTOP: The eroded summit of Roundtop; compare with aerial photo of Isanotski taken at same time (8/22/42) and presumably same altitude.

ROUNDTOP

Eastern Aleutian Islands

Type: *Stratovolcano*
Lat/Long: *54.80°N, 153.60°W*
Elevation: *1,871 m*
Eruptive History: *Holocene; pre-Holocene*
Composition: *Unknown*

Roundtop is an eroded and glaciated stratovolcano, located ~13 km southwest of the village of False Pass. In the 1930s warm springs were found on its slopes. Recent field study indicates the presence of Holocene pyroclastic flows and a group of domes south of Roundtop, as well as a thick silicic pre-Holocene flow.

How to get there: *Unimak Island is part of the Aleutian Islands National Wildlife Refuge. Re-*

quest permit from US Fish and Wildlife Service, Cold Bay. Roundtop may be accessible by foot from the village of False Pass (air service by Peninsula Airways, Cold Bay). Be prepared for bears.

References

Finch, R. H., 1934, Shishaldin Volcano: *Proc. Fifth Pacific Sci. Conf. Canada 3*, 2369-2376.
Miller, T., 1988, Personal communication.

John Fournelle

COLD BAY

Alaska Peninsula

Type: *Overlapping stratovolcanoes*
Lat/Long: *55.00°N, 162.75°W*
Elevation and Height: *1,920 m*
Volcano Dimensions: *18 × 24 km*
Eruptive History:
Summit Cone: Late Pleistocene to Recent
Frosty Peak Volcanics: Mid to late Pleistocene
Morzhovoi Volcanics: late Pliocene to early Pleistocene
Composition: *Basalt to andesite*

The extinct Cold Bay volcanic center is at the western end of the Alaska Peninsula, immediately southwest of the town of Cold Bay. Episodic late Pliocene to Recent volcanism has produced a large volcanic complex consisting of three overlapping collapsed stratovolcanoes and an intact central summit cone. The total estimated eruptive volume is ~100 km^3. The oldest lavas (**Morzhovoi** Volcanics) constitute a late Pliocene to early Pleistocene basaltic and andesitic stratovolcano. Volcanism culminated in the formation of a large collapsed crater. Though presently ice-free, extensive Pleistocene glaciation has eroded most original stratovolcano features, including the actual crater rim, leaving long U-shaped valleys. The younger, snow- and ice-covered lavas to the north (**Frosty Peak** Volcanics) consist entirely of mid to late Pleistocene andesites. Two adjacent, steep-walled craters indicate at least two cycles of stratovolcano formation and collapse. A large valley glacier has breached the western wall of the northern crater, while a symmetrical, late Pleistocene to Recent andesitic summit cone

COLD BAY: Oblique view of the Cold Bay volcanic center looking from the southwest toward the Frosty Peak Volcanics. Outer rims of two craters are in the foreground and the central summit cone rises behind.

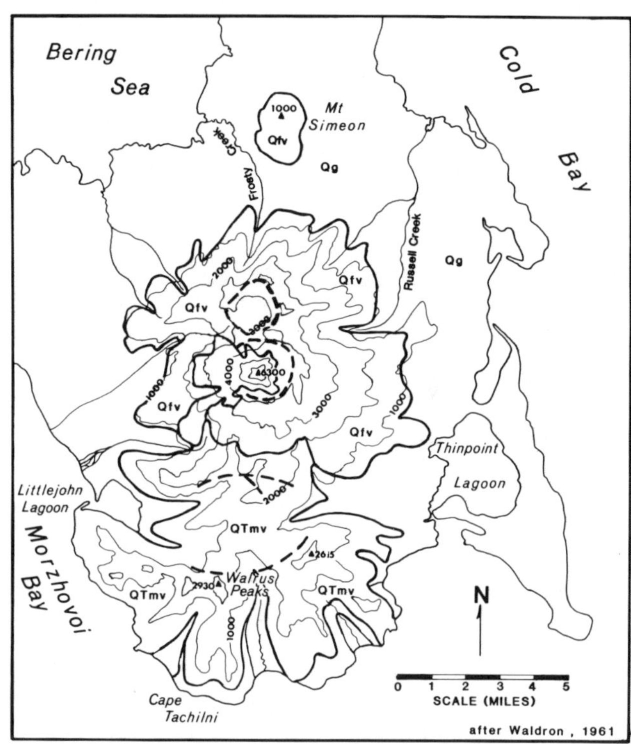

COLD BAY: QTmv = Morzhovoi Volcanics, Qfv = Frosty Peak Volcanics.

rises ~600 m above the floor of the southern crater. Few geologic investigations have been reported other than reconnaissance mapping and petrologic studies.

How to get there: *The Cold Bay volcanic center is ~8 km southwest of the town of Cold Bay, which is accessible only by sea or air. A rough dirt road from Cold Bay provides access to the northernmost margin of the volcanic complex. Vehicles may be rented in Cold Bay. The southern mar-*

gins are accessible only by boat or seaplane, both of which may be chartered at Cold Bay. The entire volcanic center lies within the Izembek National Wildlife Refuge; applications for entry permits may be obtained from the Refuge Manager at Cold Bay.

References

Waldron, H. H., 1961, Geologic reconnaissance of Frosty Peak Volcano and vicinity, Alaska: *USGS Bull. 1028-T*, 677-708.

Brophy, J. G., 1987, The Cold Bay Volcanic Center, Aleutian volcanic arc. II. Implications for fractionation and mixing mechanisms in calc-alkaline andesite genesis. *Cont. Min. Petr. 97*, 378-388.

James G. Brophy

AMAK
Aleutian Islands, Alaska

Type: *Stratovolcano*
Lat/Long: *55.42°N, 163.15°W*
Elevation: *513 m*
Eruptive History: *1700-1710, 1796*
Composition: *Basaltic andesite, andesite*

Amak is a small (~1 km^3), young volcano in the Bering Sea some 50 km north of Frosty Peak volcano at Cold Bay at the western tip of the Alaska Peninsula. It is unusual in its position, which is significantly north of the main Aleutian volcanic front; Bogoslof, some 250 km west, is the only other such Aleutian volcano. In overall character, Amak is much like a large volcanic dome except that it has a well-formed crater from which granular, blocky leveed flows have erupted in historic times. The earlier volcanism, perhaps some 4,000-5,000 yr ago, consisted mostly of thin (~3 m) platy to massive andesite. U-shaped valleys in the older series of flows indicate significant glaciation during the latest ice phase some 6,700 yr BP. The southern margin of Amak is a grassy, apparently wave-cut alluvial plain, which contains a flat-bottomed crater with a distinct upturned rim, which may be a maar. Amak lavas are similar in overall composition to those of the Cold Bay volcanic center except in a small but significant higher concentration of potash in Amak rocks.

How to get there: *Commercial air service is available daily from Anchorage to Cold Bay.*

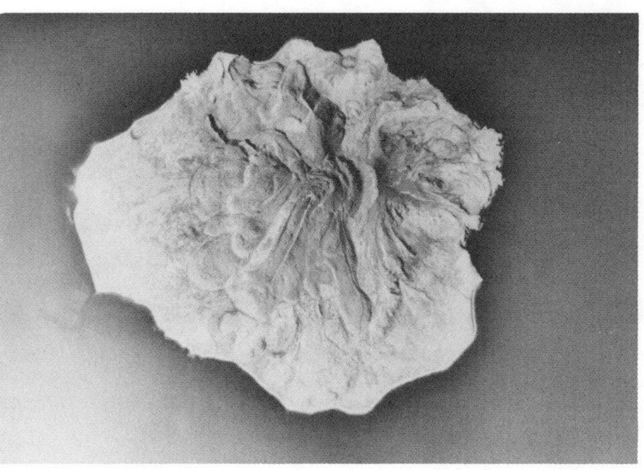

AMAK: Aerial view of Amak Island (north to top), showing older series of flows on margins and north, and most recent flows. Possible maar is tangential to southwest coast (photograph: US Coast and Geodetic Survey).

Private sea transportation must be found at Cold Bay; aircraft landing is prohibited on Amak, and permission to visit the island must be obtained from the US Fish and Wildlife Service in Cold Bay.

References

Marsh, B. D., and R. E. Leitz, 1979, Geology of Amak Island, Aleutian Islands, Alaska. *J. Geol. 87*, 715-723.

Morris, J. D., and S. R. Hart, 1983, Isotopic and incompatible element constraints on the genesis of island arc volcanics from Cold Bay and Amak Island, and implications for mantle structure. *Geochim. Cosmochim. Acta 47*, 2015-2030.

Bruce D. Marsh

DUTTON
Alaska Peninsula

Type: *Stratovolcano with central domes*
Lat/Long: *55.18°N; 162.27°W*
Elevation: *1,506 m*
Height: *760 m*
Volcano Diameter: *6 km*
Volcano Volume: *7-15 km^3*
Eruptive History: *No historic activity*
Composition: *Andesitic basalt to dacite*

Mount Dutton is a little-studied snow- and ice-covered Quaternary volcano near the tip of the Alaska Peninsula ~2,000 km southwest of

DUTTON: View of Mount Dutton volcano and associated summit dome. View is looking southwest from the rim of Emmons Lake Caldera.

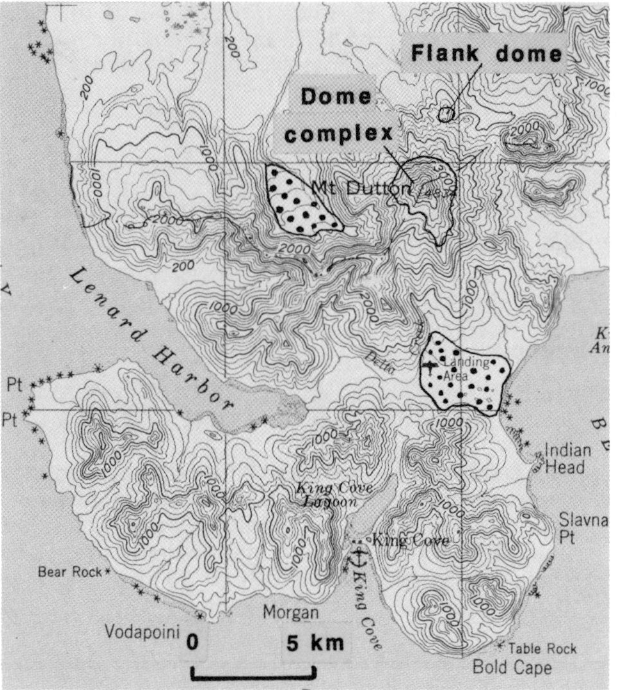

DUTTON: Map from USGS Cold Bay 1:250,000-scale quadrangle map; contour interval: 200 feet. Dot pattern marks debris avalanche deposits.

Anchorage and 14 km north of the village of King Cove. The volcano lies on the crest of the Aleutian Range; the north half of the volcano is in the Izembek Wilderness Area and the south half is in the Alaska Peninsula Wildlife Refuge.

The volcano is built on an east-sloping basement of Miocene volcanic rocks and consists of a central dome complex in which successive dacitic domes displaced earlier domes and the enclosing cone-building andesitic volcanic rocks. The dome-building activity and associated massive slope failure has caused extensive destruction of the earlier andesitic flows and, to a lesser extent, the later domes, resulting in thick debris flows that surround the central dome complex. The west side of the summit area is a steep, amphitheater-like avalanche crater ~300 m high and inclined to the northwest about 45°. Debris avalanches from this and lower areas moved down either side of the east–west range crest, covering an area of about 11 km^2. An andesitic dome of Holocene age occurs 3.5 km north-northeast of the summit of Mount Dutton at an elevation of 520 m.

Although Mount Dutton has seen no historic eruptive activity, earthquake swarms occurred beneath the volcano in 1984-85 and in 1988. Cone-building volcanism is assumed to be chiefly late Pleistocene in age based on similarity to nearby Quaternary volcanoes. At least some of the central dome complex and most of the debris avalanches and flows are Holocene, as is a dome on the northeast flank of the volcano.

How to get there: *The King Cove airstrip is 7 km from the volcano resulting in ready access by air and foot.*

References: None

Thomas P. Miller

EMMONS AND HAGUE

Alaska Peninsula

Type: *Stratovolcano with caldera*
Lat/Long: *55.57°N, 162.03°W*
Elevation: *1,463 m (caldera rim)*
Height: *850 m*
Caldera Depth: *1,159 m*
Volcano/Caldera Diameter: *30/18 × 11 km*
Eruptive History: *Cone building assumed to be Plio-Pleistocene; No historic activity*
Composition: *Basalt to rhyolite*

The **Emmons Lake** volcanic center is the site of a large stratovolcano with a pre-caldera volume of 300-400 km^3 and one of the largest calderas in the Aleutian arc. The center is very complex; at least two large caldera-forming eruptions have occurred in Late Quaternary time, and voluminous post-caldera volcanism continues to the present. The entire area has been subjected to multiple glaciations, and ice

EMMONS AND HAGUE: Interior view of southern caldera rim and Emmons Lake.

fields presently occur in the caldera. A small lake in the southwest corner of the caldera drains to the Pacific Ocean through a breach in the southeast rim of the caldera.

The pre-caldera Emmons Lake volcano was built on the northeast flank of a topographic basement high composed of Tertiary volcanogenic and intrusive rocks that are exposed at an elevation of over 900 m in the caldera wall west of Emmons Lake. This period of volcanism consisted of an outpouring of andesitic basalt and volcaniclastic rocks over an area of 700 km^2 or more. Following deep erosion and glaciation, a major caldera-forming eruption occurred, resulting in the deposition of welded ash-flow tuffs of rhyolitic composition on both the Bering Sea and Pacific Ocean flanks of the volcano. In some cases, these welded tuffs filled deep valleys in the pre-caldera cone. At least some post-caldera basaltic volcanism appears to have occurred at this time, principally associated with Mount Emmons.

A second caldera-forming eruption, probably in late Wisconsin time, resulted in the emplacement of non-welded rhyolitic ash-flow tuffs in all quadrants around the volcano to distances >30 km from the caldera rim.

Renewed post-caldera volcanism appears to have been chiefly basaltic in character. Within the caldera this activity has been characterized by aa and block lava flows emitted from several small cinder cones and vents. Some of the young Holocene flows moved through the breach in the southern caldera wall and are found within a kilometer of the Pacific Ocean.

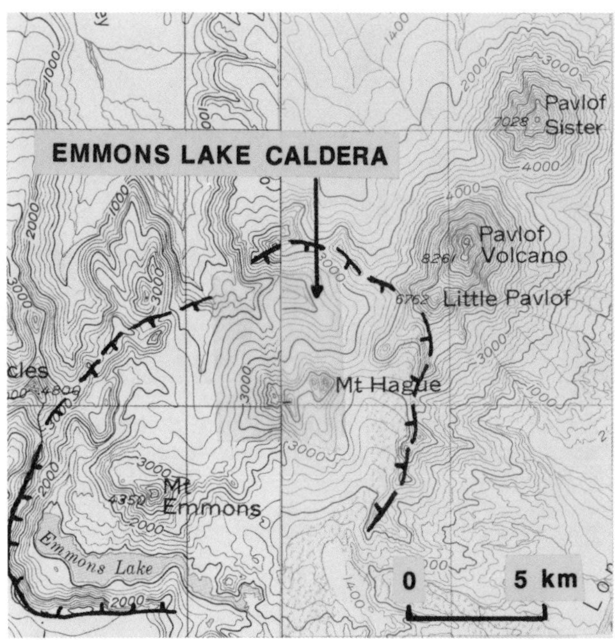

EMMONS AND HAGUE: Map from USGS Port Moller and Cold Bay 1:250,000-scale quadrangle maps; contour interval: 200 ft.

A large fumarolic area occurs on the south side of Mount Hague, which is near the eastern margin of the caldera.

How to get there: *The Emmons Lake volcanic center is located near the tip of the Alaska Peninsula about 55 km east of Cold Bay. There are no roads in the vicinity of the volcano and access is by air from Cold Bay to Emmons Lake and then by foot.*

References

Miller, T. P., and Smith, R. L., 1987, Late Quaternary caldera forming eruptions in the eastern Aleutian arc, Alaska: *Geology 15*, 434-438.

Kennedy, G. C., and Waldron, H. H., 1955, Geology of Pavlof volcano and vicinity, Alaska: *USGS Bull. 1028-A*, 1-20.

Thomas P. Miller

PAVLOF AND PAVLOF SISTER
Alaska Peninsula

Type: *Stratovolcanoes*
Lat/Long: *Pavlof: 55.42°N, 161.90°W*
Pavlof Sister: 55.45°N, 161.95°W

PAVLOF: East-looking view of Pavlof Sister (left) and Pavlof (right). Active vent and recent ash are visible on Pavlof.

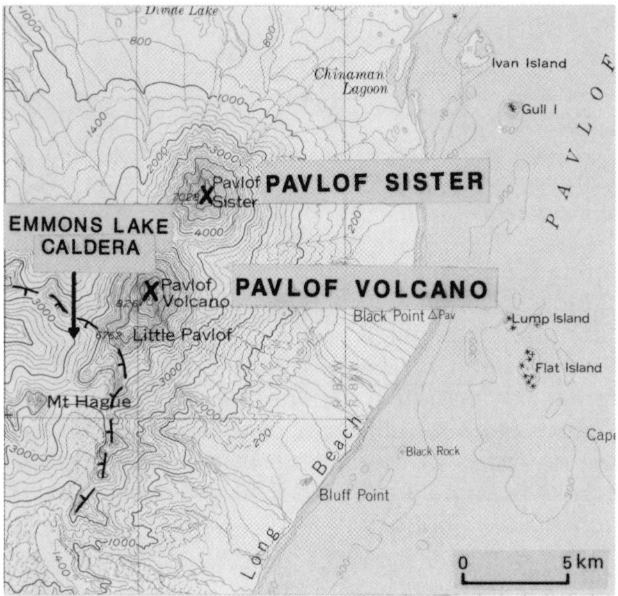

PAVLOF: Map from USGS Port Moller 1:250,000-scale quadrangle; contour interval: 200 ft.

Elevation: *P: 2,519 m; PS: 2,143 m*
Height: *P: 1,100 m; PS: 1,100 m*
Volcano Diameter: *P: 5 km; PS: 6 km*
Eruptive History: *Holocene, with 39 Pavlof eruptions from 1790 to 1986-87*
Composition: *Basalt*

Immediately east of Emmons Lake caldera are the prominent twin stratovolcanoes, Pavlof and Pavlof Sister. Although these symmetrical young volcanoes are very conspicuous, they have relatively small volumes (<10 km^3), as they are built high on an older volcanic rock basement related to the Emmons Lake center. Historic eruptions are chiefly from Pavlof volcano and typically are strombolian, occurring sporadically over a several-month period. One historic eruption has been reported for Pavlof Sister in 1762, but Pavlof is probably the most consistently active volcano in the Aleutian arc, with nearly 40 historic eruptions.

How to get there: *The volcanoes are located near the tip of the Alaska Peninsula about 55 km east of Cold Bay. There are no roads in the vicinity and access is by air from Cold Bay to the beach below the volcanoes and then by foot.*

Reference

Kennedy, G. C., and Waldron, H. H., 1955, Geology of Pavlof volcano and vicinity, Alaska: *USGS Bull. 1028-A*, 1-20.

Thomas P. Miller

DANA
Alaska Peninsula

Type: *Dome complex*
Lat/Long: *55.63°N, 161.22°W*
Elevation: *1,354 m*
Height: *1,036 m*
Crater Diameter: *2 km*
Volcano Diameter: *5.5 km*
Eruptive History: *3,840 yr BP: block-and-ash-flow eruption; no known historic eruptions*
Composition: *Andesite to dacite*

Mount Dana is a small calc-alkaline volcanic center consisting largely of volcaniclastic debris surrounding a central dome or domes, reminiscent of Augustine volcano. The volcano rests on relatively undeformed Jurassic and Cretaceous marine sandstone and shale. Mesozoic sedimentary rocks, dipping steeply southwest, form the southwest crater rim and are exposed in the canyon at the crater outlet. Remnants of a high-silica andesite dome are exposed on the west crater rim and in a small mound on the east side of Knutson Lake. A block-and-ash flow erupted 3,840 yr BP fills valleys south and west of the crater. A 200-m-wide tufa mound and several cold springs occur at elevations of 490 to 520 m on the southwest flank.

How to get there: *Mount Dana, 900 km southwest of Anchorage, is within the Alaska Penin-*

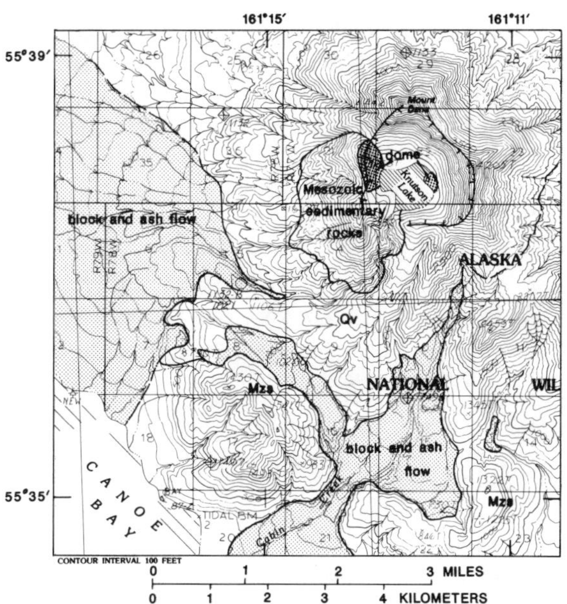

DANA: Annotated topographic map (USGS Port Moller C-4 quadrangle).

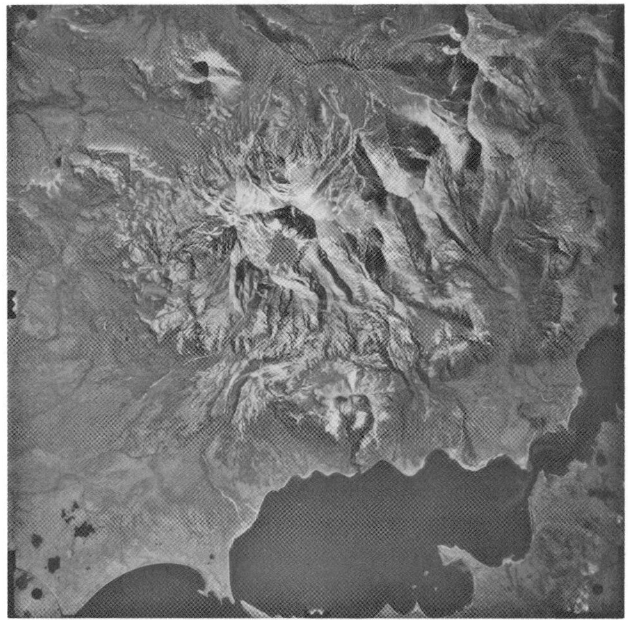

DANA: USGS aerial view; north toward upper left.

sula National Wildlife Refuge. Amphibious planes can be chartered from Cold Bay, 108 km southwest of Dana, landing in Canoe Bay immediately south of the volcano. Passage by boat from the fishing community of Sand Point, 56 km southeast of Dana, might be possible to arrange depending on the time of year. Dana is readily accessible on foot from Canoe Bay.

References

Burk, C. A., 1965, Geology of the Alaska Peninsula-island arc and continental margin: *Geol. Soc. Am. Mem. 99*, 250 pp (see especially Plate 6).

Jagger, T. A., 1929, Mapping the home of the great brown bear: Adventures of the National Geographic Society's Pavlof Volcano expedition to Alaska: *Nat. Geogr. 55*, 109-134.

M. Elizabeth Yount

KUPREANOF (STEPOVAK BAY)
Alaska Peninsula

Type: *Line of stratovolcanoes*
Lat/Long:

Volcano 1:	*55.87°N, 160.09°W*
Volcano 2:	*55.91°N, 160.04°W*
Volcano 3:	*55.93°N, 160.00°W*
Volcano 4:	*55.95°N, 159.96°W*
Kupreanof volcano:	*56.01°N, 159.79°W*

Elevation: *1,320–1,895 m from SW to NE*
Eruptive History: *No historic activity*
Composition: *Basalt, andesite and dacite*

The Stepovak Bay group is a chain of five volcanoes at the southwest end of a N40°E oriented linear segment of the Aleutian arc on the Alaska Peninsula. This same segment includes the better known Veniaminof and Aniakchak calderas. Three of the Stepovak Bay volcanoes [2, 4, and Kupreanof (K on map)] have clearly had Holocene eruptions, resulting in three small debris flows filling late Pleistocene glacial valleys, and a small cinder cone and associated lava flow. The other two volcanoes (1 and 3) do not show unmistakable evidence of Holocene activity. They have ice-filled summit craters 500 m (1) and 300 m (3) in diameter that may be late Pleistocene age. These volcanoes have contributed to extensive late Tertiary and Quaternary lava flows, some extending near sea level.

Kupreanof is the largest and best known volcano of the group. In 1982, reconnaissance mapping located the three southernmost volcanoes. Later work further delineated these three, and located an additional center (4). Mapping and K-Ar dating indicate the area has been the locus of

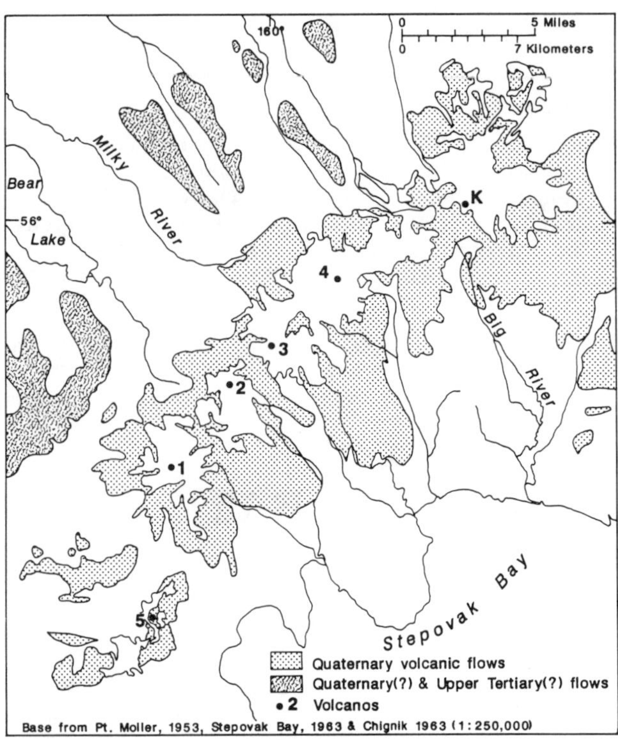

KUPREANOF: Map of volcano line.

voluminous volcanic activity for approximately the last 4 million years. A fumarole occurs at Kupreanof volcano; another has been reported on volcano 4. Air photos from the late 1940s also suggest the presence of a fumarole on volcano 1.

Northwest of the line of volcanoes a projected fault vertically offsets the volcanoes from flows that are related to the early history of the centers. These flows. which dip away from the centers, yield K-Ar ages between 3.9 and 1.7 Ma. Along the trend, and 7 to 12 km southwest of the volcanoes, is another area of early (?) Quaternary flows possibly related to a yet unrecognized center (5 on map); outcrops of hypabyssal rocks in a snowfield among these flows may be this center. The group of volcanoes overlies sedimentary rocks of Oligocene to late Miocene age; the southern centers also overlie thick tuff deposits of uncertain but presumably late Tertiary or early Quaternary age.

How to get there: *The Stepovak Bay group of volcanoes is located on the Alaska Peninsula 70 km northwest of the city of Sand Point, and 40 km east of the cannery at Port Moller. Access is difficult, requiring a charter flight via helicopter or bush plane. Commercial service is available to Sand Point, but as of 1988 there were no charter operators in Sand Point. There are no roads or trails in the area, or lakes suitable for floatplane landings within 12 km of the volcanoes.*

References

Eakins, G. R., 1970, Mineralization near Stepovak Bay, Alaska Peninsula, Alaska: *Alaska Div. Mines Geol. Sp. Rpt. 4*, 12 pp.

Yount, M. E., Wilson, F. H., and Miller, J. W., 1985, Newly discovered Holocene volcanic vents, Port Moller and Stepovak Bay quadrangles, Alaska Peninsula, *in* Bartsch-Winkler, S., and Reed, K. M. (eds.), The United States Geologic Survey in Alaska: Accomplishments in 1983: *USGS Circ. 945*, 60-62.

Frederic H. Wilson

VENIAMINOF

Alaska Peninsula

Type: *Stratovolcano with caldera*
Lat/Long: *56.18°N, 159.38°W*
Elevation: *2,507 m*
Height: *2,200 m*
Caldera Diameter: *9 km*
Volcano Diameter: *35 km*
Eruptive History: *3,700 yr BP: caldera collapse; 1830-38, 1852?, 1874?, 1892, 1930?, 1939, 1944, 1956, 1983-84*
Composition: *Tholeiitic basalt to dacite*

Mount Veniaminof is one of the highest peaks on the Alaska Peninsula and one of the Aleutian volcanic arc's most voluminous centers (>400 km^3). Deposits from its 3,700-yr-BP climactic eruption extend from coast to coast across the peninsula. A 9-km-wide, ice-filled caldera truncates the summit.

Two-thirds of the caldera is enclosed by steep walls rising as much as 520 m above the intracaldera ice. Within the caldera the ice surface ranges in elevation from 2,000 m to 1,750 m and slopes to the south, where it drapes and obscures the south caldera rim as it merges with a 220-km^2 ice field on the south flank. Alpine glaciers totaling another 60 km^2 radiate from the north and west rims, in some cases cutting significant notches in the rim. Cone Glacier drains west from the caldera and has cut a notch in the caldera wall probably down to the level of

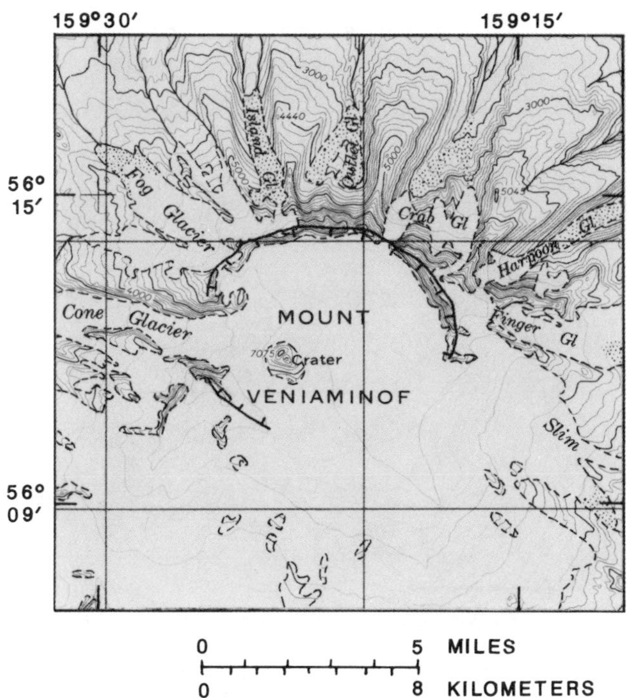

VENIAMINOF: USGS topographic map.

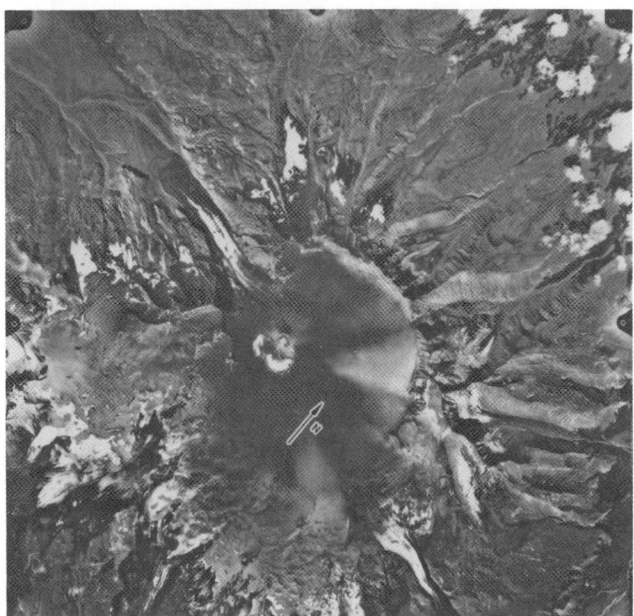

VENIAMINOF: USGS aerial photograph, August, 1983.

the floor. A subglacial, east-facing breach of the caldera rim has been postulated based on ice fracture patterns but remains unconfirmed.

A 60-km-long belt of basaltic cinder and scoria cones, most if not all of Holocene age, extends from the Bering Sea coast across the caldera and down the Pacific flank. Two of the cinder cones are within the caldera, one protruding 330 m above the ice and the other almost entirely covered; the more prominent of the two has been the locus of eruptive activity since at least 1930.

VENIAMINOF: Space Shuttle picture, November–December, 1983; frame width ~65 km. S09-40-2651.

One of the principal hazards from an eruption of Mount Veniaminof is a jokulhlaup from melting of the intracaldera glacier. During the 1983-84 eruption ~0.15 km^3 of ice melted, presenting the possibility that subglacially ponded water could be catastrophically released from the caldera. Although no outburst floods were observed in 1983-84 on any of the rivers and creeks draining from the caldera, the possibility of future jokulhlaups cannot be discounted.

How to get there: *Mount Veniaminof is in the Alaska Peninsula National Wildlife Refuge, 100 km southwest of Port Heiden and 115 km northeast of Sand Point, both of which can be reached by scheduled airline from Anchorage. Smaller aircraft can be chartered to Perryville, an Aleut village ~35 km south of Veniaminof, from which the volcano can be reached either on foot or by an all-terrain vehicle (ATV). Lodging is available in Perryville by prior arrangement with the Village Corporation.*

References

Miller, T. P., and Smith, R. L., 1987, Late Quaternary caldera-forming eruptions in the eastern Aleutian arc, Alaska: *Geology 15*, 434-438.

Yount, M. E., Miller, T. P., Emanuel, R. P., and Wilson, F. H., 1985, Eruption in an ice-filled caldera, Mount Veniaminof, Alaska Peninsula, *in* Bartsch-Winkler, S., and Reed, K.M. (eds.), USGS in Alaska: Accomplishments during 1983: *USGS Circ. 945*, 58-60.

M. Elizabeth Yount

BLACK PEAK
Alaska Peninsula

Type: *Stratovolcano with caldera*
Lat/Long: *56.53°N, 158.62°W*
Elevation: *1,019 m (caldera rim); 1,032 m (intercaldera dome)*
Height: *1,019 m*
Caldera Relief: *330 m*
Volcano/Caldera Diameter: *8/3.5 km*
Eruptive History: *Cone-building assumed to be Plio-Pleistocene; caldera formation: 3,600-4,700 ^{14}C yr BP; no historic activity*
Composition: *Andesite to dacite*

Black Peak is a deeply eroded, highly altered stratovolcano/dome complex with a small caldera at its eastern edge. The caldera is ice-free and contains two small lakes; most of the interior of the caldera is occupied by a complex of nested dacitic domes. The volcano rests on a north-dipping basement of Pliocene volcanogenic non-marine sedimentary rocks.

The flows, domes, and volcaniclastic rocks that make up the pre-caldera cone range in composition from andesite to dacite. Dacitic ash-flow tuffs and coarse block-and-ash flows fill the Bluff and Ash Creek valleys to as much as 100 m on the north and west sides of the volcano. The pyroclastic flows that deposited these tuffs had limited distribution and appear to have been relatively sluggish; they also appear to have had a large air-fall component, judging from the crude stratification in the ash-flow tuff. In spite of the caldera's small diameter, the widespread climactic air fall and the thickness of the ash-flow tuffs suggest a bulk eruption volume of >10 km^3.

How to get there: *Black Peak is located on the central Alaska Peninsula ~740 km southwest of Anchorage. There are no roads in the area; access is by air from King Salmon to one of the nearby lakes and then by foot.*

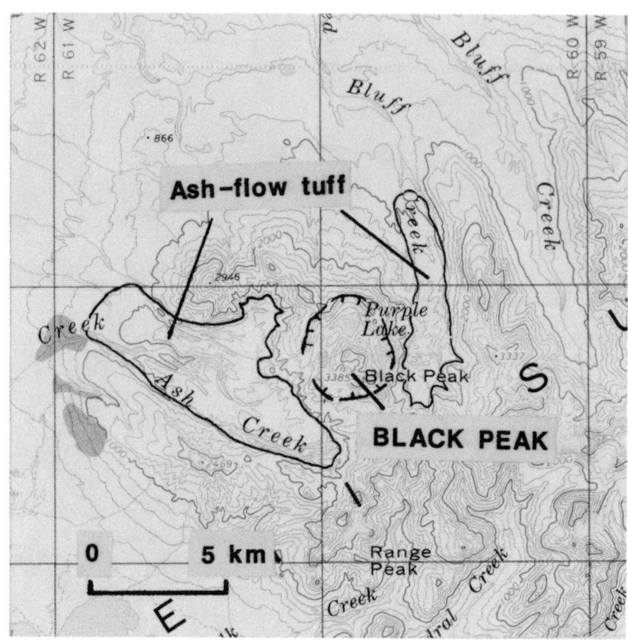

BLACK PEAK: Map from USGS Chignik 1:250,000-scale topographic map series; contour interval: 200 ft.

BLACK PEAK: View looking southwest of flat-topped ashf-low tuffs in Ash Creek adjacent to Black Peak volcano. Veniaminof volcano in background.

References

Detterman, R. L., Miller, T. P., Yount, M. E., and Wilson, F. H., 1981, Geologic map of the Chignik and Sutwik Island quadrangles, Alaska: *USGS Misc. Invest. Ser. Map I-1229*, scale 1:250,000.

Miller, T. P., and Smith, R. L., 1987, Late Quaternary caldera forming eruptions in the eastern Aleutian arc, Alaska: *Geology 15*, 434-438.

Thomas P. Miller

ANIAKCHAK: USGS low oblique view eastward across caldera.

ANIAKCHAK
Alaska Peninsula

Type: *Stratovolcano with summit caldera*
Lat/Long: *56.88°N, 158.17°W*
Elevation: *1,341 m (caldera rim)*
Height: *~650 m*
Caldera Relief: *1,028 m*
Volcano /Caldera Diameter: *~20/10 km*
Eruptive History: *Pleistocene cone-building (K-Ar ages range from 0.661-0.242 Ma); caldera formation ~3,400 ^{14}C yr BP; ash eruption and possible dome emplacement in 1931*
Composition: *Basalt to dacite (rhyolite)*

Aniakchak caldera is one of the most spectacular volcanic landforms in the Aleutian arc. This circular steep-walled ice-free caldera was formed about 3,400 ^{14}C yr BP by collapse of a deeply eroded and glaciated andesitic stratovolcano following the catastrophic eruption of >50 km^3 (bulk volume) of material.

The volcano was built on a west-dipping erosion surface, and basement rocks consisting of Mesozoic and Tertiary sediments are exposed in the east and south walls of the caldera to elevations of about 610 m. The original stratovolcano probably had a volume of about 75 km^3 prior to caldera collapse. The cone-building flows and volcaniclastic rocks range in composition from andesitic basalt to dacite but are generally andesitic, with SiO_2 contents ranging from 51-61%.

Ash-flow tuff, generally non-welded and deposited during the caldera-forming eruption, occurs in thick continuous exposures surrounding the stratovolcano and in isolated outcrops as far away as 50 km. The tuff is compositionally zoned: basal silicic tuffs are overlain by andesitic tuffs, and SiO_2 ranges from 57.1 to 67.9%. The distribution of the tuff suggests the

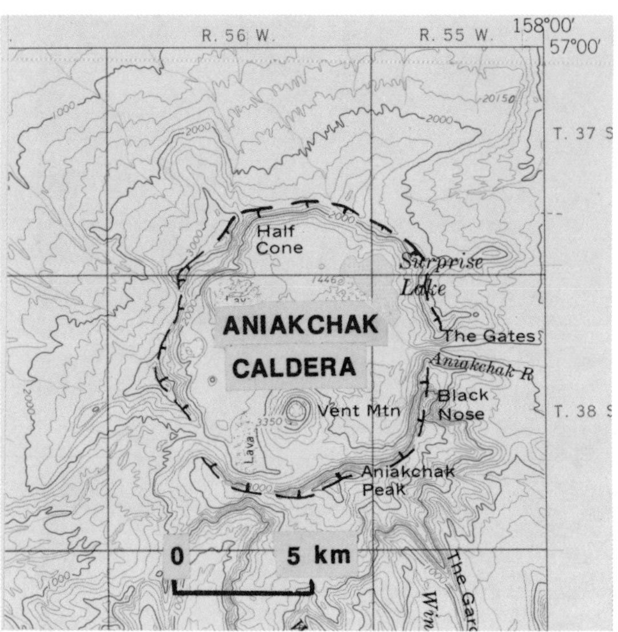

ANIAKCHAK: Map from USGS Chignik 1:250,000-scale quadrangle map; contour interval: 200 ft.

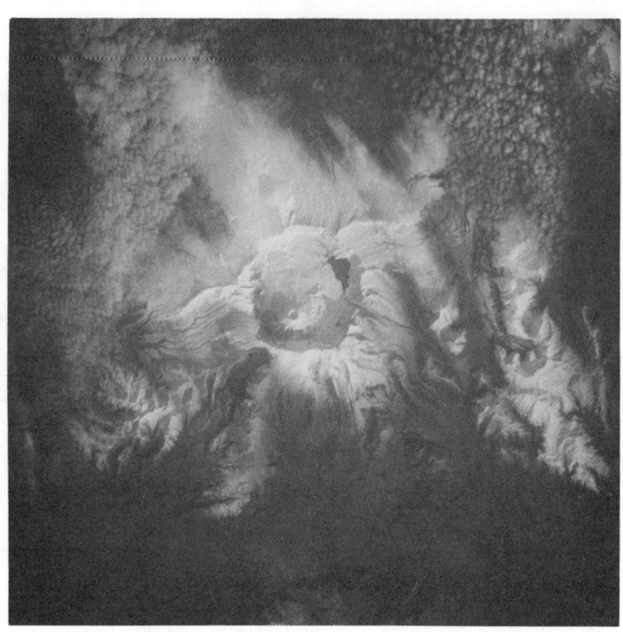

ANIAKCHAK: Space Shuttle photograph, October, 1984. Frame width ~65 km. S17-33-043.

pyroclastic flows originally covered an area of at least 2,500 km^2. The pyroclastic flows that deposited these tuffs were extremely mobile and crossed over topographic barriers of 500 m relief at distances of 30 km from the caldera rim. As much as 7 cm of tephra from the caldera-forming eruption have been found 1,100 km to the north on the Seward Peninsula.

Post-caldera activity has resulted in a 1,000-m-high spatter cone (**Vent Mountain**), numerous domes, tuff cones, maars, lava flows, and cinder cones inside the caldera indicating over 20 major post-caldera events. Surprise Lake, a small lake covering <5% of the northeast caldera floor, is drained through The Gates, a breach in the east wall.

The only documented historic eruption occurred in 1931 from a vent on the west side of the caldera floor. The eruption resulted in an ash fall over much of southwestern Alaska and the possible emplacement of a dome. Thermal areas inside the caldera in 1985 included warm springs north of Surprise Lake and an area having ground temperatures of 85°C at depths of 25 cm near Half Cone.

How to get there: *Aniakchak caldera is ~650 km southwest of Anchorage in Aniakchak National Monument on the central Alaska Peninsula. No roads occur in the area and access is generally by floatplane from King Salmon into Surprise Lake.*

References

Miller, T. P., and Smith, R. L., 1977, Spectacular mobility of ash flows around Aniakchak and Fisher calderas, Alaska: *Geology 5*, 173-176.

Miller, T. P., and Smith, R. L., 1987, Late Quaternary caldera-forming eruptions in the eastern Aleutian arc, Alaska: *Geology 15*, 434-438.

Thomas P. Miller

YANTARNI
Alaska Peninsula

Type: *Stratovolcano with domes*
Lat/Long: *57.02°N, 157.20°W*
Elevation: *1,336 m*
Height: *730 m*
Eruptive History: *Stratovolcano construction in Pleistocene; no historic activity*
Composition: *Andesite to dacite*

Yantarni volcano is an andesitic stratovolcano comprising ~3.5 km^3 located adjacent to, and within the same arc segment as, Mount Chiginagak. The volcano was not discovered until 1979, owing to its modest summit elevation, re-

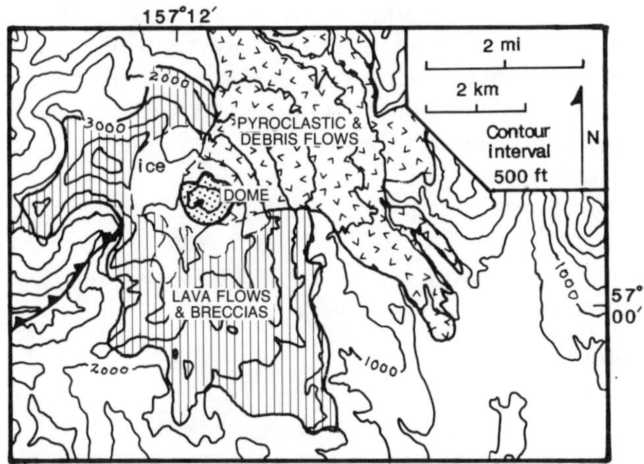

YANTARNI: Simplified geologic map of Yantarni volcano on the central Alaska Peninsula, 640 km southwest of Anchorage.

YANTARNI: View to west showing central dome (center, 5 km distant), breached cone (on both sides of dome), and flat-topped pyroclastic-flow deposits (foreground).

mote location, and lack of documented historic activity. Its name is taken from the adjacent bay, named on a Russian chart as "Z. Yantarniy" for amber purportedly found there. First mapped at 1:250,000 scale, the volcano has since been mapped at 1:63,360 and its eruptive history and chemistry determined in greater detail.

The volcano formed near a high-angle fault at a site of previous Tertiary magmatism. The current cycle of eruptive activity began in middle Pleistocene time with extrusion of andesitic lava flows, perhaps from multiple vents. By the late Pleistocene, central-vent volcanism had initiated construction of a small stratovolcano. The cone was breached in late Holocene time (between 2,000 and 3,500 yr ago?), forming a debris-avalanche deposit which was possibly accompanied by a directed blast and closely followed by emplacement of a dome and pyroclastic flows. The pyroclastic flows are about 1 km^3 in volume and extend 4 km down-valley.

Disequilibrium phenocryst assemblages, a lack of correlation between phenocryst assemblages and whole-rock compositions, eruption of different compositions of magma in close succession, and the small volume of eruptive products suggest Yantarni is an immature volcano lacking a large, shallow magma chamber.

How to get there: *Yantarni volcano is in a remote part of the Aleutian Range and can be reached only by aircraft; there are no prepared landing sites within 10 km.*

References

Orth, D. J., 1967, Dictionary of Alaska place names: *USGS Prof. Pap. 567*, 1084.

Riehle, J. R., Yount, M. E., and Miller, T. P., 1987, Petrography, chemistry, and geologic history of Yantarni volcano, Aleutian volcanic arc, Alaska: *USGS Bull. 1761*, 27 pp.

James R. Riehle

CHIGINAGAK

Alaska Peninsula

Type: *Stratovolcano with domes*
Lat/Long: *57.13°N, 156.98°W*
Elevation: *2,067 m*
Height: *1,390 m*
Eruptive History: *Stratovolcano construction in Pleistocene; 1929?, 1972?*
Composition: *Andesite, dacite*

Chiginagak is a typical calc-alkaline stratovolcano comprising ~25 km^3 located midway in an arc segment which extends from Aniakchak caldera northeast to Peulik volcano. Having a high summit and conical shape, the volcano has been recognized since at least the time of Russian occupation. The name is native and first appeared in writing on an 1827 Russian map showing adjacent Chiginagak Bay.

An active fumarole at an elevation of about 1,600 m on the north flank typically produces a small cloud. As no tephra beds at the ground surface are known beyond the barren slopes of the volcano, the reported historic activity may

CHIGINAGAK: View to northeast of Chiginagak volcano, about 10 km distant. Outward-dipping lava flows and breccias and the profile of the summit crater are readily visible. Curving levees of Holocene(?) lava flows extend toward the viewer at the bottom center of the cone.

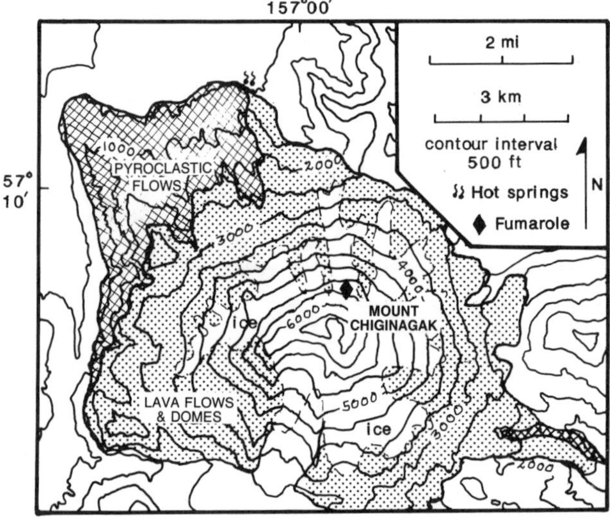

CHIGINAGAK: Simplified geologic map of Chiginagak volcano on the central Alaska Peninsula, 620 km southwest of Anchorage.

be only fumarolic. Thermal springs occur just beyond the cone deposits downslope from the fumarole. Although exposures are limited by snow and ice, Chiginagak appears to be a stratovolcano having a summit crater and one or more domes, built mainly during Pleistocene time. The oldest known deposits are block-and-ash flows of low-silica dacite composition (64% SiO_2) low on the northwest flank. Most of the cone-forming lavas are andesite (56%-61% SiO_2). The most recent deposits are an unglaciated lava flow and an overlying block-and-ash flow (59% SiO_2) that extend east from the cone onto the floor of a glacial valley. The pyroclastic flows of Chiginagak comprise a total of about 4 km^3, and all probably resulted from dome collapse; the source of the Holocene pyroclastic flow is probably a dome at 1,687 m elevation on the cone, about 1 km southeast of the summit crater.

How to get there: *Mount Chiginagak occupies a remote location in the Aleutian Range and can be reached only by aircraft; there are no prepared landing sites nearby.*

References

Detterman, R. L., Wilson, F. H., Yount, M. E., and Miller, T. P., 1987, Quaternary geologic map of the Ugashik, Bristol Bay, and western part of Karluk quadrangles, Alaska: *USGS Misc. Invest. Ser. Map I-1801.*

Motyka, R. J., Moorman, M. A., and Liss, S. A., 1981, Assessment of thermal springs sites, Aleutian arc, Atka Island to Becherof [sic] Lake-preliminary results and evaluation: State of Alaska, Dept. Natural Resources, *ADGGS Open-file Rpt 144*, 173 pp.

James R. Riehle

KIALAGVIK
Southeast Alaska

Type: *Stratovolcano with dome*
Lat/Long: *57.20°N, 156.75°W*
Elevation: *1,575 m*
Height: *900 m*
Eruptive History: *Holocene: dome emplacement, block-and-ash flows(?); no historic activity*
Composition: *Andesite, dacite*

Kialagvik volcano is a small (~5 km^3), poorly known central-vent volcano adjacent to, and in the same arc segment as Chiginagak volcano. The volcano is informally named after the Eskimo word for the adjacent Wide Bay. The published geologic map is at 1:250,000 scale; owing to remoteness and extensive snow and ice cover, little additional work has been done.

The volcano is constructed partly on and adjacent to older, early Quaternary(?) or Tertiary hypabyssal and extrusive rocks from other vents; consequently, the exact extent of the deposits properly assigned to the Kialagvik vent is uncertain. Probable Kialagvik deposits comprise lava flows of andesitic composition and

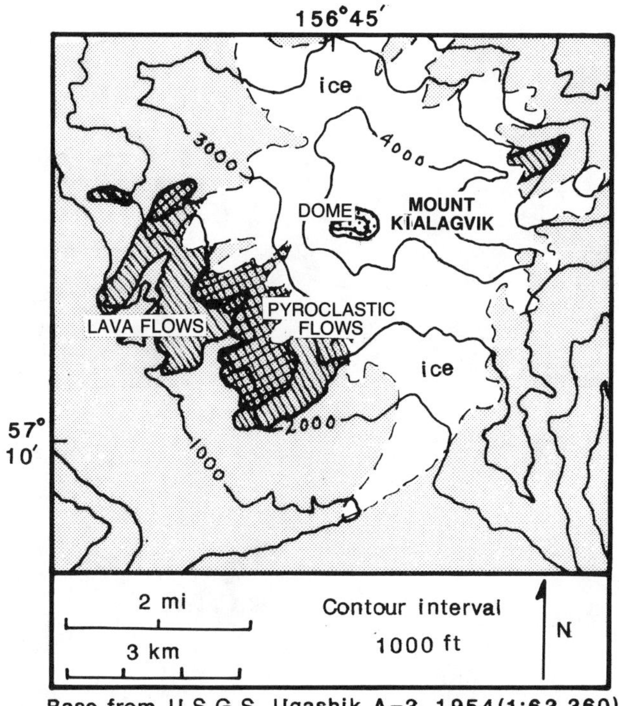

KIALAGVIK: Simplified geologic map of Kialagvik volcano on the central Alaska Peninsula, 600 km southwest of Anchorage. Pre-Quaternary rocks not covered by snow or ice are lightly shaded.

KIALAGVIK: View to northwest of central dome of Kialagvik volcano. The dome rises about 350 m above the surrounding ice field.

overlying dacitic block-and-ash flow deposits similar in composition to the Holocene(?) dome. Except for lacking an exposed stratovolcano edifice, Kialagvik has probably been most similar in eruptive style to Yantarni volcano to the southwest.

How to get there: *Kialagvik volcano is in a remote part of the Aleutian Range and can be reached only by air. There are no prepared landing sites nearby.*

Reference

Detterman, R. L., Wilson, F. H., Yount, M. E., and Miller, T. P., 1987, Quaternary geologic map of the Ugashik, Bristol Bay, and western part of Karluk quadrangles, Alaska: *USGS Misc. Invest. Ser. Map I-1801.*

James R. Riehle

UGASHIK AND PEULIK

Alaska Peninsula

Type: *Stratovolcano with caldera and parasitic stratovolcano (Peulik)*
Lat/Long: *Ugashik: 57.70°N, 156.35°W; Peulik: 57.75°N, 156.35°W*
Elevation: *Ugashik: 521 m; Peulik: 1,474 m*
Height: *Ugashik: 300 m; Peulik: 650 m*
Volcano Diameter: *Ugashik: 6-8 km; Peulik: 53.5 km*
Caldera Diameter: *4.6 km*
Eruptive History: *Ugashik cone-building: 0.17 Ma; caldera-formation: >30,000 yr BP; Peulik: 1814, 1852.*
Composition: *Ugashik: dacite to rhyolite; Peulik: basalt to dacite*

Ugashik caldera is a remnant of a small, deeply eroded stratovolcano (pre-caldera volume probably <15 km^3) with a 5-km-wide summit caldera. Only a few hundred meters of dacitic flows and volcaniclastic rocks remain perched high on the caldera rim, and a single K-Ar age of 0.17 Ma has been obtained from this unit. Most of the caldera walls consist of sandstone basement rocks of the Jurassic Naknek Formation. At least 5 non-glaciated (Holocene) domes of dacite to rhyolite composition occupy the interior of the caldera. Ash-flow tuffs associated with the caldera-forming eruption have not been found outside the caldera, but reworked pumice fragments occur in surficial deposits around the volcano. Consequently, the caldera is thought to have formed in late Wisconsin time, possibly when most of the volcano was surrounded by a large glacial lake. Pyroclastic flows associated with the caldera-forming eruption may have moved into the lake.

UGASHIK–PEULIK: North-looking view of Peulik volcano showing summit crater and plug dome.

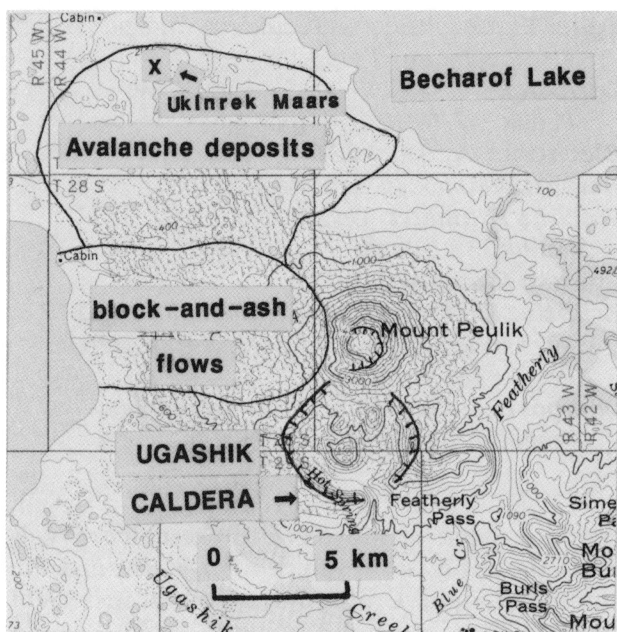

UGASHIK–PEULIK: Map from USGS Ugashik 1:250,000-scale quadrangle map; contour interval: 200 ft.

Peulik is a small (3 km^3), non-glaciated Holocene parasitic stratovolcano developed on a topographic basement high (820 m) 2.5 km north of the caldera. Flows from the summit area of Peulik have breached the north rim of the caldera. Basaltic flows from flank eruptions of Peulik cover ~8 km^2 north of the volcano, extending as far as Becharof Lake. The Peulik summit crater is breached to the west and contains a large dacite dome; a smaller dome occurs high on the east side of the cone. Block-and-ash-flow deposits resulting from summit dome-building activity cover a 40 km^2 area west of the breached cone extending down to Ugashik Lake. Avalanche deposits whose source is uncertain underlie an area of 75 km^2 northwest of the volcano.

UGASHIK–PEULIK: Close-up view of Peulik volcano showing dacite dome. View from southwest. Bechanof Lake is in the background. Photograph by D. J. Lalla.

How to get there: *Peulik and Ugashik volcanoes are located on the Alaska Peninsula about 5-10 km south of Becharof Lake. They can be reached only by floatplane from King Salmon to Becharof Lake and then by foot, as there are no roads in the area.*

References

Detterman, R. L., Wilson, F. H., Yount, M. E., and Miller, T. P., 1987, Quaternary geologic map of the Ugashik, Bristol Bay, and western part of Karluk quadrangles, Alaska: *USGS Map I-1801.*

Miller, T. P., and Smith, R. L., 1987, Late Quaternary caldera-forming eruptions in the eastern Aleutian arc, Alaska: *Geology 15*, 434-438.

Thomas P. Miller

UKINREK
Alaska Peninsula

Type: *Two maars*
Lat/Long: *57.83°N, 156.51°W*
Elevation: *70 m*
Volcano Dimensions:
East Maar: 300 m diameter, 70 m deep;
West Maar: 105 × 170 m, 35 m deep
Eruptive History: *Monogenetic formation, March 30-April 9, 1977*
Composition: *Weakly undersaturated alkali olivine basalt*

During ten days of phreatomagmatic activity in early April 1977, two maars formed 13 km inboard of the Aleutian arc. The eruption site is 13 km northwest of Peulik volcano, a parasitic stratovolcano on the rim of Ugashik caldera. The maars formed in terrain that was heavily glaciated in late Pleistocene/early Holocene time. The groundwater contained in the underlying till and silicic volcanics from nearby Peulik volcano controlled the dominantly phreatomagmatic course of the eruption. The maars have been named "Ukinrek Maars," which means "two holes in the ground" in Yupik Eskimo. (In Yupik Eskimo, Peulik means "one with smoke"; for the location of the Ukinrek Maars, see map that accompanies Ugashik and Peulik description.)

The Ukinrek eruption began at **West Maar** on the crest of a low ridge in glacial till deposits near Becharof Lake. During the most violent phase of the eruption, steam and ash plumes reached maximum heights of ~6 km. A thin blanket of ash was dispersed to a distance of at least 160 km north and east of the vent, covering an area of ~25,000 km^2. Three days later, **East Maar** began to erupt, forming a new vent 400 m away. At the same time, activity ceased at West Maar. This new crater quickly grew by strong phreatomagmatic explosions and on day four lava first appeared in the center of East Maar.

The fresh maars were steep-sided with near vertical walls, blasted out by strong phreatomagmatic explosions. East Maar was excavated below the water table and is now filled by a lake 50 to 60 m deep. The steep sides of the fresh maars are in striking contrast to older maars that typically have shallow dish shapes because of modification by slumping and erosion.

UKINREK: Both maars on day 3 of the eruption when the activity shifted from West Maar (foreground) to East Maar, 400 m away (view to east). Black area in front of West Maar is a hot ejecta blanket of juvenile basaltic cinders. Steam rises from a hot water lake in West Maar. Phreatomagmatic activity is beginning to excavate the East Maar. Photograph by L. Conyers.

UKINREK: East Ukinrek Maar, fall, 1977. The maar is starting to fill with water. Steam rises from a lava cupola in the bottom of the vent. A thick (26 m maximum) ejecta blanket built the maar's rim. Meter-size blocks of glacial till can be seen in the ejecta blanket.

The ejecta blanket surrounding the two maars has a maximum thickness of 26 m on the rim of East Maar and rapidly decreases to a few centimeters 3 km from the vents. Deposits within a 5-km radius of the maars have a bulk volume of at least 26 million m^3. They consist of three types: (1) Aprons of glacial till and bedrock blocks with juvenile bombs (~60-80% of the deposit); (2) ash and lapilli fallout lobes associated with strombolian pulses; and (3) minor vesiculated tuffs associated with base surge activity.

The primitive chemistry of the basalt erupted at Ukinrek suggests a deep mantle origin. The dikes through which the basalt rose to the surface were probably controlled by a deep crustal fracture, the Bruin Bay fault.

How to get there: *The Ukinrek craters are an exceptionally fresh example of maar volcanism. The maars can be reached by air from King Salmon, 100 km to the north. Access is by floatplane to Gas Rocks on the southern shore of Becharof Lake. From there it is a 2-km hike southward to the maars. Be prepared for bears. The maars are in the Becharof National Wildlife Refuge.*

References

Kienle, J., Kyle, S., Motyka, R. J., and Lorenz, V., 1980, Ukinrek Maars, Alaska, I. April 1977 eruption sequence, petrology and tectonic setting: *J. Volc. Geotherm. Res. 7*, 11-37.

Self, S., Kienle, J., and Huot, J-P., 1980, Ukinrek Maars, Alaska, II. Deposits and formation of the 1977 craters: *J. Volc. Geotherm. Res. 7*, 39-65.

Jürgen Kienle

KEJULIK
Alaska Peninsula

Type: *Dissected stratovolcano?*
Lat/Long: *58.05°N, 155.50°W*
Elevation: *1,517 m*
Volcano Diameter: *15 km*
Eruptive History: *No historic activity*
Composition: *Andesite*

Kejulik is a remnant of a volcano in the southwestern Kejulik Mountains. A radial system of glacially carved valleys surrounds the present topographic high. Extensive glacial erosion has removed most of the lavas, but remnants of lava flows and lahars are preserved on the radial ridges. A series of dikes intrudes the near-horizontal Mesozoic sedimentary rocks beneath the flows. A zone of hydrothermally altered breccia near the present summit marks the former vent of the volcano.

Little information has been published on Kejulik. All of the analyzed samples are porphyritic andesites. No recent activity is known.

KEJULIK: Kejulik Mountains in background with snow and ice fields. Foreground hills are near-horizontal Mesozoic sediments.

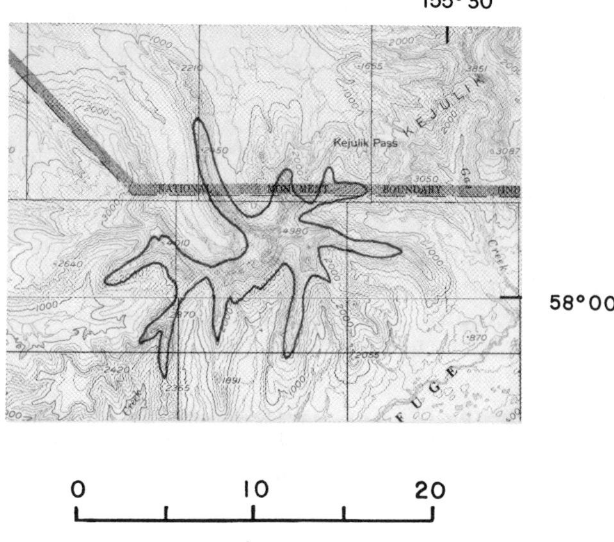

KEJULIK: USGS topographic map.

How to get there: *The Kejulik Mountains are in a remote area without nearby bodies of water suitable for floatplanes. The best access is via commercial air to King Salmon or Kodiak, then by helicopter charter to the mountains.*

References

Kienle, J., and Swanson, S. E., 1983, Volcanism in the eastern Aleutian arc: Late Quaternary and Holocene centers, tectonic setting and petrology: *J. Volc. Geotherm. Res. 17*, 393-432.

Kienle, J., Swanson, S. E., and Pulpan, H., 1983, Magmatism and subduction in the eastern Aleutian arc: *in* Shimozuru, D., and Yokoyama, I. (eds.), *Arc Volcanism: Physics and Tectonics*, D. Reidel, Boston, 191-224.

Samuel E. Swanson

MARTIN: Summit of Mount Martin showing fumarole activity in crater.

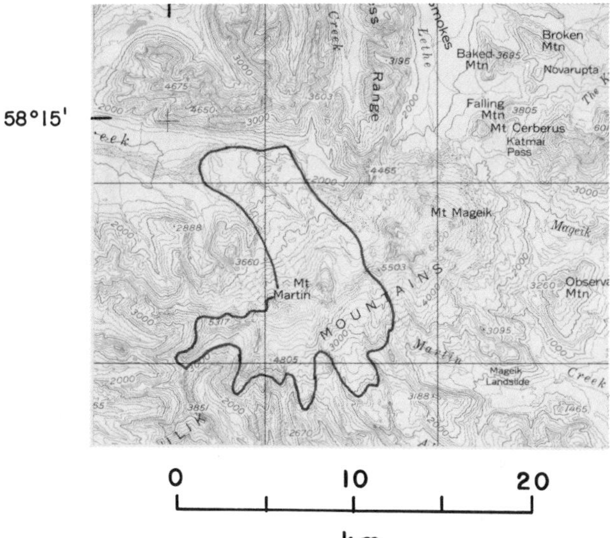

MARTIN: USGS topographic map.

MARTIN
Alaska Peninsula

Type: *Stratovolcano*
Lat/Long: *58.15°N, 155.38°W*
Elevation: *1,862 m*
Volcano Diameter: *15 km*
Eruptive History: *1951?, 1953?*
Composition: *Andesite to dacite*

Mount Martin is a large stratovolcano mostly covered by glacial ice. Despite the extensive glacial covering, the volcano is only moderately dissected. A large post-glacial lava flow extends 12 km from the summit of Mount Martin to the northwest. A summit crater is kept free of ice by vigorous solfatara activity which feeds an almost constant steam plume. A small yellow lake occupied the floor of the crater in 1980 and 1981.

Historic eruptions of Mount Martin are limited to two questionable events. However, the very active fumarole field and the large post-glacial lava flow suggest recent activity.

Only reconnaissance geologic surveys are available for Mount Martin. The lavas are porphyritic, high-silica andesite and dacite.

How to get there: *Access to Mount Martin is difficult. The most direct ground access is through the Valley of Ten Thousand Smokes past Mount Mageik.*

References

Keller, A. S., and Reiser, H. N., 1959, Geology of the Mount Katmai area, Alaska: *USGS Bull. 1058-G*, 261-298.

Kienle, J., Swanson, S. E., and Pulpan, H., 1983, Magmatism and subduction in the eastern Aleutian arc: *in* Shimozuru, D., and Yokoyama, I. (eds.), *Arc Volcanism: Physics and Tectonics*, D. Reidel, Boston, 191-224.

Samuel E. Swanson

MAGEIK
Alaska Peninsula

Type: *Stratovolcano*
Lat/Long: *58.2°N, 155.25°W*
Elevation: *2,165 m*
Relief: *~1,565 m*
Volcano Diameter: *9 km*
Eruptive History: *Cone mostly Holocene; no historic activity; sulfur-rich fumaroles in and around Crater Lake*
Composition: *Silicic andesite to dacite*

Mount Mageik (pronounced Muh-GEEK) faces Trident across Katmai Pass, both centers standing atop the drainage axis of the Alaska Peninsula, at the head of the **Valley of Ten Thousand Smokes.** The edifice is moderately symmetrical, but its broad ice-capped summit is marked by 4 discrete knobs, each a center of spatter and lava flow emission having 50-250 m of local relief. Nestled between two of them is a

MAGEIK: Mounts Mageik (left) and Martin (distant right) from the Valley of Ten Thousand Smokes. Photograph by J. Fierstein, USGS.

young explosion crater 300 m wide and 100 m deep, reamed through a massive dacite dome and now containing a shallow turbulent lake of yellow-green water at 70°C. When Fenner visited the crater in 1923, the fumaroles were active but no lake was present.

Half of the cone is ice-covered, but glacial dissection is not severe. Glaciers on the north flank have retreated nearly 1 km since 1912. Eroded lavas of probable Pleistocene age rest on Jurassic basement at the southwest and northwest bases of the cone, but most exposed lavas appear to have erupted in Holocene time. The present volume of the volcano is 30-35 km^3, predominantly lava flows of pyroxene andesite and dacite. A small debris-avalanche deposit fills a deglaciated valley just south of the cone. A dacite pumice fall as thick as 85 cm occurs on Mageik's north slope and probably erupted there; upon it is developed a soil dated at 2140 ^{14}C yr. Thirty-seven samples from the summit knobs and the north and east flanks fall in the limited range 60-65% SiO_2 · K_2O values are 1.3% at 60% SiO_2 and 2.0% at 65%.

No deposits have been found to substantiate reports of historic eruptions. These reports were all based on remote sightings, probably misinterpretations of the persistent vapor plume above the crater.

How to get there: *A long day's walk up the Valley of Ten Thousand Smokes from the Overlook roadhead. Rope and crampons to the summit crater. Crevasses, weather, and bears are all dangers. The area is uninhabited wilderness.*

References

Fenner, C. N., 1930, Mount Katmai and Mount Mageik: *Zeitschrift fur Vulkanologie 13*, 1-24.

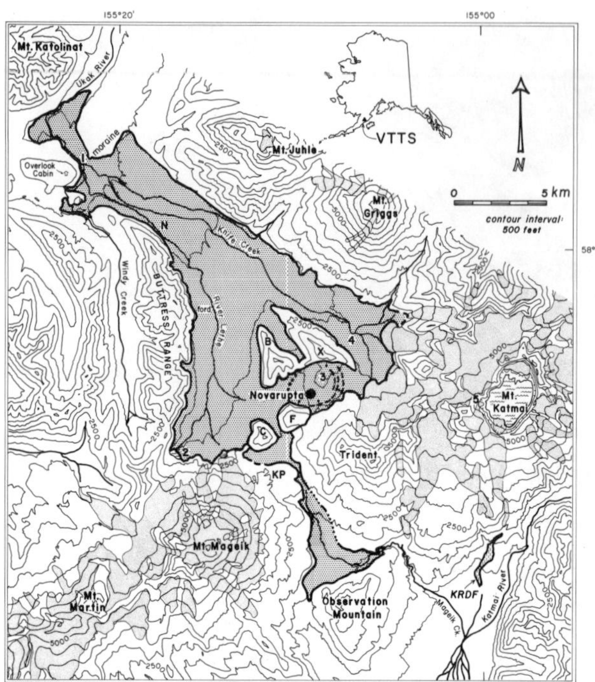

MAGEIK TO GRIGGS: USGS topographic map with Valley of Ten Thousand Smokes ash flow indicated.

Hildreth, W., 1987, New perspectives on the eruption of 1912 in the Valley of Ten Thousand Smokes, Katmai National Park, Alaska: *Bull. Volc. 49*, 680-693.

Wes Hildreth

TRIDENT
Alaska Peninsula

Type: *Stratovolcano cluster*
Lat/Long: *58.23°N, 155.1°W*
Elevation: *1,864 m*
Relief: *~965-1,265 m*
Volcano Diameter: *7 km for cluster*
Eruptive History: *New cone built 1953-63; minor activity to 1968; blocky lava flows 1953-60; vigorous fumaroles still active on new cone*
Composition: *Andesite to dacite*

When initially named, Trident consisted of 3 adjacent andesite-dacite cones, each a discrete center of lava and ejecta emission, and each gutted and sharpened by glacial erosion. Two gla-

TRIDENT: Volcano cluster seen from the south.

TRIDENT: Oblique aerial view showing the young flank cone that has developed in a series of eruptions since 1953. View is to the south. Fresh black andesite lava flows were erupted between 1953 and 1960. Photograph by J. Kienle.

ciated domes on the south flank completed the the cluster. Beginning in 1953, however, a fourth cone (now having 400-800 m of local relief) was constructed on the steep southwest flank of the older complex. Trident adjoins Mount Mageik to the southwest across Katmai Pass, its main vents lie only 3-5 km southeast of Novarupta, and it is continuous to the northeast with a similar cone cluster that composed pre-caldera Mount Katmai.

The 1953 eruption opened with a vapor and ash plume 9 km high, issuing from the site of a pre-existing fumarole at 1,100 m. Four complexly lobate tongues of blocky two-pyroxene andesite (60-64% SiO_2) as long as 4.5 km were emplaced between 1953 and 1960. The vent cone of scoria, crusted bombs, coarse angular blocks, and thin proximal lava sheets continued to grow through 1963, with minor emission as late as 1968. Its shallow crater is today 250 m wide, breached to the southwest. Cone structure and mode of formation appear similar for older components of the cluster.

Numerous fumaroles and steaming ground remain active on the upper outer slopes of the new cone. All major fumaroles had orifice temperatures of 97°C (boiling) in 1979; those on the northwest side were near-neutral, but those on the northeast and southeast were SO_2-rich, sulfur-depositing, and yielded condensates with pH ~1. Dozens of near-neutral warm springs at the south extremity of the young lava flows emerge at 12°-50°C.

Chemically analyzed lavas and ejecta (n = 40) representing all members of the Trident cluster yield a range of 57-65% SiO_2; magmatic inclusions in some flows are as mafic as 55% SiO_2. K_2O content is 1.0% at 57.5% SiO_2, 1.3% at 60%, and 1.6-1.8% at 64%. Major element variation trends are essentially identical to those for the adjoining Mageik and Katmai centers. Trident's present volume is 10-15 km^3, but glaciation has probably removed at least half as much more. The volume erupted in 1953-63 was slightly <0.5 km^3.

How to get there: *A long day's walk from a rough beach landing at Katmai Bay or, alternatively, up the Valley of Ten Thousand Smokes from the Overlook roadhead. Weather, glaciers, bears, and river crossings are dangers. The area is uninhabited wilderness.*

References

Snyder, G.L., 1954, Eruption of Trident volcano, Katmai National Monument, Alaska, February-June 1953: *USGS Circ. 318*, 7 pp.

Hildreth, W., 1987, New perspectives on the eruption of 1912 in the Valley of Ten Thousand Smokes, Katmai National Park, Alaska: *Bull. Volc. 49*, 680-693.

Wes Hildreth

NOVARUPTA: Dome-filled Novarupta with Mount Griggs in the background. Photograph by J. Fierstein, USGS.

NOVARUPTA, FALLING MOUNTAIN, AND CERBERUS
Alaska Peninsula

Type: *Lava domes and caldera*
Lat/Long: *58.27°N, 155.17°W*
Elevation: *841 m, 1,160 m, 1,100 m, respectively*
Relief: *70 m, 430 m, 365 m, respectively*
Volcano Diameter: *395 m, 1,700 m, 1,600 m, respectively*
Eruptive History: *Falling and Cerberus possibly early Holocene; Novarupta extruded in 1912 after voluminous pyroclastic eruptions; weak fumaroles persist*
Composition: *Rhyolite > dacite > andesite*

Novarupta was the source of the world's most voluminous 20th-century eruption, the most extensive historic ignimbrite, and of the only Quaternary high-silica rhyolite on the Alaska Peninsula. The 60-hour eruptive sequence of 6-8 June 1912 included (1) an initial

NOVARUPTA: Falling Mountain and Mount Cerberus from the northwest. West Trident is behind in the clouds.

rhyolite plinian fall with near-vent blasts, (2) an 11-km^3 zoned ignimbrite sheet filling the **Valley of Ten Thousand Smokes,** and (3) a complex series of plinian dacite falls. Total ejecta volume was 30-35 km^3; magma volume was 12-15 km^3. The ignimbrite vent is marked by a 2-km-wide backfilled depression, near the middle of which is nested the 500-m-wide vent of the post-ignimbrite plinian dacite eruptions. A blocky rhyolite lava dome (Novarupta, s.s.) subsequently plugged the inner vent.

Three discrete magmas erupted together, in part mingling syneruptively to yield an abundance of banded pumice: quartz-hypersthene rhyolite (77.0% SiO_2), pyroxene dacite (64.5 - 66% SiO_2), and black pyroxene andesite (58.5-61.5% SiO_2). More than half was rhyolite. Because the three vented concurrently and repeatedly after eruptive pauses hours in duration, the compositional gaps between them are thought intrinsic to the reservoir, not merely effects of withdrawal dynamics. K_2O contents are 1.3-1.4% at 60% SiO_2, 1.8-1.9% at 65%, and 3.2% at 77%. Major element variation trends are nearly identical to those of Mount Katmai and Trident. Linear fractures between Novarupta and Trident strike perpendicular to the volcanic front, raising the possibility that some or all of the 1912 magma was transferred to the 1912 vent from reservoir components beneath Trident; these, in turn, may have withdrawn magma from beneath Mount Katmai, permitting its collapse.

The "ten thousand smokes" in the ash-flow sheet mostly died out by 1930, and the few fumaroles remaining today are odorless wisps that issue from Novarupta and from intravent fractures nearby. Although orifice temperatures were as high as 645°C in 1919, none are hotter than 90°C at present. Many fossil fumarole deposits, both pipelike and fissure-controlled, retain spectacular zoning in color, mineralogy, and composition.

Falling Mountain and Mount Cerberus are twin domes of pyroxene dacite (64% SiO2) that frame the entrance to Katmai Pass. They are similar in volume (each 0.25 km^3), composition, lithology, and weathering. Although undated, they are thought to be of early Holocene age because their carapaces are degraded but not significantly glaciated. The two domes are compositionally similar (but not identical) to 1912 dacite. Together with the rhyolitic Novarupta dome, they define a 3-km-long line parallel to, but 4 km behind, the volcanic front. Falling Mountain stands on the edge of the 1912 vent depression, and its northeast face was truncated during the 1912 eruption. Its location and its apparent youth suggest either that the voluminous rhyolite of 1912 had not yet evolved or that the rhyolite was stored elsewhere at the time of the Falling Mountain extrusion.

How to get there: *A long day's walk up the Valley of Ten Thousand Smokes from the Overlook roadhead. Weather, bears, and river crossings can be dangerous. The area is uninhabited wilderness.*

References

Hildreth, W., 1983, The compositionally zoned eruption of 1912 in the Valley of Ten Thousand Smokes, Katmai National Park, Alaska: *J. Volc. Geotherm. Res. 18*, 1-56.

Hildreth, W., and Fierstein, J., 1987, Valley of Ten Thousand Smokes, Katmai National Park, Alaska: *Geol. Soc. Am. Centennial Field Guide – Cordilleran Section*, 425-432.

Wes Hildreth

KATMAI
Alaska Peninsula

Type: *Stratovolcano cluster with caldera*
Lat/Long: *58.27°N, 154.98°W*
Elevation: *2,047 m (~2,290 before 1912 collapse)*
Volcano Diameter: *9 km*
Eruptive History: *Caldera collapse in 1912 at time of Novarupta flank eruption 10 km to west; dacite dome on caldera floor in 1912(?); no other historic eruptions verified*
Composition: *Basalt to rhyolite*

KATMAI: Katmai caldera with Trident and snow-clad Mageik in upper left; and Novarupta, Falling Mountain, and Mount Cerberus in center distance, at the margin of the Valley of Ten Thousand Smokes. Photograph by A. Post, USGS.

Before 1912, Mount Katmai was a cluster of 3 or 4 small stratovolcanoes, contiguous and similar to those of Trident. The great flank eruption at Novarupta in June 1912 led to engulfment of all but one of Katmai's cones (Peak 6128), probably owing to magma withdrawal by way of a complex reservoir beneath Trident. The caldera has steep walls 500-1,050 m high, areas of 4.2 km^2 at lake level and 7 km^2 at the rim, a collapse volume of 4-5 km^3, and a blue-green lake 250 m deep and rising.

The truncated edifice is about half ice-covered, is the most severely dissected of the Katmai group volcanoes, and is heavily mantled with Novarupta fallout. Its slopes consist mostly of pyroxene andesite-dacite lavas and breccias, and the caldera is partly rimmed by dense sheets of pre-1912 agglutinate. Basaltic and rhyolitic blocks have been found on the rim but nowhere in situ. Analyzed samples (n = 25) define a continuous range of 49-66.5% SiO_2 plus the rhyolite blocks at 72%. K_2O contents and major-element variations are essentially identical to those of Trident. Present volume of Mount Katmai is ~25 km^3.

Nothing is known to have erupted from Mount Katmai itself at the time of its collapse. Before investigators first reached the rim in 1916, however, a small dacite dome (the **Horseshoe Island**) had been emplaced on the caldera floor and centrally reamed out by subsequent explosions; the dome has been submerged since the late 1920s. Sulfurous fumaroles and mudpots, once abundant on the caldera floor and the fringing talus, were also drowned by the lake, but yellow-green plumes can still be seen in the lake and sulfurous odors persist.

How to get there: *A long day's walk from a rough beach landing at Katmai Bay or, alternatively, up the Valley of Ten Thousand Smokes from the Overlook roadhead. Rope and crampons to the caldera. Weather, crevasses, bears, and river crossings are dangerous. The area is uninhabited wilderness.*

References

Hildreth, W., 1987, New perspectives on the eruption of 1912 in the Valley of Ten Thousand Smokes, Katmai National Park, Alaska: *Bull. Volc. 49*, 680-693.

Hildreth, W., and Fierstein, J., 1987, Valley of Ten Thousand Smokes, Katmai National Park, Alaska: *Geol. Soc. Am. Centennial Field Guide - Cordilleran Section*, 425-432.

Wes Hildreth

GRIGGS
Alaska Peninsula

Type: *Stratovolcano with 3 nested craters*
Lat/Long: *58.35°N, 155.1°W*
Elevation: *2,317 m (base at 550-1,200 m)*
Dimensions: *7 × 11 km*
Eruptive History: *Most of exposed cone is Holocene; no historic eruptions; vigorous sulfur-depositing fumaroles in summit crater and on upper southwest slope*
Composition: *Mafic to silicic andesite*

A spectacular, little-dissected cone, Mount Griggs towers 1,700 m above the north margin of the Valley of Ten Thousand Smokes, its summit only 10 km north of Novarupta. Uniquely among the stratovolcanoes of the Katmai district, it lies 10 km *behind* (northwest of) the remarkably linear (N66°E-trending) volcanic front defined by the Martin, Mageik, Trident, Katmai, and Snowy Mountain centers.

Griggs's truncated summit is an expression of 3 concentric craters, the outermost of which is 1.5 km wide, breached to the southwest, and filled by a semi-annulus of ice wrapped around a young central cone containing the nested inner craters. The main outer crater probably originated by early Holocene collapse, and formation of a 1-km^3-scale debris avalanche, remnants of which survive on the lower west-southwest flank and across the valley on Broken Mountain. Much of the amphitheater was subsequently filled by the nested inner cone. Outer slopes of the main and inner cones expose a few scoria flows and local scoria falls but principally consist of complexes of the thin brecciated lava tongues, overlapping and bifurcating in the manner so characteristic of summit-fed andesites on steep slopes. The total volume of Mount Griggs is ~25 km^3.

Griggs's products are chiefly olivine-pyroxene andesites; 27 samples range continuously from 54.5 to 63.5% SiO_2. They are consistently more potassic than products of the other Katmai cluster volcanoes, their K_2O contents being 1.3-1.4% at 57.5% SiO_2 and 1.7-1.9% at 60%. Isotopic data also suggest a source mixture and plumbing system independent of the other nearby centers.

Numerous SO_2-rich fumarolic jets occur between 1,940-m and 2,180-m elevation atop the inner cone and along a steep chute on its upper southwest slope. Orifice temperatures (measured by Dave Johnston in 1978-9) range from 96° to 108°C, and their condensates have pH ~1. Each orifice is constructing its own mound of sulfur sublimate. Downhill fumaroles are hottest and loudest, their roar commonly audible from the valley floor.

How to get there: *A long day's walk up the Valley of Ten Thousand Smokes from the Overlook roadhead. Weather, bears, and river crossings are dangerous. The area is uninhabited wilderness.*

References

Hildreth, W., 1987, New perspectives on the eruption of 1912 in the Valley of Ten Thousand Smokes, Katmai National Park, Alaska: *Bull. Volc. 49*, 680-693.

Kosco, D. G., 1981, Characteristics of andesitic to dacitic volcanism at Katmai National Park, Alaska: Ph.D. Thesis, Univ. Calif., Berkeley, 249 pp.

Wes Hildreth

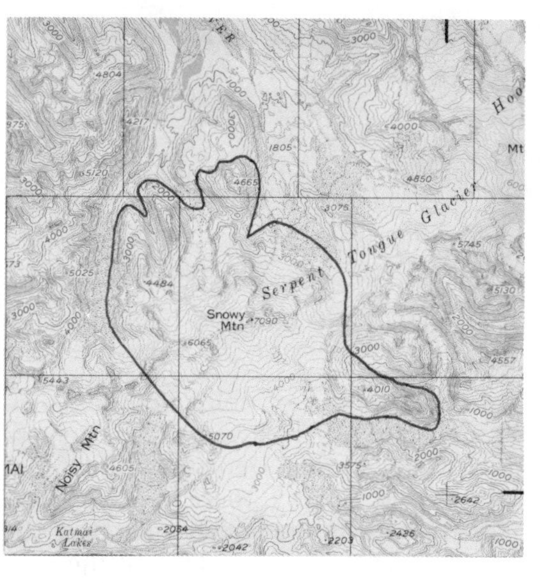

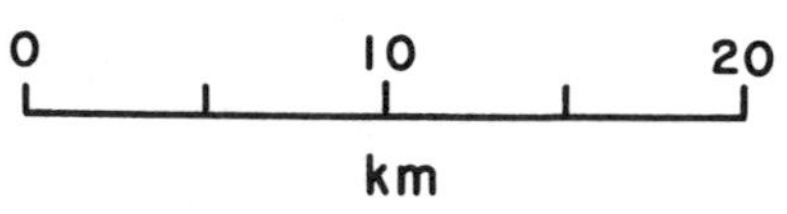

SNOWY MOUNTAIN: USGS topographic map.

SNOWY
Alaska Peninsula

Type: *Stratovolcano*
Lat/Long: *58.33°N, 154.73°W*
Elevation: *2,161 m*
Volcano Diameter: *21 km*
Eruptive History: *No historic activity*
Composition: *Andesite*

Snowy Mountain is a complex volcanic center almost entirely covered by the Serpent Tongue Glacier. Three peaks extending above the ice along a northeast-southwest trending ridge define the summit of Snowy Mountain. Outcrops of lava are sparse and include isolated peaks and ridges surrounded by glacial ice. Some lava appears to crop out below the ice on the north flank of the glacier, but this has not been verified on the ground. The mountain retains a conical form despite the extensive ice cover, suggesting a relatively undissected volcano beneath the glacier.

An active fumarole field on the summit of the tallest peak has melted holes through the ice. A zone of active shallow-level seismicity beneath Snowy Mountain is probably a manifestation of a hydrothermal system. This, together with its youthful, undissected form, suggests that volcanism at Snowy Mountain is recent. But no historic activity has been reported.

SNOWY: Summit dome (?) and erosional remnants of lava flows in ridges.

How to get there: *Access to Snowy Mountain is via chartered floatplane into either Geographic or Hidden Harbor, followed by a 20-25-km hike. Commercial air service is available to Kodiak or King Salmon, where a floatplane may be chartered.*

References

Keller, A. S., and Reiser, H. N., 1959, Geology of the Mount Katmai area, Alaska: *USGS Bull. 1058-G*, 261-298.

Kienle, J., Swanson, S. E., and Pulpan, H., 1983, Magmatism and subduction in the eastern Aleutian arc: *in* Shimozuru, D., and Yokoyama, I. (eds.), *Arc Volcanism: Physics and Tectonics*, D. Reidel, Boston, 191-224.

Samuel E. Swanson

DENISON, STELLER, AND KUKAK
Alaska Peninsula

Type: *Stratovolcanoes?*
Lat/Long:
D: 58.42°N, 154.45°W
S: 58.43°N, 154.40°W
K: 58.47°N, 154.35°W
Elevation: *D: 2,318 m, S: 2,272 m, K: 2,040 m*
Volcano Diameter: *6 km*
Eruptive History: *No historic activity*
Composition: *Andesite to dacite*

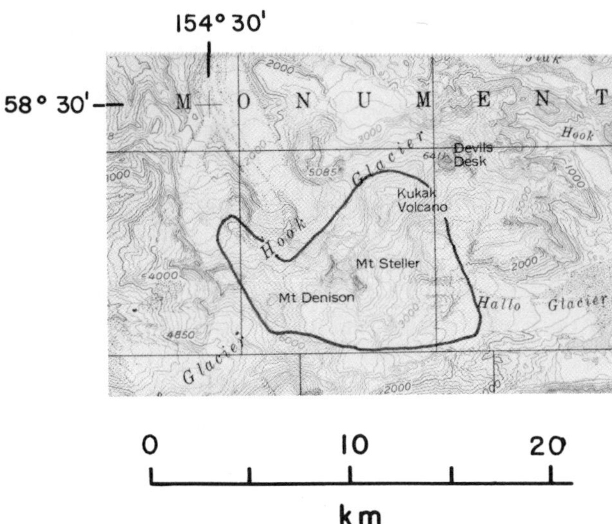

DENISON: USGS topographic map of three ice-clad volcanoes: Denison, Steller, and Kukak.

DENISON: East side of Kukak volcano showing lava flows.

DENISON: Summit of Mount Steller showing lava flows dipping to left (northwest).

Numerous outcrops of volcanic rock protrude above the ice of Hook and Hallo Glaciers in the vicinity of Mounts Denison and Steller and Kukak volcano. Clearly, there is at least one volcano *somewhere* under all this ice, but the location of the vent(s) and relations of the outcrops are unclear. Some authors consider each of these peaks a volcano; others identify only 1 or 2 volcanic vents in this area. Orientation of lava flows and a thick, cross-bedded tephra outcrop are consistent with a volcanic vent near Mount Denison. Aside from its elevation, nothing suggests Mount Steller is a separate volcanic vent; it probably is an erosional remnant.

Kukak volcano is almost completely covered by glacial ice. A vigorous fumarole field on Kukak's northern peak keeps that area free of ice and reveals the volcanic character of Kukak. Rocks exposed around the fumaroles show intense hydrothermal alteration. These fumaroles are the only evidence of recent volcanic activity in the area.

This is a complex area that is known only from reconnaissance studies of a few exposures. More detailed investigations may reveal additional volcanoes or result in a consolidation into fewer actual vents. Lava found in this complex is high-silica andesite and dacite.

How to get there: *The Denison-Steller-Kukak complex can be reached via a 15-20-km foot traverse from Hallo Bay. Floatplanes can be chartered in either King Salmon or Kodiak for flights to the bay. Commercial air service is available to Kodiak or King Salmon.*

References

Keller, A. S., and Reiser, H. N., 1959, Geology of the Mount Katmai area, Alaska: *USGS Bull. 1058-G*, 261-298.

DENISON: Summit of Mount Denison showing cross-bedded tephra.

Kienle, J., Swanson, S. E., and Pulpan, H., 1983, Magmatism and subduction in the eastern Aleutian arc: *in* Shimozuru, D., and Yokoyama, I. (eds.), *Arc Volcanism: Physics and Tectonics*, D. Reidel, Boston, 191-224.

Samuel E. Swanson

DEVILS DESK
Alaska Peninsula

Type: *Stratovolcano*
Lat/Long: *58.48°N, 154.30°W*
Elevation: *1,954 m*
Volcano Diameter: *5 km*
Eruptive History: *No historic activity*
Composition: *Andesite*

Devils Desk is a volcanic neck of a former stratovolcano now completely surrounded by Hook Glacier. Total relief from the top of Hook Glacier to the top of Devils Desk is ~450 m. Extensive erosion has removed most of the stratovolcano. Three radial dikes (now exposed as ridges) extend outward from the Desk toward the east.

How to get there: *Devils Desk is in an isolated part of Katmai National Park and Preserve. Commercial air service is available to Kodiak or King Salmon. A floatplane can then be chartered for the trip to Hallo Bay. Access to Devils Desk is then via a 13-km foot traverse.*

References

Keller, A. S., and Reiser, H. N., 1959, Geology of the Mount Katmai area, Alaska: *USGS Bull. 1058G*, 261-298.

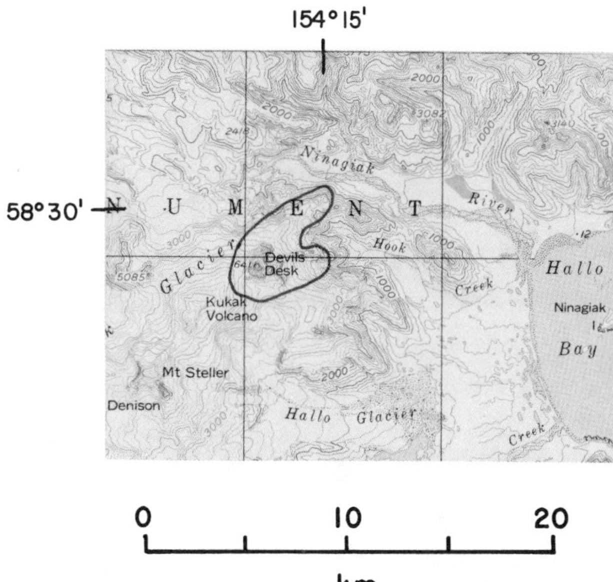

DEVILS DESK: USGS topographic map.

DEVILS DESK: Radial dike in the foreground.

Kienle, J., Swanson, S. E., and Pulpan, H., 1983, Magmatism and subduction in the eastern Aleutian arc: *in* Shimozuru, D., and Yokoyama, I. (eds.), *Arc Volcanism: Physics and Tectonics*, D. Reidel, Boston, 191-224.

Samuel E. Swanson

KAGUYAK
Alaska Peninsula

Type: *Stratovolcano with caldera*
Lat/Long: *58.62°N, 154.08°W*
Elevation: *901 m*
Volcano Diameter: *11 km*
Caldera Diameter/Depth: *2.5 / 0.73 km*

KAGUYAK: Summit crater surrounded by vegetated pyroclastic flow deposits. Photograph by J. Kienle.

Eruptive History: *No historic activity*
Composition: *Andesite to dacite*

Kaguyak is a stratovolcano abbreviated by a caldera. The highest point on the caldera rim (901 m) is over 550 m above a crater lake (depth >180 m) that partially fills the caldera. The lake is 2.5 km in diameter. Exposed in the walls of the caldera are pre-caldera agglomerates, lava flows, dikes, and some small plugs. A saddle on the northwest rim of the caldera represents a pre-caldera stream valley that was beheaded by caldera formation.

An extensive apron of pyroclastic flow deposits surrounds Kaguyak crater. These deposits consist of white dacitic tephra and pumice and are the result of the caldera-forming eruptions. Immediately after eruption, the pyroclastic flows filled the stream valleys around the volcano. Subsequent stream erosion has provided exposures up to 30 m thick of the pyroclastic flows, but the base of the deposits has not been observed. A paleosol immediately on top of the pyroclastic flow deposits yields a ^{14}C date of 1,060 yr BP; this is a minimum age for the caldera-forming eruption. The caldera has not been modified by glacial activity, thus making its origin post-glacial.

Both pre- and post-caldera domes are found at Kaguyak Crater. A pre-caldera dome (elevation 706 m) on the southeast flank of the volcano is heavily mantled by caldera-related pumice, while a post-caldera dome (elevation 616 m) on the east flank does not have any pumice mantle. Within the caldera two large post-caldera domes (elevation 614 m, 1 km diameter) have coalesced on the southwest edge of the caldera. A smaller dome forms an island rising ~10 m above the center of the lake. Weak solfataric activity has been reported from this dome, but was not seen in 1982.

Kaguyak is better known than some of the other Katmai volcanoes. Caldera formation resulted from the eruption of a dacite with phenocrysts of augite, hypersthene, hornblende, quartz, and plagioclase. The extensive pyroclastic deposits surrounding Kaguyak suggest that tephra from the eruption might be widespread, and, indeed, possible correlative tephras have been found 100 km from the volcano. Work is in progress to better define the timing of caldera formation and the petrochemical evolution of the Kaguyak system.

How to get there: *A floatplane can land on the lake in Kaguyak Crater. Commercial air service*

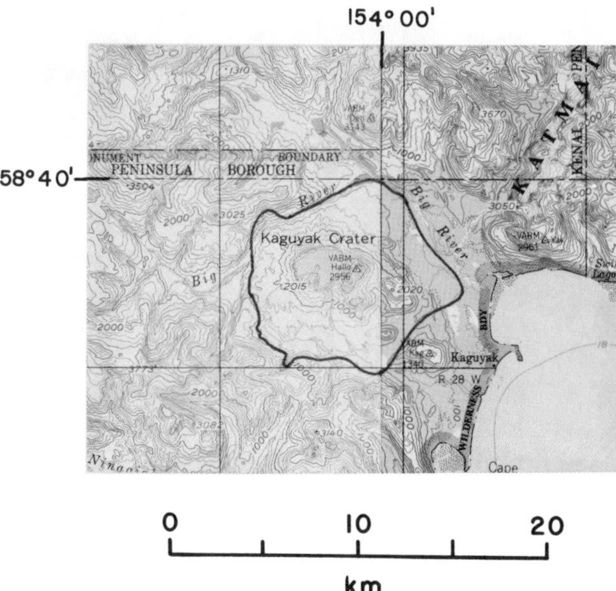

KAGUYAK: USGS topographic map.

is available to Kodiak or King Salmon, where a suitable aircraft can be chartered for the trip to Kaguyak.

References

Kienle, J., Swanson, S. E., and Pulpan, H., 1983, Magmatism and subduction in the eastern Aleutian arc: *in* Shimozuru, D., and Yokoyama, I. (eds.), *Arc Volcanism: Physics and Tectonics*, D. Reidel, Boston, 191-224.

Swanson, S. E., Kienle, J., and Fenn, P. M., 1981, Geology and petrology of Kaguyak Crater, Alaska: *EOS, Trans. Am. Geophys. Un.* 62, 1062.

Samuel E. Swanson

FOURPEAKED
Alaska Peninsula

Type: *Stratovolcano*
Lat/Long: *58.77°N, 153.68°W*
Elevation: *2,104 m*
Volcano Diameter: *18 km*
Eruptive History: *No historic activity*
Composition: *Andesite*

Fourpeaked Mountain consists of small isolated volcanic exposures surrounded by the Fourpeaked Glacier. The exposures are found along ridge crests and cliff faces on the sides of ridges that radiate out from the ice-covered summit. Lava flows are interlayered with volcanic agglomerate in the isolated exposures.

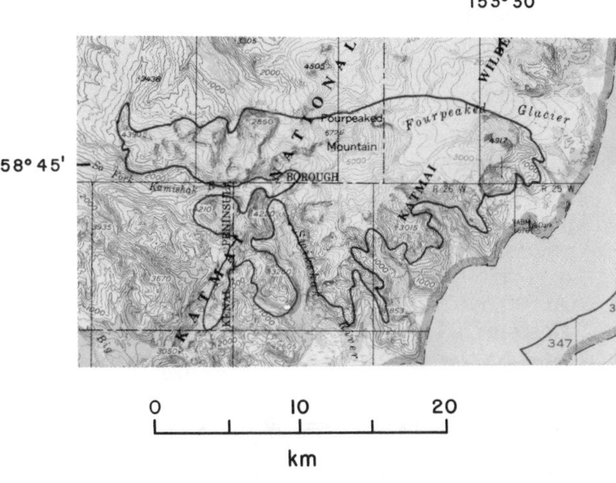

FOURPEAKED: USGS topographic map.

FOURPEAKED: Summit of Fourpeaked Mountain surrounded by Fourpeaked Glacier.

Orientation of lava flows suggests the present summit of Fourpeaked is probably the vent for Fourpeaked volcano. Extensive hydrothermal alteration of rocks in this area is consistent with this vent location.

Fourpeaked is known only from limited reconnaissance studies. The lavas are porphyritic andesite. No recent volcanic or hydrothermal activity has been identified.

How to get there: *The isolated exposures and extensive ice cover make access to Fourpeaked volcano difficult. The most obvious access is by floatplane from Kodiak or King Salmon to the shore of Shelikof Strait southwest of Fourpeaked and then foot traverse to the volcano.*

Commercial air service is available to Kodiak or King Salmon.

References

Keller, A. S., and Reiser, H. N., 1959, Geology of the Mount Katmai area, Alaska: *USGS Bull. 1058-G*, 261-298.

Kienle, J., Swanson, S. E., and Pulpan, H., 1983, Magmatism and subduction in the eastern Aleutian arc: *in* Shimozuru, D., and Yokoyama, I. (eds.), *Arc Volcanism: Physics and Tectonics*, D. Reidel, Boston, 191-224.

Samuel E. Swanson

DOUGLAS
Alaska Peninsula

Type: *Stratovolcano*
Lat/Long: *58.87°N, 153.55°W*
Elevation: *2,140 m*
Volcano Diameter: *35 km*
Eruptive History: *No historic activity*
Composition: *Andesite*

Mount Douglas is a dissected stratovolcano covered by ice of the Spotted Glacier. The summit of the volcano is marked by a crater with a small (160 m wide) crater lake. An active fumarole field on the north side of the crater keeps the area free of ice. A black scum floating on the lake in 1980 was probably sulfide minerals of some sort (consistent with a lake temperature of 25°C and a pH of 1 measured in 1982).

Much of the volcano has been subjected to glacial erosion, but a ramp of lava flows on the northwest flank is relatively uneroded. Most of the volcano is ice-covered, but isolated outcrops of lava flows (high-silica andesite) are found within the ice. Reconnaissance geologic surveys suggest that the lavas extend to elevations lower than the glacier's, but this has not been confirmed on the ground.

No historic activity has been reported for Mount Douglas; however, the presence of unglaciated lava flows and the active fumaroles indicate recent activity.

How to get there: *Mount Douglas forms a point of land between Shelikof Strait and Cook Inlet and is thus accessible from two coastlines. Floatplane landings can be made on Sukoi and Kamishak Bays. Shallow water and high tidal range can present problems in Kamishak Bay, however. Floatplanes can be chartered at Kodiak or King Salmon, both accessible by commercial air service.*

DOUGLAS: Summit of Mount Douglas with crater lake.

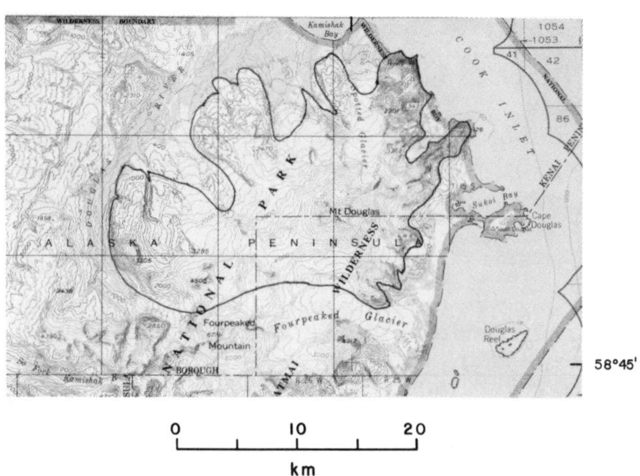

DOUGLAS: USGS topographic map; outline shows limit of volcano.

References

Keller, A. S., and Reiser, H. N., 1959, Geology of the Mount Katmai area, Alaska: *USGS Bull. 1058G*, 261-298.

Kienle, J., Swanson, S. E., and Pulpan, H., 1983, Magmatism and subduction in the eastern Aleutian arc: *in* Shimozuru, D., and Yokoyama, I. (eds.), *Arc Volcanism: Physics and Tectonics*, D. Reidel, Boston, 191-224.

Samuel E. Swanson

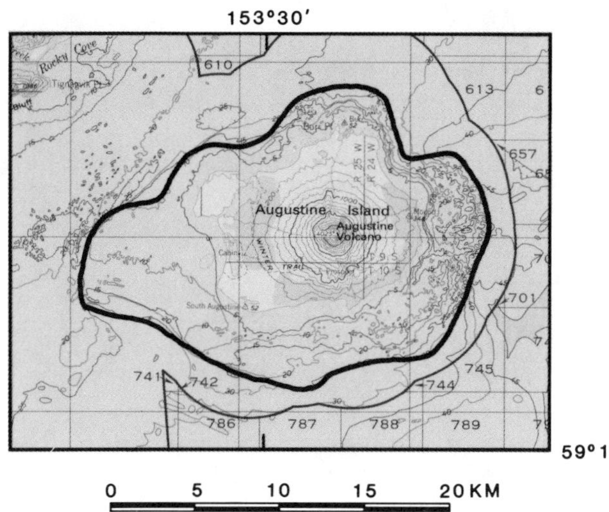

AUGUSTINE: USGS topographic map.

AUGUSTINE: From the southeast; Aleutian Range in the background. The island has a diameter of about 10 km. The smooth ridges on the lower left of the photograph are part of the uplifted basement of sediments. The steep central part of the volcano consists of historic lava domes. The wave-cut sea cliffs in the foreground are in debris avalanche deposits. Flat-lying debris avalanche lobes can also be seen in the background to the left and right of the central cone. The black deposits on the snow are pyroclastic flows from eruptive activity in January 1976, one week before this photograph was taken. Photograph by R. E. Wilson.

AUGUSTINE
Cook Inlet, Alaska

Type: *Dome cluster*
Lat/Long: *59.37°N, 153.42°W*
Elevation/Height: *1,282 m*
Eruptive History: *1812, 1883, 1935, 1963-4, 1976, 1986*
Composition: *Andesite to dacite, minor basalt*

The name Mount St. Augustine was in use during much of the 19th and early 20th centuries, following the naming of the volcano by Captain James Cook on St. Augustine's day (May 26) in 1778. During the Russian occupation of Alaska, the volcano was known as Chernabura, a local corruption of the Russian name (Ostrov) Chernoburoy, meaning black-brown (island).

Augustine volcano, a postglacial island volcano in lower Cook Inlet, is part of the eastern Aleutian arc. The undissected, symmetrical volcano is built on an uplifted basement of Jurassic to Tertiary sedimentary rocks that are exposed to 320-m altitude on the south flank of the volcano.

Structurally, Augustine consists of a central complex of young (mostly historic) lava domes and steep, short lava flows. The center is surrounded by an apron of debris avalanches and pyroclastic flows that form the bulk of Augustine Island. Debris avalanches occur on all flanks of the volcano indicating repeated collapse of the growing central dome complex. Bathymetry offshore from Augustine Island defines at least 8 lobate hummocky avalanche deposits that represent submarine extensions of debris avalanches. Debris avalanche activity on Augustine Island presents a special hazard in the lower Cook Inlet region – the impact of such rock slides into the sea can create tsunamis (as during a historic eruption in 1883). It appears that late Holocene lateral growth of Augustine Island was mainly achieved by debris avalanche activity.

Historic eruptions of Augustine volcano have been similar in terms of eruptive style, distribution of products, and magma composition. The abundance of pyroclastic products on Augustine Island points to the generally explosive nature of Augustine's eruptions. Augustine magmas are rather viscous due to their high degree of crystallinity and a dry rhyolitic groundmass. Interaction of this high viscosity magma with groundwater can produce highly explosive eruptions, usually during the initial phase of an eruptive cycle, that result in widespread tephra fall. During the past six historic eruptions much of the Cook Inlet region has been affected by these regional ash falls. Less explosive, extrusive dome growth has marked the later phases of all six historic eruptions.

The total volume of volcanic products making up the visible cone of Augustine volcano is

AUGUSTINE: From the north, July 14, 1986. The 1986 lava dome occupies the center of the photograph. Fresh pyroclastic flow deposits erupted between March and April 1986 are in the foreground. The lava flow on the right (casting a deep shadow) is from the 1883 eruption. Above it, on the skyline, are two small prehistoric domes. The 1935 dome lies to the right of the steam plume.

15 km^3. The offshore debris avalanches constitute another 1 or 2 km^3. Typical eruptions involve <0.5 km^3 of expanded juvenile material.

At present the summit of the volcano is breached to the north. This breach has funneled many of the pyroclastic flows northward during the four eruptions that occurred after 1883, when the breach formed by edifice collapse. Future pyroclastic flow activity is likely to be affected by the present geometry of the summit vent.

How to get there: *Augustine Island can be reached in a 45-minute floatplane ride from Homer on the Lower Kenai Peninsula. Commercial air service is available to Homer from Anchorage. A climb to the summit can be made in one day from the lagoon on the west side of Augustine Island where floatplanes can land.*

References

Kienle, J., Swanson, S. E., and Pulpan, H., 1983, Magmatism and subduction in the eastern Aleutian arc: *in* Shimozuru, D., and Yokoyama, I. (eds.), *Arc Volcanism: Physics and Tectonics*, D. Reidel, Boston, 191-224.

Swanson, S. E., and Kienle, J., 1988, The 1986 eruption of Mount St. Augustine: Field test of a hazard evaluation, *J. Geophys. Res. 93*, 4500-4520.

Jürgen Kienle

ILIAMNA
Cook Inlet, Alaska

Type: *Stratovolcano*
Lat/Long: *60.035°N, 153.17°W*
Elevation: *3,054 m*
Height: *1,750 m*
Volcano Diameter: *10 km*
Eruptive History: *Cone-building: Plio-Pleistocene. No historic activity.*
Composition: *Andesitic basalt to andesite*

Iliamna, a prominent stratovolcano on the west side of Cook Inlet, consists of an intercalated sequence of strongly altered hypersthene-augite andesite flows and volcaniclastic rocks. It is not a simple symmetrical cone; rather, the present summit peak is the northernmost of 4 peaks which form a 5-km-long ridge 2,200 to 2,400 m high and rising almost 1,800 m above the adjacent basement rocks. Most of the volcano is covered by snow and ice, and at least 10 glaciers radiate from its summit area. Little of the original constructional surface of the cone remains as the volcano has been deeply eroded and glaciated. The volcano was constructed on a basement that ranges in elevation from 1,200-1,400 m and consists chiefly of Jurassic granodiorite of the Alaska–Aleutian Range batholith; Lower Jurassic volcanic rocks occur beneath the southeastern part of the volcano.

Cone-building activity began prior to Wisconsin glaciation and probably has occurred within the past 10^6 yr. Regional tephrochronology study of the Cook Inlet region has not identified any major Holocene tephra with Iliamna volcano although more recent work suggests at least one mildly explosive Holocene event. Historic activity other than steam emissions has not been documented for Iliamna volcano. Two large fumarolic areas occur at 2,600-2,750 m elevation on the south and east sides of the ice-capped summit. These sulfur-rich fumaroles are inaccessible; under favorable meteorological conditions, the plumes from these fumaroles extend 700 m or more above the summit.

How to get there: *Iliamna volcano is 30 km west of Cook Inlet and ~210 km southwest of Anchorage, entirely within Lake Clark National Park and Wilderness. The volcano can be*

ILIAMNA: North-looking view, with Redoubt volcano in right background. North and South Twins in right foreground may represent old vent areas.

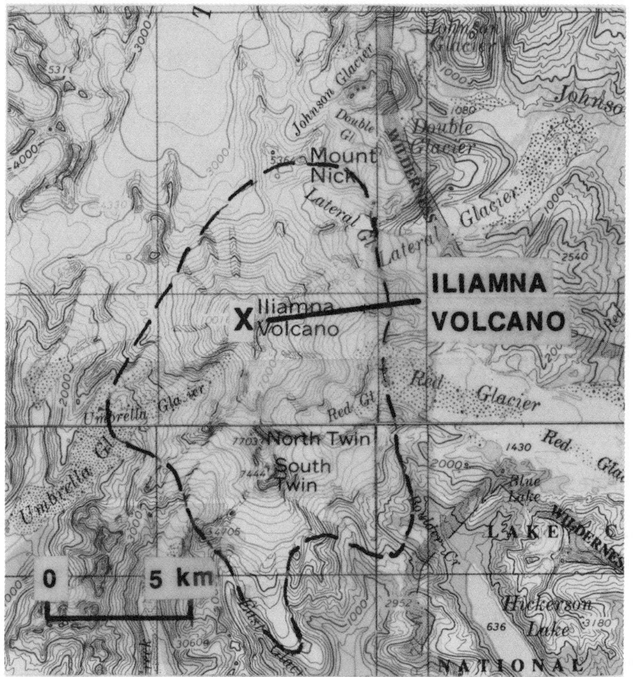

ILIAMNA: Map from USGS Iliamna, Lake Clark, Seldovia and Kenai 1:250,000-scale quadrangle maps; contour interval: 200 feet.

reached only by boat or air travel from Anchorage or Kenai to the beach below the volcano and then by foot.

References

Juhle, R. W., 1955, Iliamna volcano and its basement: *USGS Open-file Rpt. 55-77*, 74 pp.

Riehle, J. R., 1985, A reconnaissance of the major Holocene tephra deposits in the upper Cook Inlet region, Alaska: *J. Volc. Geotherm. Res. 26*, 37-74.

Thomas P. Miller

REDOUBT
Cook Inlet, Alaska

Type: *Stratovolcano*
Lat/Long: *60.48°N, 152.75°W*
Elevation: *3,108 m*
Crater Diameter: *2 km*
Volcano Diameter: *11 km*
Eruptive History: *1902, 1966, 1967?, 1968, 1989 (pre-1902 and 1933 "eruption" reports probably represent fumarole activity only)*
Composition: *Calc-alkaline basalt to dacite*

Redoubt, one of the easternmost volcanoes of the Aleutian arc, erupted through and rests on Mesozoic granitoid rocks of the Alaska Range batholith. Volcanic activity may have begun as early as 0.9 Ma with explosive eruptions of high-silica andesite to dacite pyroclastic debris, now preserved as small outcrops low on the

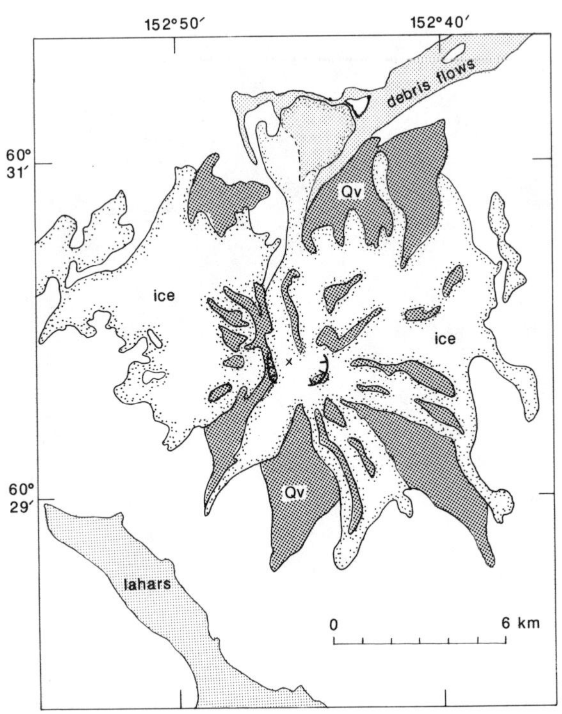

REDOUBT: Schematic geologic map showing Quaternary volcanics (Qv), debris flows, and lahars.

flanks of the mountain. Approximately 0.2 Ma the present cone began to be built during a period dominated by eruption of basalt to basaltic andesite flows. A third eruptive period capped the cone with thick andesite flows and flanking pyroclastic debris aprons. At least 30 Holocene eruptions of Redoubt are recorded in ash stratigraphy of the Cook Inlet basin.

Redoubt is heavily covered by glaciers, and Holocene deposits are dominated by lahars and debris flows. A lahar dated at ^{14}C 3,500 yr BP fills the Crescent River valley south of the volcano for 25 km to Cook Inlet; it dams Crescent Lake and has a volume of ~0.4 km^3. North of the volcano, deposits of numerous younger and smaller debris flows fill the Drift River canyon and undoubtedly contribute large volumes of sediment to the Drift River delta.

The ice-filled summit crater is the accumulation basin for a large glacier which drains north around the most recently active vent. Approximately 0.06 km^3 of the upper glacier was destroyed in 1966 during an eruption-induced jokulhlaup which disconnected the accumulation area from the piedmont lobe of the glacier. When the glacier reconnected 10 yr later, a kinematic wave and associated glacial surge threat-

REDOUBT: Debouching debris flows; seen from the north in 1981.

ened to temporarily dam the Drift River. The river, confined against the north side of its valley by the glacier, is susceptible to blockage by glacial advance or emplacement of a volcanic debris dam.

How to get there: *Redoubt volcano, 170 km southwest of Anchorage, is within Lake Clark National Park and Preserve. Access to the region is almost exclusively by small aircraft. Wheeled planes may land on open gravel bars of the Drift River in the summer. Planes on skis may land on gravel bars (in winter and early spring) and glaciers. Glaciers east and northeast of Redoubt are suitable for landings at about the 1,200-m level. During summer, potential landing spots are reduced due to softening of the glacial surfaces. Helicopters are prohibited from landing within the park except under the terms of a permit issued by the superintendent. Further advice on access and a list of air taxi operators can be obtained from Lake Clark National Park and Preserve, 701 C St., Box 61, Anchorage, AK 99513; (907) 271-3751.*

References

Sturm, M., Benson, C., and MacKeith, P., 1986, Effects of the 1966-1968 eruptions of Mount Redoubt on the flow of Drift Glacier: *J. Glaciol. 32*, 355-362.

Till, A. B., Yount, M. E., and Riehle, J. R., in prep., Volcanic hazards of Redoubt Volcano, Cook Inlet, Alaska: *USGS Bull.*

M. Elizabeth Yount

SPURR
Cook Inlet, Alaska

Type: *Stratovolcano*
Lat/Long: *61.30°N, 152.26°W*
Elevation: *3,374 m*
Height: *~2,500 m*
Volcano/Caldera Diameter: *19/5x6 km*
Caldera Depth: *600 m*
Eruptive History: *Late Pleistocene cone-building, late Pleistocene or early Holocene avalanche caldera. Holocene growth of a central dome and a flank stratovolcano; major ash eruption in 1953*
Composition: *Basaltic andesite and andesite*

The Spurr volcanic complex consists of an ancestral volcano whose growth was terminated by the formation of an avalanche caldera. The present Mount Spurr then grew in the center of the caldera and Crater Peak grew in the breach. The ancestral Mount Spurr is an andesitic stratovolcano some 19 km in diameter which was constructed in late Pleistocene time. It overlies Late Cretaceous and Paleocene plutonic rocks and Eocene sedimentary rocks. Basement elevation ranges from <150 m in the south to at least 2,000 m to the northwest. Snow and ice obscure much of the northern flanks of the volcano, but the southern flank is well exposed. The basal portion of the volcano is dominated by thick ashflows and other pyroclastic deposits which contain abundant dikes and sills. The upper part is dominated by lava flows, most of which are between 150,000 and 50,000 yr old. These flows are mostly andesite of uniform composition, although a few more mafic flows are present.

Cone growth was terminated during the very late Pleistocene or early Holocene by avalanche caldera formation. The caldera is 5 × 6 km, has a present maximum rim altitude of 3,000 m, and is breached to the south. The caldera contains an ice field which feeds glaciers that drain in all directions; thus the rim is highly dissected.

Caldera formation resulted in the emplacement of a volcanic debris avalanche with a volume of a few cubic kilometers which lies between Straight Creek and the Chakachatna River. The distal end of the avalanche deposit is at least 25 km from the center of the ancestral Mount Spurr. The debris avalanche is overlain by ash-flow tuffs which are more silicic than typical ancestral Mount Spurr andesites.

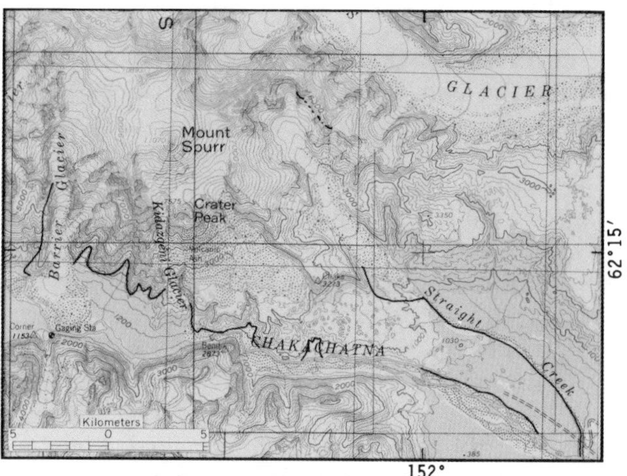

SPURR: A portion of the USGS 1:250,000-scale Tyonek quadrangle showing Mount Spurr, Crater Peak, and the ancestral volcano and the areal extent of their products. The debris avalanche deposit lies between Straight Creek and the Chakachatna River.

SPURR: From the south. The highest elevation is the post-caldera central dome. Crater Peak, the site of the 1953 eruption, is in front of and lower than the summit and occupies the breach created during the caldera-forming eruption. The small rock rib immediately to the right of Crater Peak is post-caldera in age and compositionally similar to Crater Peak. Remnants of the ancestral volcano are to the left and right of Crater Peak and define the rim encircling the summit ice field. The debris avalanche deposit is off the lower right-hand edge of the photo. Photograph by A. Post.

Subsequent to caldera formation a dome grew in the center of the caldera to form the present Mount Spurr. Mount Spurr is largely ice-covered but apparently similar in composition to the ash flows overlying the debris avalanche. It is currently the highest part of the volcano, with a summit elevation of 3,374 m. Exposed ground has diffuse fumaroles near the boiling point. A small summit crater produces enough heat to periodically reduce the thickness of accumulated ice. During historic time before 1953, this vent was the site of fumarolic activity, which was the only volcanic manifestation at the volcano.

A second cone, **Crater Peak**, grew in the breach of the ancestral volcano. The older part of this cone grew at the same time as the Mount Spurr dome. Crater Peak lavas are more mafic than the typical andesites of the ancestral Mount Spurr. The present summit elevation is 2,309 m, and the crater contains a small, warm lake and zones of vigorous fumarolic activity. Crater Peak was the locus of the only major historic eruption. On July 9, 1953, Crater Peak erupted with no warning. The eruption cloud reached in excess of 21 km and deposited over 6 mm of ash on Anchorage, 125 km to the east. Prior to the eruption, Crater Peak's crater was ice-filled and there were no active fumarole fields.

How to get there: *Mount Spurr is 125 km west of Anchorage and is accessible only by chartered aircraft. Landing areas on the southern flank of Crater Peak are suitable for small airplanes equipped for bush landings.*

References

Capps, S. R., 1929, The Mount Spurr region, Alaska. *USGS Bull. 810-C*, 140-172.

Nye, C. J., 1987, Stratigraphy, petrology and geochemistry of the Spurr volcanic complex, eastern Aleutian arc, Alaska. *Univ. Alaska Fairbanks Geophys. Inst. Rpt. UAG R-311*, 135 pp.

Christopher J. Nye

HAYES

Cook Inlet, Alaska

Type: *Stratovolcano*
Lat/Long: *61.62°N, 152.48°W*
Elevation: *1,970 to 2,788 m,*
Eruptive History: *No historic activity*
Composition: *Dacite*

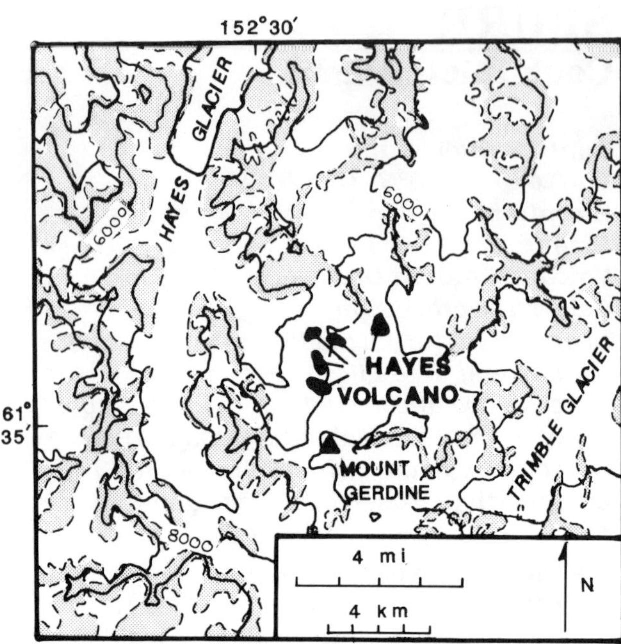

HAYES: Location of Hayes volcano in the Tordrillo Mountains, 150 km west-northwest of Anchorage. Pre-Quaternary rocks not covered by snow or ice are lightly shaded.

Hayes volcano was discovered in 1975 and is informally named for the adjacent Hayes Glacier. Hayes is almost totally ice-covered. No fumaroles have been observed during site visits in three different years. Little is known of the volcano except that it is the site of catastrophic eruptions that deposited six regional tephra beds to the south, east, and northeast between about 3,500 and 3,800 yr ago. The average volume of each of the six beds is about 2.4 km^3. The tephra-forming eruptions may have been accompanied by emplacement of a dome, although no such feature has been positively identified. A thin, poorly preserved tephra bed northeast of the vent may be evidence of late Holocene activity subsequent to the catastrophic eruptions.

How to get there: *Hayes volcano is located in a remote and rugged part of the Alaska Range and can be reached only by aircraft; there are no prepared landing sites nearby, and the adjacent Hayes Glacier is heavily crevassed.*

References

Miller, T. P., and Smith, R. L., 1976, "New" volcanoes in the Aleutian volcanic arc, *in* Cobb, E.H. (ed.), The USGS in Alaska: Accomplishments during 1975: *USGS Circ. 733*, p. 11.

HAYES: View to southeast of Hayes volcano. Mount Gerdine is the high peak on the skyline at right. Breccias comprising the remains of the volcano are the four small outcrops surrounded by ice in the center one-half of the view and the first small peak on the skyline to the left of the four small outcrops.

Riehle, J. R., 1985, A reconnaissance of the major Holocene tephra deposits in the upper Cook Inlet region, Alaska: *J. Volc. Geotherm. Res. 26*, 37-74.

James R. Riehle

BUZZARD CREEK
Central Alaska

Type: *Two tuff rings*
Lat/Long: *64.07°N, 148.42°W*
Elevation: *830 m*
Diameters: *300 m, 66 m*
Rim Heights: *10 m*
Eruptive History: *Formed 3,000 yr BP*
Composition: *Transitional tholeiitic/alkaline basalt*

The Buzzard Creek craters are two tuff rings at the headwaters of Buzzard Creek, a tributary of the Totatlanika River, near Healy at the northern foot of the central Alaska Range. The two craters are shallow and contain small lakes. Rim ejecta contain 80% country rock fragments and 20% juvenile basaltic material, suggesting a phreatomagmatic origin. The basalt rests on the youngest glacial terraces on Buzzard Creek, which are correlated with the Riley Creek glaciation that ended ~10,000 yr BP. Three ^{14}C dates from charcoal samples above and below the basaltic ejecta give an age of eruption of ~3,000 yr BP, in agreement with stratigraphic evidence of a Holocene age. A 300-m-wide ejecta blanket associated with the larger crater can be traced 1.6 km from the vent. The total volume of the ejecta probably does not exceed 1 million m^3, of which only 20% is volcanic, consisting of vesicular basaltic lapilli and small bombs.

BUZZARD CREEK: Larger of the two Buzzard Creek craters. Helicopter is parked on the tuff ring rim.

The Buzzard Creek craters, though insignificant in the volume of ejecta, are of regional tectonic interest because they occur on trend with the Aleutian arc structure and are situated directly over the northernmost corner of the subducting Pacific plate. The easternmost known volcano of the Aleutian arc is Hayes volcano, 320 km southwest of the Buzzard Creek craters. Whether the craters are tectonically linked to the Aleutian subduction zone is not clear.

How to get there: *The Buzzard Creek craters can be reached on foot in a 15-km hike from the Parks Highway near Healy.*

References

Pewe, T., Wahrhaftig, C., and Webber, F., 1966, Geologic map of the Fairbanks quadrangle, Alaska: *USGS Misc. Geol. Invest. Map I-455.*

Albanese, M., 1980, The geology of three extrusive bodies in the central Alaska Range, Alaska: M.S. Thesis, Univ. Alaska, Fairbanks, 104 pp.

Jürgen Kienle

DRUM

Wrangell Mountains, Alaska

Type: *Stratovolcano*
Lat/Long: *62.12°N, 144.64°W*
Elevation: *3,661 m*
Height: *2,716 m*
Eruptive History: *No historic activity*
Composition: *Andesite, dacite, and rhyolite*

Mount Drum, the westernmost volcano in the Wrangell volcanic field, was formed between ~0.65 and 0.24 Ma during at least two cycles of cone-building and ring-dome extrusion. The first cycle began with the construction of a cone consisting chiefly of andesite and dacite lava flows, breccias, lahars, and tuffs, and culminated with the emplacement of a series of rhyolite ring domes around the cone's southeast flank. The second cycle of activity, following without an apparent time break, continued to build the cone but with more dacitic flows and fewer pyroclastic and volcaniclastic deposits. This stage was followed by the emplacement of at least nine dacite domes that lie on 270° of arc, crudely defining a circle ~12-13 km in diameter centered approximately at the present summit of Mount Drum. The rhyodacite dome of **Snider Peak** and its massive dacite flows erupted late in the second cycle, probably marking the end of major constructive activity. Following the second cycle, paroxysmal explosive activity, probably from the central vent area, destroyed the south half of the stratovolcano and deposited ~7 km^3 of hot and cold avalanche debris over an area >200 km^2.

How to get there: *Mount Drum volcano is ~47 km east of Glennallen, Alaska, in the Wrangell–St. Elias National Park in southcentral Alaska. The volcano is not accessible by road, but can best be viewed from the Glenn Highway between ~15 km west of Glennallen and Chistochina.*

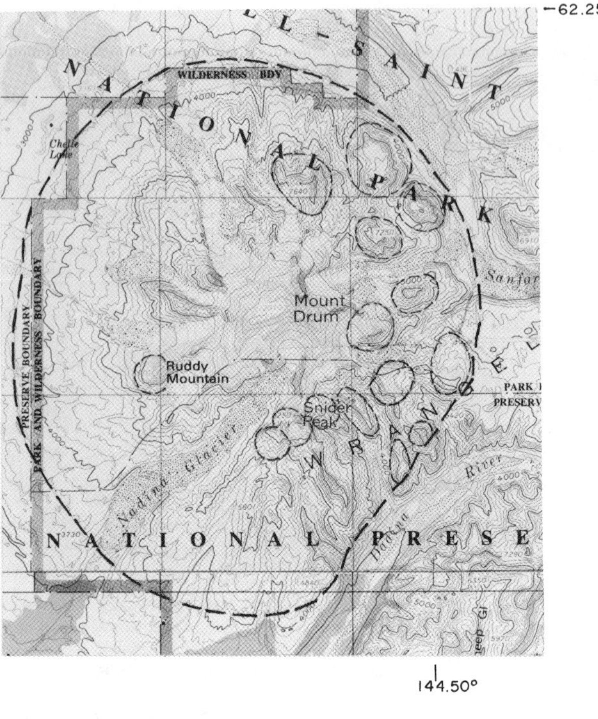

DRUM: Topographic map showing ring domes (thin dashed lines) and approximate extent of lava flows (heavy dashed line).

DRUM: Viewed from the east. Rhyolite and dacite domes of both the first and second cycles form light colored areas surrounding the volcano. Snider Peak, a late second-stage dome, is silhouetted to the left of the summit. Photograph by C. Nye.

Reference

Richter, D. H., Smith, R. L., Yehle, L. A., and Miller, T. P., 1979, Geologic map of the Gulkana A-2 quadrangle, Alaska: *USGS Geol. Quad. Map GQ-1520*, scale 1:63,360.

Donald H. Richter

SANFORD
Wrangell Mountains, Alaska

Type: *Shield volcano?*
Lat/Long: *62.66°N, 144.12°W*
Elevation: *4,949 m*
Height: *3,577 m*
Volcano Diameter: *34 km*
Eruptive History: *Base construction 0.85-0.32 Ma; upper (younger) structure age unknown; no historic activity*
Composition: *Andesite and basaltic andesite, minor basalt and dacite, one known rhyodacite flow*

Mount Sanford, a very large dissected shield with an impressive bulbous top, is the highest volcano in the Wrangell volcanic field. Most of the upper part (>2,500 m) is covered by perennial snow and ice, making study and observation difficult. The principal "window" through the ice cap is the great amphitheater at the head of the Sanford Glacier which rises more than 2,400 m in less than 1,500 m. Data from unpublished geologic mapping around the volcano's base, and fly-by observations of the great amphitheater and other cirques, indicate that the upper part of Mount Sanford is a young feature, possibly Holocene in age, that developed on a base of at least three coalescing andesitic shield volcanoes, referred to as the north, west, and south Sanford eruptive centers. The centers may contain shallow dacite and andesite intrusives, dike complexes, vent deposits, and may be the locus of linear rift vents; all centers are marked by topographic highs.

How to get there: *Mount Sanford volcano is ~75 km east-northeast of Glennallen, Alaska, in the Wrangell–St. Elias National Park in south-central Alaska. The volcano is not accessible by road, but can be viewed best from the Glenn Highway between Gakona and Slana.*

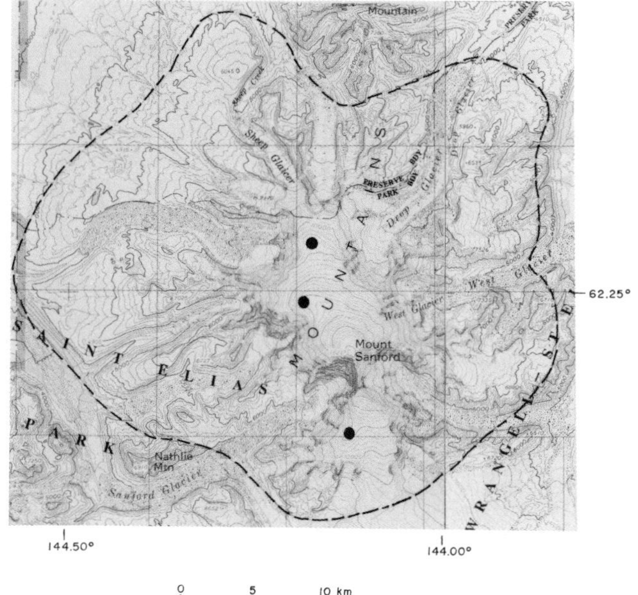

SANFORD: Topographic map showing north, west, and south eruptive centers (dark circles) and approximate extent of lava flows (heavy dashed line).

SANFORD: From the southwest. The south and west eruptive centers are on the skyline to the right and left of the amphitheater, respectively. Photograph by C. Nye.

Reference

Richter, D. H., Ratte, J. C., Schmoll, H. R., Leeman, W. P., Smith, J. G., and Yehle, L. A., 1988, Geologic map of the Gulkana B-1 quadrangle, Alaska: *USGS Geol. Quad. Map GQ-1655*, scale 1:63,360.

Donald H. Richter

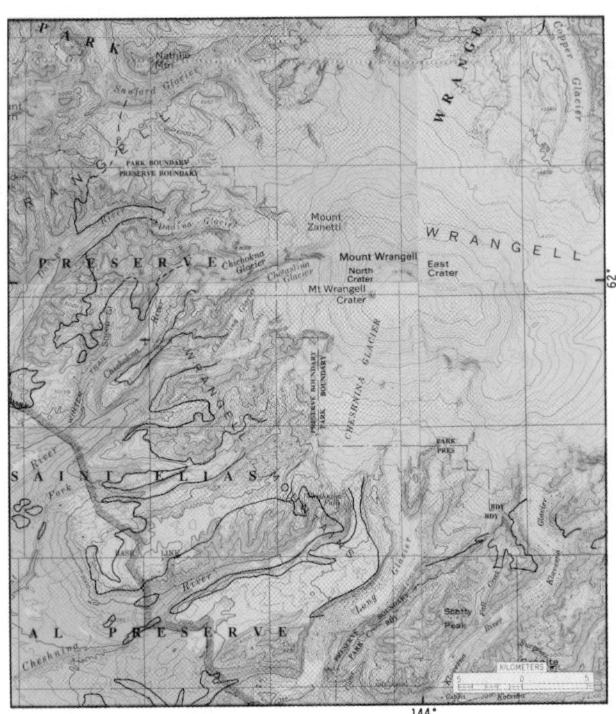

WRANGELL: Portions of the USGS 1:250,000-scale Gulkana, Valdez, McCarthy, and Nabesna quadrangles showing Mount Wrangell and the extent of its deposits (from USGS geologic maps). Distal lava flows and debris flows extend south and southwest of the map area shown.

WRANGELL
Wrangell Mountains, Alaska

Type: *Shield volcano*
Lat/Long: *62.00°N, 142.00°W*
Elevation: *4,317 m*
Height: *3,550 m*
Volcano/Caldera Diameter: *40 km/4 × 6 km*
Caldera Depth: *1,000 m*
Eruptive History: *Shield-building: 0.75 Ma to early Holocene; historic activity limited to minor phreatic eruptions in 1819 and 1884*
Composition: *Basaltic andesite through dacite; mostly andesite*

Mount Wrangell is one of the few large andesitic shield volcanoes, one of the world's largest andesite volcanoes (900 km^3), and the only active center in the western Wrangell Mountains. It is constructed on a basement of late Paleozoic and early Mesozoic limestone, argillite, and basalt. The basement has irregular relief with an upper surface between 750 and 1,800 m in elevation. The modern Mount Wrangell was preceded by a mid-Pleistocene andesitic volcano with a volume of ~450 km^3 whose vent area was ~15 km to the southwest. Remnants of the older volcano are exposed up to the ice limit at 3,200 m on the southern flank of the present shield. Wrangell lavas are mostly andesite, although basaltic andesite and dacite have also erupted. Most of the shield is composed of flows rather than pyroclastic units, and has been built since 0.75 Ma. Eruption volumes were generally high compared to other andesitic volcanoes at convergent margins.

WRANGELL: Seen from Glennallen, 80 km to the west. The outlines of cinder cones on the caldera rim can be seen. The flank cone Mount Zanetti is on the left, slightly behind a sharper mountain, which is one of the late-stage rhyodacite domes near Mount Drum.

The shield morphology is noteworthy because it suggests that Wrangell lavas were very fluid. Wrangell lavas are dominantly silicic andesite similar in composition to other circum-Pacific andesite which, because it is normally viscous, generally forms steep, stubby flows. Wrangell lavas, in contrast, formed long flows; in one extreme case lava flowed 35 km across topography with very low slope. The fluidity of Wrangell magmas does not appear to have resulted from anomalously high temperature or alkali content, low phenocryst content, or mechanical assistance during flow by water or ice. The long flows may have formed due to high eruption rates in a manner similar to that which produced high fluidity of Columbia River basalt.

A flank cone, Mount **Zanetti**, is 6 km northwest of the summit. It is steep-sided, 3,965 m tall, and completely ice-covered.

The present Wrangell summit is a 4 × 6-km caldera ~1 km deep. The caldera is ice-filled and drained to the southeast by Long Glacier. Three cinder cones, each 0.5 to 1 km in diameter, sit on the northern and western rims of the caldera. Portions of these cones are kept ice-free by volcanic heat. Increased heat flux to the northern crater has resulted in the melting of 43×10^6 m^3 of ice since the late 1950s. The westernmost cone has been the source of historic eruptions, which produced plumes of steam or steam and ash a few kilometers in maximum height. The present caldera lies on the southern edge of a circular subglacial feature which may be an older caldera some 20 km in diameter.

Large debris flows lie southwest of Wrangell. These deposits consist of multi-hued clasts, some as much as 90 m in diameter, set in a sandy-to-silty matrix. They are exposed discontinuously over some 850 km^2, and as far as 70 km south of the summit. They are overlain by lava flows as old as 0.43 Ma.

How to get there: *Glennallen, 250 km east-northeast of Anchorage and 80 km west of Mount Wrangell, is the nearest town. The Richardson and Edgerton highways skirt the Wrangell Mountains on the west and south, but the Copper River prevents closer access. Direct access to the volcano is by chartered helicopter or bush airplane from Glennallen. A few landing areas suitable for use by small single-engine airplanes equipped for bush operation exist on the west and south flanks of the volcano.*

References

Benson, C.S, and Motyka, R.J., 1977, Glacier–volcano interactions on Mount Wrangell, Alaska, *in* Univ. Alaska Fairbanks Geophys. Inst. *Annual Report* 1975-1976, 1-25.

Yehle, L.A., and Nichols, D.R., 1980, Reconnaissance map and description of the Chetaslina volcanic debris flow (new name), southeastern Copper River Basin and adjacent areas, south-central Alaska. *USGS Misc. Field Studies Map MF-1209*, scale 1:250,000.

Christopher J. Nye

CAPITAL

Wrangell Mountains, Alaska

Type: *Shield volcano*
Lat/Long: *62.43°N, 144.10°W*
Elevation: *2,356 m*
Height: *1,290 m*

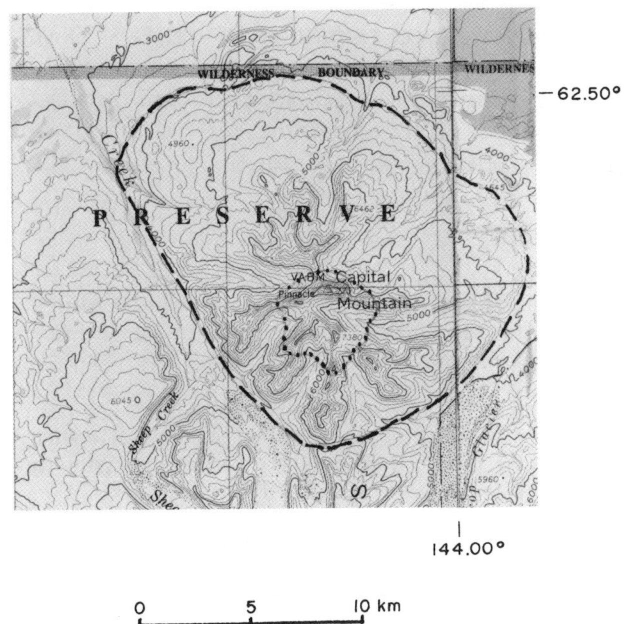

CAPITAL: Topographic map showing caldera (dotted line) and approximate extent of lava flows (heavy dashed line).

Diameter: *18 km*
Eruptive History: *Shield and intracaldera lavas and lake dikes 1-1.1 Ma; no historic activity*
Composition: *Predominantly andesite and basaltic andesite, minor basalt and dacite, rare rhyolite*

Capital Mountain is a relatively small andesitic shield volcano with a roughly circular summit caldera 4 km in diameter. The shield consists chiefly of lava flows and subordinate volcaniclastic rocks that dip 3° to 25° away from the summit area. The caldera, apparently of non-explosive origin, is filled with thick, flat-lying flows. Talus, flow breccias, and pillow lavas occur locally between the caldera wall and intracaldera flows. A prominent andesite plug, 100 m high, marks the general center of an area of post-caldera-fill activity and is the locus of a spectacular radial dike swarm. Shield and intracaldera lavas are chiefly hypersthene andesite, but shield lavas range in composition from basalt to dacite. Dikes are also chiefly andesite; one prominent rhyolite dike originating from a small rhyolite laccolith extends almost completely across the volcano.

How to get there: *Capital Mountain is ~ 33 km southeast of Chistochina, in the Wrangell–St. Elias National Park in southcentral Alaska. The volcano, not accessible by road, can be viewed from the Glenn Highway between Chistochina and Slana.*

References

Richter, D. H., Smith, J. G., Ratte, J. C., and Leeman, W. P.,1984, Shield volcanoes in the Wrangell Mountains, Alaska, *in* The USGS in Alaska: Accomplishments during 1982: *USGS Circ. 939*, 71-75.

Richter, D. H.,, Ratte, J. C., Schmoll, H. R., Leeman, W. P., Smith, J. G., and Yehle, L.A., 1988, Geologic map of the Gulkana B-1 quadrangle, Alaska: *USGS Geol. Quadr. Map GQ-1655*, scale 1:63,360.

Donald H. Richter

TANADA
Wrangell Mountains, Alaska

Type: *Shield volcano*
Lat/Long: *62.30°N, 143.50°W*
Elevation: *2,852 m*
Height: *1,630 m*
Volcano Diameter: *22 km*
Eruptive History: *Shield and intracaldera? lavas 1-2 Ma; no known historic activity*
Composition: *Predominantly andesite; late lavas and dikes are dacitic*

Tanada Peak is the erosional remnant of an andesitic shield volcano that contained an oval-shaped summit caldera, ~8 km long × 6 km wide. Very little of the original shield, which covered more than 400 km^2, remains; rugged Tanada Peak, highest point on the volcano, is composed entirely of flat-lying intracaldera? flows. Tanada was built on a thick sequence of andesitic flows, flow breccias, and laharic deposits probably of late Pliocene to early Pleistocene age. A riftlike chain of younger basalt and basaltic andesite cones mantles part of the northern flank of the shield, and much of the volcano's southern flank is covered by flows and pyroclastic deposits from younger unmapped eruptive centers.

Tanada shield lavas consist chiefly of thin (<10 m) andesite flows that dip 1° to 20° away from the summit area. The summit caldera, of apparent non-explosive origin, is filled with >900 m of massive, flat-lying andesite flows and dacitic agglutinates. The dacitic rocks are restricted to the uppermost part of the caldera fill, suggesting a temporally related change in magma chemistry. A few post-caldera dikes are either andesitic or dacitic in composition. The caldera wall dips 20° to 50° inward and is mantled locally by thin pyroclastic beds and rubbly

TANADA: View from north to Tanada (foreground) and Sanford (background) in the new Wrangell–St. Elias National Park. National Park Service photograph by M. Woodbridge Williams.

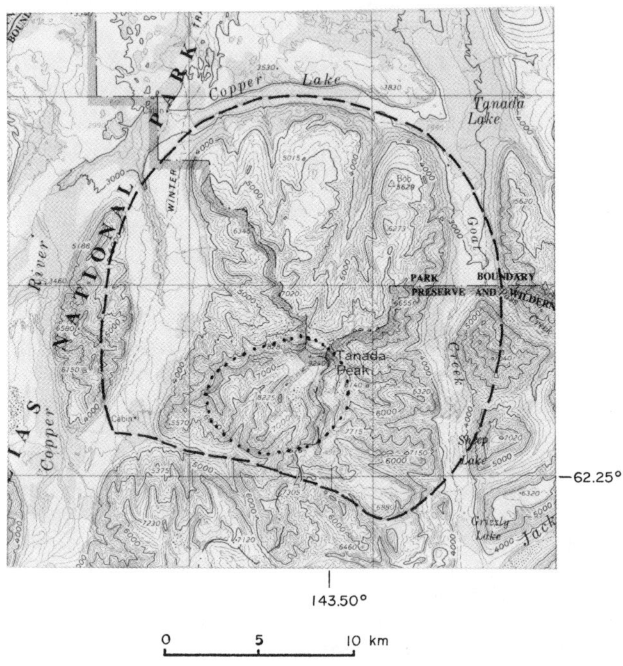

TANADA: Topographic map showing caldera (dotted line) and approximate extent of lava flows (heavy dashed line).

breccias. Most of the remaining exposed wall is composed of the older andesite sequence; only on the east side of the volcano, and locally elsewhere, are shield lavas still present in the wall.

How to get there: *Tanada volcano is ~51 km southeast of Slana in the Wrangell–St. Elias National Park in southcentral Alaska. The volcano is not accessible by road, but can best be viewed from the unpaved Nabesna Road between ~Mile 10 and Mile 30.*

Reference

Richter, D. H., Smith, J. G., Ratte, J. C., and Leeman, W. P., 1984, Shield volcanoes in the Wrangell Mountains, Alaska, *in* The USGS in Alaska: Accomplishments during 1982: *USGS Circ. 939*, 71-75.

Donald H. Richter

EASTERN WRANGELL
Wrangell Mountains, Alaska

Type: *Stratovolcanoes, shield volcanoes, domes, and cinder cones*
Lat/Long: *61.3° -62.5°N, 141°-144°W*
Elevation: *900 -5,005 m*
Eruptive History: *Miocene to Holocene; no historic activity*
Composition: *Basalt to rhyolite*

The Wrangell volcanic field extends from the Copper River Basin east-southeast across southcentral Alaska into the Yukon Territory of Canada, a distance of more than 250 km. At its west end the field includes the Quaternary volcanoes of Mount Wrangell, Mount Drum, Mount Sanford, Tanada, and Capital Mountain, which are described separately. In Alaska the field includes many of the higher elevations of the Wrangell Mountains and contiguous parts of the eastern Alaska Range and the St. Elias Mountains; it is in large part snow- and ice-covered.

The eastern part of the field is characterized chiefly by extensive and thick sequences of relatively flat-lying andesite and basaltic andesite flows, probably mostly from a number of coalescing shield volcanoes. Thick dacite flows and dacite and rhyolite domes, pyroclastic rocks, and subvolcanic plutons occur at and around

Volcano	Lat N/Long W	Age (Ma)
Jarvis	62.03°, 143.63°	1-2
Skookum Creek	62.40°, 143.13°	2-4
Castle	61.67°, 142.00°	8-10
Solo Creek	61.78°, 141.80°	3-10
Sonya Creek	61.93°, 141.25°	18-20

some of the known eruptive centers. Available radiometric dates indicate that, in general, volcanic activity proceeded from east to west across the field.

Exceptions to the general east to west progression of eruptive activity include numerous Quaternary basalt and basaltic andesite cinder cones throughout the western half of the field, and the Holocene White River Ash source vent near the Yukon border. The cinder cones, of both Pleistocene and Holocene age, occur along the north side of the field between Mount Drum and the Nabesna Glacier River system, where they are superimposed on older volcanic rocks. Many of the cones retain their original constructional form, and most are <1 km in diameter and <100 m in height. The most prominent cone is Mount **Gordon**, a composite cinder-lava cone, 5 km in diameter and 625 m high. It has an elevation of 2,755 m and is located at 62.13°N, 143.08°W. The young White River Ash was apparently erupted from a vent now covered by the Klutlan Glacier in the eastern part of the field. In addition, two prominent peaks in the eastern part of the field - Mount **Churchill** (4,766 m high, 61.25°N, 141.70°W) and Mount **Bona** (5,005 m, 61.25°N, 141.70°W) - appear to be relatively young constructional volcanic forms. Mount Churchill also apparently contains an ice-filled summit caldera 3 km wide.

How to get there: *The Eastern Wrangell field is accessible via expeditionary-scale bushwhacking from the Nabesna Road to the north or the McCarthy Road to the south. Alternatively, small, single-engine airplanes equipped for bush operation can land on some gravel bars, glaciers, and small landing strips. Many of the landing areas shown on maps have been blockaded recently.*

References

Lowe, P. C., Richter, D. H., Smith, R. L., and Schmoll, H. R., 1982, Geologic map of the Nabesna B-5 quadrangle, Alaska: *USGS Geol. Quad. Map GQ-1566*, scale 1:63,360.

Richter, D. H., Smith, J. G., Ratte, J. C., and Leeman, W. P., 1984, Shield volcanoes in the Wrangell Mountains, Alaska, *in* The USGS in Alaska: Accomplishments during 1982: *USGS Circ. 939*, 71-75.

Donald H. Richter

WHITE RIVER ASH

Wrangell Mountains, Alaska, and Yukon Territory, Canada

Type: *Tephra deposit*
Eruptive History: *~1885 yr BP, ~1260 yr BP*
Composition: *Dacite*

White River Ash covers most of the southern Yukon of Canada and part of eastern Alaska. It is the product of two separate Plinian eruptions, the combined tephra volume likely exceeding 30 km^3. The tephra is a calc-alkaline, hornblende-rich dacite, although its pumiceous glass has a rhyolitic composition. The source vent is located on the northwestern flank of Mount **Bona** in the Wrangell Mountains. Eleven radiocarbon age determinations on each lobe give a weighted mean age and uncertainty of 1,885 yr BP and 1,265 yr BP for the northern and eastern lobes, respectively.

References

Downes, H., 1985, Evidence for magma heterogeneity in the White River Ash (Yukon Territory): *Can. J. Earth Sci. 22*, 929-934.

Lerbekmo, J. F., Westgate, J. A., Smith, D. G. W., and Denton, G. H., 1975, New data on the character and history of the White River volcanic eruption, Alaska: *in* "Quaternary Studies," Suggate, R. P., and Cresswell, M. M. (eds.), *Royal Soc. New Zealand Bull. 13*, 203-209.

John Westgate

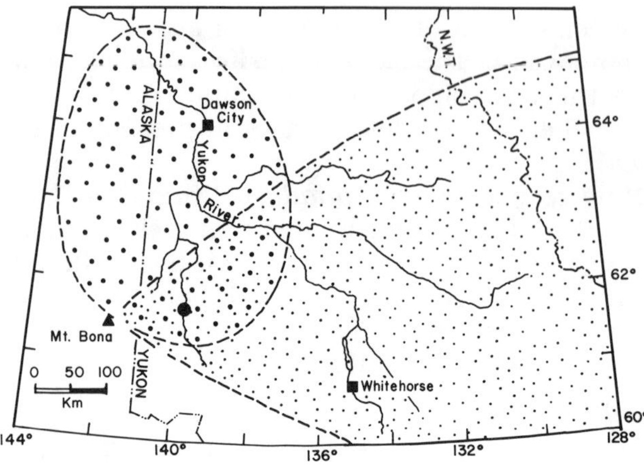

WHITE RIVER ASH: Distribution of the bi-lobate White River Ash. The northern and eastern lobes have axis lengths in excess of 500 and 1,000 km, respectively.

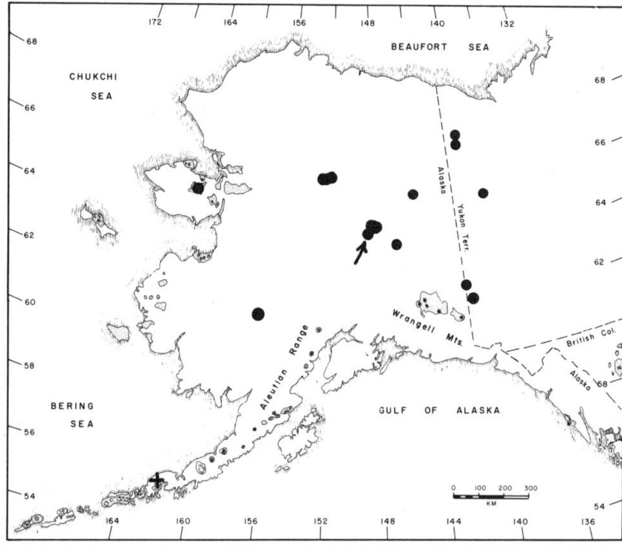

OLD CROW TEPHRA: Location of sites where the tephra has been found. Arrow shows site in photograph; cross shows location of Emmons Lake caldera.

OLD CROW TEPHRA

Eastern Alaska and Canada

Type: *Tephra deposit*
Eruptive History: *0.15 Ma*
Composition: *Rhyolite*

Old Crow Tephra, a two pyroxene, calc-alkaline rhyolite, is widely distributed across Alaska and the western Yukon. Its source vent is unknown, although petrographic and geochemical characteristics indicate a location in the eastern Aleutian arc. **Emmons Lake** caldera has been suggested as a possible source volcano. The tephra is fine-grained (MdØ) = 5.33 ± 0.32) at all the observed sites, and bed thickness values suggest a tephra volume exceeding 200 km^3.

The weighted mean age of Old Crow Tephra is 149 ka, which supports the view that the tephra was deposited during the cold interval that immediately preceded the last interglacial, dated in the marine record at 125 ka. The distinc-

OLD CROW TEPHRA: Interbedded with loess at Mile 329.5 on George Parks Highway, 35 km west of Fairbanks, Alaska. Thickness varies from 15 to 30 cm. Scale in cm.

tive characteristics and widespread distribution of this tephra have encouraged its use as a stratigraphic marker horizon, greatly facilitating correlation of late Pleistocene sediments across Alaska and the Yukon Territory.

References

Westgate, J. A., 1988, Isothermal plateau fission-track age of the late Pleistocene Old Crow Tephra, Alaska: *Geophys. Res. Lett.* 15, 376-379.

Westgate, J. A., Walter, R. C., Pearce, G. W., and Gorton, M. P., 1985, Distribution, stratigraphy, petrochemistry and paleomagnetism of the Old Crow tephra in Alaska and the Yukon: *Can. J. Earth Sci.* 22, 893-906.

John Westgate

EDGECUMBE
Southeastern Alaska

Type: *Shield volcano complex*
Lat/Long: *57.05°N, 135.75°W*
Elevation: *975 m*
Eruptive History: *Pleistocene: construction of shield volcano began ~600 ka; domes and stratovolcano constructed chiefly in late Pleistocene time. Holocene: extensive pyroclastic eruptions 9 to ~13 ka; two phreatomagmatic eruptions between 4 and 6 ka; no historic activity*
Composition: *Calc-alkaline and marginally tholeiitic basalt, basaltic andesite, andesite, dacite, and low-silica rhyolite*

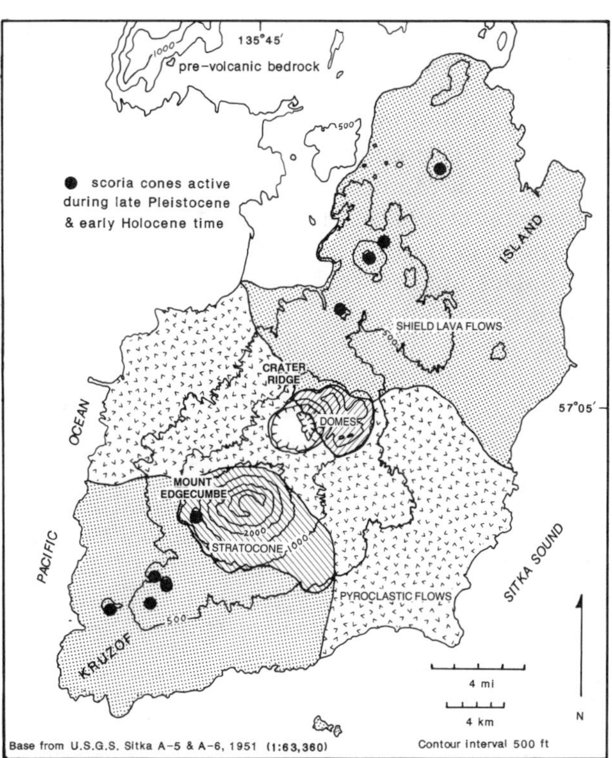

EDGECUMBE: View to north of Mount Edgecumbe volcanic field showing stratovolcano of Mount Edgecumbe (foreground), crater and domes of Crater Ridge (middle), and scoria cones (background). Photograph by D. Grybeck, 1981.

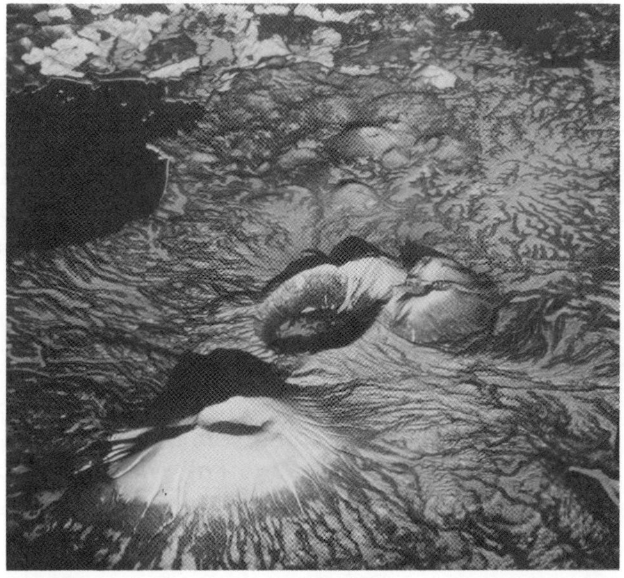

EDGECUMBE: Simplified geologic map of the Mount Edgecumbe volcanic field on Kruzof Island, ~20 km west of Sitka in southeastern Alaska

The Edgecumbe volcanic field on southern Kruzof Island is on the North American plate 10-15 km inboard of the Queen Charlotte–Fairweather transform fault. The Edgecumbe volcanic field is dominated by the symmetric stratovolcano of Mount Edgecumbe and the domes and crater of adjacent **Crater Ridge**. Mount Edgecumbe was named by Captain James Cook in 1778. Despite the fresh constructional morphology of the cone, there is no evidence for historic eruptions in Russian documents or in native oral traditions. Thus, the last certain activity of the Edgecumbe volcanic field was two minor tephra-forming eruptions between 4 and 6 ka. The basal shield comprises ~35 km^3 and consists of basalt, basaltic andesite, and andesite lava flows and breccias. The composite cone of Mount Edgecumbe is dominantly of andesite composition and has a volume of ~3.5 km^3. The low-silica rhyolite domes of Crater Ridge also contain ~3.5 km^3 of magma.

The latest significant eruptive activity was postglacial and produced voluminous pyroclastic deposits (7.6 km^3 dense-rock equivalent). The main geomorphic features of the Edgecumbe volcanic field were formed during this activity and include basaltic andesite scoria cones, a crater explosively reamed from the Crater Ridge domes during eruptions of rhyolitic pyroclastic flows, and eruption of andesite and dacite tephra during dome emplacement and crater formation on the Mount Edgecumbe cone. Tephra deposits produced by the late Pleistocene–early Holocene activity of the Edgecumbe volcanic field have been found as far away as Juneau and Lituya Bay, 200 km to the north. Vents active during the pyroclastic eruptions have a northeast–southwest alignment that probably marks a regional fissure.

How to get there: *The Edgecumbe volcanic field is 15 km west of Sitka in southeastern Alaska. Sitka has scheduled ferry and airline service; there are no roads to the city. Visitors desiring to visit the Edgecumbe area can charter a boat, airplane, or helicopter in Sitka. There are no prepared landing sites on Kruzof Island.*

References

Riehle, J. R., and Brew, D. A., 1984, Explosive Holocene activity of the Mount Edgecumbe volcanic field, Alaska, *in* Reed, K. M., and Bartsch-Winkler, S. (eds.), USGS in Alaska: Accomplishments during 1982: *USGS Circ. 939*, pp.111-115.

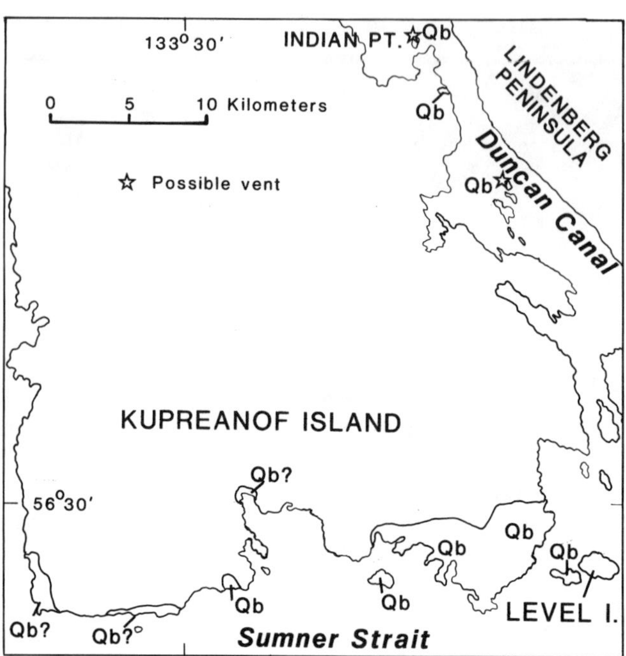

DUNCAN CANAL: Sketch map of location of Duncan Canal volcanics (stars = vents, Qb = flows).

Riehle, J. R., Brew, D. A., and Lanphere, M. A., *in press*, Geologic map of the Mount Edgecumbe volcanic field, Kruzof Island, southeastern Alaska: *USGS Misc. Invest. Series Map I-1983*, scale 1:63,360.

James R. Riehle

DUNCAN CANAL
Southeastern Alaska

Type: *Low gradient flows, two possible vents*
Lat/Long: *56.5°N, 133.1°W*
Elevation: *Sea level to ~15 m*
Dimensions: *~40 ×40 km*
Eruptive History: *Post-glacial: ~270 ka*
Composition: *Olivine basalt*

Olivine-bearing tholeiitic basalt with average K-content, together with minor sodic alkalic basalts, overlie volcanic-rich gravel and glacial till along the southern coast of Kupreanof Island west of Duncan Canal (Qb on map). The flows are recognized in the field by their conspicuous fresh-appearing pahoehoe and aa surfaces. Two probable vents lie within the Mesozoic and Paleozoic rocks of the Duncan Canal

fault zone north of the main exposures; the northern one contains scattered peridotite nodules. The relations of the main exposures to the subjacent middle Tertiary volcanic rocks of the Kuiu–Etolin volcanicplutonic belt are not clear at the western end of the field because the Holocene basalt is difficult to differentiate from the adjacent older basalt.
How to get there: *Duncan Canal can be reached by charter boat, helicopter, or float-equipped fixed-wing aircraft from Petersburg or Wrangell, Alaska.*

Reference

Brew, D. A., Karl, S .M., and Tobey, E. F., 1985, Reinterpretation of age of Kuiu–Etolin belt volcanic rocks, Kupreanof Island, southeastern Alaska, *in* Bartsch-Winkler, S. (ed.), The USGS in Alaska: Accomplishments during 1983: *USGS Circ. 945*, 86-88.

David A. Brew

TLEVAK STRAIT AND SUEMEZ ISLAND
Southeastern Alaska

Type: *Low gradient lava flows*
Lat/Long: *55.25°N, 133.30°W*
Elevation: *Sea level to 50 m*
Dimensions: *Scattered flows over 25 × 10 km*
Eruptive History: *No historic activity; probably post-glacial*
Composition: *Olivine basalt*

Flat-lying sodic alkalic olivine basalt flows with fresh pahoehoe surfaces occur in the vicinity of Tlevak Strait west of Prince of Wales Island, at Trocadero Bay on Suemez Island, and in valleys on northern Dall Island. The outcrop areas are indicated by Qb on the map. The flows unconformably overlie Paleozoic sedimentary rocks and on Suemez Island are associated with rhyolite and dacite flows that are interpreted to be Miocene and Late Oligocene in age.
How to get there: *The Alexander Archipelago can be reached by charter boat or float-equipped fixed-wing aircraft from Craig or Ketchikan, AK. Helicopters can be chartered in Ketchikan.*

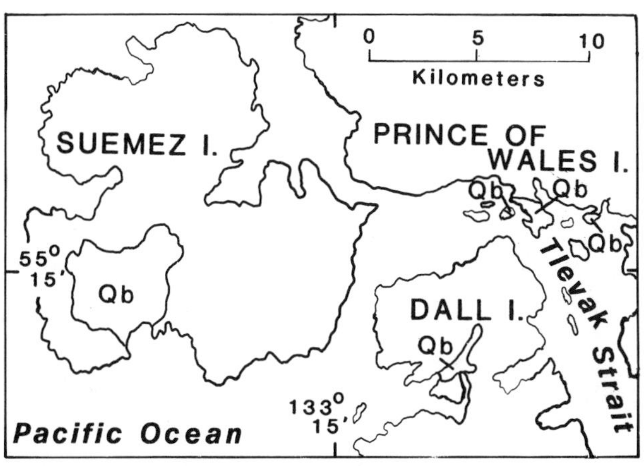

TLEVAK STRAIT-SUEMEZ ISLAND: Sketch map of location of volcanics (Qb = flows).

References

Eberlein, G. D., and Churkin, M., Jr., 1970, Tlevak basalt, west coast of Prince of Wales Island, southeastern Alaska, *in* Cohee, G. V., Bates, R. G., and Wright, W. B. (eds.), Changes in stratigraphic nomenclature by the USGS, 1968: *USGS Bull. 1294-A*, A25-A28.

Eberlein, G. D., Churkin, M., Jr., Carter, C., Berg, H. C., and Ovenshine, A. T., Geology of the Craig quadrangle, Alaska: *USGS Open-file Rpt. 83-91*, 4 sheets, scale 1:250,000, 52 pp.

David A. Brew

BEHM CANAL AND RUDYERD BAY
Southeastern Alaska

Type: *Cinder cones and low gradient flows*
Lat/Long: *55.5°N, 131.0°W*
Elevation: *Sea level to 500 m*
Dimensions: *Scattered flows in area 40 × 25 km*
Eruptive History: *Possibly two periods: Older ~5 Ma, and younger ~1 Ma to 0.5 Ma*
Composition: *Olivine basalt, andesite, and trachyandesite*

Potassic alkali olivine basalt cinder cones and columnar-jointed flows occur as separate entities in the field, which is more or less centered on the Coast Range megalineament. Minor andesite and trachyandesite flows are locally present. There are also thin patches of ash and lapilli layers. The lava flows unconformably overlie glacial–fluvial sand and silt deposits

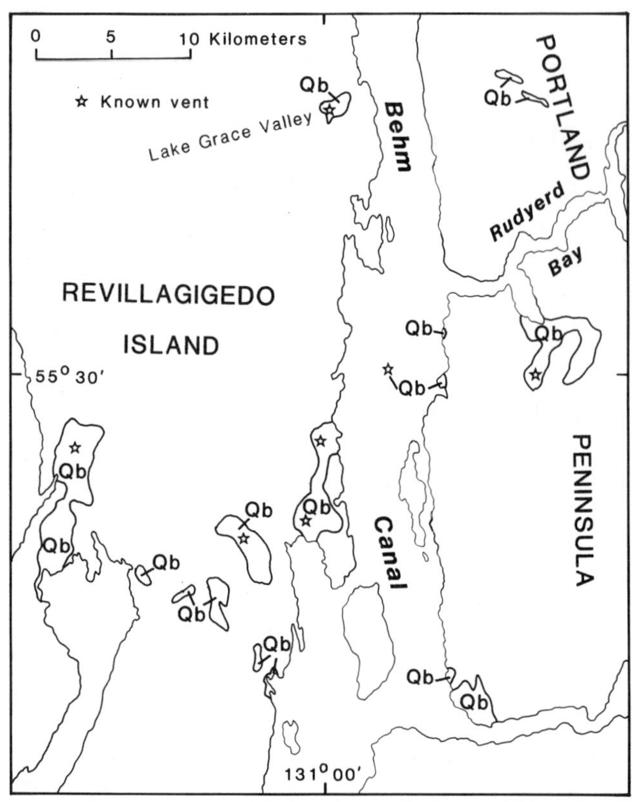

BEHM CANAL–RUDYERD ISLAND: Sketch map of location of volcanics (star = vent, Qb = flows).

and granitic gneiss of the Coast Mountains plutonic-metamorphic complex. Glacial striations and grooves indicate that some of the flows pre-date the last glaciation, but undisturbed cinder cones and flow surface features indicate that most of the activity was post-glacial. Flows are locally columnar-jointed, and some tuff and breccia deposits form void-rich spires and cliffs.

How to get there: *The Behm Canal–Rudyerd Bay Field can be reached by charter boat or float-equipped fixed-wing aircraft from Ketchikan, Alaska.*

References

Berg, H. C., Elliott, R. L., Smith, J. G., and Koch, R. L., 1978, Geologic map of the Ketchikan and Prince Rupert quadrangles, Alaska: *USGS Open file Rpt. 78-73A*, 1 sheet, scale 1:250,000.

Wanek, A. A., and Callahan, J. E., 1971, Geologic reconnaissance of a proposed power site at Lake Grace, Revillagigedo Island, southeastern Alaska: *USGS Bull. 1211-E*, 24 pp.

David A. Brew

ST. PAUL

Bering Sea, Alaska

Type: *Monogenetic volcano field*
Lat/Long: *57.17°N, 170.27°W*
Elevation: *0 to 204 m*
Field Area: *~110 km²*
Eruptive History: *Intermittent activity since at least 0.3 Ma*
Composition: *Alkali olivine basalt, basanite*

St. Paul is the largest of the Pribilof Islands, the emergent parts of a lava-topped structural high on the western edge of the Bering Sea shelf. St. George Island is described elsewhere in this volume; **Otter** and **Walrus** Islands and **Sea Lion Rock** are smaller volcanic islands of unknown age in the group. Most of St. Paul consists of coalescing small volcanoes, each composed of a central cinder cone and a surrounding shield of Quaternary lava flows. Most flows are of the pahoehoe type, but at least three are aa.

There is no evidence of glaciation on St. Paul, but the youthful volcanic topography has been modified by frost action, deposition of wind-blown sand, faulting, and, perhaps, folding. A lava flow near Tolstoi Point has been broken up by closely spaced and gaping tectonic faults and fissures as much as 30 m deep, 50 m wide, and hundreds of meters long.

The lava flows are interbedded with marine sediments, and pillow lava is exposed in Einahnuhto Bluffs (western St. Paul). Eolian and colluvial sediments are also interbedded with or cover many of the flows. Most of the northern and eastern parts of the island are covered by wind-blown sand.

The cinder cones or cone complexes rise 30-100 m above their bases; 12-15 cones or cone complexes exceed 0.5 km in basal diameter. The cones are composed of pyroclastic material that includes primarily basaltic tuff; there is one rare occurrence of rhyolitic pumice in the central eastern part of the island. A maar is exposed in cross-section at Black Bluffs (east of the village of St. Paul), and many flows are patchily zeolitized, a process that may have occurred during initial cooling. A tuffaceous marine gravel at the base of Black Bluffs that hosts Pliocene diatom flora indicates a history of Miocene or Pliocene volcanism.

ST. PAUL: Panorama of southwestern St. Paul Island from Bogoslof Hill (central St. Paul Island). Photograph by D. M. Hopkins, 1965.

Lava flows mapped on St. Paul are all <0.3 Ma. Based on available data, ~1 eruption occurred per 5,000 yr during the past 0.12 Ma. The most widely exposed are pahoehoe lavas of Bogoslof Hill (~8 km^2) and the aa basalt flow of Rush Hill (~2.5 km^2). The **Bogoslof Hill** lavas erupted from an east–west trending series of vents marked by collapse craters and a spatter cone near the center of the island. The originally smooth surface of this flow has been disrupted by frost riving. The **Rush Hill** flow erupted from a northeast-trending row of cinder cones near the west coast of St. Paul. Exposed flow units are as much as 40 m thick. The flow rocks are nepheline-normative alkali olivine basalt and basanite with porphyritic and aphyric textures. The lavas have a limited SiO_2 range compared to those from nearby St. George Island. Mafic inclusions in some of the flows are evidently derived from the mantle. Granite inclusions and pebbles on the beaches suggest the presence of a quartz-monzonite pluton beneath the basalts of St. Paul.

How to get there: *St. Paul Island, part of the Pribilof Islands National Preserve for seal and otter, is in the western Bering Sea, 420 km northwest of the Aleutian Island of Unalaska. The island can be reached by scheduled air service from Anchorage (~1,600 km) by way of Cold Bay (520 km) or by charter plane from Dutch Harbor.*

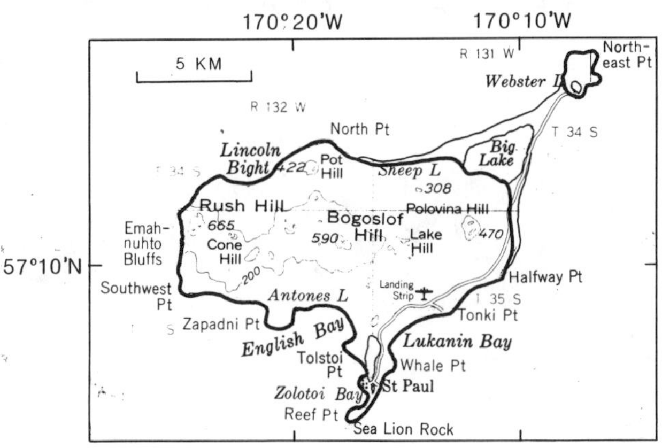

ST. PAUL: Extent of lavas on St. Paul Island indicated by heavy outline. Mapped by D.M. Hopkins and Th. Einarsson, 1965. Contour interval, 200 feet (~61 m).

References

Cox, A., Hopkins, D. M., and Dalrymple, D. B., 1966, Geomagnetic polarity epochs-Pribilof Islands, Alaska: *Geol. Soc. Am. Bull. 77*, 883-910.

Lee-Wong, F., Vallier, T. L., Hopkins, D. M., and Silberman, M. L., 1979, Preliminary report on the petrography and geochemistry of basalt from the Pribilof Islands and vicinity, southern Bering Sea: *USGS Open-file Rpt. 79-1556*, 51 pp.

Florence Lee-Wong

ST. GEORGE
Bering Sea, Alaska

Type: *Monogenetic volcano field*
Lat/Long: *56.57°N, 169.63°W*
Elevation: *0 to 289 m*
Field Area: *~90 km^2*
Eruptive History: *Intermittent activity from 2.2 to 1.0 Ma*
Composition: *Alkali olivine basalt, basanite*

St. George is the second largest of the Pribilof Islands, a group of late Tertiary and Quater-

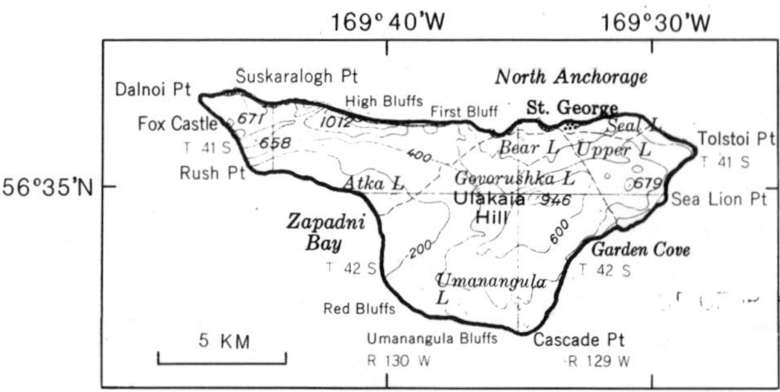

ST. GEORGE: Extent of lavas on St. George Island indicated by heavy outline. Mapped by D. M. Hopkins and Th. Einarsson, 1965. Contour interval, 200 ft.

ST. GEORGE: St. George Village, northern St. George Island. Fault scarp in left background. Middle Pleistocene glacial moraine in right background. Photograph by D. M. Hopkins, 1970.

ternary volcanic piles on the western edge of the Bering Sea shelf. The volcanic rocks on St. George are Pliocene and early Pleistocene in age and are interbedded with marine sand and gravel, glacially derived sediments, frost breccia, and wind-blown sediments. Along the southeastern coast, the St. George volcanic field has an exposed basement of serpentinized peridotite intruded by quartz diorite. The lavas on St. George are older than those on nearby St. Paul, and the original volcanic topography has been subdued by weathering and later redefined by faulting, uplift, and glaciation. Vestiges of cinder cones, with filled craters and slopes reduced by frost and mass-wasting processes, occur as flat-topped or domical hillocks. Apparently alternating lava flows and pyroclastic eruptions in the eroded volcanic cones are actually evidence of much fountaining, which produced near-vent pumice interbedded with some flows in one eruptive event.

Sea cliffs provide the best exposures of the products of repeated eruptions; for example, the cliff at Tolstoi Point on the eastern tip of St. George is made up of seven flows as much as 7.6 m thick, with associated pyroclastic layers. Similar to those on St. Paul Island, the St. George lavas consist of alkali olivine basalt and basanite, but the St. George samples and those from submarine dredge hauls near St. George have a broader range in SiO_2 and include hypersthene-normative as well as nepheline-normative chemistries. Mafic inclusions are common, and some flows contain granitic inclusions. Many flows display a patchy zeolitization, which may have resulted from alteration during cooling shortly after emplacement. Several flows that erupted during high sea-level episodes display thick and extensive pillow breccias and palagonite tuff sequences.

How to get there: *St. George Island, part of the Pribilof Islands National Preserve for seal and otter, is in the western Bering Sea 350 km northwest of the Aleutian Island of Unalaska. The island can be reached by scheduled air service from Anchorage (~1,600 km) by way of Cold Bay (455 km) or by charter plane from Dutch Harbor, all cities in Alaska.*

References

Cox, A., Hopkins, D. M., and Dalrymple, D. B., 1966, Geomagnetic polarity epochs-Pribilof Islands, Alaska: *Geol. Soc. Am. Bull.* 77, 883-910.

Lee-Wong, F., Vallier, T. L., Hopkins, D. M., and Silberman, M. L., 1979, Preliminary report on the petrography and geochemistry of basalt from the Pribilof Islands and vicinity, southern Bering Sea: *USGS Open-file Rpt. 79-1556*, 51 pp.

Florence Lee-Wong

NUNIVAK

Bering Sea, Alaska

Type: *Shield?*
Lat/Long: *60.15°N, 166.30°W*
Elevation: *510 m*
Island Size: *115 × 60 km*
Eruptive History: *5 broad periods of activity:*

6.1 –4.0 Ma	*>20 km³*	*Thol. + Alk. Bst.*
3.4 –3.1 Ma	*>65 km³*	*Thol.*
1.7 –1.5 Ma	*210 km³*	*Thol.*
0.9 –0.3 Ma	*130 km³*	*Thol. + Alk. Bst.*
0.3 –0 Ma	*2.3 km³*	*Alk. Bst.*

Composition: *Tholeiite and alkali basalt (basanite)*

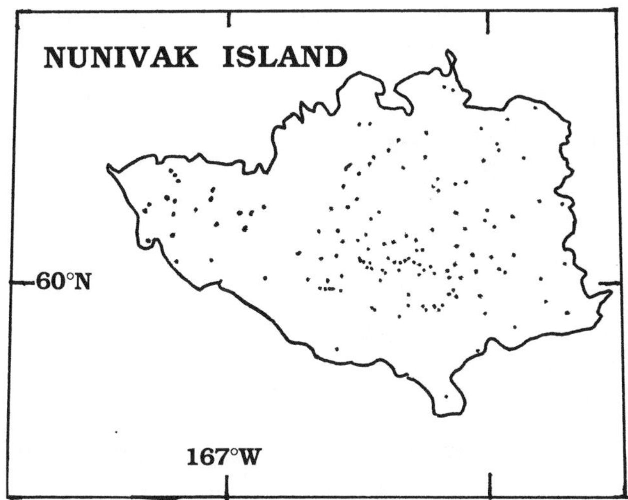

NUNIVAK: Sketch map of vents (dots) and distribution of feldspathic basalt (pattern) and basalt (plain). Modified from Luedke and Smith, USGS Map I-1091-F, 1986.

The volcanic carapace of Nunivak Island is built on Cretaceous sedimentary rocks in the eastern Bering Sea, within 50 km of western Alaska. The surface of the island is dominated by broad thin pahoehoe lava flows, with subsidiary alkalic basalts forming small lava flows, ~60 cinder cones and 4 maars. The pahoehoe flows are 3-15 m thick and build small shield volcanoes. Exposures are best at sea cliffs and among the young flows of the eastern part of the island, which is otherwise covered with tundra and shallow lakes. Permafrost is ubiquitous.

Alkali basalt lava flows are <3.5 km long, and most erupted from cinder cones. Some flows contain collapsed lava tubes and rafted segments of cinder cones. Although many true cinder cones occur on Nunivak Island, some alkali basalt cones (e.g., **Twin** Mountain) have small summit craters and steep flanks, much more like small stratovolcanoes. Alkali basalts also erupted phreatomagmatically, forming 4 maars in an east–west line: **Binalik, Ahkiwiksnuk, Nanwaksjiak**, and **"385,"** which is the elevation in feet of the crater lake of an otherwise unnamed maar. The latter 2 maars have ~200 m of relief, with floors near sea level. Although various types of nodules are found in many of the alkali basalt cones, ejecta from Nanwaksjiak includes lherzolite xenoliths within *tholeiite* boulders, an apparently unique occurrence.

Detailed radiometric dating demonstrated that volcanism occurred in distinct episodes, and systematically migrated eastward (~50 km/ 5 Ma). Additionally, tholeiites and alkali basalts erupted nearly simultaneously. Five pulses of volcanism (some combined in listing above), each lasting ~0.1-0.2 m.y., and separated by quiescent intervals of 1-2 m.y., occurred between 6 and 1.5 Ma. Since 0.9 Ma volcanism has been frequent, with both tholeiite and alkali basalt common, but since 0.3 Ma only alkali basalt has been erupted. The overlapping occurrence of the two magma types differs from the more familiar pattern of tholeiite followed by alkali basalt in Hawaii, and suggests that at Nunivak the latter is not derived from the former.

How to get there: *The village of Mekoryuk on the north shore of Nunivak Island is served by commercial airlines from Bethel on the mainland. Chartered floatplanes from Bethel can deliver geologists to lakes in the interior of Nunivak, and small boats can be rented at Mekoryuk for exploration of coastal exposures. The best weather is in late May and early June.*

References

Hoare, J. M., Condon, W. H., Cox, A., and Dalrymple, G. B., 1968, Geology, paleomagnetism and potassium-argon ages of basalts from Nunivak Island, Alaska: *in* Coats, R. R., Hay, R .L., and Anderson, C. A. (eds.), *Studies in Volcanology-A Memoir in Honor of Howel Williams: Geol. Soc. Am. Mem. 116*, 377-414.

Lucchitta, I., 1968, 1968 expedition to Nunivak Island, Alaska: *Unpublished USGS internal report*, 7 pp, Flagstaff, AZ.

Charles A. Wood

YUKON DELTA
Western Alaska

The Yukon Delta volcanic fields are part of a late Cenozoic subalkaline to alkaline basalt province in the Bering Sea region. Little geologic work has been published on these fields. The following descriptions are based primarily on interpretation of 1:63,360 scale topographic maps, aerial photos, and regional geologic maps. All of the basalt fields in the Bering Sea region are composed of rocks with primitive whole-rock compositions characterized by high MgO and low SiO_2 contents. The suites show de-

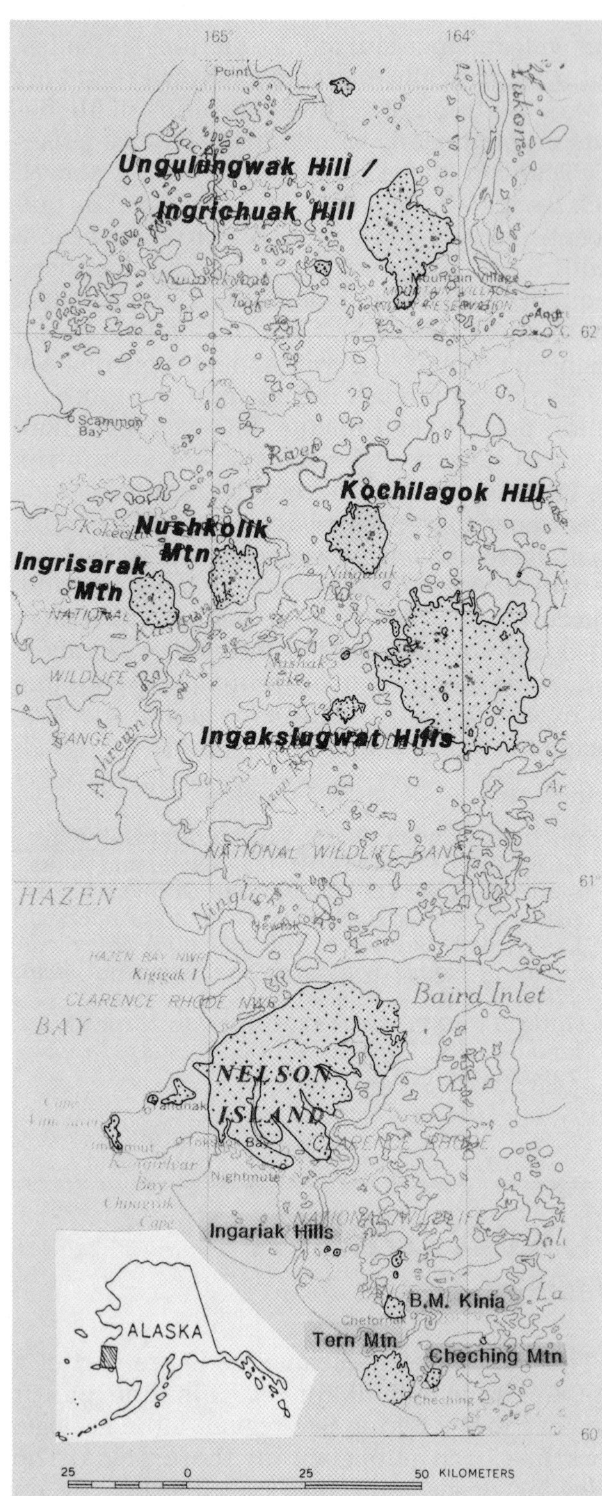

YUKON DELTA: Map showing locations of late Cenozoic basalt fields. Based on Luedke and Smith, USGS Map I-1091-F, 1986.

creasing alkalies with increasing SiO_2, similar to rocks from Hawaii.

UNGULUNGWAK HILL–INGRICHUAK HILL

Type: *Lava shield field*
Lat/Long: *61.17°N, 164.25°W*
Elevation: *15 to 188 m*
Volcanic Field Dimensions: *20 × 29 km*
Eruptive History: *No historic activity*
Composition: *Olivine basalt*

This field consists of numerous lava flows 1.5 to 4.5 m thick, and at least five low shield volcanoes from 30 to 188 m high (four are 166 to 188 m high) and 5 to 11 km across. Although the volcanoes are modified by erosion and widely covered with silt, they are inferred to be Quaternary in age based on their physiographic expression. The volcanoes consist of vesicular and scoriaceous olivine basalt.

How to get there: *The Ungulungwak Hill–Ingrichuak Hill volcanic field is located 5 to 15 km west of the Yukon River. There are no roads or trails in the area. The field can be reached by floatplane from the village of St. Mary's (55 km to the east-southeast) or by charter helicopter from Bethel (220 km to the southeast).*

Reference

Hoare, J. M., and Condon, W. H., 1966, Geologic map of the Kwiguk and Black quadrangles, Alaska: *USGS Misc. Geol. Invest. Map I-469*, scale 1:250,000.

Betsy Moll-Stalcup

INGAKSLUGWAT HILLS

Type: *Monogenetic volcanic field*
Lat/Long: *61.38°N, 164.00°W*
Elevation: *8 to 200 m*
Volcanic Field Dimensions: *33 × 33 km*
Eruptive History: *No historic activity, probably Holocene*
Composition: *Alkali olivine basalt, basanite and nephelinite*

The Ingakslugwat Hills volcanic field consists of >32 small cinder cones and 8 larger craters with associated flows, all covering an area of

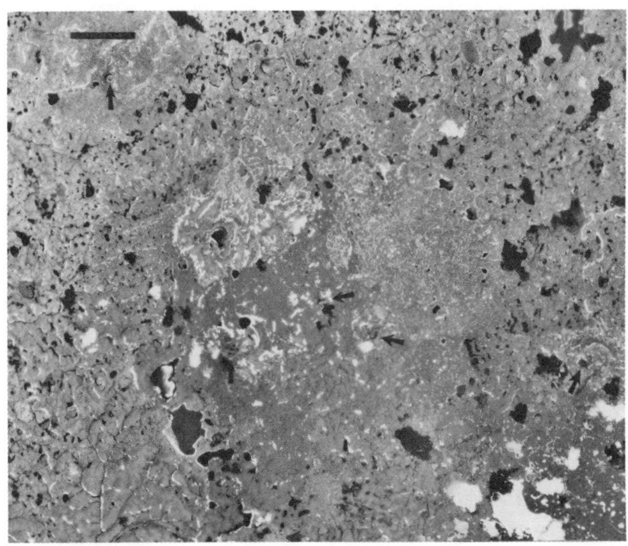

INGAKSLUGWAT HILLS: NASA vertical air view of central part of field. Arrows identify 5 cones; bar is 2 km long.

>500 km^2. Older vents are low, wide cones with saucer-shaped craters up to 1.5 km in diameter; younger eruptions formed relatively small, steep cones 30 to 90 m high and 90 to 150 m wide. Numerous small spatter cones and cinder cones ~30 m high occur on the northwest side of the volcanic field. Some cones are aligned west-northwest, apparently defining a fracture. One low cone with a lake ~400 m across may be a maar. The rocks are chiefly alkali olivine basalt with lesser amounts of basanite and nephelanite. The eruptive history of this field is unknown, but the well-preserved volcanic morphology suggests that some of the activity occurred in the Holocene.

Inclusions of lherzolite, layered gabbro, and granular gabbro occur in nephelnite ash and in an alkali basalt flow erupted from a cone in the southwest part of the Ingakslugwat Hills.

How to get there: *The Ingakslugwat Hills field is in the Yukon Delta between the Yukon and Kuskokwin Rivers. It can be reached by charter floatplane from the village of St. Mary's (85 km northeast) or by charter helicopter from the town of Bethel (140 km southeast).*

Reference

Hoare, J. M., and Condon, W. H., 1971, Geologic map of the Marshall quadrangle, western Alaska: *USGS Misc. Geol. Invest. Map I-668*, scale 1:250,000.

Betsy Moll-Stalcup

KOCHILAGOK HILL

Type: *Cinder cone and lava flow*
Lat/Long: *61.63°N, 164.42°W*
Elevation: *8 to 70 m*
Volcanic Field Dimensions: *13 × 15 km*
Eruptive History: *No historic activity*
Composition: *Olivine basalt*

A low vent and associated flow occur at Kochilagok Hill. Geologic maps designate the volcano as chiefly basalt scoria.

How to get there: *The Kochilagok Hill volcano is 25 km northwest of the Ingakslugwat Hills volcanic field. It can be reached by charter floatplane from the village of St. Mary's (75 km northeast) or charter helicopter from the town of Bethel (140 km southeast).*

Reference

Hoare, J. M., and Condon, W. H., 1971, Geologic map of the Marshall quadrangle, western Alaska: *USGS Misc. Geol. Invest. Map I-668*, scale 1:250,000.

Betsy Moll-Stalcup

NUSHKOLIK MOUNTAIN

Type: *Shield volcano?*
Lat/Long: *61.55°N, 164.92°W*
Elevation: *8 to 195 m*
Volcanic Field Dimensions: *12 × 14 km*
Eruptive History: *No historic activity*
Composition: *Olivine basalt*

Nothing is known of this field beyond the information above.

How to get there: *The Nushkolik Mountain volcanic field is located along the Aphrewn River and Nanvaranak Lake and is accessible by boat. The river or lake can be reached by floatplane from St. Mary's (105 km northeast) or directly by charter helicopter from Bethel (190 km southeast).*

Reference

Hoare, J. M., and Condon, W. H., 1971, Geologic map of the Marshall quadrangle, western Alaska: *USGS Misc. Geol. Invest. Map I-668*, scale 1:250,000.

Betsy Moll-Stalcup

INGRISARAK MOUNTAIN

Type: *Cinder cone*
Lat/Long: *61.50°N, 165.23°W*
Elevation: *8 to 112 m*
Volcanic Field Dimensions: *10 × 11 km*
Eruptive History: *No historic activity*
Composition: *Olivine basalt*

Ingrisarak Mountain is a volcanic vent, composed entirely of red and black basalt scoria.
How to get there: *The Ingrisarak Mountain volcanic region can be reached by charter floatplanes from the village of St. Mary's (124 km northeast) or by chartered helicopters from the town of Bethel (200 km southeast).*

Reference

Hoare, J. M., and Condon, W. H., 1968, Geologic map of the Hooper Bay quadrangle, Alaska: *USGS Misc. Geol. Invest. Map I-523,* 1:250.000 scale.

Betsy Moll-Stalcup

NELSON ISLAND

Type: *Lava flow field?*
Lat/Long: *60.67°N, 164.75°W*
Elevation: *0 to 450 m*
Volcanic Field Dimensions: *42 × 38 km*
Eruptive History: *No historic activity*
Composition: *Olivine basalt?*

Reconnaissance geologic mapping of Nelson Island revealed basaltic lava flows ~30 m thick dipping gently eastward. The topography of the island suggests it is highly dissected and probably late Tertiary rather than Quaternary in age. Younger looking isolated cones are evident to the south at **Ingariak Hills, Tern** Mountain, **Cheching** Mountain, and benchmark **Kinia**.
How to get there: *Nelson Island is located in the western Yukon–Kuskokwim Delta near the Kolavinarak and Ninglick Rivers. The eastern and southern parts of the field are accessible by boat. Rivers or nearby lakes can be reached by charter floatplanes from Bethel (160 km to the east). Chartered helicopters are also available in Bethel.*

References

Coonrad, W. L., 1957, Geologic reconnaissance in the Yukon–Kuskokwim Delta region, Alaska: *USGS Misc. Geol. Invest. Map I-223.,* 1:500,000 scale.

Moll-Stalcup, E. J.,1989, Latest Cretaceous and Cenozoic magmatism in mainland Alaska, *in* Plafker, G., and Jones, D. L. (eds.), The Cordilleran Orogene–Alaska: *DNAG Spec. Pub.,* Geol. Soc. Am., Boulder, CO.

Betsy Moll-Stalcup

TOGIAK
Western Alaska

Type: *Lava field*
Lat/Long: *59.2°N, 160.1°W*
Elevation: *Flows: 10-50 m; Tuya: 353 m*
Eruptive History: *<0.76 Ma*
Composition: *Tholeiitic and alkali-olivine basalt*

The Togiak volcanics are a late Pleistocene set of thin (<20 m thick) lava flows underlying ~450 km^2 of the Togiak River valley in southwest Alaska. The flows are largely covered by glacial deposits, but excellent exposures occur in sea cliffs on Togiak Bay and along cutbanks of various rivers. Three probable vents have been detected; the most interesting is a 300-m-high tuya or subglacial volcano. Togiak Tuya is 6 km long × 2.5 km wide and is elongated parallel to regional faults and the flow of glacial ice. The elongated shape suggests formation in an actively flowing glacier; most other tuyas apparently formed in stagnant ice. Sideromelane tuff and pillow basalts make up the main mass of the tuya, but the top 40-50 m are subaerial flows, formed where the volcano grew above the glacier. It has been speculated that the tuya may have formed during a glacial period ~39,000 yr ago, and a single K-Ar age of 0.76 Ma is recorded for the underlying basalts, all of which have normal magnetic polarity. The lavas are confined to a graben between northeast-trending faults, and appear to have erupted from vents along the faults.
How to get there: *The Togiak River valley can be reached by bush plane from Dillingham in southwest Alaska. Commercial air service is available to Dillingham from Anchorage.*

References

Hoare, J. M., and Coonrad, W. L., 1978: A tuya in Togiak Valley, southwest Alaska. *USGS J. Res. 6,* 193-201.

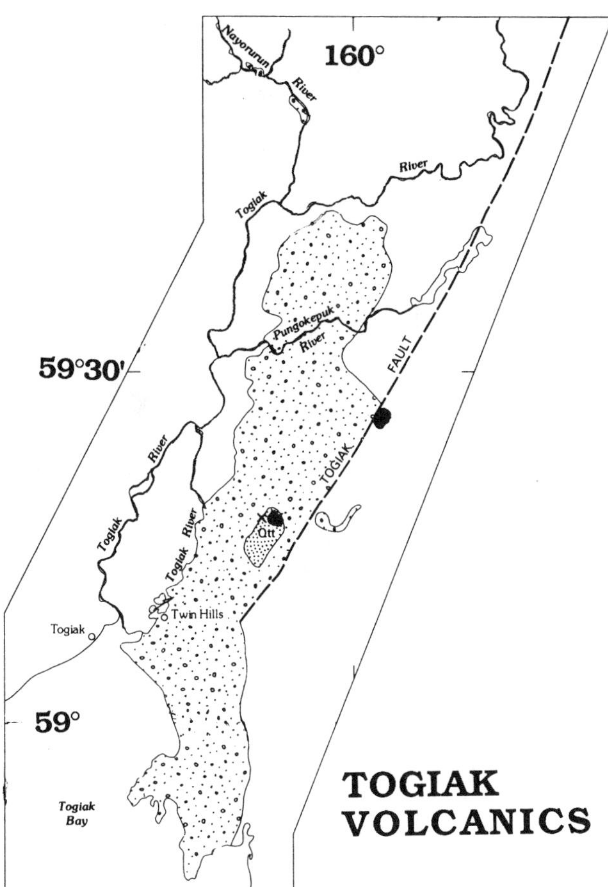

TOGIAK: Map of the extent of Togiak lava flows; map from Hoare and Coonrad, 1980. Two vents marked in black, with tuya deposits indicated by dense stipple pattern (Qtt) near center of field.

Hoare, J. M., and Coonrad, W. L., 1980: The Togiak Basalt, a new formation in southwestern Alaska. *USGS Bull. 1482-C*, 11 pp.

Charles A. Wood

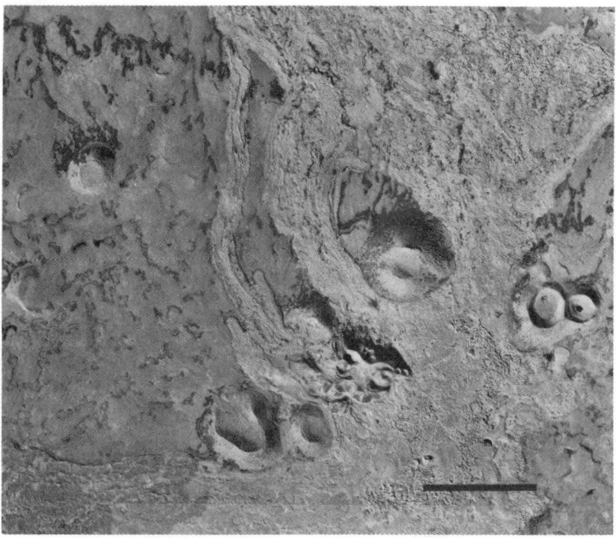

ST. MICHAEL: US Navy aerial photograph of Holocene lava flows and several generations of basanite and alkali basalt cones in "The Sisters" region of the St. Michael volcanic field. Scale bar = 1 km. North up.

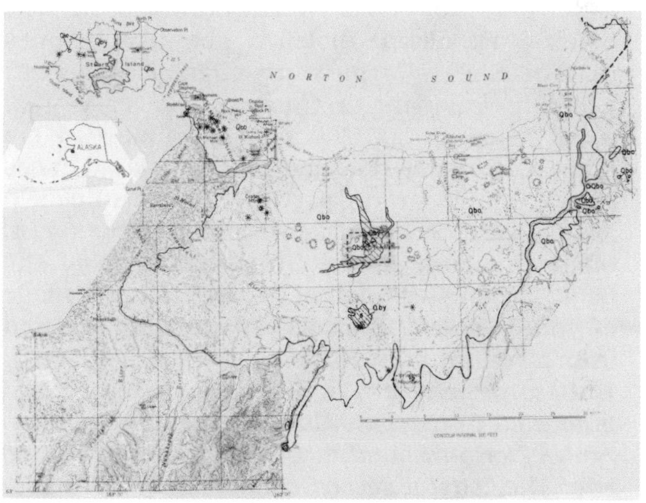

ST. MICHAEL: Schematic map showing outline of field lavas, with area of accompanying photograph outlined. Map a composite from two references and additional mapping by Patton, Moll-Stalcup, and Murphy.

ST. MICHAEL
Western Alaska

Type: *Monogenetic cone field*
Lat/Long: *63.13°-63.63°N, 160.92°-162.67°W*
Elevation: *0 to 715 m*
Volcanic Field Dimensions: *Crescent-shaped, ~40 × 78 km*
Eruptive History: *Five K-Ar dated flows : 3.25 Ma, 2.80 Ma, 1.99 Ma, 1.39 Ma, 0.19 Ma; no historic activity*
Composition: *Tholeiite, olivine tholeiite, alkali-olivine basalt, hawaiite, and basanite*

The St. Michael volcanic field is one of a number of late Tertiary and Quaternary, alkaline to subalkaline basaltic volcanic fields in the Bering Sea region of western Alaska. The volcanic field covers all of St. Michael and

Stuart islands, extending inland as far as the Golsovia and Kogok rivers. It consists of >55 cones and craters and numerous flat-lying flows and tuffs covering >3,000 km^2. The section of flows has a maximum thickness of 40 m exposed in sea cliffs along Norton Sound, but is probably much thicker in the interior of the field. Most of the volcanic field consists of flat-lying tholeiitic or alkali-olivine basalt flows which appear to be erupted from broad shield volcanoes, such as those comprising St. Michael Mountain and **Stuart** Mountain. These older dissected shield volcanoes are ~4 km across and 150 m high. Young steep-sided cinder cones consisting of alkali-olivine basalt, basanite or hawaiite overlie the flat-lying flows. The cones and flows surrounding **Crater** Mountain, in the western part of the field, consist of highly alkalic basanite containing lherzolite nodules. Most of the cones in **"The Sisters"** region are basanite or alkali olivine basalt, although a small group of cones ~4 km northeast of "The Sisters" are olivine tholeiite. The alkalic cones in "The Sisters" region are aligned east–west, possibly following a fracture. These young, small cones are 200-700 m across and ~40-75 m high. A Holocene flow and two associated cones in the south-central part of the volcanic field are tholeiite. Seven maar volcanoes occur on St. Michael Island in the northwest part of the field. The western half of Stuart Island is made of young tholeiitic aa flows, probably erupted from a vent at West Hill.

How to get there: *The St. Michael volcanic field is located in western Alaska along the southern rim of Norton Sound in the easternmost Bering Sea. The coastal section can be reached by boat from the villages of St. Michael or Unalakleet, ~50 km north. A small airstrip exists at the village of St. Michael; the rest of the area is not easily accessible without charter helicopter.*

References

Hoare, J. M., and Condon, W. H., 1971, Geologic map of the St. Michael quadrangle, Alaska: *USGS Misc. Geol. Invest. Map I-682*, scale 1:250,000.

Patton, W. W., Jr., and Moll, E. J., 1985, Geologic map of the northern and central parts of Unalakleet quadrangle: *USGS Misc. Field Studies Map MF-1749*, scale 1:250,000.

Betsy Moll-Stalcup

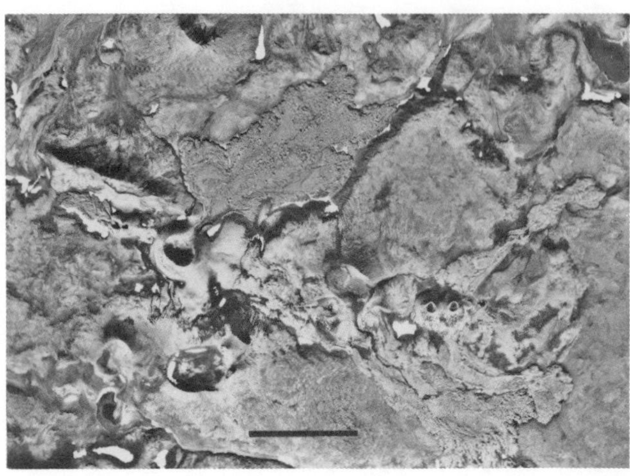

KOOKOOLIGIT: US Navy aerial photo of several cones and Recent flows (chiefly basanite) along the east-west highlands in the Kookooligit Mountains. Scale bar = 1 km. North up.

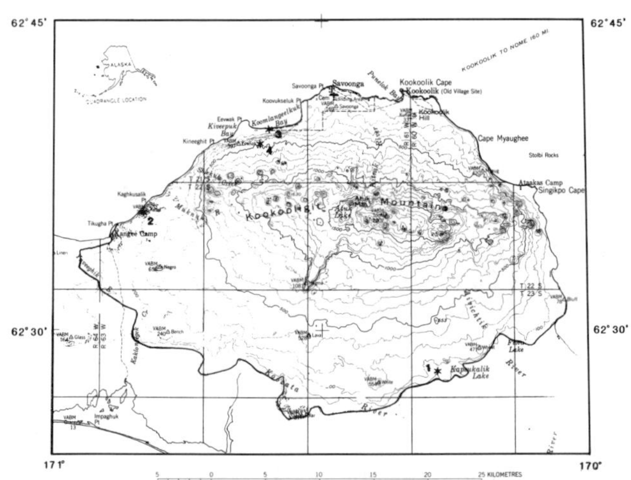

KOOKOOLIGIT: Topographic map showing east-west trending highlands covered by numerous cinder cones. Photo location just south of "S" in "Mountains." Numbers 1-4 indicate locations of K-Ar ages listed in statistics table. Geology from Patton and Csejtey, 1980. Contour interval 100 ft.

KOOKOOLIGIT

Bering Sea, Alaska

Type: *Shield volcano with cinder cones*
Lat/Long: *63.58°N, 170.43°W*
Elevation: *680 m*
Volcano Diameter: *30 × 40 km*
Eruptive History: *Four K-Ar dated flows: (4) 1.46 Ma, (3) 0.65 Ma, (2) 0.36 Ma, (1) 0.24 Ma; no historic activity*

Composition: *Olivine tholeiite, alkali-olivine basalt, hawaiite, basanite, and nephelenite*

The Kookooligit volcanic field, on northern St. Lawrence Island, is one of a number of basalt fields in the Bering Sea region. The field consists of an elongate shield volcano over 500 m high composed of massive columnar-jointed basalt flows overlain by >100 small cones (20 to 60 m high) aligned east–west along a volcanic highlands. The older volcanic flows are chiefly alkali olivine basalts and olivine tholeiite. The younger cones and flows are dominantly basanite and alkali olivine basalt with subordinate hawaiite and nephelanite. Most of the basanite and nephelanite flows and cones contain inclusions of deformed peridotite or gabbro; some also contain megacrysts of anorthoclase.

Most of the young volcanic rocks have primitive whole-rock compositions characterized by high MgO and low SiO_2 contents. The suite is characterized by decreasing total alkalies with increasing SiO_2, similar to the trend observed in Hawaiian lavas.

How to get there: *The Kookooligit volcanic field is located on St. Lawrence Island in the Bering Sea between the Soviet Far East and western Alaska. There are no roads or trails in the area. A landing strip at the village of Savoonga occurs at the northernmost flank of the volcanic field. Much of the northern and some of the southern margin of the field is accessible by boat along the Bering Sea coast and along the Koozata and Aivichtik Rivers. Several lakes in the Kookooligit Mountains are probably large enough to land a floatplane. All of St. Lawrence Island is native-owned, and permission to visit the island must be granted by the native corporation.*

References

Patton, W. W., Jr. and Csejtey, B, Jr., 1980, Geologic map of St. Lawrence Island, Alaska: *USGS Misc. Geol. Invest. Map I-1203*, scale 1:250,000.

Moll-Stalcup, E. J., 1989, Latest Cretaceous and Cenozoic magmatism in mainland Alaska, *in* Plafker, G., and Jones, D. L. (eds.), The Cordilleran Orogene–Alaska: *DNAG Spec. Pub.*, Geol. Soc. Am., Boulder, CO.

Betsy Moll-Stalcup

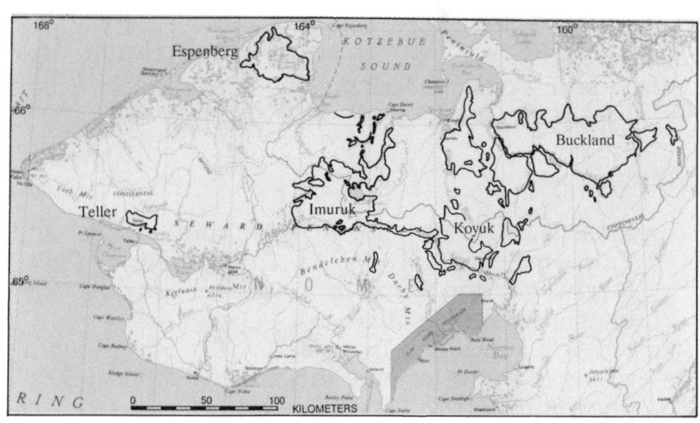

TELLER: Seward Peninsula basalt fields. Base map from USGS National Atlas; geologic boundaries from Luedke and Smith (1986).

TELLER: Volcanic necks north of Teller. Note subdued topography and the extent of erosion of cinder cones. Photograph by S. E. Swanson.

TELLER
Seward Peninsula, Alaska

Type: *Monogenetic volcano field*
Lat/Long: *65.4°N, 166.3°W*
Elevation: *175 m*
Field Size: *120 km^2*
Eruptive History: *4 dated flows, 2.8 - 2.5 Ma*
Composition: *Alkali basalt*

The Teller basalt field is a small, poorly known field of cinder cones and flows on the western Seward Peninsula. A geologic sketch map shows 30 vents, most of which are small cinder cones. Flow and cone morphology is presumably similar to the better known Imuruk basalt field. Reported ages range from 2.8 to 2.5 Ma, which are similar to the Imuruk Series of the Imuruk basalt field. According to unpublished data from S.E. Swanson the basalts are alkalic.

How to get there: *The western edge of the Teller basalt field is <5 km overland from Brevig Mission, which can be reached by scheduled air service from Nome.*

References

Hopkins, D. M., Rowland, R. W., Echols, R. E., and Valentine, P. C., 1974, An Anvilian (Early Pleistocene) marine fauna from western Seward Peninsula, Alaska: *Quat. Res. 4*, 441-470.

Turner, D. L., and Swanson, S. E., 1981, Continental rifting – a new tectonic model for geothermal exploration of the central Seward Peninsula: *in* Westcott, E., and Turner, D. L. (eds.), Geothermal Reconnaissance Survey of the Central Seward Peninsula. *Univ. Alaska Fairbanks Geophys. Inst. Rpt. UAG-R284*, 7-36.

Christopher J. Nye

ESPENBERG
Seward Peninsula, Alaska

Type: *Maars and basalt field*
Lat/Long.: *66.3°N, 164.4°W*
Elevation: *<250 m*
Field Size: *850 km*^{2}
Eruptive History: *Pleistocene shield formation; maars date from >0.12 Ma to ~7,000 yr BP*
Composition: *Basalt*

These are among the northernmost volcanoes in North America, being just south of the Arctic Circle. The basalt field contains 5 maars and 5 small shield-shaped volcanoes. The shields are of Pleistocene age and form tundra-covered hills as tall as 240 m. The maars are younger than the shields and range in age from >0.12 Ma to ~7,000 yr. The maars are 2 to 5 km in diameter and are filled by lakes with surfaces 60 to 80 m below surrounding topography and as much as 30 m deep. Tholeiitic and alkalic basalt of the maars contains up to several tens of percent of xenoliths of basement metamorphic and sedimentary rocks as well as masses of unconsolidated Quaternary sediments which must have been frozen at the time of eruption.

How to get there: *The Cape Espenberg area can be reached via chartered bush aircraft from Kotzebue or Nome. There are no roads or maintained airstrips in the area.*

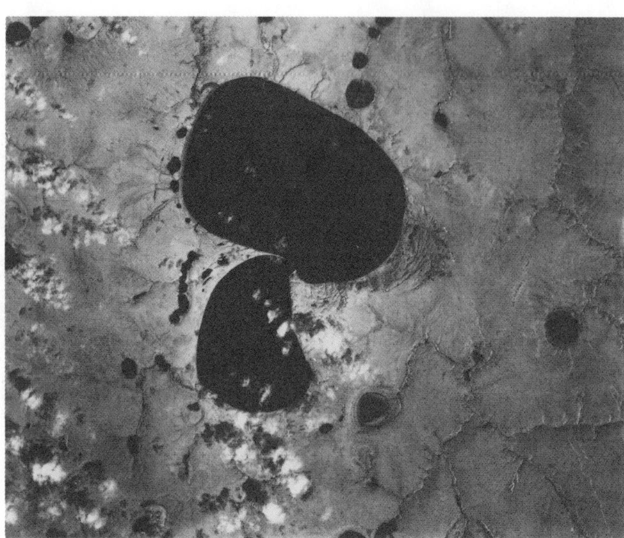

ESPENBERG: Devil Mountain Lakes, which occupy two of the five maars in the Cape Espenberg area. North at top. Larger lake is 5 km from northwest to southeast. The smaller lakes in the photograph are thaw-lakes in the permafrost. Vertical air photograph by NASA.

Reference

Hopkins, D. M., 1988, The Espenberg maars: A record of explosive volcanic activity in the Devil Mountain-Cape Espenberg area, Seward Peninsula, Alaska: *in* Schaaf, J. (ed.), The Bering Land Bridge National Preserve: An Archaeological Survey. *US Natl. Park Service Alaska Region Research Management Rpt. AR-14*, 262-321.

Christopher J. Nye

IMURUK
Seward Peninsula, Alaska

Type: *Monogenetic volcano field*
Lat.Long.: *65.6°N, 163.5°W*
Elevation: *480 m*
Field Size: *2,280 km*^{2}
Eruptive History: *Oligocene, Pliocene through Holocene basalt extrusion*
Composition: *Alkali and tholeiitic basalt*

The Imuruk basaltic volcanic field consists of ~75 vents (mostly small cones) surrounded by lava flows. Cones generally fed a single flow, and range in height from as little as 3 m to as much as 30 m. Some cones are little more than spatter ramparts at the highest parts of flows.

IMURUK: Oblique aerial view of Lost Jim cinder cone and associated lava flows. View is in the direction of Imuruk Lake. Photograph by J. Kienle.

The largest, and most recent, cone is the Holocene **Lost Jim** cone which is 30 m high and surmounted by a crater 30 m in diameter and 12 m deep. There are a few small shield-shaped volcanoes which were also probably monogenetic. Representative of these are the **Twin Calderas**, which are 120 m high with craters 500 to 750 m in diameter and 15 to 35 m deep. Flows are generally pahoehoe and have surface relief averaging only a few meters. There are a few aa and block lava flows. Lava flows are a few to a few tens of meters thick at their distal ends and up to a few tens of kilometers long. Only two flows are young enough that their geometry can be recognized clearly: the Late Pleistocene **Camille** flow extends 39 km west from its vent, and the Holocene Lost Jim flow extends 35 km west and 9 km north of its vent and covers ~230 km^2.

Four major stratigraphic units have been recognized based on degree of weathering and nature of overlying deposits. From youngest to oldest these are: the Holocene Lost Jim Flow, the Pleistocene Camille Flow, and the Pliocene and early Pleistocene Gosling and Imuruk volcanics. The Lost Jim Flow erupted 1,655 yr BP, and the **Gosling** volcanics have ages of 0.9 and 0.8 Ma. The Imuruk volcanics have been dated at 5.7 to 2.2 Ma. A fifth unit, the Kugruk volcanics, is Oligocene and probably unrelated to late Cenozoic volcanism. Ages of the Gosling and Imuruk volcanics are typical of other basalt fields in the Bering Sea region. The Imuruk volcanics have the greatest areal extent of the lava units and cover nearly 2,300 km^2. The Imuruk basalt field lies in an area of broad structural warping accompanied by normal faulting. Fault scarps as high as 30 m, and as long as 5 km, trend northwest and north-northeast cutting units as young as the Gosling volcanics.

Geochemical data are sparse but suggest that among Quaternary basalts the older, more voluminous Imuruk volcanics are dominantly subalkaline while the younger, less voluminous basalts are mainly alkalic. The latter contain ultramafic nodules.

How to get there: *The Imuruk basalt field can be reached via chartered bush aircraft from Kotzebue or Nome. There are no roads or maintained airstrips in the area.*

References

Hopkins, D. M., 1963, Geology of the Imuruk Lake area: Seward Peninsula, Alaska: *USGS Bull. 1141-C*, 101 pp.

Turner, D. L., and Swanson, S. E., 1981, Continental rifting – a new tectonic model for geothermal exploration of the central Seward Peninsula: *in* Westcott, E., and Turner, D. L. (eds.), Geothermal Reconnaissance Survey of the Central Seward Peninsula. *Univ. Alaska Fairbanks Geophys. Inst. Rpt. UAG-R284*, 7-36.

Christopher J. Nye

KOYUK–BUCKLAND
Seward Peninsula, Alaska

Type: *Monogenetic volcano field*
Lat/Long: *65.2°-66.1°N, 159.0°-162.0°W*
Elevation: *550 m*
Area: *~5,890 km^2*
Eruptive History: *Unknown*
Composition: *Basalt*

The Koyuk–Buckland monogenetic volcano field is the largest basalt field in Alaska, and among the least known. It is mostly contained within the Candle quadrangle. The field consists of basalt flows, with local aggregate thickness of 150 m, which have erupted from ~46 cinder cones and small shield volcanoes. Most of the basalts are flat-lying, but some have been gently folded. The age of the basalts is unknown, but they are thought to be correlative with the Imuruk basalt field 100 km to the west. The composition of the field is also unknown, but

the presence of ultramafic nodules suggests that at least some flows are alkaline.
How to get there: *The town of Buckland, which is accessible from Kotzebue, lies on the northern margin of the field. Transportation to other parts of the field requires chartered aircraft. Some stretches of some rivers may also be navigable by small boats.*

Reference

Patton, W. W., Jr., 1967, Regional geologic map of the Candle quadrangle, Alaska: *USGS Misc. Geol. Invest. Map I-492.*

Christopher J. Nye

PORCUPINE
Northeastern Alaska

Type: *Lava field*
Lat/Long: *67.30°N, 142.70°W*
Area: *~400 km²*
Eruptive History: *Tertiary to Quaternary?*
Composition: *Basanite?*

The Porcupine Basalt is the northernmost young volcanic field known in Alaska. The field contains no known cinder cones. The lavas occur along the Porcupine River, south of it, and along the Black River tributary. Based on nearly 30-year-old notes from Earl Brabb (USGS), the lavas apparently have no physiographic expression, and are covered by river terrace deposits. The lavas are visible along the river banks. Original estimates of the basalt age based on geomorphology suggested that it was Pliocene or Pleistocene. Recent K-Ar ages, however, show it to be ~16 Ma. The basalt is definitely not Holocene because of its cover by river terrace material. A single analysis gives 45.8% SiO_2 and 2.0% K_2O. More recent isotopic studies illustrate that the basalts differ from ones of similar age on the Seward Peninsula and Nunivak Island in being more like ocean island basalts or the least contaminated continental basalts.
How to get there: *This remote and little visited area is accessible by riverboat from Ft. Yukon. The Dalton Highway passes through parts of the field.*

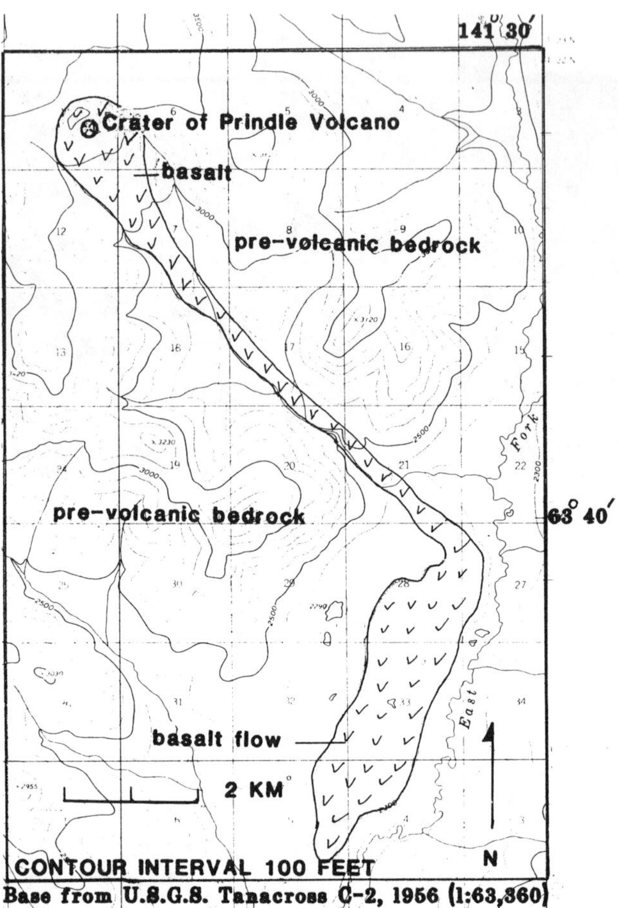

PRINDLE: Location and extent of Prindle lava flow.

References

Brabb, E. E., 1970, Preliminary geologic map of the Black River quadrangle, east-central Alaska: *USGS Misc. Invest. Map I-60.*

Plumley, P. W., and Vance, M., 1988, Porcupine River basalt field, northeast Alaska: Age, paleomagnetism and tectonic significance (abstract): *EOS Trans. Am. Geophys. Un. 69*, 1458.

Charles A. Wood

PRINDLE
Eastern Alaska

Type: *Cinder cone and lava flow*
Lat/Long: *63.72°N, 141.62°W*
Elevation: *1,000-1,250 m,*
Volcano Diameter: *0.95 km*
Eruptive History: *Pleistocene*
Composition: *Basanite*

PRINDLE: View north of Prindle volcano showing breached rim of crater on the cone's south side.

Prindle volcano is a small isolated basaltic cone in the midst of the metamorphic and granitic terrane of the Yukon-Tanana upland, east-central Alaska. The cone is ~900 to 1,000 m in diameter at its base and has a crater ~90 m deep, which is breached on the south. A lava flow extends from the breached crater ~6.4 km to the southeast, where it turns southwest and continues an additional 4.8 km in a river valley.

The cone and lava flow are vesicular basanite, rich in phenocrystal and xenocrystal olivine as well as inclusions of peridotite ranging up to 13 cm in diameter. Fragments of crystalline schists of the granulite facies with gneissose structure also occur as inclusions, but are less abundant than peridotite inclusions. Prindle volcano and its lava flow probably formed in post-early Pleistocene, but before 1,900 yr BP.

How to get there: *Prindle volcano is located in east-central Alaska ~80 km northeast of the town of Tok, and ~31 km west of the Canadian border. Prindle can be reached by charter helicopter from Tok or can be flown over by fixed wing planes chartered from Tok or Northway. Overland it is ~32 km east from the Taylor Highway and not accessible by roads or trails.*

References

Foster, H. L., Forbes, R. B., and Ragan, D. M., 1966, Granulite and peridotite inclusions from Prindle Volcano, Yukon-Tanana Upland Alaska: *USGS Prof. Pap. 500-B*, B115-B119.

Foster, H. L., 1970, A minimum age for Prindle volcano, Yukon–Tanana Upland, *in* Albert, N. R. D., and Hudson, T. (eds.), USGS in Alaska: Accomplishments during 1979: *USGS Circ. 823-B*, B37-B38.

Helen L. Foster

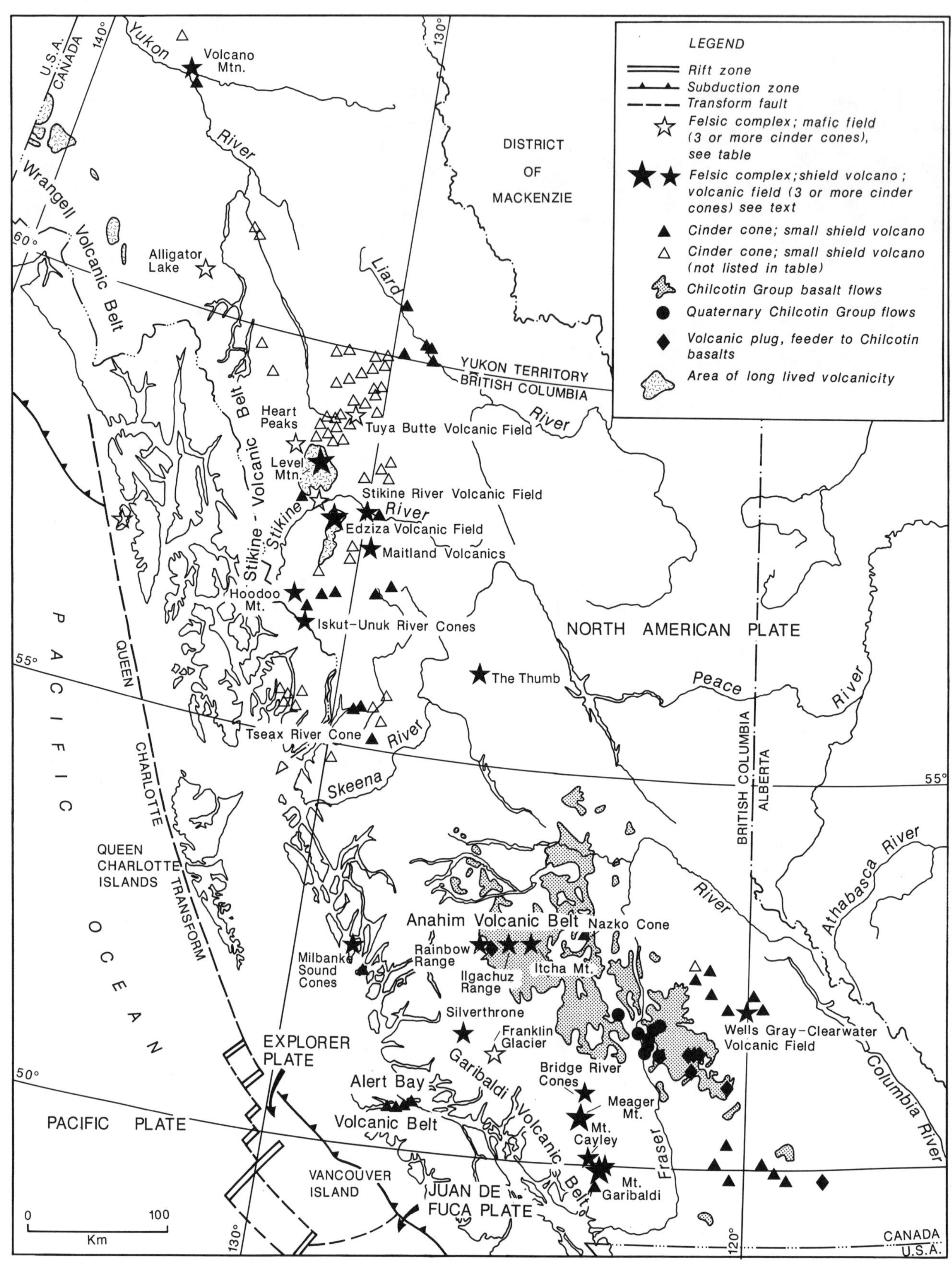

CANADA FIGURE 1: Canadian volcanic fields <5 m.y. (compiled by C. J. Hickson).

VOLCANO TECTONICS OF CANADA

J. G. Souther

Canada is commonly thought to occupy a gap in the Pacific Ring of Fire between the Cascade volcanoes of the western United States and the Aleutian volcanoes of Alaska, yet the Cordillera of British Columbia and Yukon includes more than 100 separate volcanic centers that have been active during the Quaternary. Many of these are small monogenetic cones but others, such as the vast shield and dome complex of Mount Edziza, are products of repeated eruptions that began in the Miocene and continued episodically into Recent time. Others, such as Mount Garibaldi and Meager Mountain, are northern manifestations of Cascade volcanism which differ from the High Cascade volcanoes of the United States only in their more advanced stage of glacial erosion. Because volcanic activity in western Canada was contemporaneous with the ebb and flow of Cordilleran glaciations, many of the volcanoes display icecontact features. Mount Garibaldi itself is a supraglacial volcano which erupted onto a regional ice sheet. Others, such as Hoodoo Mountain, were contained within basins thawed in the ice and assumed the flat–topped form of tuyas. Still others, such as the subglacial mounds of the Clearwater Field, were erupted under glacial ice to form piles of pillow lava and hyaloclastite. Although no historic eruptions have been recorded in Canada, native legends of lava flows destroying villages and killing fish are confirmed by radiocarbon dates of 200–250 yr BP on carbonized wood beneath basalt flows in the Stikine Volcanic Belt.

Neogene and Quaternary igneous rocks in the Canadian Cordillera are concentrated along five principal volcano-tectonic belts that are closely related to the modern tectonic regime (Fig. 1) (Souther, in press). In southwestern British Columbia, the Pemberton and Garibaldi volcanic belts and the Chilcotin Group plateau basalt define an arc–backarc pair related to subduction of the Juan de Fuca and Explorer plates under the continental margin. The Anahim Volcanic Belt, which extends from the coast near Bella Bella easterly across central British Columbia to the eastern boundary of the Intermontane Belt, is interpreted to be the trace of a mantle hot spot. The Stikine Volcanic Belt, which forms a broad zone through northwestern British Columbia and the southern Yukon, is thought to be a zone of extension developed in response to shear along the adjacent, transcurrent boundary between the continent and Pacific crust. The Wrangell Volcanic Belt, extending from Alaska into southwestern Yukon, is part of a continental arc related to convergence between the Pacific and northern North American plates. Neogene volcanic activity outside the five main volcano-tectonic belts was limited to the eruption of small basaltic centers in the Clearwater–Quesnel and McConnell Creek areas near the suture bounding the eastern edge of Quesnellia, and to the eruption of basaltic to rhyolitic volcanics and hypabyssal rocks of the Alert Bay Belt possibly coincident with the subducted boundary between the Explorer and Juan de Fuca plates.

Southern British Columbia Subduction Complexes

At least four volcanic elements are associated with the converging plate boundary between the Juan de Fuca–Explorer plate system and the continental margin. Calc-alkaline volcanic fronts of the Pemberton and Garibaldi volcanic belts and transitional (alkali-tholeiitic) plateau lavas of the Chilcotin Group form an arc–backarc pair which persisted through Neogene and Quaternary time (Fig. 1). The Alert Bay Volcanic Belt on Vancouver Island is probably a zone of basaltic and felsic volcanism related to the subducted Juan de Fuca–Explorer plate edge.

The Pemberton Volcanic Belt (Souther, 1975, 1977; Berman and Armstrong, 1980) extends west-northwest from south-central British Columbia to Queen Charlotte Islands. In the south it is defined by a group of epizonal plutons and a few erosional remnants of eruptive rock. Farther north it includes the intracaldera complexes of Mount Silverthrone and Franklin Glacier and the Masset Formation on Queen Charlotte Islands. Although widely separated from one another, all of the Pemberton Belt rocks are of similar age (late Miocene) and calc-alkaline affinity, and are believed to be the products of arc volcanism related to subduction of the Farallon Plate. By late Pliocene time the Farallon Plate had been greatly reduced in size, and the remnants reorganized to form the present Juan de Fuca–Explorer system. With this change in plate configuration volcanic activity shifted to the more northerly trending Garibaldi Volcanic Belt.

The Garibaldi Volcanic Belt (Mathews, 1958; Green et al., 1988) is a northern extension of the Cascade Volcanic Belt in the western United States. It is offset to the west of the main Cascade trend and is a composite of at least three en echelon, north-trending segments, referred to here as the southern, central, and northern segments. The Garibaldi Volcanic Belt intersects the older Pemberton Volcanic Belt at a low angle near Meager Mountain where Garibaldi Group lavas rest on uplifted and deeply eroded remnants of Pemberton Volcanic Belt subvolcanic plutons. North of Meager Mountain the Garibaldi and Pemberton volcanic belts appear to merge into a single belt. Mount Silverthrone, farther to the northwest, was episodically active during both Pemberton and Garibaldi stages of volcanism.

The principal volcanoes in the southern segment of the Garibaldi Volcanic Belt are Mount Garibaldi, Mount Price, and the Black Tusk. Black Tusk, the oldest volcano in the southern segment, is a composite pile formed during two distinct stages of magmatic activity, the first between 1.1 and 1.3 Ma and the second between 0.17 and 0.21 Ma. Mount Garibaldi, a moderately dissected composite Pelean cone, was built during the waning stages of the last major glaciation (Mathews, 1952, 1958; Green, 1981). Mount Price, a composite volcano much smaller than Mount Garibaldi, formed during three distinct periods of activity beginning at 1.2 Ma and culminating with the eruption of Clinker Peak ~0.3 Ma (Green, 1981). In addition to the large, central andesite-dacite volcanoes, the southern Garibaldi Belt is the site of numerous remnants of basalt and basaltic andesite flows and pyroclastic rocks. These include valley flows interbedded with till containing wood which yielded a ^{14}C age of ~34,000 yr.

The central segment of the Garibaldi Belt (Souther, 1980) is defined by a group of eight volcanoes on the height of land east of Squamish River, and by remnants of basaltic flows preserved in the adjacent Squamish valley. Mount Cayley, the largest and most long-lived center (3.8 to 0.31 Ma) is a multiple plug dome of dacite and minor rhyodacite from which most of the original, outer cone of pyroclastic material has been eroded away. Mount Fee, a narrow elliptical spine of rhyodacite ~1 km long and 0.25 km across, rises 150 m above the ridge. Complete denudation of the central spine as well as the absence of till under the Mount Fee flows suggest a preglacial age. The other volcanoes of the central Garibaldi Belt (Ember Ridge, Pali Dome, Cauldron Dome, Slag Hill, and Crucible Dome) are intraglacial, tuya-like forms with oversteepened, ice-contact margins.

The northern segment of the Garibaldi Volcanic Belt includes the Meager Mountain complex and several remnants of basaltic and andesitic piles which extend north of Meager Mountain almost to the Interior Plateau. Meager Mountain (Read, 1978; Lewis and Souther, 1978) is a complex of at least four overlapping composite dacite to rhyodacite volcanoes that become progressively younger from south to north, ranging in age from ~2 Ma to ~2,490 yr BP. North of Meager Mountain, the Salal Glacier Volcanic Complex (Lawrence et al., 1984) and the Bridge River cones (Souther, this volume) comprise remnants of both andesitic and alkali basalt cones and flows. These range in age from ~1 to 0.5 Ma and commonly display ice-contact features. The alkaline affinity of many of these lavas contrasts with the calc-alkaline nature of the larger central volcanoes of the Garibaldi Volcanic Belt. Their proximity to the volcanic front suggests a discontinuity in the subducted plate, possibly a subducted plate edge analogous to that proposed for the Alert Bay Volcanic Belt, or magma generation at the northern end of the active arc (Rogers, 1985).

The Alert Bay Volcanic Belt (Armstrong et al., 1985) extends from Brooks Peninsula northeastward across Vancouver Island to Port McNeil (Fig. 1). It encompasses several separate remnants of late Neogene volcanic piles and related intrusions ranging in composition from basalt to rhyolite and in age from ~8 Ma in the west to ~3.5 Ma elsewhere. Major element analyses of Alert Bay volcanic and hypabyssal rocks suggest two different basalt-andesite-dacite-rhyolite suites with divergent fractionation trends. The first coincides with the typical calc-alkaline, Cascade trend, whereas the other is more alkaline and more Fe-enriched following a trend which straddles the calc-alkaline-tholeiite boundary.

The western end of the Alert Bay Volcanic Belt is now ~80 km northeast of the Nootka Fault Zone, which separates the Explorer and Juan de Fuca plates. However, at the time of its formation the volcanic belt may have been coincident with the subducted plate boundary. Also, the timing of volcanism corresponds to shifts of plate motion and changes in the locus of volcanism along the Pemberton and Garibaldi volcanic fronts. This brief interval of plate motion adjustment at ~3.5 Ma may have triggered the generation of basaltic magma along the descending plate edge.

The Chilcotin Group (Bevier, 1983; Mathews 1989) of flat-lying, Mio-Pliocene basalt flows covers 50,000 km^2 of British Columbia's interior plateau. They consist of thin, crudely columnar flows which are inferred to form a series of coalesced, low shield volcanoes erupted from central vents. Anahim Peak, near the eastern flank of the Rainbow Range, was one source, and several gabbroic and basaltic plugs (Farquharson, 1973) which cut the Chilcotin flows are also presumed to be vents. These centers define a northwest trend along the axis of the volcanic plateau, parallel with the Pemberton Volcanic Belt.

Chilcotin Group basalt is coeval with, and ~150 km inland from, arc volcanoes of the Pemberton–Garibaldi volcanic front. Most of the Chilcotin basalt is coeval with volcanic and subvolcanic plutons of the Pemberton Volcanic Belt, but some of the youngest flows are coeval with early stages of Garibaldi volcanism. They are mainly olivine-bearing, transitional basalt believed to have been generated by partial melting in the upper mantle, due to asthenospheric upwelling in a backarc setting, above the subducting Juan de Fuca Plate. Silicic tuff, interbedded with the basalt, probably originated from calc-alkaline arc volcanoes in the Pemberton Volcanic Belt and was preserved between successive flows of basalt in the Chilcotin, backarc lava plain.

Central and Northern British Columbia Plume and Rift Complexes

Two well-defined belts of Neogene volcanoes extend into the continent from near the transcurrent boundary between the North American and

Pacific plates (Fig. 1). The east-trending Anahim Volcanic Belt is defined by a chain of Neogene volcanoes, plutons and dike swarms which extend across the Coast Belt and Interior Plateau of central British Columbia. The Stikine Volcanic Belt of northern British Columbia and southern Yukon encompasses a series of en echelon, north-trending segments which occupy a broad zone curving northwestward, subparallel with the continental margin. Volcanoes in both of these belts are distinctly alkaline.

The Anahim Volcanic Belt (Souther, 1977, 1986; Bevier et al., 1979) extends from coastal British Columbia across the Coast Mountains into the Interior Plateau. Its western end is defined by alkaline intrusive and comagmatic volcanic rocks of the Bella Bella–King Island complex, exposed in fjords and islands of the western Coast Mountains (Souther, 1986). The central part of the belt consists of three composite, peralkaline shield volcanoes, the Rainbow, Ilgachuz, and Itcha ranges. These moderately dissected shields are superimposed on the northern end of the area flooded by Chilcotin plateau lava, and distal flows at the margins of the shields merge imperceptibly with flat-lying flows of the Chilcotin Group. Unlike the Chilcotin basalt, which is not associated with any felsic derivatives, the volcanoes of the central Anahim Volcanic Belt are markedly bimodal, comprising a mixed assemblage of basalt and peralkaline silicic rocks. A cluster of postglacial basaltic cones in the Nazko area west of Quesnel forms the youngest and most easterly part of the belt. Tephra from the Nazko center is intercalated with peat having a ^{14}C date of ~7,200 yr BP (Souther et al., 1987).

A systematic decrease in the age of volcanism from west to east along the Anahim Belt suggests that it may be the trace of a mantle hot spot over which the continent moved during the past ten to fifteen million years (Bevier et al., 1979; Souther, 1986; Hickson, 1986).

The Stikine Volcanic Belt includes scores of relatively small pyroclastic cones and related lava fields, and three large, compositionally diverse volcanic complexes: Level Mountain (Hamilton, 1981), the Edziza–Spectrum Complex (Souther, 1989), and Hoodoo Mountain. In the south the belt is relatively narrow and traverses diagonally across the northwesterly structural trend of the Coast Mountains. Farther north it is less clearly defined, forming a broad arch that swings westward through the central Yukon. Volcanoes within the Stikine Volcanic Belt are disposed along several short, northerly trending en-echelon segments which, in the southern part of the belt, are clearly associated with north-trending extensional structures including synvolcanic grabens and half-grabens. This extensional regime is believed to be related to incipient gash fractures formed in response to dextral plate motion along transcurrent faults to the west.

The compositions of lavas in the Stikine Volcanic Belt are consistent with an environment of continental extension. Mantle-derived alkali olivine basalt, lesser hawaiite, and basanite form the large shield volcanoes and small pyroclastic cones. Many of them contain inclusions of lherzolite. Felsic rocks which form the large central volcanoes comprise mostly trachyte, pantellerite, and comendite (Souther, 1989). These peralkaline end members are believed to have formed by fractionation of primary alkali basalt magma in crustal reservoirs. A region of crustal extension such as the Stikine Volcanic Belt would favor the development of high-level reservoirs of sufficient size and thermal capacity to sustain prolonged fractionation.

Northwestern British Columbia and Southwestern Yukon Subduction-Related Arc Volcanics

The Wrangell Volcanic Belt (Souther and Stanciu, 1975) comprises scattered remnants of upper Tertiary subaerial lavas and pyroclastic rocks which are preserved along the entire eastern fringe of the Saint Elias Mountains. This calc-alkaline assemblage is believed to be the product of arc volcanism along a volcanic front related to the convergence of Pacific crust with the northern North American continent. Over large areas extrusive rocks lie in flat undisturbed piles on a Tertiary surface of moderate relief. Locally, however, strata of the same age have been affected by a late pulse of tectonism, during which they were faulted, contorted into tight symmetrical folds, or overridden by pre–Tertiary basement rocks along southwesterly dipping thrust faults. Considerable recent uplift, accompanied by rapid erosion, has reduced once vast areas of upper Tertiary volcanic rocks to small isolated remnants.

The Wrangell Volcanic Belt of northern Canada is coextensive with active Quaternary volcanoes in adjacent Alaska. Although no centers younger than late Miocene are known in Canada, the eruption of rhyolite pumice, White River Ash, from a vent near the head of Klutlan Glacier 24 km west of the Alaska–Yukon border, blanketed large areas of northwestern Canada with tephra which yielded ^{14}C dates of 1,990 to 1,200 yr BP (Lerbekmo and Campbell, 1969).

Volcanic Rocks of East-Central British Columbia

Late Neogene and Quaternary basaltic cones, necks, and associated intravalley lava flows are distributed in the highlands east of the Interior Plateau. These centers, which include the Clearwater–Quesnel and McConnell Creek volcanic provinces, are outside the main volcanic belts, and their tectonic affiliation is not clear. They lie near and parallel to the eastern edge of Quesnel Terrane and may be a manifestation of small magma batches generated in a zone of incipient extension related to normal faulting (Hickson, 1986).

References

Armstrong, R. L., Muller, J. E., Harakal, J. E., and Muehlenbachs, K.,1985, The Neogene Alert Bay Volcanic Belt of northern Vancouver Island, Canada: descending-plate-edge volcanism in the arc-trench gap: *J. Volc. Geotherm. Res. 26*, 75-97.

Berman, R. G., and Armstrong, R. L., 1980, Geology of the Coquihalla Volcanic Complex, southwestern British Columbia: *Can. J. Earth Sci. 17*, 985-995.

Bevier, M. L., 1983, Implications of chemical and isotopic composition for petrogenesis of Chilcotin Group basalts, British Columbia: *J. Petr. 24*, 207-226.

Bevier, M. L., Armstrong, R. L., and Souther, J. G., 1979, Miocene peralkaline volcanism in west-central British Columbia – its temporal and plate-tectonic setting: *Geology 7*, 389-392.

Farquharson, R. B., 1973, The petrology of late Tertiary dolerite plugs in the South Cariboo region, British Columbia: *Can. J. Earth Sci. 10*, 205-225.

Green, N. L., 1981, Geology and petrology of Quaternary volcanic rocks, Garibaldi Lake area, southwestern British Columbia: *Geol. Soc. Am. Bull. 92*, 697-702, and 1359-1470.

Green, N. L., Armstrong, R. L., Harakal, J. E., Souther, J. G., and Read, P. B., 1988, Eruptive history and K-Ar geochronology of the Garibaldi volcanic belt, southwestern British Columbia: *Geol. Soc. Am. Bull. 100*, 563-579.

Hamilton, T. S., 1981, Late Cenozoic alkaline volcanics of the Level Mountain Range, northwestern British Columbia: Geology, petrology and paleomagnetism: Ph.D. Thesis, University of Alberta, Edmonton.

Hickson, C. J.,1986, Quaternary volcanics of the Wells Gray–Clearwater area, east central British Columbia: Ph.D. Thesis, University of British Columbia, Vancouver.

Lawrence, R. B., Armstrong, R. L., and Berman, R. G., 1984, Garibaldi Group volcanic rocks of the Salal Creek area, southwestern British Columbia: alkaline lavas on the fringe of the predominantly calc-alkaline Garibaldi (Cascade) volcanic arc: *J. Volc. Geotherm. Res. 21*, 255-276.

Lerbekmo, J. F., and Campbell, F. A., 1969, Distribution, composition, and source of the White River Ash, Yukon Territory: *Can. J. Earth Sci. 6*, 109-116.

Lewis, T. J., and Souther, J. G., 1978, Meager Mountain, B.C. – a possible geothermal energy resource: Earth Physics Branch, *Geothermal Ser. No. 9*.

Mathews, H. W., 1952, Mount Garibaldi, a supraglacial Pleistocene volcano in southwestern British Columbia: *Am. J. Sci. 250*, 81-103.

Mathews, H. W., 1958, Geology of the Mount Garibaldi map-area, southwestern British Columbia, Canada: *Geol. Soc. Am. Bull. 69*, 179-198.

Mathews, H. W., 1989, Neogene Chilcotin Basalts in south-central British Columbia: geology, ages and geomorphic history; *Can. J. Earth Sci.*, in press.

Read, P. B., 1978, Geology, Meager Creek geothermal area, British Columbia: Geol. Surv. Can., *Open-file Rpt. 603*.

Rogers, G. C., 1985, Variation in Cascade volcanism with margin orientation: *Geology 13*, 495-498.

Souther, J. G., 1975, Geothermal Potential of Western Canada, *in Proceedings*, 2nd United Nations Symposium on the Development and Use of Geothermal Resources, San Francisco, May 1975, 259-267.

Souther, J. G., 1977, Volcanism and tectonic environments in the Canadian Cordillera – a second look: *in* Baragar,W. R. A., Coleman, L. C., and Hall, J. M. (eds.), *Volcanic Regimes in Canada*, Geol. Assoc. Canada, *Spec. Pap. 16*, 3-24.

Souther, J. G., 1980, Geothermal reconnaissance in the central Garibaldi Belt, British Columbia: *in* Current Research, Part A, *Geol. Surv. Can. Pap. 80-1A*, 1-11.

Souther, J. G., 1986, The western Anahim Belt, root-zone of a peralkaline magma system: *Can. J. Earth Sci. 23*, 895-908.

Souther, J. G., in press, The late Cenozoic Mount Edziza Volcanic Complex, British Columbia: *Geol. Surv. Can. Mem. 420*.

Souther, J. G., in press, Igneous Assemblages, *in* Chapter 10, Neogene Assemblages, in Gabrielse, H., and Yorath, C. J. (eds.), The Cordilleran Orogen: Canada, Geol. Surv. Can., *Geology of Canada, no. 4.* (also Geol. Soc. Am., *The Geology of North America, no. G-2*).

Souther, J. G., and Stanciu, C., 1975, Operation Saint Elias, Yukon Territory: Tertiary volcanic rocks, *in* Report of Activities, Part A, *Geol. Surv. Can. Pap. 75-1A*, 63-70.

Souther, J. G., Clague, J. J., and Mathews, H. W., 1987, Nazko Cone, a Quaternary volcano in the eastern Anahim Belt: *Can. J. Earth Sci 24*, 2477-2485.

Canadian Cordillera: Volcano Vent Map and Table

C. J. Hickson

Volcanoes in Canada fall into three categories. Those that have been studied at least at a reconnaissance level are described individually in this chapter. Volcanic centers that have received only cursory attention are shown on Figure 1 as open stars and are listed in the accompanying table. Cones marked on Figure 1, but not listed in the table, have never been studied. Their composition and distribution are taken from 1:250,000 map sheets.

J. G. Souther (Geological Survey of Canada) and W. H. Mathews (University of British Columbia) have both contributed extensively to this compilation. Their long association and interest in volcanic rocks of the Canadian Cordillera have sparked many studies, but much more needs to be done. This table is intended to provide a starting point for future studies of poorly known volcanic centers in the Canadian Cordillera.

References for Table 1

Armstrong, R. L., Muller, J. E., Harakal, J. E., and Muehlenbachs, K., 1985, The Neogene Alert Bay Volcanic Belt of northern Vancouver Island, Canada: Descending-plate-edge volcanism in the arc-trench gap: *J. Volc. Geotherm. Res. 26*, 75-97.

Bevier, M. L., 1983, Regional stratigraphy and age of Chilcotin Group basalts, south-central British Columbia, Canada: *Can. J. Earth Sci. 20*, 515-524.

Casey, J. J., and Scarfe, C. M., 1980. Summary of the petrology of the Heart Peaks volcanic centre, northwestern British Columbia: Current Research, Part A, *Geol. Sur. Can. Pap. 80-1A*, 356.

Carter, N. C., 1974. Geology and geochronology of porphyry copper and molybdenum deposits in west-central British Columbia: Ph.D. Thesis, University of British Columbia, Vancouver.

Church, B. N., 1980, Geology of the Kelowna Tertiary outliers: Geol. Branch, Brit. Col. Min. Energy, Mines Petrol. Res., *Prelim. map 39.*

Church, B. N., and Suesser, U., 1983, Geology and magnetostratigraphy of Miocene basalts of the Okanagan Highland, British Columbia: Geol. Branch, Brit. Col. Min. Energy, Mines Petrol. Res., *Pap. 1983-1*, 32-36.

Eiche, G., 1986, Petrology of Quaternary alkaline lavas from the Alligator Lake Volcanic Complex, Yukon Territory, Canada: U M.Sc. Thesis, McGill Univ., Montreal.

Farquharson R. B., and Stipp, J. J., 1969, Potassium-argon ages of dolerite plugs in South Cariboo region, British Columbia: *Can. J. Earth Sci. 6*, 1468-1470.

Green, N. L., Armstrong, R. L., Harakal, J. E., Souther, J. G., and Read, P. B., 1988, Eruptive history and K-Ar geochronology of the Garibaldi volcanic belt, southwestern British Columbia: *Geol. Soc. Am. Bull. 100*, 563-579.

Jackson, L. 1989, Pleistocene subglacial volcanism near Fort Selkirk, Yukon Territory: Current Research, Part H, *Geol. Sur. Can. Pap. 89-1H* (in press).

Klassen, R. W.,1986, Tertiary-Pleistocene stratigraphy of the Liard Plain, southeastern Yukon Territory: *Geol. Sur. Can. Pap. 86-17*, 16.

Littlejohn, A. L., and Greenwood, H. J., 1974, Lherzolite nodules in basalts from British Columbia, Canada: *Can. J. Earth Sci. 11*, 1288-1308.

Mathews, H. W., 1988, Neogene geology of the Okanagan Highlands, British Columbia: *Can. J. Earth Sci. 25*, 725-731.

Mathews, H. W., 1989, Neogene Chilcotin Basalts in south-central British Columbia: geology, ages and geomorphic history: *Can. J. Earth Sci.* (in press).

Stevens, R. D., Delabio, R. N., Lachance, G. R., 1982, Age determinations and geological studies; K-Ar Isotopic Ages, Report 15: *Geol. Sur. Can. Pap. 81-2*, 15.

Sutherland Brown, A., 1969, Aiyansh lava flow British Columbia: *Can. J. Earth Sci. 6*, 1460-1468.

Table 1. *Additional Volcanic Areas in Canada*

Rock Type	Lat N/Long W	NTS	Age	Comments	Reference
GARIBALDI VOLCANIC BELT					
Franklin Glacier					
Andesite to dacite	51°20' 125°24'	92M	Plio	Cauldron subsidence	Green et al., 1988
ALERT BAY VOLCANIC BELT					
Rhyolite	50°28' 127°28'	92L	3.0	N-NE Jeune Landing	Armstrong et al., 1985
Andesite	50°30' 127°15'	92L	3.0	Twin Peaks Mountain	Armstrong et al., 1985
Felsite, K-poor Dacite	50°34' 127°10'	92L	2.5	Cluxewe Mountain	Armstrong et al., 1985
Felsite, K-poor Dacite	50°36' 127°02'	92L	3.7	Haddington Islan	Armstrong et al., 1985
CHILCOTIN GROUP BASALTS					
Olivine basalt	49°53' 118°32'	82E	2.5	Lightning Pk (nodules)	Mathews, 1988
Basalt/Gabbro plug	51°09' 120°26'	92P	LM	Skoatl Point	Farquharson & Stipp, 1969
Basalt/Gabbro plug	51°21' 121°11'	92P	LM	Tin Cup Mt	Farquharson & Stipp, 1969
Basalt/Gabbro plug	51°29' 121°23'	92P	7.8	Mt. Begbie	Farquharson & Stipp, 1969
Basalt/Gabbro plug	51°33' 121°08'	92P	8.3	Forestry Hill	Farquharson & Stipp, 1969
Basalt/Gabbro plug	51°33' 121°13'	92P	6.0	Lone Butte	Farquharson & Stipp, 1969
Basalt/Gabbro plug	52°34' 125°37'	93C	6.7	Anahim Peak	Bevier, 1983
Quaternary Chilcotin Flows					
Basalt	49°47' 119°04'	82E	0.24	E. of Hydraulic Lake	Mathews, 1988
Basalt	51°23' 121°58'	92P	0.72	Big Bar 82-1	Mathews, 1989
Basalt	51°25' 122°12'	92O	1.14	Crows Bar	Mathews, in prep.
Basalt	51°25' 122°15'	92O	0.78	Browns Lake	Mathews, 1989
Basalt	51°35' 122°15'	92O	1.1	Dog Creek	Mathews, 1989
Basalt	51°48' 122°09'	92O	0.70	Alixton Creek	Mathews, 1989
Basalt	51°56' 122°45'	92O	1.0	Thaddeus Lake	Mathews, 1989
VALLEY BASALTS, ISOLATED CONES and FLOWS					
Basalt	49°48' 120°32'	92H	0.04	Missezula Lake	Mathews, in prep.
Basalt	49°52' 119°32'	82E	0.81	W. Kelowna	Mathews, 1988
Olivine Basalt	49°57' 119°33'	82E	0.76	Near Kelowna	Church and Suesser, 1983
Basalt	50°07' 120°30	92I	0.04	Quilchena Creek	Mathews, in prep.
Basalt	50°81' 120°42'	92I	0.6	Chester Ranch, Nicola	Church, 1980
Basalt	52°07' 120°55'	93A	Quat	Boss Mt. (nodules)	Littlejohn & Greenwood, 1974
Basalt	52°03' 121°10'	93A	Quat	Jacques L. (nodules)	Littlejohn & Greenwood, 1974
Basalt	52°39' 120°59'	93A	0.17	Quesnel Lake	Mathews, in prep.
STIKINE VOLCANIC BELT					
Basalt	55°25' 129°25'	103P	1.7	Alice Arm area	Carter, 1974
Basalt	55°28' 129°20'	103P	1.1	Alice Arm area	Carter, 1974
Tseax River Cone					
Basalt	55°07' 128°54'	103O	Holo	S. Nass River	Sutherland Brown, 1969
Basalt	56°52' 129°37'	104A	Quat	Bowser Basin cone	R.F. Gerath pers. comm.
Basalt	56°53' 12°22'	104A	Quat	Bowser Basin cone	R.F. Gerath pers. comm.
Basalt	56°54' 129°22'	104A	1.6	Bowser Basin cone	G.E. Eisbacher unpub. data
Basalt	57°02' 129°00'	104A	Quat	Bowser Basin cone	R.F. Gerath pers. comm.
Stikine River Volcanic Field					
Basanite	57°53' 129°49'	104H	4.6	Near Thatue Mt	Stevens et al., 1982
Porph. basalt	57°58' 129°59'	104H	4.9	Near Mount Tsaybahe	Stevens et al., 1982
Porph. basalt	58°01' 129°46'	104I	5.3	Near Stikine River	Stevens et al., 1982
Porph. basalt	58°04' 129°53'	104I	5.4	Near Stikine River	Stevens et al., 1982
Porph. basalt	58°05' 130°00'	104I	4.8	Near Stikine River	Stevens et al., 1982
Tuya Butte Volcanic Field					
Basalt (?)	60°05' 128°54'	105A	0.604	Liard Plain	Klassen, 1986
Basalt (?)	60°09' 129°41'	105A	0.232	Liard Plain	Klassen, 1986
Basalt	60°14' 129°06'	105A	0.545	Liard Plain	Klassen, 1986
Basalt (?)	60°14' 128°49'	105A	0.765	Liard Plain	Klassen, 1986
Basalt	60°15' 129°0.7'	105A	Pleist	Rancheria Ck	J. Westgate pers. comm.
Alligator Lake Volcanic Complex					
Alk. Ol. basalt	60°25' 135°25'	105D	Quat	lherzolite xenoliths	Eiche, 1986
Basalt	62°45' 137°16'	115I	Pleist	Ne Ch'e Ddhawa Tuya	Jackson, 1989
Basalt	62°49' 137°27'	115I	Pleist	Selkirk Volcanics	J. Westgate pers.comm.

[a]Geographic place names can be found on the figure; rock compositions, in most cases, are not known precisely.
[b]Geographic place names in this column can be found by referring to the corresponding 1:250,000 NTS map sheet or reference. In the age column LM = Lower Miocene.

VOLCANOES OF CANADA

Volcano Mountain, 118
Tuya Butte, 119
Level Mountain, 121
Edziza, 124
Maitland, 126
Hoodoo, 127
Iskut–Unuk River Cones, 128
The Thumb, 129
Milbanke Sound Cones, 130
Rainbow Range, 131
Ilgachuz Range, 132
Itcha Range, 134
Nazko, 135
Chilcotin Basalt, 136
Wells Gray–Clearwater, 137
Silverthrone, 138
Bridge River Cones, 139
Meager Mountain, 141
Cayley, 142
Garibaldi Lake, 143
Garibaldi, 144

VOLCANO MOUNTAIN
Yukon, Canada

Type: *Monogenetic volcanic field*
Lat/Long: *62.93°N, 137.38°W*
Elevation: *1,238 m*
Relief: *325 m*
Volcano Diameter: *1.5 km*
Eruptive History: *1 Ma to historic?*
Composition: *Olivine nephelinite*

Volcano Mountain is the youngest vent of the **Fort Selkirk** alkaline volcanic complex at the junction of the Yukon and Pelly Rivers in the Central Yukon. Two styles of volcanism occurred in the Fort Selkirk complex. One was characterized by the effusive eruption of relatively fluid magma to produce two sequences of smooth-topped, valley-filling lavas, which blocked the Yukon River. These valley-filling lavas evolved in composition from basanite to transitional alkaline olivine basalt with time. The second style of volcanism was characterized by more explosive eruptions of relatively viscous olivine nephelinite magma, which constructed three pyroclastic to composite cones with aprons of aa lava flows. The olivine nephelinite eruptions span the history of the Fort Selkirk complex. The Fort Selkirk olivine nephelinite vent predates the Pelly Sequence valley-filling lavas, **Wootton's Cone** postdates the Wolverine Sequence valley-filling lavas, and Volcano Mountain ends volcanic activity at Fort Selkirk.

The olivine nephelinite, basanite, and alkaline olivine basalt lavas represent three distinct alkaline magma series that have evolved along diverging fractionation paths. They cannot be related by low-pressure crystal fractionation, and systematic isotopic differences make it difficult to derive them via variable degrees of melting of a common mantle source. When the effects of differential olivine fractionation are ignored, however, the compositional spectrum of the Fort Selkirk lavas defines a binary mixing line between transitional alkaline basalt and olivine nephelinite. A population gap along this mixing line located between the compositions of the nephelinitic and basanitic lavas coincides with the compositions of amphibole-clinopyroxene-garnet assemblages observed in mantle xenoliths. This compositional gap may

VOLCANO MOUNTAIN: Looking northeast from the end of the trap line trail leading from the old Dawson–Whitehorse stage road.

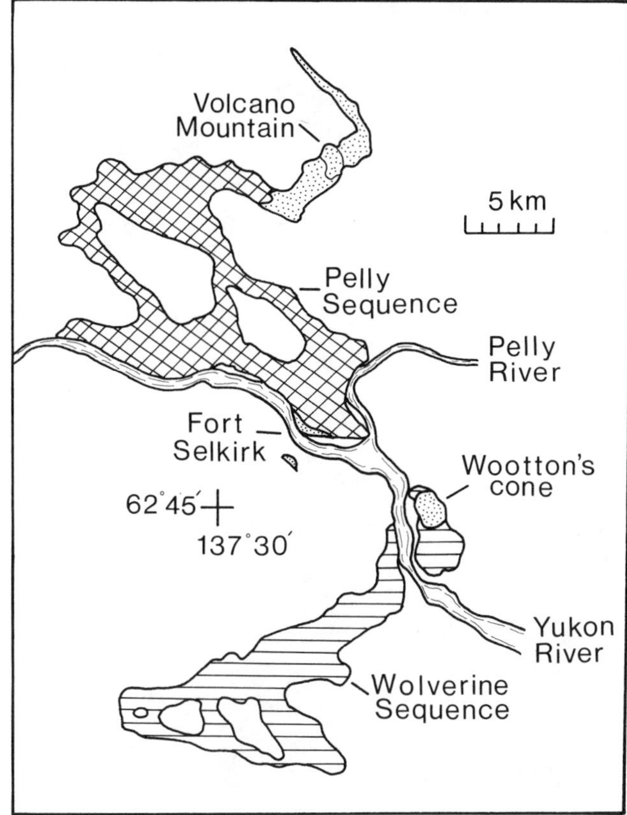

VOLCANO MOUNTAIN: Simplified geologic map of the Fort Selkirk volcanic complex. The stippled pattern represents the olivine nephelinite vents, the lined pattern the Wolverine Sequence of valley-filling lavas, and the crossed line pattern the Pelly Sequence of valley-filling lavas.

represent a thermal divide separating two minimum melt compositions in an amphibole pyroxenite-veined lithosphere source.

How to get there: *Volcano Mountain can be reached by foot from the Pelly Ranch by following the old Whitehorse–Dawson stage road for 8 km and then a trap line trail leading to the northeast for 2.5 km. These trails are indicated on the 1:50,000 NTS Map 115 I/14. Pelly Ranch is accessible by car via an unmarked dirt road which leaves the present Whitehorse–Dawson Highway from the north side of the Pelly Crossing bridge.*

Reference

Francis, D., and Ludden, J., 1989, The petrogenesis of alkaline magma series at Fort Selkirk, Yukon, Canada: submitted to *J. Petrol.*

Don Francis

TUYA BUTTE
Canada

Type: *Subglacial volcanic field*
Lat/Long: *59°N, 131°W*
Elevation: *1,200 to 2,290 m*
Eruptive History: *Late Pleistocene*
Composition: *Basalt*

Ten flat-topped volcanoes and eight volcanic cones, all attributed to subglacial eruption, occur in north-central British Columbia. These basaltic mountains, collectively called "tuyas," are part of the Stikine Volcanic Belt associated with an array of north–south trending normal faults dated as Tertiary or younger. During the Pleistocene this region was completely covered by continental glaciers, perhaps several times. The most recent deglaciation occurred ~10,000 years ago. The bulk of the preserved eruptive products were formed beneath glacial ice, but several small post-glacial flows have also been identified in the area.

The best preserved of the ten flat-topped volcanoes in the region is Tuya Butte, which is 4 × 2.5 km at its base and rises 390 m above the adjacent lake. The slopes are composed of crudely layered pillow fragments and sideromelane tuff which has been partially altered to palagonite, zeolites, and clay. Tuya Butte is capped by columnar-jointed olivine basalt flows ranging

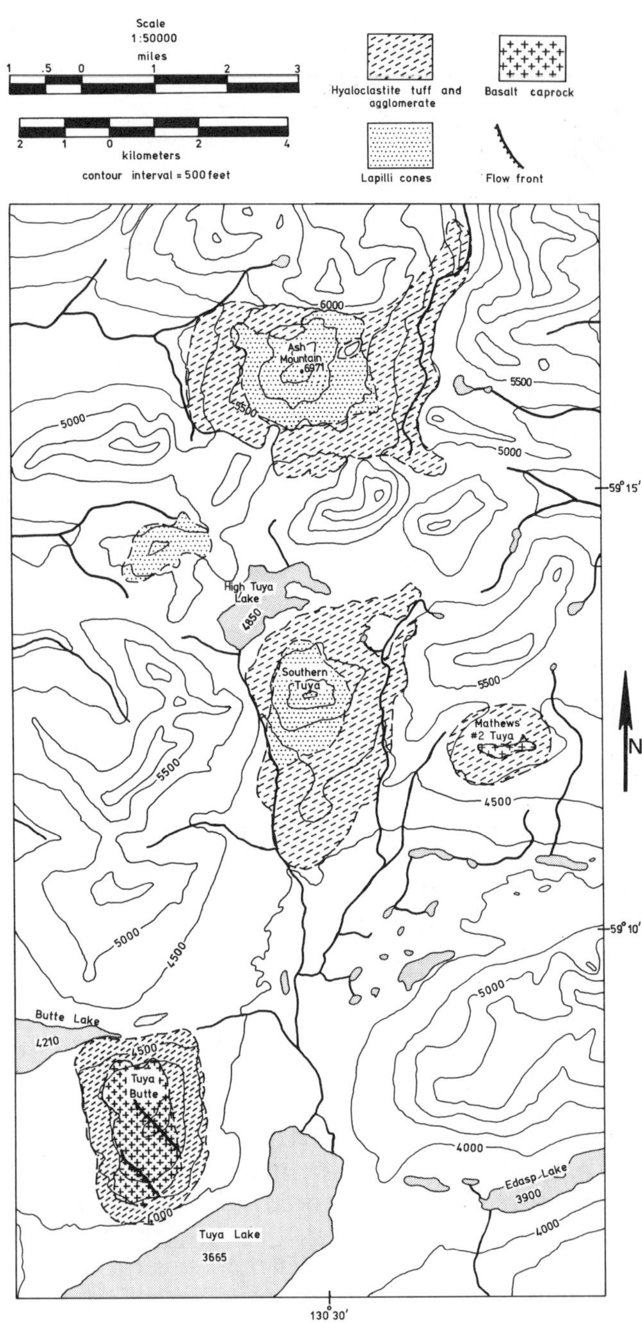

TUYA BUTTE: Map showing hyaloclastite tuff and agglomerate (diagonal dash pattern), basalt caprock (pluses), and lapilli cones (dots). Flow fronts indicated on Tuya Butte by heavy lines. North up, contour interval 500 ft., map width 13 km.

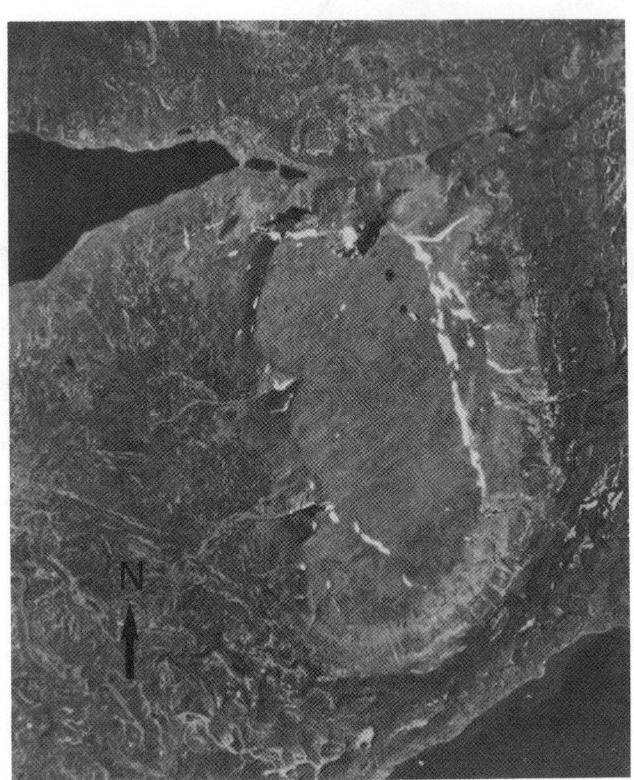

TUYA BUTTE: Vertical air photograph of Tuya Butte.

TUYA BUTTE: Ash Mountain seen from the south.

from 5 to 40 m thick. Following eruption, the mountain was completely covered by glacial ice as indicated by numerous erratics on the caprock and a large cirque at the north end.

The largest of the eight conical tuyas, **Ash** Mountain, rises 646 m above nearby High Tuya Lake. The upper cone, 3 km across, rests on a 4 × 6-km-wide base of pillow basalts with minor interbedded tuff. The upper conical unit is a complex composite of palagonized tuff, ash, and black cinders. It is deeply eroded, and much of the ash and cinder has been reworked by streams. Ash Mountain shows no evidence of a crater or a caprock, although the cone is buttressed by long northeast–southwest trending dikes.

How to get there: *The tuya region is totally roadless. Access is by foot or floatplane (~80 km from the town of Dease Lake). High Tuya Lake and Butte Lake provide the most convenient landing sites.*

References

Allen, C. C., Jercinovic, M. J., and Allen, J. S. B., 1982, Subglacial volcanism in north-central British Columbia and Iceland: *J. Geol. 90*, 699-715.

Mathews, W. H., 1947, "Tuyas," flat-topped volcanoes in northern British Columbia: *Am. J. Sci. 245*, 560-570.

Carl Allen

LEVEL MOUNTAIN
Canada

Type: *Composite central volcano (shield, stratovolcano, domes)*
Lat/Long: *58.42°N, 131.35°W*
Elevation: *760 to 2,190 m*
Eruptive History: *Three principal periods of eruptive activity:*

14.9 to 6.9 Ma	*Shield*
7.1 to 5.3 Ma	*Stratovolcano*
4.5 to 2.5 Ma	*Domes*

Composition: *Sodic alkalic basalt series, ankaramite to comendite*

The Level Mountain Range is the largest (area ~1,800 km^2, volume ~860 km^3), most persistent eruptive center of the Stikine Volcanic Belt, and the earliest locus of volcanism in the region. Level Mountain straddles two major northwest trending faults active in Mesozoic/Cenozoic time, the King Salmon and the Nahlin; the latter is the Stikine/Atlin tectonostratigraphic terrane boundary.

Resting on an erosional surface marking extensive Neogene regional uplift, Level Mountain has two principal components. The lower, more extensive component is a cliff-bounded volcanic plateau (average thickness 750 m) which rises to an average elevation of 1,400 m as a constructional basaltic shield. The shield is built of 4 distinctive stratigraphic units dominated by thin basalt flows (1 to 8 m thick) separated by

LEVEL MOUNTAIN: View south from dome on Kakuchuya Beatty divide. Late domes (central) cut glacially eroded core of the stratovolcano. East–West cliffs of breached stratovolcano visible at upper left. Plateau margin (near) and Coast Ranges (far) at upper right.

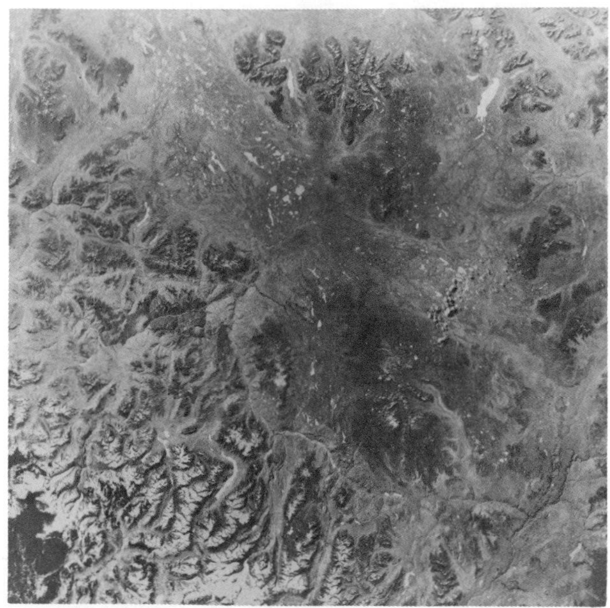

LEVEL MOUNTAIN: Landsat 5 TM image, band 7, taken July 2, 1987. Elevation and snow help define the geology of the region. The Coast Range to the southwest has heavy snow cover. Intermontane lowland lakes show as irregular white patches, Tertiary to Recent volcanic centers of the Stikine stand in relief. Central, and largest, is the treeless plateau of Level Mountain whose stratovolcano and domes are etched out by a light snowfall. The southern portion of the Kawdy Peaks cinder cone field, representing 35 km^3 *of basic lava is visible 20 km to the north. The dispersed style of the small vents is typical of the 50+ Plio-Pleistocene centers of the Stikine-Atlin region.*

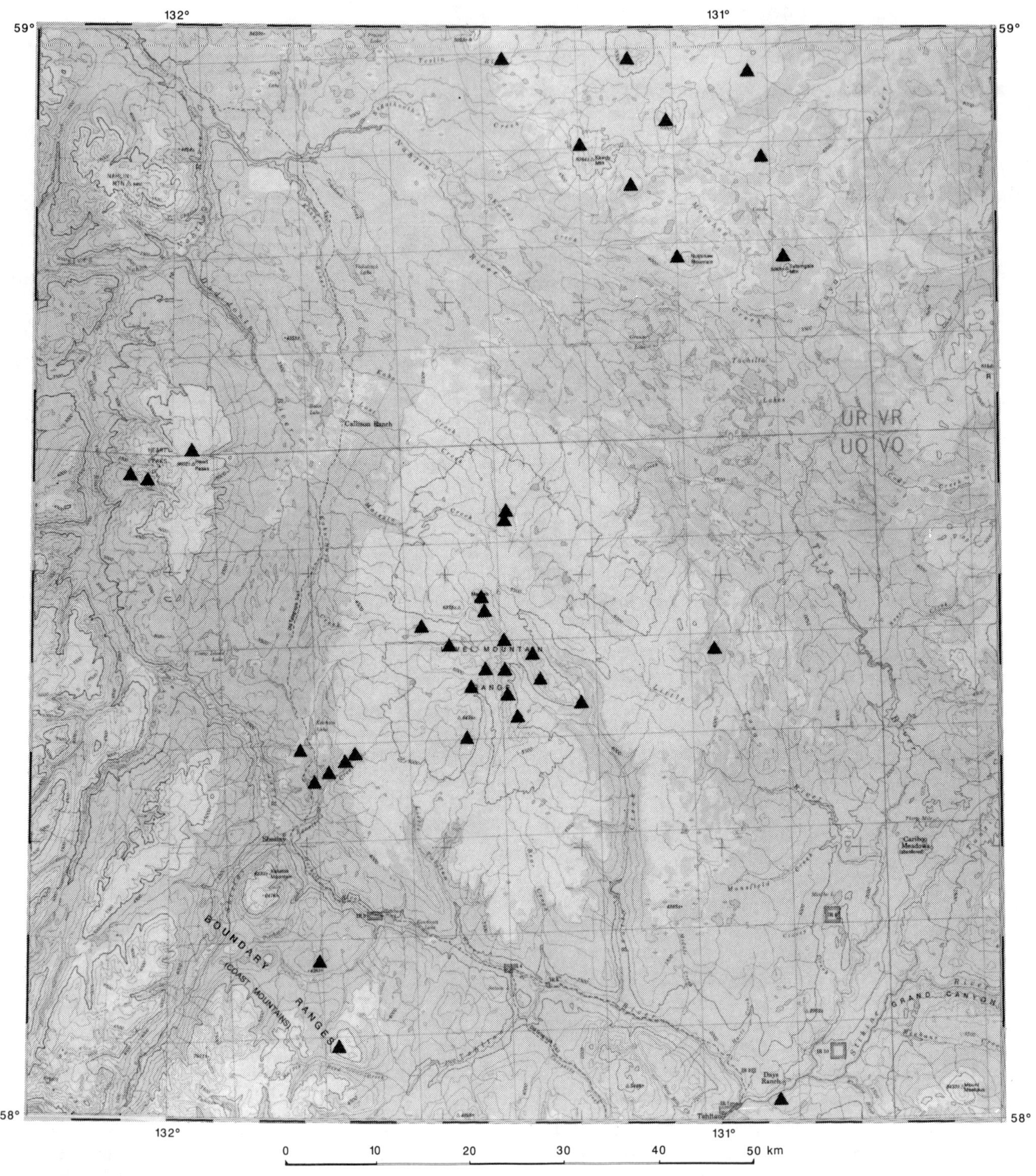

LEVEL MOUNTAIN: Composite topographic map from portions of 1:250,000 NTS sheets 104J Dease Lake and 104K Tulsequah, covering the region of Tertiary to Recent volcanism in the northern Stikine. Eruptive centers, vents, cinder cones, domes, and isolated flows are indicated by triangles. Concentrations of vents correspond to Level Mountain (central), Kawdy Peaks cinder cones and flows (north), and Heart Peaks (west).

thin discontinuous breccias, sporadic tuff horizons, and local lenses of fluvial, lacustrine, and glacial sediment. Most strata are undeformed. Pahoehoe flows predominate over blocky aa flows, breccias, and tuffs, attesting to a fluid and effusive character for the volcanism. Radial flow directions and quaquaversal dips suggest that most flows originated from central vents. Dikes are rare, thin, radially oriented, and intermittent features related to high-level volcanotectonic processes within the pile.

The upper component of Level Mountain is a bimodal stratovolcano cap centrally located atop the shield and comprising several eruptive centers (some formerly >2,500 m elevation). The stratigraphy of the stratovolcano is complex due to the influence of adjacent vents and to the tremendous variation in the erupted magmas (oversaturated, undersaturated, peralkaline, and metaluminous). Five bimodal packages of flows and ejecta comprise the stratovolcano. Felsic compositions dominate (>80% by volume) with peralkaline trachyte and comendite the most common.

Peralkalinity had remarkable effects on lava flow morphology and mineralogy. Low volume flows (7 km long × 3 to 8 m thick), with small-scale flow folds and lava tubes (1 to 2 m diameter) and thin dikes (1 m), owed their fluidity to liquidus temperatures in excess of 1,200°C and viscosities as low as 10^5 poise. These physical properties are atypical for melts of such high silica content.

Level Mountain records glaciation extending back to the Pliocene, affording numerous examples of the interaction of volcanic and glacial processes. Widespread evidence for contemporaneous volcanism and glaciation is present in the upper part of the shield and in the stratocone. This evidence includes tuyas (on the uppermost surface of the shield and as outliers), tills and glacial erratics at the base of flows and tuffs, till cemented by siliceous sinter, lahars comprised of till and agglomerate, volcano-glacial tuff breccias, interlayered unconsolidated fluvioglacial and tuffaceous deposits, and freshwater pillow basalts. The shield margins have been sculpted by continental ice, and its uppermost member includes volcano-glacial facies. Radially directed alpine glaciers have dissected the stratovolcano into a series of U-shaped valleys with intervening ridges that constitute the Level Mountain Ranges proper. This offers a unique opportunity to view cross sections through an effusive peralkaline volcano.

Post-dating the alpine valley system are high-level faulting and the extrusion of a series of salic domes (with volumes up to 9.4×10^7 m^3). Headward erosion has further modified the volcano by incising youthful V-shaped stream canyons into the plateau margin. This is the local expression of continuing regional uplift with canyon development on all of the adjacent major river systems, including the Stikine with its prominent section of Tertiary basalts.

From trace elements and isotope systematics, the petrogenesis of the Level Mountain Formation requires an upper mantle source for both the basaltic and the salic lavas. Refractory xenocrysts (olivine, orthopyroxene, spinel) from Level Mountain basalt and spinel lherzolites from nearby vents in the Stikine Canyon and at Castle Rock support this contention. The bimodal sodic alkalic lava suite at Level Mountain is typical of within-plate volcanism (hot spots) and volcanism associated with lithosphere in tension (e.g., continental rift and leaky transform settings).

How to get there: *The closest approach to Level Mountain via graded road (Cassiar–Stewart (Highway 37A) requires a 30-km+ hike from the Tahltan village at Telegraph Creek or Day's Ranch. Charter helicopter service from Dease Lake provides direct access to the central vents. Alternatively, fixed-wing landings are possible on peripheral low-lying lakes, and at the Sheslay air strip for a 15-km trek traversing a cross section of this major volcano.*

References

Hamilton, T. S., and Evans, M. E. 1983. A magnetostratigraphic and secular variation study of Level Mountain, northern British Columbia. *Geophys. J. R. Astr. Soc. 73*, 39-49.

Scarfe, C. M., and Hamilton, T. S. 1980. Viscosity of lavas from the Level Mountain volcanic centre, northern B.C.: *Carnegie Inst. Washington Yearbook 79*, 318-320.

T. S. Hamilton

EDZIZA
Canada

Type: *Complex of shields, domes, stratovolcanoes, calderas, and cinder cones*
Lat/Long: *57.7°N, 130.7°W*
Elevation: *816 to 2,786 m*
Eruptive History: *Episodic activity from 7.5 Ma to <1,300 yr BP*
Composition: *Alkali basalt, hawaiite, trachyte, and comendite*

The Mount Edziza volcanic complex covers an area of ~1,000 km^2 in the central part of the Stikine Volcanic Belt. The complex includes a group of overlapping basaltic shields, felsic stratovolcanoes, domes, small calderas, and monogenetic cones, ranging in age from 7.5 Ma to <1,300 yr BP. Four central felsic volcanoes, Armadillo Peak, Spectrum Range, Ice Peak, and Mount Edziza, lie along the northerly trending axis of an elliptical, composite basaltic shield. The shield forms a broad intermontane plateau, dotted with monogenetic cinder cones and bounded by steep escarpments which expose tiers of columnar basalt flows interlayered with distal clastic and pyroclastic deposits derived from the central, felsic volcano.

The volcanic assemblage is chemically bimodal, comprising voluminous alkali olivine basalt and hawaiite, which form most of the composite shield, and a felsic suite of mainly peralkaline trachyte and comendite which is confined mainly to the central volcanoes and associated lava dome. A relatively small volume of intermediate rocks was erupted from Ice Peak.

The complex is the product of five cycles of magmatic activity, each of which began with the effusion of alkali olivine basalt and culminated with the eruption of felsic magma. This cyclical behavior is attributed to the episodic rise of primitive, mantle-derived alkali basalt both to the surface and partly into crustal reservoirs where the felsic, peralkaline magmas were produced by prolonged crystal fractionation.

→

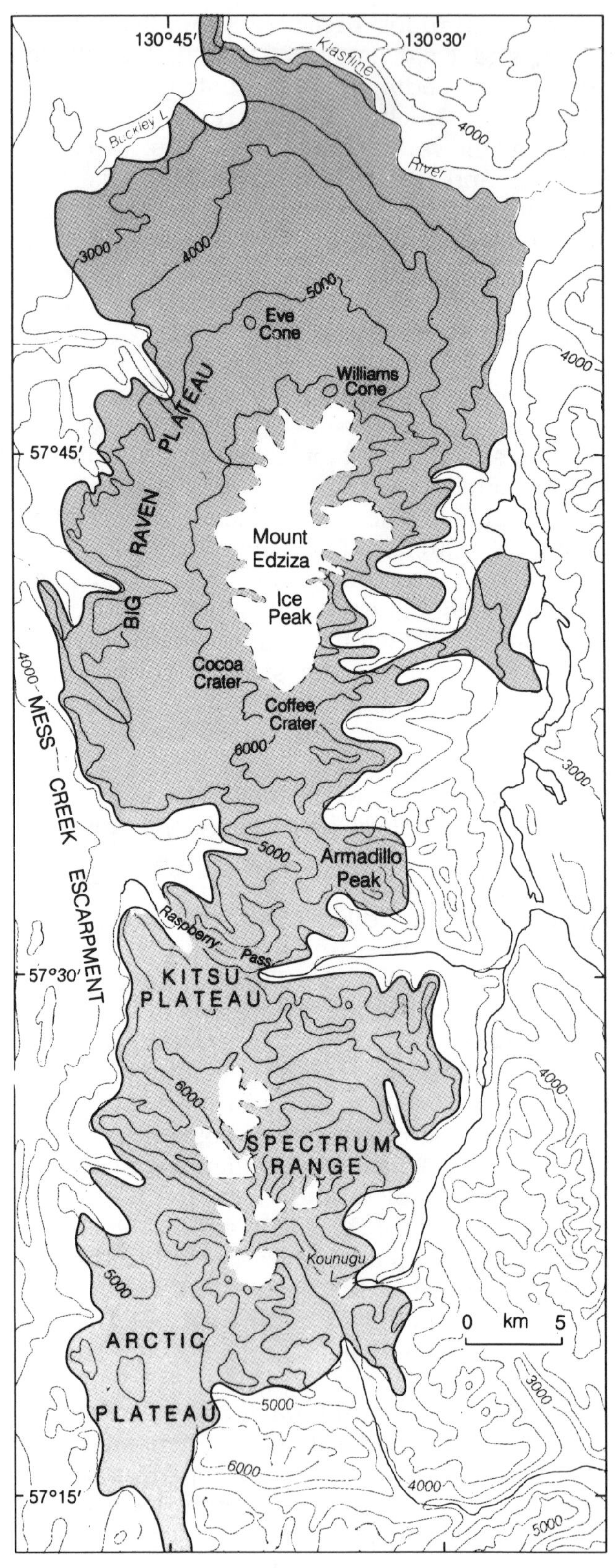

EDZIZA: Map showing the extent of the Mount Edziza volcanic complex and the locations of the four central volcanoes.

The oldest of the central volcanoes, **Armadillo Peak** (7 to 6 Ma) is the dissected remnant of a small caldera flanked by several steep-sided satellitic domes, and a thick pile of interlayered felsic lava flows, pyroclastic flows, air fall pumice, and epiclastic deposits. The 2,194-m summit of Armadillo Peak is capped by 180 m of trachyte flows which were ponded within the caldera.

The southern flank of the Armadillo pile is overlapped by distal flows from the younger **Spectrum Range** (3 to 2.5 Ma), which consists of a nearly circular, felsic lava dome over 10 km across and up to 650 m thick. Deeply incised radial valleys expose sections through the massive comendite and trachyte flows of the dome, as well as the bounding faults of a buried, cogenetic caldera ~4.5 km across.

Ice Peak (1.6 to 1.5 Ma), which overlaps the northern flank of the Armadillo pile, is a composite stratovolcano that produced both basic and intermediate to felsic lavas and pyroclastic rocks. The basic lava spread to the flanks of the cone, where it merges with and forms part of the surrounding shield. More viscous, intermediate, and felsic lava accumulated around the vents and forms most of the steep, upper slopes of the volcano. The present 2,500-m-high summit is a remnant of the western rim of a small summit caldera, which has been almost completely destroyed by alpine glaciation.

Mount **Edziza** (1.0 to 0.9 Ma), the youngest and most northerly of the four central volcanoes, is a composite, trachytic cone associated with several satellitic domes. It is superimposed on, and covers most of the north flank of, Ice Peak. Edziza's smooth northern and western slopes, only slightly channeled by erosion, curve up to a circular, 2,700-m summit ridge which surrounds a central, ice-filled crater 2 km in diameter. Active cirques on the east side have breached the crater rim, exposing the remnants of several lava lakes which ponded in the crater and rest on hydrothermally altered breccia of the main conduit.

The history of the Mount Edziza volcanic complex includes at least two periods of regional glaciation and numerous lesser advances of alpine glaciers. Piles of pillow lava and hyaloclastite, formed by subglacial eruptions, are found on the flanks of both Ice Peak and Mount

EDZIZA: View of the southwest face of Mount Edziza. Escarpment in the foreground exposes tiers of basalt flows which form the older shield. The glacier-covered trachyte dome of Mount Edziza in the background is flanked by numerous basaltic cinder cones.

EDZIZA: ***Eve Cone****, on the northern flank of Mount Edziza, is one of the youngest and best preserved of >30 monogenetic cones within the complex.*

Edziza as well as on the surface of the surrounding shield. The eruption of Quaternary basalt, from satellitic vents along the flanks of the central volcanoes, began when remnants of ice were still present and continued into post-glacial time. The initial flank eruptions, quenched by glacial meltwater, formed hyaloclastite tuff rings, whereas later activity produced >30 subaerial pyroclastic cones surrounded by blocky lava fields and tephra deposits. Charred willow stems, still rooted in a paleosol under 2 m of loose basaltic tephra, gave a ^{14}C date of 1,300 yr BP. That event was followed by at least two younger, but still undated eruptions. ***How to get there:*** *The Cassiar-Stewart Highway and a spur road from Dease Lake to Telegraph Creek provide the closest road access. From Kinaskan Lake, on the Cassiar-Stewart Highway, a poorly maintained trail extends west for 30 km into the central part of the complex. From Telegraph Creek a fairly good trail extends east for 25 km to the north slope of Mount Edziza.*

References

Souther, J. G., Armstrong, R. L., and Harakal, J., 1984, Chronology of the peralkaline, late Cenozoic Mount Edziza Volcanic Complex, northern British Columbia, Canada: *Geol. Soc. Am. Bull. 95*, 337-349.

Souther, J. G., in press, The late Cenozoic Mount Edziza Volcanic Complex, British Columbia: *Geol. Surv. Can. Mem. 420.*

Jack G. Souther

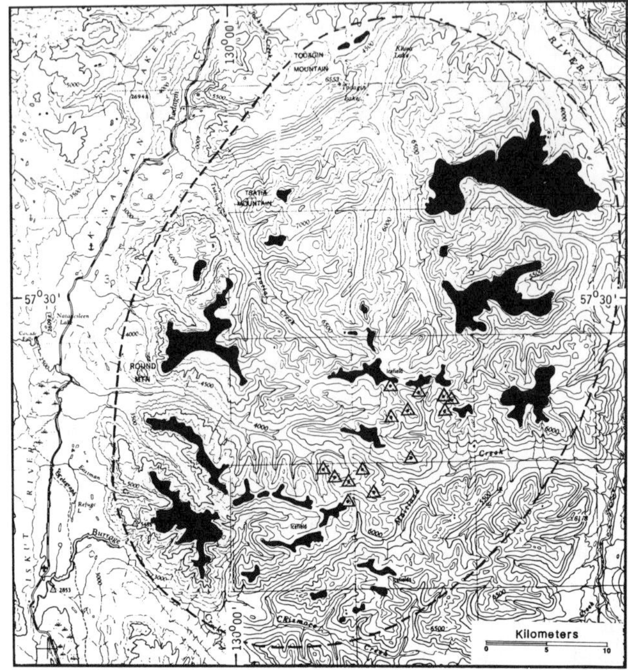

MAITLAND: Map showing lava flow remnants (black) and necks (triangles). Dashed line indicates the inferred extent of the original shield.

MAITLAND
Canada

Type: *Shield*
Lat/Long: *57.4°N, 129.7°W*
Elevation: *1,828 to 2,514 m*
Eruptive History: *5 to 4 Ma*
Composition: *Alkali basalt and hawaiite*

The Maitland volcanics consist of basalt necks and flat-lying basalt flows in the western Cassier Mountains of northern British Columbia. They are remnants of a composite shield volcano which was built onto a mature, Tertiary erosion surface between 5 and 4 Ma. The shield, which formerly covered an area of >900 km^2, has been reduced by erosion to scattered, cliff-bounded remnants which cap the higher mountains. Intervening valleys, deeply incised into the underlying Jurassic shale and sandstone, are as much as 600 m below the base of the shield. Individual lava caps are up to 230 m thick and contain from 1 to 20 flows of columnar basalt separated by beds of scoriaceous flow-topped breccia and tephra.

Near the center of the shield a cluster of 14 volcanic necks forms prominent steep-sided monoliths of basalt that rise 60 to 120 m above the surrounding shale slopes. These nearly circular conduits, each ~30 m across, are the only known feeders for the flows of the shield.

MAITLAND: Basalt neck cutting Jurassic shale. Flat-lying basalt flows capping the mountain in the right background are remnants of a once contiguous shield volcano.

The Maitland volcanics include both alkali olivine basalt and hawaiite. Most samples contain phenocrysts of plagioclase and/or olivine.

How to get there: *Follow Highway 37 north from New Hazelton to Iskut River, where remnants of the Maitland volcanics can be seen capping ridges along the east side of the valley. From Iskut River valley the only access to the flows and necks is by strenuous hiking or by charter helicopter.*

Reference

Souther, J. G., and Yorath, C. J., in press, Chapter 10, Neogene Assemblages, *in* The Cordilleran Orogen: Canada: Gabrielse, H., and Yorath, C. J. (eds.), *Geol. Surv. Can. No. 4* (also Geol. Soc. Am., *The Geology of North America, No. G-2*).

Jack G. Souther

HOODOO
Canada

Type: *Tuya and valley flows from post-glacial flank eruptions*
Lat/Long: *56.76°N, 131.30°W*
Elevation: *150 to 1,700 m*
Eruptive History: *0.11 to 0.09 Ma and earlier*
Composition: *Peralkaline trachyte and comendite, minor hawaiite*

Hoodoo Mountain lies west of the main axis of the Stikine Volcanic Belt. It consists of a symmetrical lava dome, ~6 km in diameter, surrounded on three sides by alpine glaciers. Only its southern slope, which extends down to the floodplain of Iskut River, is ice-free. Hoodoo's steep sides and nearly flat 900-m summit suggest it formed as a subglacial tuya when regional ice sheets covered all but the highest peaks of the northern Coast Mountains. Subaerial lava flows which rest on glacial till along Iskut River indicate that volcanic activity continued after retreat of the ice. Radiometric dates of 0.11 and 0.09 Ma are consistent with the age of other ice-contact features in the Stikine Volcanic Belt.

Pantellerite and comendic trachyte are the principal rock types. Most specimens are porphyritic with phenocrysts of sanidine and sodic pyroxene in a groundmass similar to oversaturated peralkaline rocks elsewhere in the Stikine Volcanic Belt.

How to get there: *Bob Quinn Lake, on the Cassiar–Stewart Highway 60 km northeast of Hoodoo Mountain, is the closest point accessible by road. From there the mountain can be reached by charter helicopter or by trekking across extremely difficult terrain. Alternatively, it can be reached by boat from Wrangell, Alaska, up the swift and shifting channels of Stikine and Iskut rivers.*

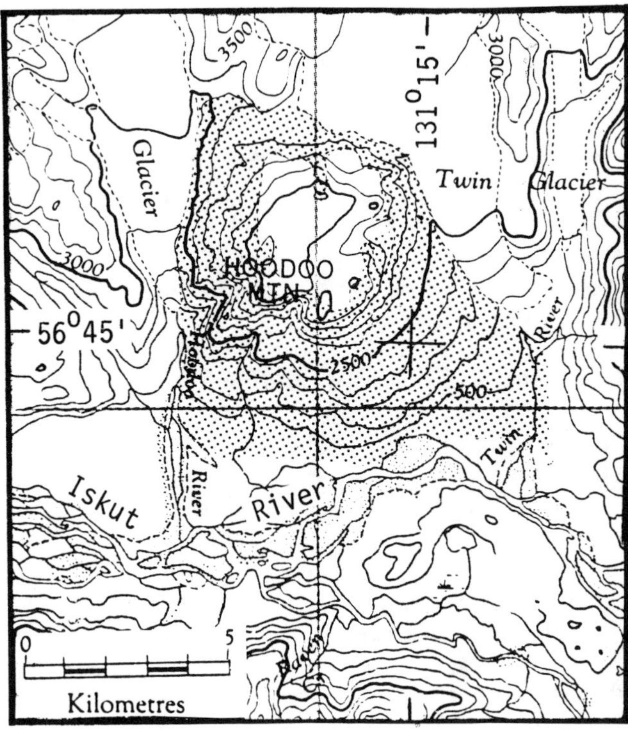

HOODOO: Map.

HOODOO: Viewed from the south across the Iskut River valley (photograph by R. G. Anderson).

Reference

Souther, J. G., and Yorath, C. J., *in press*, Chapter 10, Neogene Assemblages, *in* The Cordilleran Orogen:

Canada: Gabrielse, H., and Yorath, C. J. (eds.), *Geol. Survey Can., No. 4* (also Geol. Soc. Am., *The Geology of North America, No. G-2*).

Jack G. Souther

ISKUT-UNUK RIVER CONES

Canada

Type: *Monogenetic cones and small stratovolcanoes*
Lat/Long: *56.52°N, 130.33°W*
Elevation:

Iskut Canyon Cone:	*914 m*
Cinder Mountain:	*1,820 m*
Snippaker Creek Cone:	*1,370 m*
King Creek Cone:	*1,828 m*
Canyon Creek Cone:	*762 m*
Lava Fork:	*1,200 m*

Eruptive History: *All probably active between 8,780 and 360 yr BP*
Composition: *Alkali basalt*

The Iskut River cones comprise a group of six small basaltic centers in the southern part of the Stikine Volcanic Belt. The largest, **Iskut Canyon Cone** – a stratovolcano ~1 km across and 200 m high – is exposed on the steep southern slope of Iskut valley near its junction with Forrest Kerr Creek. This cone is the source of at least 10 thick lava flows which initially dammed Iskut River and are now exposed in a box canyon extending ~20 km downstream from the cone. The flows are divided into two groups separated by an erosion surface. A lower group of dark-colored alkali basalt flows predates wood that gave a ^{14}C date of 8,780 yr BP. Wood from a gravel lens in an upper group of light-colored flows yielded a ^{14}C date of 3,660 yr BP. Drilling to assess the hydroelectric potential of the canyon revealed that the subaerial flows are underlain by hyaloclastite resting on pre-eruption river gravel.

Cinder Mountain, a partly eroded composite cone at the head of Snippaker Creek, is the source of a basalt flow that extends 4 km north into Copper King Creek, where it rests on an alpine moraine. The edifice, which rises ~300 m above the surface of surrounding alpine glaciers, consists mostly of pillow lava and hyalo-

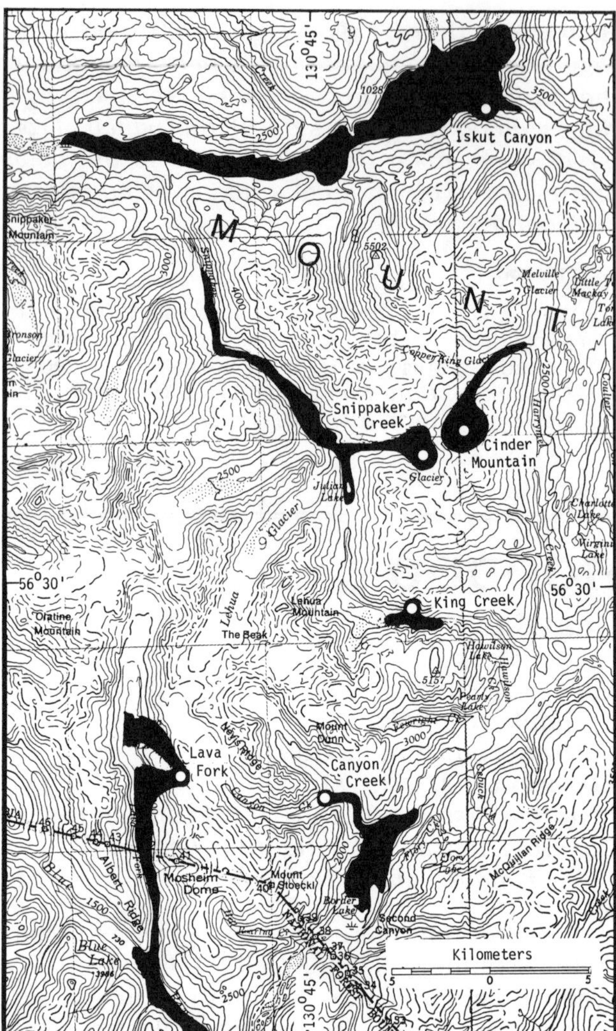

ISKUT-UNUK RIVER: Map showing locations of the Iskut River cones and lava flows.

ISKUT-UNUK RIVER: View across basalt flows to lava-dammed lake at head of Lava Fork valley.

clastite breccia. These ice-contact deposits probably formed during a period of Holocene glacial advance when alpine glaciers extended to the limit of their present trimlines. The initial accumulation of hyaloclastite in a subglacial meltwater basin was followed by failure of the ice dam and eruption of the subaerial flow which entered Copper King valley after the meltwater had drained.

An isolated pile of subaerial basalt flows and associated pillow lava rest on varved clay and till in **King Creek**. This material probably originated from a small subglacial center that was roughly coeval with Cinder Mountain.

A pyroclastic cone ~300 m high (**Snippaker Creek Cone**) near the west flank of Cinder Mountain is the source of a narrow, levee-bounded flow that descended Snippaker valley for ~20 km, almost to its junction with Iskut valley.

About one kilometer north of the Alaska border, Unuk River valley is occupied by a blocky lava field that has forced the river against its eastern bank. The source of this flow is a cluster of dissected cinder cones in **Canyon Creek**, a small steep-sided tributary valley on the east side of Unuk River valley.

The flow in **Lava Fork** valley is probably the youngest in Canada. It issued from a vent near the crest of a ridge on the north side of the valley, where ropy pahoehoe lava is associated with irregular mounds of fire-fountain deposits, open lava tubes, and steep-walled lava troughs. No cone is present in the vent area, but thick deposits of loose tephra still cling to ledges on the surrounding granite peaks and form lenses within some of the adjacent glaciers. From the vent, at an elevation of ~1,200 m, the flows cascade down >1,000 m of steep granite cliffs to the broad valley of Lava Fork, where two small lakes are ponded above the flow. The main flow extends south along Lava Fork for 12 km to Blue River valley in Alaska, and along Blue River valley for another 9 km, where it spreads into a broad terminal lobe on the flat alluvial plain of Unuk River. Several successive overlapping lava flows with varying degrees of reforestation suggest intermittent eruptive activity separated by quiescent periods. Carbonized wood on one of the older flows yielded a ^{14}C date of 360 yr BP.

All analyzed samples of lava from the Iskut cones are alkali olivine basalt.

How to get there: *No roads are within practical walking distance of any of the Iskut River cones. Cinder Mountain and the King Creek Cone can be reached on foot from an abandoned airstrip on Snippaker Creek. The lake at the head of Lava Fork is large enough for a floatplane. Access to the other centers requires a helicopter.*

References

Elliot, R. L., Koch, R. D., and Robinson, S. W., 1981, Age of basalt flows in the Blue River valley, Bradfield Canal quadrangle, *in* The United States Geological Survey in Alaska: Accomplishments during 1979: *USGS Circ. 823-B*, B115-B116.

Grove, E. W., 1974, Deglaciation – A possible triggering mechanism for recent volcanism: Internat. Assn. Volc. Chem. Earth's Interior, *Proc. Symp. Andean Antarctic Volcanology Problems, Santiago Chile, Sept. 1974.*

Jack G. Souther

THE THUMB
Canada

Type: *Basaltic neck*
Lat/Long: *56.16°N, 126.70°W*
Elevation: *1,840 m*
Relief: *180 m*
Eruptive History: *Quaternary (age unknown)*
Composition: *Alkali olivine basalt*

The Thumb is the largest among a group of at least seven volcanic necks which are associated with dikes, flows, and remnants of pyroclastic cones in the Omineca Mountains of northeastern British Columbia. Although they have not been dated, the presence of loose scoria with open vesicles and the relationship of intravalley flows to the present topography suggest a Quaternary age.

The steep-sided edifice of The Thumb forms a prominent landmark which rises 180 m above gently sloping benchland near the crest of the Connelly Range. The Thumb consists mostly of columnar basalt surrounded by vesicular breccia containing accidental clasts of the underlying Paleocene sandstone.

Alkali olivine basalt from The Thumb and other Quaternary necks in the Omineca Mountains commonly contains phenocrysts of labradorite, clinopyroxene, and olivine. These volcanic centers lie outside the main volcanic belts

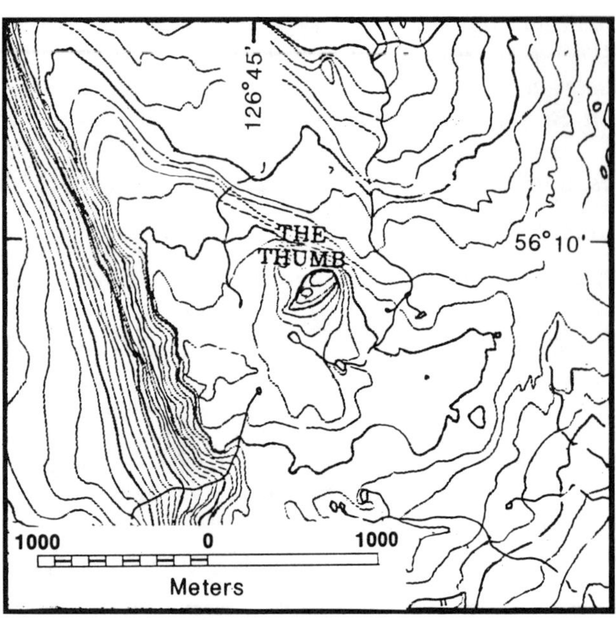

THE THUMB: Topographic map.

THE THUMB: View from northwest.

of British Columbia, and their tectonic affiliation is not clear.

How to get there: *From Vanderhoof on Highway 16 a railway grade extends north to Bear Lake, which is 5 km west of and 1000 m below The Thumb. Construction of the grade has been suspended and, although parts of it are used as a logging road, vehicular access is restricted.*

Reference

Lord, C. S., 1948, McConnell Creek Map Area, Cassiar District, British Columbia: *Geol. Surv. Can. Mem. 251.*

Jack G. Souther

MILBANKE SOUND CONES

Canada

Type: *Monogenetic cones*
Lat/Long: *52.3°N, 128.5°W*
Elevation: *0 to 385 m*
Eruptive History: *Post-glacial (age unknown)*
Composition: *Alkali basalt*

Quaternary volcanic activity in the outer islands of coastal British Columbia was confined to a group of four small monogenetic cones in the Milbanke Sound area. The best preserved, **Kitasu Hill**, is a symmetrical pyroclastic cone that rises from the southern shoreline of Swindle Island to the rim of a crescent-shaped summit crater at an elevation of 250 m. The edifice consists mostly of subaerial tephra and bombs enclosing remnants of proximal olivine basalt flows. The cone is underlain by till which rests on glacially scoured bedrock.

Parts of Lake Island and adjacent Lady Douglas Island are covered by moderately welded basaltic tuff breccia that rests on either glaciated granitic rock or locally on unconsolidated beach gravel. The breccia originated from a central vent on Lake Island, where a pile of welded blocks and anastomosing feeder dikes form **Helmet Peak**, a steep-sided mound that rises from sea level to an elevation of 335 m. Blocks of basement granodiorite, some up to 2 m across, are randomly suspended within the breccia. The Lake Island basalt is characterized by abundant 0.5 to 1 cm phenocrysts of labradorite and clinopyroxene.

Cones on Price and Dufferin islands have been reduced to structural mounds covered by mature forest. Basaltic flows from both centers rest on beach deposits on the adjacent shorelines.

The Milbanke Sound cones are near remnants of an older (14 to 12 Ma) peralkaline complex, which is believed to define the western end of a hot spot trace, the Anahim Volcanic Belt, through central British Columbia. Like anomalously young volcanoes on other hot spot traces, the relationship of these cones to the Anahim Volcanic Belt is controversial.

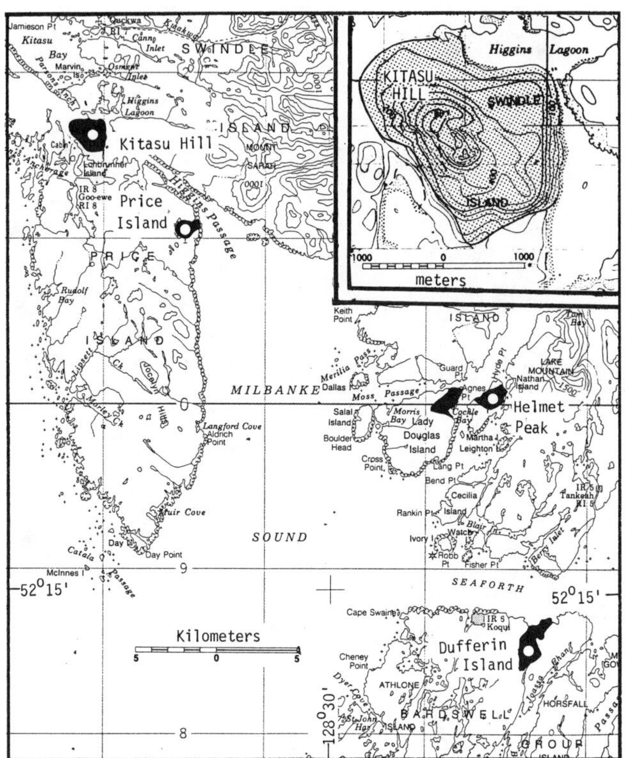

MILBANKE SOUND: Map showing locations of the Milbanke Sound cones and detail of Kitasu Hill (inset).

MILBANKE SOUND: Helmet Peak, remnant of a Quaternary pyroclastic cone on Lake Island.

How to get there: *With the exception of Kitasu Hill, which is the site of the weather station and fire lookout, the Milbanke Sound cones are on uninhabited islands of the British Columbia coast. The nearest settlement is Bella Bella, where small boats or float-equipped aircraft may be chartered.*

References

Baer, A. J., 1973, Bella Coola–Laredo Sound map area, British Columbia: *Geol. Surv. Can. Mem. 372*, 38-39.

Souther, J. G., 1986, The western Anahim Belt: Root zone of a peralkaline magma system: *Can. J. Earth Sci. 23*, 895-908.

Jack G. Souther

RAINBOW RANGE
Canada

Type: *Shield*
Lat/Long: *52.73° N, 125.28° W*
Elevation: *1,890 to 2,500 m*
Eruptive History: *8.7 to 6.7 Ma*
Composition: *Oversaturated alkaline to peralkaline suite*

The Late Miocene Rainbow Range is one of three large peralkaline shield volcanoes that lie on an east–west trend in west-central British Columbia. These shield volcanoes, along with other peralkaline extrusive and intrusive centers and many small basaltic centers, comprise the Anahim Volcanic Belt. Excellent exposure of the interior of the Rainbow Range shield volcano is afforded today because of incised alpine glacial valleys that radiate from the center of the volcano.

During a period of ~2 m.y., extrusion of highly fluid basic and silicic lava flows built up the gently sloping (5°-8°) flanks of the shield volcano. The stratiform flank zone surrounds a central complex of small domes, short, stubby flows, and small intrusive bodies. Alkaline and peralkaline lava flows from four volcanic episodes make up an 845-m composite section on the north flank of the shield volcano. Basal comenditic trachyte flows are unconformably overlain by flows and flow breccias of mugearite. A sequence of 40-60-m-thick columnar-jointed comendite flows blankets the underlying units and gives the volcano its shieldlike form. Hawaiite dikes, plugs, and minor capping flows are scattered over the north flank. Comendite flows, which commonly have a glassy selvage at the base, account for 75% of the lavas in the flank zone. **Anahim Peak**, a small center on the northeast flank of the main volcano,

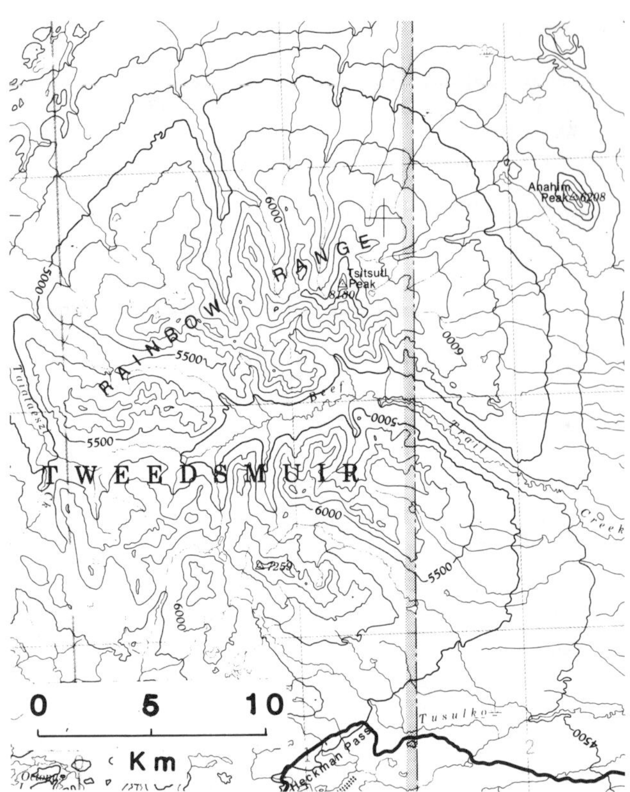

RAINBOW RANGE: Topographic map of the Rainbow Range shield volcano [from Canadian National Topographic Service Map 93C, 1:250,000 scale (Anahim Lake sheet)].

RAINBOW RANGE: Oblique air photo of the north flank of the Rainbow Range shield volcano, looking east. Anahim Peak is partially obscured by the clouds in the middle background, and the Ilgachuz Range, another peralkaline shield volcano of the Anahim belt, is visible in the background. Province of British Columbia photograph B.C.461:23, XL48:32.

consists of a small (<2 km^2) pile of 7 thick hawaiite flows cut by a trachyte plug.

The unique characteristic of the Rainbow Range is that, although the flank zone is composed dominantly of silicic lavas (64-71% SiO_2), the form of the volcanic edifice is a shield volcano. The apparent morphological paradox of a silicic shield volcano, made of flows that should be too viscous to form such a shape, is also found in the Kenya rift. It occurs because the peralkaline nature of the silicic lavas decreases the viscosity of the flows a minimum of 10-30 times over that of calc-alkaline silicic flows.

How to get there: *The Rainbow Range is on the boundary between the Coast Mountains and Chilcotin Plateau, ~440 km north of Vancouver, B.C. The volcano is best observed from Heckman Pass, ~38 km west of Anahim Lake, B.C., on Highway 20, a good gravel road, where on a clear day one sees a spectacular profile view of the multicolored volcano. The volcano itself is accessible by foot only, through Tweedsmuir Provincial Park, via a hiking trail that leaves Highway 20 from the picnic site on the East Branch of Young Creek, ~46 km west of Anahim Lake.*

References

Bevier, M. L., Armstrong, R. L., and Souther, J. G., 1979, Miocene peralkaline volcanism in west-central British Columbia – Its temporal and plate tectonics setting: *Geology* 7, 389-392.

Bevier, M. L., 1981, The Rainbow Range, British Columbia: a Miocene peralkaline shield volcano: *J. Volc. Geotherm. Res. 11*, 225-251.

Mary Lou Bevier

ILGACHUZ RANGE
Canada

Type: *Felsic shield*
Lat/Long: *52.75° N, 125.30° W*
Elevation: *1,220 to 2,400 m*
Volcano Diameter: *25 km*

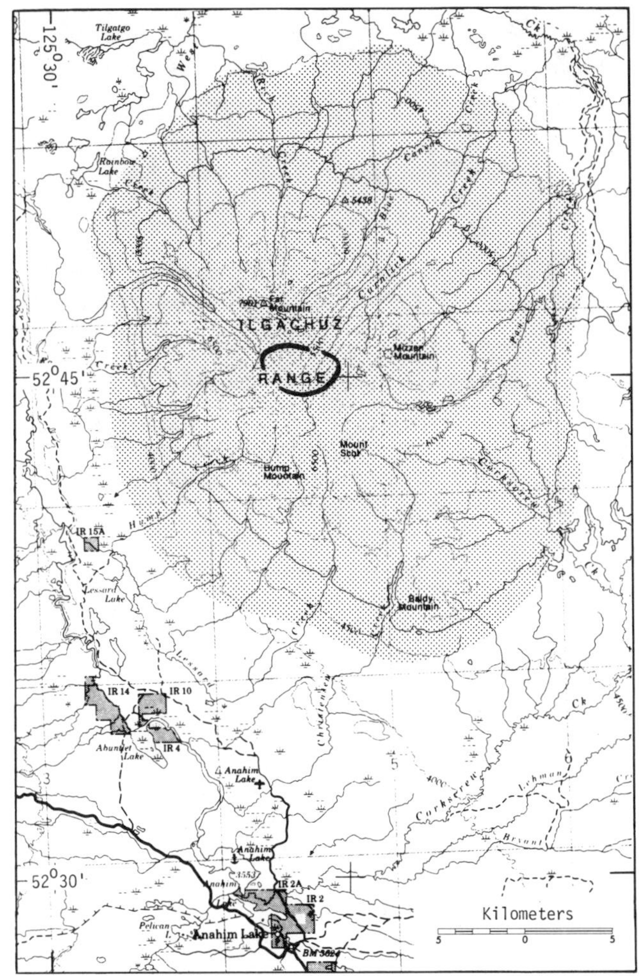

ILGACHUZ: Topographic map showing the extent of the shield (stipples) and the location of the summit caldera (dashed line).

Eruptive History:
6 to 5 Ma: Shield-forming assemblage
5 to 4 Ma: Post-caldera assemblage
Composition: *Trachyte, alkali basalt, hawaiite, comendite*

The Ilgachuz Range, near the center of the Anahim Volcanic Belt, is a symmetrical shield volcano ~25 km across. The surface of the outer shield, cut by deeply incised radial valleys, rises to a complexly dissected summit area where remnants of a small central caldera and numerous felsic domes and basaltic feeders are exposed. The outer flanks of the shield merge with flat-lying basalt of the Chilcotin Group, which covers much of the surrounding interior plateau of central British Columbia.

ILGACHUZ: View south across the central Ilgachuz Range. Basalt dike in foreground cuts trachyte flows of the outer shield. Flat-topped mountain in background is a remnant of the intracaldera assemblage comprising a thick ponded flow of trachyte resting on lacustrine tuff.

The oldest rocks exposed in the Ilgachuz Range, the 6-to-5-Ma lower shield assemblage, are hydrothermally altered breccias, domes, and lava flows of alkali rhyolite and trachyte. These are overlain by the upper shield assemblage, which forms the bulk of the Ilgachuz shield and consists of moderately to strongly peralkaline comendite and pantellerite flows and domes interlayered locally with lenses of basalt.

Formation of the shield was followed by collapse of a central caldera ~3 km in diameter that later filled with epiclastic breccia, lacustrine tuff, and ponded trachyte lava flows.

A 5-to-4-Ma post-caldera assemblage, consisting of alkali basalt and hawaiite flows, issued from satellitic vents on the flanks of the shield and from a few small vents within the caldera.

The Ilgachuz suite is chemically bimodal, consisting of voluminous oversaturated alkaline to peralkaline rocks (rhyolite, comendite, and pantellerite), minor trachyte, and a large volume of alkali basalt and hawaiite. The relationship between distal flows from the Ilgachuz shield and basalt of the Chilcotin Group is unknown, but the two are chemically and probably petrogenetically distinct. The alkaline to mildly peralkaline lavas of the Ilgachuz Range are believed to be the product of a mantle plume which generated a hot spot trace defined by the

Anahim Volcanic Belt, whereas transitional basalt of the Chilcotin Group is the product of back-arc extension related to subduction of the Juan de Fuca Plate.

How to get there: *The Ilgachuz Range is ~30 km north of Anahim Lake, a small town on the Chilcotin Highway (Highway 20). Several pack-horse trails lead from the vicinity of Anahim Lake onto the plateau surface of the outer shield.*

References

Souther, J. G., 1984, The Ilgachuz Range, a peralkaline shield volcano in central British Columbia: *in* Current Research, Part A, *Geol. Surv. Can. Pap. 84-1A*, 1-10.

Souther, J. G., 1986, The western Anahim Belt: Root zone of a peralkaline magma system: *Can. J. Earth Sci. 23*, 895-908.

Jack G. Souther

ITCHA RANGE
Canada

Type: *Shield*
Lat/Long: *52.71° N, 124.85° W*
Elevation: *2,370 m*
Height: *690 m*
Volcano Diameter: *15 km*
Eruptive History:
3.8-3.0 Ma: Early trachytic shield flows
2.2-0.8 Ma: Late capping basalts
Composition: *Trachyte, hawaiite to basanite*

The Itcha Mountain Range is the easternmost and youngest of three felsic shield volcanoes that compose the Anahim belt in central British Columbia, which is interpreted to represent the east–west trending trace of a mantle hot spot. The dissected Itcha shield is dominated by undersaturated trachyte as opposed to the oversaturated felsic lavas that characterize the more western shields of Ilgachuz and the Rainbow Ranges. Mildly peralkaline, aphyric trachyte and some commendite occur at the base of the volcanic pile and are overlain by metaluminous feldspar-phyric trachyte.

The broad scale morphology of the shield is given by the extensive lateral distribution of thick (70-150 m) felsic flows which erupted from fissures around a central vent to cover an area of 330 m^3. The early shield rocks are unconformably capped by a series of small cinder cones and thin flows (1-4 m) of mafic alkaline lava. At least 30 cinder cones of ~1-2 km diameter are randomly distributed on the eastern half of the complex. Hawaiite is the dominant rock

ITCHA: Looking northwest from the eastern edge of the complex. Note the sharp contact between the dark hawaiitic scoria and cinder cone and the earlier, underlying thick trachytic flow.

ITCHA: Topographic map. Ruled areas show the distribution of the capping alkaline basaltic flows across the shield.

type, but compositions ranging from alkaline olivine basalt to spinel lherzolite-bearing basanite (the most primitive lavas have 10% MgO and are Si-undersaturated – 15.33% norm Ne) are found in the Itcha Mountain Range. These alkaline mafic flows appear to merge laterally with the less Si-undersaturated (Hy norm) lavas of the Chilcotin flood basalt that surrounds the Anahim belt. The exact nature of the relationship between the Anahim belt and the Chilcotin basalt is, however, unknown.

How to get there: *The Itcha Mountain Range is 47 km northeast of Nimpo Lake in British Columbia. There is no road access to the Range and one must fly by fixed wing from Nimpo to Itcha Lake, 4 km northeast of Itcha Mountain, or by helicopter from Bluff Lake, 100 km south of the Range.*

References

Souther, J. G., 1984, The Ilgachuz Range, a peralkaline shield volcano in central British Columbia: *In* Current Research, part A. *Geol. Surv. Can. Pap. 84-1A*, 1-10.

Souther, J. G., 1986, The western Anahim Belt: Root zone of a peralkaline magma system: *Can. J. Earth Sci. 23*, 895-908.

Anne Charland

NAZKO
Canada

Type: *Subglacial mound and composite pyroclastic cone*
Lat/Long: *52.90°N, 123.73°W*
Elevation: *1,230 m*
Height: *120 m*
Eruptive History: *0.34 Ma to 7,200 yr BP*
Composition: *Alkali basalt, basanite*

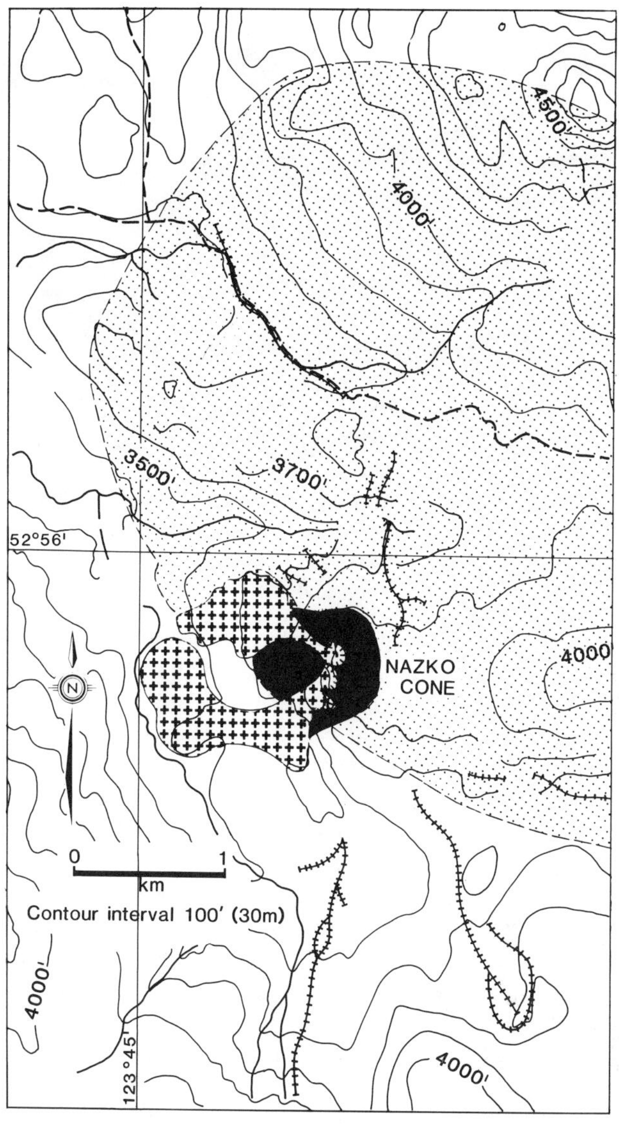

NAZKO: Map showing composite pyroclastic cone (black), post-glacial lava flows (crosses), and tephra plume (stipple).

Nazko Cone is the youngest and most easterly eruptive center in the Anahim Volcanic Belt. It consists of a small composite cone which rises ~120 m above the surrounding Interior Plateau and appears to be the product of three episodes of activity. The oldest unit, a thick subaerial flow of alkali olivine basalt (K-Ar age 0.34 Ma), issued onto the plateau during an ice-free, interstadial period. It is overlain by a conical mound of quenched basalt, broken pillows, and sideromelane tuff breccia which forms the central core of the edifice. These ice contact deposits comprise a subglacial mound believed to have formed during the last major Pleistocene glaciation. A third eruptive episode occurred soon after the retreat of Pleistocene ice. During this final event the subglacial mound was partly enveloped by a cluster of three pyroclastic cones which merge to form the crescent-shaped summit ridge. A breach on the west side of the cone is occupied by a narrow tongue of lava bounded by steep levees. At the base of the cone these flows merge into two blocky, steepsided lobes which extend west for ~1 km onto the surface of

NAZKO: View of southwestern side of Nazko Cone showing central subglacial mound partially enveloped by younger pyroclastic cones. Steep front of blocky lava flow forms the forested margin of swamp in middle foreground.

the surrounding plateau. A tephra plume extends north and east, thinning from more than 3 m of loose lapilli at the base of the cone to a few centimeters of gritty ash at a distance of 4 km. Radiocarbon ages on peat directly above and below the tephra layer suggest that the eruption occurred ~7,200 yr BP.

The Nazko basalt, like that of other monogenetic cones in the western Anahim Belt, is silica undersaturated (10-15% normative nepheline), but – unlike the basanite of the Itcha Range – lacks any obvious lherzolite nodules.

How to get there: *From Quesnel, on Highway 97, a secondary road leads west for 45 km to the town of Nazko. Nazko cone is on an unpaved logging road ~12 km west of Nazko.*

Reference

Souther, J. G., Clague, J. J., and Mathews, R. W., 1987, Nazko cone, a Quaternary volcano in the eastern Anahim Belt: *Can. J. Earth Sci. 24*, 2477-2485.

Jack G. Souther

CHILCOTIN BASALT
Canada

Type: *Monogenetic plains basalt field*
Lat/Long: *52°N, 123°W*
Elevation: *Mostly 1,400 m but plateau surface is gently warped between 900 and 1,800 m*

CHILCOTIN: Approximately 70 m of Chilcotin basalt flows are exposed at the Painted Chasm, B.C., which is named for the brilliant red oxidized portions of the flows.

Eruptive History: *Three principal periods of activity:*
16-14 Ma
10-6 Ma (dominant volume)
3-1 Ma
Composition: *Transitional basalt (hynormative)*

Flat-lying basalt flows of the Mio-Pliocene Chilcotin Group cover ~25,000 km^2 of the interior plateau of south-central British Columbia and have a volume of ~1,800 km^3. The back-arc basalt lies east of coeval calc-alkaline volcanoes of the Pemberton Belt. Chilcotin Group basalt is representative of the basaltic plains style of volcanism and is characterized by many thin, flat-lying to gently dipping (<2°) pahoehoe flows that erupted from central vents now marked by gabbro plugs. Prior to Late Pleistocene glacial erosion these centers formed a series of coalesced, low-profile shield volcanoes. Feeder dikes for the basalt flows have not been found, although this may be a function of cover by glacial drift. The volcanic field ranges in thickness from 5-140 m, with an average value of 70 m. Two to 20 flow units are present in exposed stratigraphic sections of the Chilcotin Group, and individual flows are not generally traceable for more than a few kilometers. Most flows are 5-10 m thick and crudely columnar-jointed; however, some intracanyon flows are up to 70 m thick and have developed a lower colonnade and entablature. Individual flow lobes are very vesicular (5-20%), and in many cases are red at the base and top. Vesicle sheets and cylinders, commonly filled with chabazite, are abundant in the flow interiors. Pillow lava and pillow breccia are present in a number of exposed sections. A few air-fall tephras, containing accretionary lapilli derived from coeval arc volcanoes to the southeast, are locally preserved between successive basalt flows.

Remnants of flows from the oldest cycle of eruptive activity crop out around the margins of the present plateau, which is composed dominantly of the 6-10 Ma basalts. Pliocene flows, for which no vents have been found, occur in cliff sections along the Fraser River canyon in the southern part of the plateau. Chemically, all the lavas are hy-normative, transitional basalt, which in thin section contain only one pyroxene, titan-augite.

Although Chilcotin basalt is in part the same age as Columbia River basalt, and occurs in an analogous tectonic setting, the similarities end there. Chilcotin basalt flows are much less voluminous and not as stratigraphically continuous as the Columbia River flood basalt; they erupted from central vents rather than dike swarms, and are transitional basalt rather than tholeiite.

How to get there: *Chilcotin Group lavas form a large plateau, centered ~320 km north of Vancouver, that covers much of south-central B.C. Exposures occur in many places along Highway 97 (Cariboo Highway) between Clinton and Quesnel, and Highway 20 (Bella Coola Road, a good gravel road) between Williams Lake and Bella Coola. Two of the most accessible sections of lavas with excellent exposures of a variety of features are at Bull Canyon on the Chilcotin River (8.5 km west of Alexis Creek, B.C., on Highway 20) and the Painted Chasm (15 km north of Clinton, B.C. on Highway 97; follow signs ~5 km east to Chasm Provincial Park). Mount Begbie, a gabbro plug and probable feeder for some of the basalts, is accessible on the east side of Highway 97, ~2 km north of Eightythree Mile Creek.*

References

Bevier, M. L., 1983, Regional stratigraphy and age of Chilcotin Group basalts, south-central British Columbia: *Can. J. Earth Sci. 20*, 515-524.

Mathews, W. H., in press, Geology and age of Chilcotin Group basalts and implications for Neogene paleogeography in south-central British Columbia: *Can. J. Earth Sci.*

Mary Lou Bevier

WELLS GRAY–CLEARWATER
Canada

Type: *Monogenetic volcanic field*
Lat/Long: *52.00°N, 120.00° W*
Elevation: *600 to 2,100 m*
Eruptive History: *Pleistocene to Holocene; subhistoric activity*
Composition: *Basanite to alkali olivine basalt*

The Wells Gray–Clearwater area of east-central British Columbia contains a wealth of alkalic basaltic rocks of Quaternary age. These basalts represent small (<1 km^3 of lava) eruptions from spatially and temporally isolated vents. The deposits include subaerial, subaqueous, and subglacial material. Based on radiocarbon and whole-rock K-Ar dating, eruptions started in the early Pleistocene and continued into the Holocene. The latest eruption (**Kostal Cone**) perhaps occurred as recently as 400 years ago, based on tree-growth data.

Pleistocene valley-filling and plateau-capping lava flows have an estimated total volume of 25 km^3. The emplacement of these flows spanned at least three periods of glaciation, evidence for which is preserved in the form of tuyas, ice-ponded valley deposits, and subglacial mounds. Subaqueous explosions produced piles of hyaloclastites; interaction of groundwater and magma created several pit craters. In a few places glacial till and fluvial sands and gravels are preserved beneath the flows. Paleosols are found, but are rare. All pre-Holocene deposits are modified by glacial ice and are deeply dissected by fluvial erosion. Glaciation has left a thick blanket of till over most of the volcanic deposits; thus outcrop is largely limited to cliffforming exposures in the valleys.

Post-glacial volcanic activity occurred in three areas: Spanish Creek, Ray Lake, and Kostal Lake. Volcanism in the Spanish Creek and Ray Lake area was synglacial but continued into post-glacial time. Two cinder cone complexes,

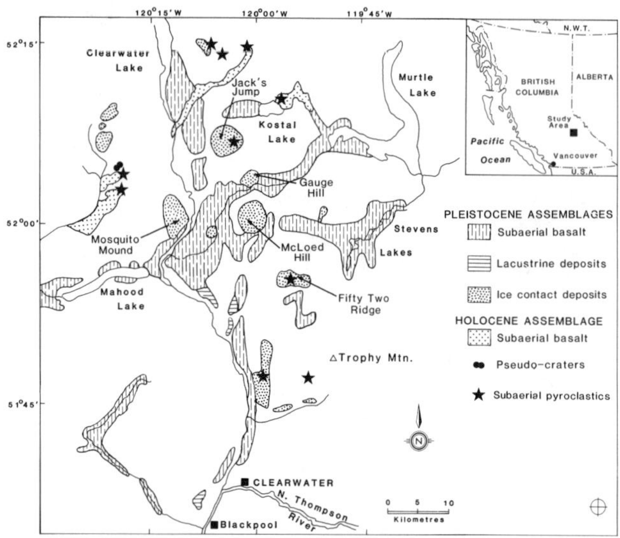

WELLS GRAY–CLEARWATER: Schematic map of Pleistocene and Holocene volcanic units.

WELLS GRAY–CLEARWATER: ***Pyramid*** *Mountain is a subglacial mound made of an accumulation of hyaloclastites. If its subglacial eruption had broken through the ice cover, the volcano would have a flat-topped tuya shape.*

the largest rising 250 m above its base, were formed in the Spanish Creek area. Eruptions near Ray Lake built a cinder cone and culminated with a 16-km-long aa lava flow. The flow is at least 15 m thick at the proximal end, but thins to 3 m at the distal end. Tree molds are preserved within the flow at the lower end.

Holocene flows are more alkalic than the Pleistocene flows and contain abundant xenoliths of Cr-spinel lherzolite, spinel clinopyroxenite, and rare ferroan websterite and spinel wehrlite. Xenoliths are not found in the older flows; however, chemical evidence suggests that all the lavas were generated in a similar manner by low degrees of partial melting. The melts originated in the upper mantle which, over time, was progressively depleted by each subsequent melting event.

How to get there: *Most of the Wells Gray–Clearwater volcanic complex lies within Wells Gray Provincial Park. The park is ~100 km north of Kamloops, the largest nearby town, and is accessible by paved highway. Once in the park, some of the complex can be reached by paved and gravel roads, but many areas are accessible only by foot or by aircraft.*

References

Hickson, C. J., 1986, Quaternary volcanism in the Wells Gray–Clearwater area, east-central British Columbia. Ph.D. Thesis, University of British Columbia, Vancouver, B.C., Canada.

Metcalfe, P., 1987, Petrogenesis of Quaternary alkaline lavas in Wells Gray Provincial Park, B.C., and constraints on the petrology of the subcordilleran mantle. Ph.D. Thesis, University of Alberta, Edmonton, Alberta, Canada.

Catherine J. Hickson

SILVERTHRONE
Canada

Type: *Caldera complex*
Lat/Long: *51.43°N, 126.30°W*
Elevation: *500 to 3,160 m*
Eruptive History:
0.75 to 0.4 Ma: Intracaldera rhyolite and dacite
Post-glacial (±1,000 yr): Marginal andesitic cones
Composition: *Calc-alkaline rhyolite, dacite, andesite and basaltic andesite*

The Silverthrone volcanic complex occupies a circular area ~20 km across in the central Coast Mountains of British Columbia. Dacite and rhyolite domes, breccia and flows, which comprise the older parts of the pile, have been deeply dissected and are now exposed in precipitous slopes extending from near sea level to elevations >3,000 m. Younger phases, consisting mostly of andesite and basaltic andesite flows

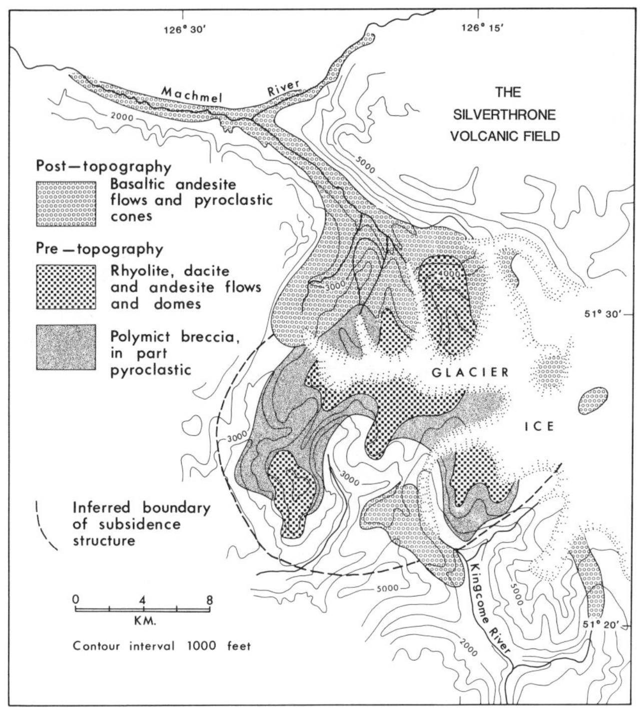

SILVERTHRONE: Geological map.

and pyroclastic breccia, postdate the present topography and have been only slightly modified by erosion.

Steep contacts between thick basal breccia of the Silverthrone pile and older crystalline rocks of adjacent peaks suggest that the breccia is part of an intercaldera succession. The presence of irregular subvolcanic intrusions and a profusion of dikes within the breccia, but not in adjacent country rock, provide further evidence of caldera structure. K-Ar dates of 0.75 Ma and 0.4 Ma on rhyolite domes above the basal breccia are consistent with the high rates of uplift and erosion recorded elsewhere in the Coast Mountains.

The younger andesitic rocks issued from a cluster of vents, now mostly ice-covered, around the periphery of the complex. At high elevations, proximal breccia and cinders from several eroded cones rest on coarse colluvium derived from the older parts of the volcanic complex. Flows from several vents on the north side of the complex extend down tributary valleys into Machmel River, where they coalesce into a broad valley flow at least 10 km long. Although radiometric dates are inconclusive, the blocky surface of the flow has been only slightly incised by high-energy streams. This, and the

SILVERTHRONE: Overlapping andesite (dark) and rhyolite (light) domes near the summit of Silverthrone Mountain.

presence of unconsolidated glacial fluvial deposits under the flow, suggest it is <1,000 years old.

How to get there: *Silverthrone Mountain is in a remote and exceptionally rugged part of the Coast Mountains of west-central British Columbia. It can be reached by charter helicopter or, with great difficulty, by trekking on foot along one of the many valleys extending from the coast or the interior plateau onto the ice fields of the Silverthrone area.*

Reference

Green, N. L., Armstrong, R. L., Harakal, J. E., Souther, J. G., and Read, P. B., 1988, Eruptive history and K-Ar geochronology of the late Cenozoic Garibaldi volcanic belt, southwestern British Columbia: *Geol. Soc. Am. Bull. 100*, 563-579.

Jack G. Souther

BRIDGE RIVER CONES
Canada

Type: *Monogenetic cones, small strato volcanoes, and tuyas*
Lat/Long: *50.8°N, 123.40°W*
Elevation: *1,520 to 2,500 m*
Eruptive History:
0.97 to 0.59 Ma: Tuyas and stratovolcanoes with ice-contact features
<1,500 yr BP: Post-glacial valley flows
Composition: *Alkaline basalt and hawaiite*

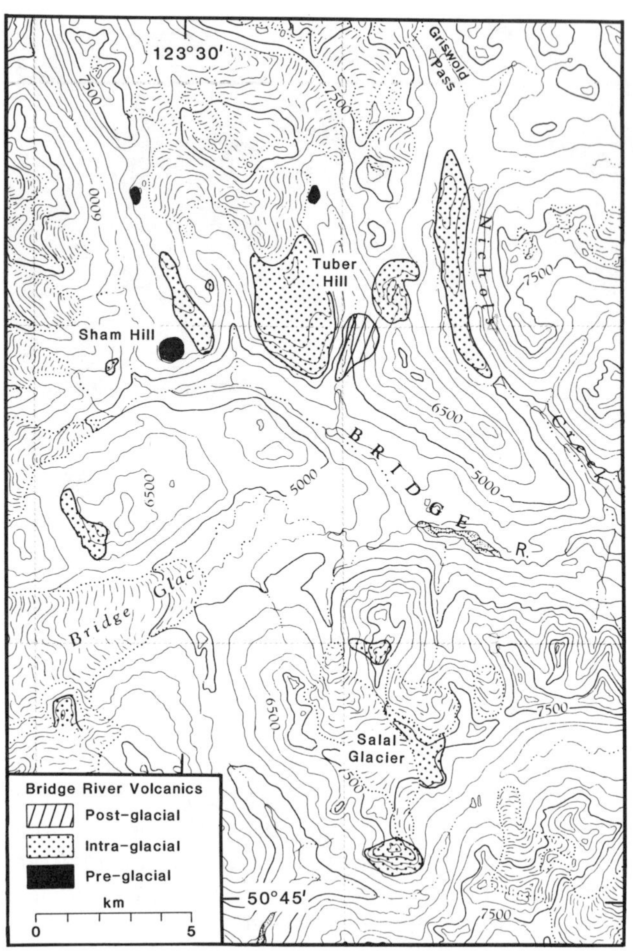

BRIDGE RIVER CONES: Map showing eruptive centers and remnants of Cenozoic volcanics in the northern Garibaldi Belt.

BRIDGE RIVER CONES: Pillow lavas and hyaloclastite resting on a thick succession of debris flows and lacustrine tuff on the southern edge of the Tuber Hill pile.

Small eruptive centers at the extreme northern end of the Garibaldi Volcanic Belt include deeply dissected pre-glacial volcanic necks and flow remnants, a group of younger intraglacial volcanoes, and a few still younger flow remnants which postdate the last retreat of ice from the main valleys.

Sham Hill (K-Ar age 1.0 Ma), a steep-sided volcanic neck ~300 m across and 60 m high, is typical of the older group. Its bare glaciated surface, strewn with glacial erratics, consists of large subhorizontal columns formed within the central conduit of an eroded stratovolcano.

The intraglacial volcanoes erupted when the major valleys were filled with pre-Wisconsin ice. Most of them include both subaerial and ice-contact deposits. In the **Salal Glacier** center (K-Ar 0.97 to 0.59 Ma) subaerial tephra and thin scoriaceous flows in the upper part of the pile are surrounded by ice-ponded flows up to 100 m thick. **Tuber Hill** (K-Ar 0.6 Ma) is a small basaltic stratovolcano erupted on the Bridge River upland when surrounding valleys were occupied by ice. Where distal flows encroached on the glaciers a marginal meltwater lake was formed in which >150 m of interbedded hyaloclastite, debris flows, and lacustrine tuff were depos-ited.

The youngest volcanic rocks in the northern Garibaldi belt are remnants of valley-filling basalt flows which rest on glacial till and clearly postdate the last major glaciation.

In contrast to the calc-alkaline character of the other volcanoes in the Garibaldi Belt, the small centers north of Meager Mountain are predominantly alkaline basalt and hawaiite, which may reflect a smaller degree of partial melting, or a descending-plate edge effect.

How to get there: *Access to the cones is by charter helicopter or on foot through difficult terrain with few trails. The cones can be reached by trekking 20 to 30 km north from a private logging road in Lillooet Valley or west for ~35 km along Bridge River from the mining camp of Gold Ridge.*

References

Lawrence, R. B., Armstrong, R. L., and Berman, R. G., 1984, Garibaldi Group volcanic rocks of the Salal Creek area, southwestern British Columbia: Alkaline lavas on the fringe of the predominantly calc-alkaline Garibaldi (Cascade) volcanic arc: *J. Volc. Geotherm. Res. 21*, 255-276.

Roddick, J. C., and Souther, J. G., 1987, Geochronology of Neogene volcanic rocks in the northern Garibaldi Belt, British Columbia: *Geol. Surv. Can. Pap. 87-2*, 21-24.

Jack G. Souther

MEAGER MOUNTAIN
Canada

Type: *Calc-alkaline stratovolcano*
Lat/Long: *50.6°N, 123.5 °W*
Elevation: *460 to 2,679 m*
Eruptive History: *Four episodes of activity:*
1.9 to 2.2 Ma
<1.9 to >0.5 Ma
1.0 to 0.5 Ma
0.15 to 0.002 Ma
Composition: *Rhyodacite to basalt*

The Meager Mountain volcanic complex is the northernmost volcano in the Garibaldi Volcanic Belt, an extension of the Cascade volcanic belt into Canada. It is a Tertiary to Quaternary edifice exhibiting at least eight vents which produced mafic to felsic rocks. Numerous feeder dikes to older units are exposed by deep erosion. The volcano is dominated by porphyritic andesite to rhyodacite lava and pyroclastic breccias, although several Quaternary basalt flows and breccias occur on the periphery. Plagioclase porphyritic andesite lava flows and breccia (0.5 to 1.0 Ma) are the most voluminous rocks, with a maximum of 1,200 m of total flow thickness. The most recent volcanic activity occurred 2,350 yr BP and produced three distinct units known as the Bridge River Assemblage. The lowest member is an air-fall deposit up to 2 m thick grading from rhyodacitic pumice blocks and lapilli to fine white-gray volcanic ash. This deposit is overlain by a pyroclastic flow up to 4 m thick and a 20-m-thick, 3-km-long rhyodacite flow. A unique breccia unit 145 m thick occurs at the toe of the flow, and appears to be disrupted flow material which filled a valley ahead of its parental lava. All units of the Bridge River Assemblage share similar mineralogy and many of the same petrographic relations (such as xenocrystic clots of biotite, plagioclase, pyroxene, and oxides, and large plagioclase phenocrysts with remelting features), suggesting that their eruption represents the emptying of a single magma reservoir.

MEAGER MOUNTAIN: The ***Devastator*** *is a dissected andesitic volcanic neck which was the source area for a thick sequence of 0.5-1.0 Ma andesite lava flows. Erosional remnants of these flows form the stratified crags of Pylon Peak (background and right). Unstable slopes of the Devastator are the source for many Recent debris flows, and consist of weak, hydrothermally altered felsic rocks. Photograph by J. K. Russell.*

A group of dissected rhyodacitic volcanic necks forms the highest peaks in the area and are flanked by their eruptive products. These necks are the upper levels of intrusions and provide a unique opportunity to study the relationships between magma chambers and their lavas. Two peripheral clusters of natural hot springs occur within the complex. For this reason the area has been investigated as a potential hydrothermal energy resource.

How to get there: *Meager Mountain lies 150 km north of Vancouver, B.C. From the town of Pemberton the paved Pemberton Meadows road followed by the graded, gravel Lillooet Forest Service Road provide excellent access to the south, east, and north parts of the complex. The Forest Service enforces a strict closure on the gravel roads from Monday to Friday, 6 a.m. to 6 p.m., for logging purposes. Access is unrestricted on weekends except for certain gated roads in the Meager Creek valley. Contact CRB logging (in Squamish) for more information and for access to gated roads before attempting a visit.*

References

Green, N. L., Armstrong, R. L., Harakal, J. E., Souther, J. G., Read, P. B., 1988, Eruptive history and K-Ar geochronology of the late Cenozoic Garibaldi volcanic belt, southwestern British Columbia: *Geol. Soc. Am. Bull. 100*, 563-579.

Read, P. B., 1977, Geology of Meager Creek geothermal area, British Columbia: *Geol. Surv. Can. Open-file 603.*

Mark Stasiuk

CAYLEY
Canada

Type: *Stratovolcano and lava domes*
Lat/Long: *50.10°N, 123.30°W*
Elevation: *260 to 2,377 m*
Eruptive History:
4 to 0.6 Ma: Stratovolcano and plug domes
0.3 to 0.2 Ma: Satellitic domes and flows
Composition: *Dacite, rhyodacite, and andesite*

Mount Cayley, the largest volcano in the central Garibaldi Belt, is a deeply dissected stratovolcano of dacite and rhyodacite. Its precipitous central edifice was formed during two stages of activity. The first (4–0.6 Ma) began with the eruption of dacite flows and pyroclastic breccia, and culminated with the emplacement of a central plug dome which forms the present summit spires of Mount Cayley. Welded vent breccia erupted during this second stage forms the present craggy summit of **Vulcan's Thumb** ridge on the south flank of Mount Cayley. Prolonged erosion, which destroyed most of the original stratovolcano, was followed by the eruption of satellitic vents. This third and final stage of activity (0.3-0.2 Ma) began with the effusion of a dacite lava flow into the present valley of Shovelnose Creek and culminated with the emplacement of two small satellitic domes.

The unstable, oversteepened southwestern flank of Mount Cayley is the source of repeated rockfalls, avalanches, and periodic debris flows which frequently damage roads and bridges in the Squamish valley.

Thermal springs associated with Mount Cayley have made it a target for geothermal exploration. Bottom hole temperatures of 50°C and thermal gradients of >100°C/km have been measured in shallow boreholes on the southwest side of the mountain.

How to get there: *From the town of Squamish, ~40 km north of Vancouver on Highway 99, a private logging road follows Squamish River north to within 2 km of Mount Cayley. Special permission is necessary to enter the logging area.*

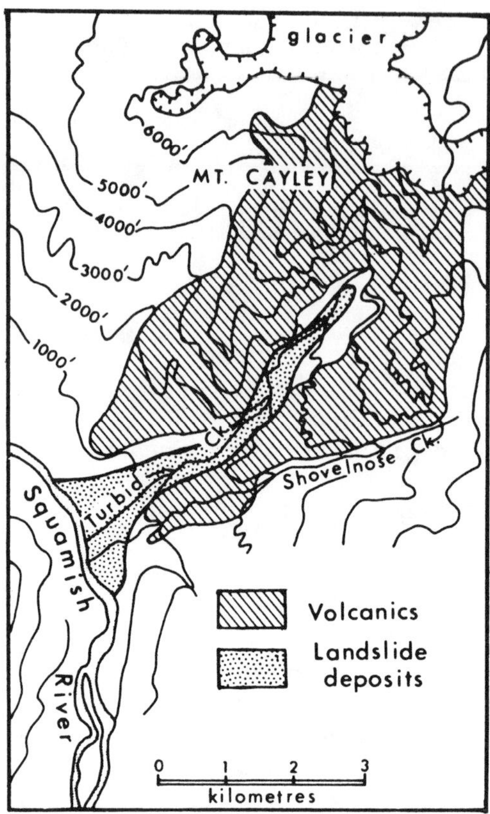

CAYLEY: Map showing volcanic rocks and recent debris flow deposits.

CAYLEY: View from northeast showing light colored breccia cut by a central spine of dacite which forms the summit ridge.

References

Souther, J. G., 1980, Geothermal reconnaissance in the central Garibaldi Belt, British Columbia: *in* Current Research, Part A, *Geol. Surv. Can., Pap. 80-1A*, 1-11.

Souther, J. G., and Dellechaie, F., 1984, Geothermal exploration at Mt. Cayley – a Quaternary volcano in southwestern British Columbia, *in Trans. Geotherm. Res. Council*, 1984 Ann. Mtg., 463-468.

Jack G. Souther

GARIBALDI LAKE
Canada

Type: *Stratovolcanoes, lavas, and a cinder cone*
Lat/Long: *49.92° N, 123.03°W*
Elevation: *290 to 2,316 m*
Eruptive History: *1.0-1.3 Ma: stratovolcanoes; 0.3-0.11 Ma: stratovolcanoes, plug domes, and flows; Early Holocene: lava flows*
Composition: *Andesite, dacite, basaltic andesite, and basalt*

Pleistocene-Holocene eruptions constructed a line of three andesite volcanoes (Black Tusk, Mount Price, and The Table) and several small basaltic andesite complexes in the Garibaldi Lake area; basaltic lavas were extruded within Cheakamus River valley, 4 km northwest of the main volcanic front.

The oldest and most striking andesite volcano is **Black Tusk**, a glacially dissected complex built during two stages of activity. The first (1.1-1.3 Ma) produced hornblende andesite flows and lithic tuffs which form bluffs northwest, southwest, and southeast of the main volcanic edifice. Prolonged erosion destroyed the original cone, and was followed by effusion of hypersthene andesite flows which locally terminate with precipitous (100 m) ice-contact margins. This eruptive activity (0.17-0.21 Ma) culminated with extrusion of an endogenous dome and related lava which form the present summit (2,316-m) spire. The late Pleistocene ice sheet subsequently incised a deep, north-trending U-shaped valley into the eastern flank of the second-stage cone. This valley was filled by 0.11-Ma basaltic andesite and 0.4-Ma basalt flows that emanated from **Cinder Cone**, 2 km east of the Black Tusk complex.

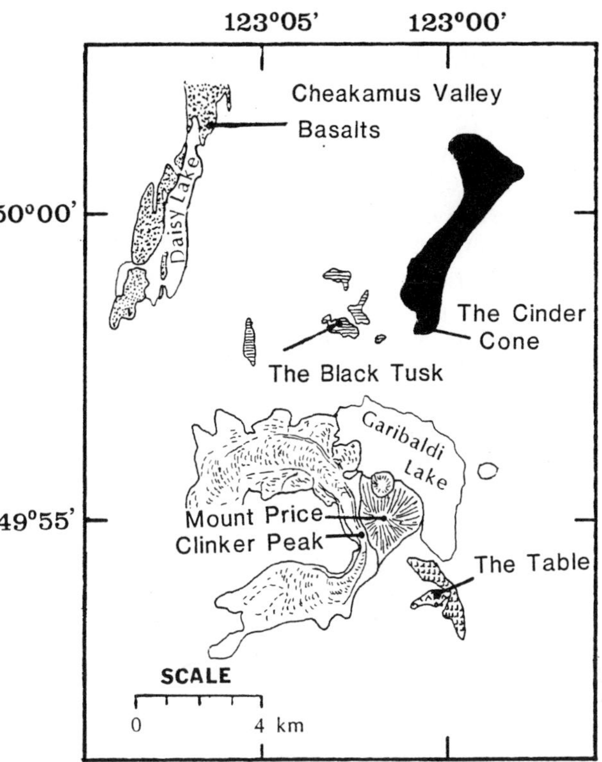

GARIBALDI LAKE: Map showing distribution of Pleistocene-Holocene volcanic rocks.

GARIBALDI LAKE: Looking east with Mount Price, satellite cone on north flank, and Barrier flow emanating from Clinker Peak summit. USGS photograph by A. Post.

The Mount **Price** complex, 5 km south of Black Tusk, formed during three periods of activity. Initial 1.2-Ma eruptions constructed a hornblende andesite stratovolcano on the drift-covered floor of a cirquelike basin. The focus of volcanic activity then shifted westward, where 0.3-Ma eruptions of andesite-dacite lavas and Pelean pyroclastic flows formed the nearly symmetrical (2,050-m) Mount Price volcano. This cone was overridden by the continental ice-sheet before eruptions at a satellite vent on its northern flanks. Possibly contemporaneous activity occurred at **Clinker Peak** on the western shoulder of Mount Price. Two hornblende-biotite andesite flows, which spread 6 km northwest and southwest from the breached lava ring, were ponded (>250 m thickness) against the continental ice sheet. These lavas postdate disappearance of the Cordilleran ice sheet from higher altitudes, but predate its disappearance from lower elevations in early Holocene time.

The Table, a hornblende andesite tuya situated ~3 km southeast of Mount Price, rises precipitously 305 m above glaciated basement rocks. The edifice formed by effusion of flat-lying flows within a pit thawed through the continental ice sheet. Absence of glacial erratics on its summit and lack of erosional features attributable to glaciation suggest eruptions occurred during early Holocene time.

How to get there: *From Squamish, ~40 km north of Vancouver, follow Highway 99 within Cheakamus River valley to Daisey Lake, and turn off to Rubble Creek parking lot in Garibaldi Provincial Park. The Garibaldi Lake area may be reached from the parking lot via a graded, 7.5-km-long hiking trail.*

References

Mathews, W. H., 1952, Ice-dammed lavas from Clinker Mountain, southwestern British Columbia: *Am. J. Sci. 250*, 553-565.

Green, N. L., 1981, Geology and petrology of Quaternary volcanic rocks, Garibaldi Lake area, southwestern British Columbia: *Geol. Soc. Am. Bull. 92*, Pt. I, 697-702; Pt. II, 1359-1470.

Nathan Green

GARIBALDI
Canada

Type: *Composite cone and domes built on glacier*
Lat/Long: *49.83° N, 123.00°W*
Elevation: *2,678 m*
Height: *~700 m*

GARIBALDI: Topographic map.

GARIBALDI: Looking north, with flat-topped ***The Table,*** *a tuya, on the north flank of Garibaldi.*

Volcano Diameter: *3 × 5 km*
Eruptive History: *Early activity 0.26 -0.22 Ma; Atwell Peak ~13,000 yr BP; Opal Cone ~post-Wisconsin glacial stage; no historic activity*
Composition: *Dacite*

Mount Garibaldi is one of the larger volcanoes (6.5 km^3) in a chain of small Quaternary volcanic piles – the Garibaldi Belt – which trend N25°W within the southern Coast Mountains of British Columbia. Mt. Garibaldi is noteworthy both for the excellent exposures of its internal structure and for its striking topographic anomalies, which can be attributed to the growth of the volcano onto a major glacial stream, part of the Cordilleran Ice Sheet, and the subsequent collapse of the flanks of the volcano with the melting of the ice.

The western slopes of the mountain reveal basement rocks, sheared and altered quartz diorite, sculptured by streams and glaciers into a rugged topography with relief up to 1,800 m. Valleys in this rough surface have been filled with 0.51 to 0.22 Ma dacite and andesite flows, tuff breccias, and domes, precursors of the activity at Mt. Garibaldi.

Eruptions from the site of the south summit (**Atwell Peak**) then created a conical pile of tuff-breccia at least 700 m thick at its apex. The tuff-breccia includes blocks of banded dacite up to 17 m across, many of which display reddened bread-crust surfaces and radial contraction cracks, showing that they were transported and deposited, still hot, by Pelean glowing avalanches. Stratification, revealed mainly by variations in block size, dips 12°-15° away from a more or less massive central core of gray-banded hypersthene dacite, which forms the south summit and the steep, highly unstable slopes leading down from it. This core is the root of a large dome, the source of the Pelean avalanches.

Unweathered glacial erratics are found resting on a remnant of the original top surface of the tuff-breccia deposit up to, but not above, the 1,660-m level, almost 300 m lower than expected had volcanism here preceded the climax of Wisconsinan glaciation. Accordingly, the Pelean eruptions are assigned to an early stage in the retreat of the Cordilleran Ice Sheet, ~15,000 to 13,000 radiocarbon years ago. Only where basement rocks extend above the 1,400-m level do the tuff-breccias lie undisturbed; below this level they are considered to have been deposited on glacier ice. With continuing retreat of the ice the support was withdrawn from those parts of the tuff-breccia cone which had been built onto the ice, causing collapse in a series of landslides which ultimately exposed the inner parts of the cone.

Later volcanism from the western summit formed lava flows which mantled the landslide headwall on the west side of the mountain. About the same time a satellite vent, **Opal Cone**, 3.5 km southeast of the summit, gave rise to a voluminous (4.5 km^3) hornblende-biotite dacite flow which moved 20 km down Ring Creek without encountering any residual glacial ice. These eruptions have been assigned to early Holocene time. There has been no subsequent eruption at Mount Garibaldi.

The close association of volcanism and glaciation in the Garibaldi volcanic belt raises the intriguing question – was glaciation an important trigger for volcanism?

How to get there: *80 km north of Vancouver, B.C. in the Garibaldi Provincial Park. Good views of Mount Garibaldi at ~25 and 55 km north of Vancouver along Highway 99. Hiking is required to reach Garibaldi itself.*

References

Green, N. L., Armstrong, R. L., Harakal, J. E., Souther, J. G., and Read, P. B., 1988, Eruptive history and K-Ar geochronology of the late Cenozoic Garibaldi Volcanic Belt: *Geol. Soc. Am. Bull. 100*, 563-579.

Mathews, W. H., 1952, Mount Garibaldi, a supraglacial Pleistocene volcano in southwestern British Columbia: *Am. J. Sci. 250*, 81-103.

William H. Mathews

WESTERN USA FIGURE 1: Volcanic areas in the continental USA younger than 5 Ma. Modified from Smith and Leudke (1984).

VOLCANO TECTONICS OF THE WESTERN USA

Charles A. Wood and Scott Baldridge*

* *Rio Grande Rift–Jemez Zone section.*

The American West is one of the most unusual volcanic provinces on Earth. It is an extraordinary wide region of contemporaneous volcanism, stretching 1,800 km from east of the Rocky Mountains to west of the Cascades (103° to 123°W) (Fig. 1). It encompasses a very wide variety of volcanic landforms and diverse rock compositions, with the only consistency being small basalt fields that are scattered throughout the whole vast region. Intriguingly, except for the Cascadian segment, there is no currently active plate subduction to drive the volcanism, although there are hot spots, persistently "leaky" faults, and rifting.

This cacophony of volcanism was nearly impossible to understand (e.g., Gilluly, 1965) until the breakthrough plate tectonic interpretation of western North America by Atwater (1970). Lipman et al. (1971) and Christiansen and Lipman (1972) immediately reinterpreted all of Cenozoic volcanism in North America in light of this new idea, creating an enduring framework for understanding American volcanism. As McKee and Noble (1986, p. 39) point out, many "later papers have refined but not fundamentally changed these early concepts."

Atwater recognized from seafloor magnetic anomalies that an oceanic spreading center was located off western North America until mid-Tertiary time and that a subduction zone must have existed with concomitant arc volcanism on the continent. Lipman, Christiansen, and colleagues showed that from ~40 to ~18 Ma subduction of the eastern limb (named the Farallon Plate) of the East Pacific Rise spreading center produced widespread calc-alkaline rocks, with very little basalt anywhere in western North America (McKee and Noble, 1986). At ~30 Ma the East Pacific Rise collided with North America in the vicinity of northern Mexico. The collision zone (which Lipman and Christiansen believe destroyed both the spreading ridge and the trench) migrated northward to its present position near Cape Mendocino, California, progressively forming the San Andreas fault near the coast and cutting off the calc-alkaline volcanism further inland. By ~18 Ma there had been a profound change in volcanism and tectonism, with arc-type volcanic activity replaced by abundant basaltic and bimodal, basaltic-rhyolitic lava fields. Today's volcanism is largely controlled by extensional tectonic regimes developed at that time.

In a sense, volcanism during the last 5 Ma in the American West is the final dribble of activity in this much grander mid-Tertiary story of large ash-flow calderas with voluminous ignimbrite eruptions, extreme crustal thinning and extension, and high mountain building. The processes that produced these epic events are still disputed (e.g., Sonder et al., 1987) nearly 20 years after the revolutionary contributions of Atwater, Lipman, and Christiansen. However, the relation of much of the younger volcanism (the subject of this book) to tectonism is generally understood, although the underlying reasons for the existence of particular tectonic structures are often uncertain. As a simplification, six interconnected volcano-tectonic provinces are described that include most of the young (<5 Ma) volcanic landforms in the western United States. These provinces combine many of 23 rectilinear volcanic zones defined by Smith and Luedke (1984).

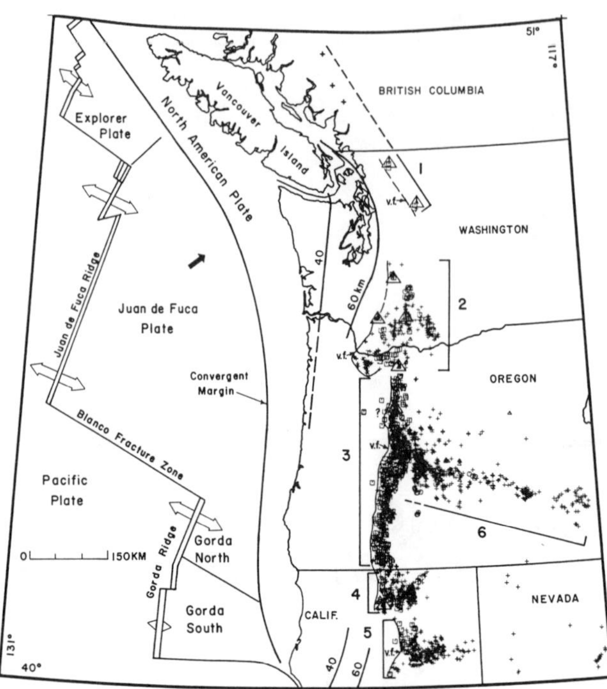

WESTERN USA FIGURE 2: Map showing <5-Ma vents in the Pacific Northwest, plate-tectonic features, volcanic segments numbered 1–6, and depth of seismicity (40 and 60 km contours). Map from Guffanti and Weaver (1988).

Cascades

This volcanic province has the best understood tectonic setting in the western USA because subduction of remnants of the Farallon Plate is apparently still driving continental arc volcanism. The Cascade region is not a typical subduction zone, however, for there is very little seismic evidence of active subduction (Weaver and Baker, 1988) and there is no trench (McBirney and White, 1982). In fact, the existence of volcanic activity in the Cascades is the best evidence for ongoing subduction.

The remaining part of the Pacific Plate currently converging with the American Northwest (Fig. 2) is the Juan de Fuca Plate, with small platelets at its northern (Explorer Plate) and southern (Gorda Plate) terminations. The Explorer Plate separated from the Juan de Fuca ~4 Ma and is apparently no longer being subducted (Hyndman et al., 1979); the Gorda split away between 18 and 5 Ma (Riddihough, 1984). The present slow rate of convergence (3-4 cm/yr) of the Juan de Fuca Plate is only about half its value at 7 Ma (Riddihough, 1984), which probably explains the reduced seismicity, lack of a trench, and debatable decline in volcanic activity.

New evidence for a possible subducting slab comes from a very large electromagnetic sounding experiment across the Juan de Fuca Plate to eastern Oregon and Washington, which detected a laterally continuous conductor extending to the High (volcanic) Cascades (EMSLAB Group, 1988). Preliminary modeling suggests an ~18° dipping slab extending from the convergence zone eastward to a depth of ~40 km under the Coast Range. From there to the High Cascades the conductor thickens (from 20 to 40 km beneath the surface) and is horizontal. This "horizontal slab" terminates abruptly against a very thick conductor associated with the Basin and Range. From this preliminary interpretation a number of obvious questions arise, chief of which is how could relatively typical volcanic arc-type magmas be generated at 40-km depth (versus 100 to 110 km as in most arcs)? Perhaps the present, geophysically inferred structure developed since the generation of the magmas that produced the Cascade volcanoes. If the Cascades were recently extinct, this might be a reasonable option, but the vigorous activity, especially during the past century, argues that magmas are still rising from source regions, apparently through the horizontal slab, to drive volcanism. Interestingly, the subducted slab is much better defined in southern British Columbia (Clowes et al., 1987).

Holocene volcanism in the Cascades extends from the Garibaldi Volcanic Belt in southern British Columbia to the Lassen volcanic complex in northern California. Pronounced differences in the nature of volcanism occur along the arc. In Washington there are five, generally large, widely spaced stratovolcanoes, with only one (Mount Adams) having significant nearby basaltic volcanics. In marked contrast, Oregon has six generally smaller stratovolcanoes, but the entire state is traversed by a 40-50-km-wide band of basaltic to andesitic lava shields, cinder cones, and smaller stratovolcanoes that the "Cascade" cones rise above. South of Crater Lake, the Cascade arc bends perceptibly toward the southeast, and continues along this trend to Lassen peak. Both Lassen and Shasta are associated with eastward extending halos of mafic shields and lava fields which, near Shasta, culminate in the huge shield volcano of Medicine Lake.

Guffanti and Weaver (1988) used the locations of 2,821 vents shown on the maps of Luedke and Smith (1981, 1982) to divide the Cascades into five segments, with a sixth extending southeastward across the High Lava Plains (Fig. 2). They note a volcano gap between Rainier and Glacier Peak which coincides with the shallowest dip (11°) of the Cascade subduction zone, and they also infer a change in the configuration of the subducting slab between Shasta and Lassen. Even though their segment boundaries differ from those of Hughes et al. (1980), Guffanti and Weaver similarly find that a number of otherwise inexplicable features of Cascade volcanism are controlled by segmentation.

Some researchers (e.g., Christiansen and Lipman, 1972) have suggested that Sutter Buttes and the Sonoma and Clear Lake volcanics, south and southwest of Lassen, are older extensions of subduction-related Cascade volcanism. This seems unlikely. If Sutter Buttes were part of a series of older Cascade stratovolcanoes abandoned due to the northward migration of the south end of Juan de Fuca Plate, the "last Cascade volcano" hypothesis would be tenable. But northward, arc volcanoes are young and active. In fact, why do the Cascades have such an abrupt southern termination?

The Clear Lake and Sonoma volcanics are the <5-Ma components of a northwesterly younging line of volcanic fields of Tertiary to Holocene age (Hearn et al., 1981). All these volcanics lie within the San Andreas fault system, which appears to have provided magma access to the surface. Hearn et al. point out that the timing of the volcanism suggests that it follows termination of subduction, as the Mendocino triple junction migrated northward. They also propose that the volcano alignment reflects an underlying hot spot. That suggestion seems inconsistent with the northward movement of the Pacific Plate which most of the volcanics ride. These volcanics are among the closest to a subduction plate boundary of any in the world and will repay closer tectonic investigation. Similarly, a tiny sliver of basalt dated at 3.57 Ma (Prowell, 1974, quoted in Luedke and Smith, 1981) occurs 45 km east of Santa Cruz, California near the Calaveras and Hayward faults. Apparently leakage of basalts along the San Andreas fault system has occurred repeatedly.

The petrology of Cascade volcanism has been reviewed by McBirney and White (1982). The large Cascade stratovolcanoes are made of basaltic andesites and andesites with smaller quantities of more silicic, calc-alkaline magmas. In Oregon, many of the large volcanoes are underlain by overlapping shields of high alumina basalt which is by far the dominant volume of erupted material. McBirney and White note that most petrologic and isotopic evidence suggests that Cascade magmas rise from the mantle with little involvement of subducted slab or sediments. This conclusion is at odds with the interpretation of the electromagetic sounding experiment (EMSLAB Group, 1988), as discussed above.

Snake River Plain–Yellowstone

An 80-km-wide swath of basaltic and rhyolitic volcanism cuts across southeast Idaho for 450 km. This Snake River Plain–Yellowstone (SRPY) volcanic province is the most dynamic area of volcanism in North America. This is not because of abundant historic eruptions – there have been none – but rather because of its rapid motion. SRPY is propagating to the northeast at

3.5 cm/yr (Armstrong et al., 1975); it will slice through Montana and be at the Canadian border in ~20 million years, if past activity is a guide. SRPY doesn't simply cover terrain with volcanic rocks, but rather the pre-existing ground subsides up to 6 km (Braile et al., 1982) between major faults (Sparlin et al., 1982) and is further churned up by the transit of magma and the formation of magma chambers. SRPY is a geologic roto-tiller.

According to the radiometric dating of Armstrong et al. (1975), SRPY activity began ~15 Ma with silicic volcanism in southern Idaho. A series of now buried rhyolitic calderas formed in a northeast progression, with abundant basaltic volcanism lagging behind by 2-5 million years. Island Park and the two Yellowstone calderas are the most recent manifestations of the silicic volcanism, and Island Park is now being colonized by the basaltic wave of magma. Braile et al. (1982) combined the age migration data with a thermal model by Brott et al. (1978) into a proposed evolutionary sequence of events, which is slightly modified here. An initial thermal perturbation causes uplift of the crust and silicic caldera formation. Rapid upwelling of basic magma from the upper mantle drives the volcanism and yields silicic magmas by partial melting of two different source materials, the upper mantle for the basalts, and perhaps lower-crust metamorphic rocks for the rhyolites (Christiansen, 1984). Northeastward movement of the underlying "hot spot" leads to cooling of the massively intruded upper crust which subsides. Basalt direct from the mantle leaks to the surface through many dikes producing the "plains style" volcanism (Greeley, 1982) characteristic of the Snake River Plain.

SRPY was an axis for rhyolitic and basaltic volcanism before 5 Ma, but mostly basalt has been erupted since then. The western Snake River Plain is tectonically related to the northwest-trending line of volcanism along the Brothers fault zone of southern Oregon. MacLeod et al. (1976) discovered a well-defined younging of silicic domes from ~11 Ma at the Oregon, Idaho, Nevada triple point to <0.5 Ma near Newberry caldera. Basalt also occurs along this northwest trend, and as in the Snake River Plain, young (<0.01-Ma) flows occur throughout the zone.

The large-scale tectonics of SRPY and its mirror image, the Brothers fault zone volcanics, are intimately related to the extension in the western United States. The two lines of volcanism form an abrupt northern boundary to the Great Basin (McKee and Noble, 1986), and volcanism along each line youngs away from their common point. Eaton et al. (1978) noted a remarkable bilateral symmetry of topography and magnetic and gravitational fields about a roughly north–south line passing through Eureka, Nevada, and extending north to the Brothers–SRPY junction. Eaton (1982) proposes that the continental lithosphere in the Basin and Range has undergone a broadly distributed rifting just as did eastern North America during the early stages of formation of the Atlantic Ocean.

Colorado Plateau

All the volcanic provinces in North America are named after regions where volcanism is concentrated, except for the Colorado Plateau, which has been a zone of volcanic avoidance since at least the Mesozoic. As summarized by Stewart (1978) the Colorado Plateau is a tectonically stable region, centered southwest of the Four Corners area, that has generally escaped serious deformation when terrain all around it was being heavily faulted and pierced by volcanism. The Colorado Plateau was uplifted, perhaps starting as early as 24 Ma, although most activity probably occurred from 10 to 5 Ma. Thompson and Zoback (1979) proposed that the uplift resulted from warming and expansion of a gently dipping subducting slab which was abandoned ~20 Ma when subduction steepened.

Much of the volcanism in Utah, Arizona, New Mexico, and Colorado occurs remarkably near the boundary of the Colorado Plateau. While it is tempting to suggest that the young (<5-Ma) volcanism occurred along peripheral fractures formed when the Colorado Plateau was uplifted in the late Cenozoic (e.g., 880 m during the last 5.5 Ma; Lucchitta, 1979), maps in Smith and Luedke (1984) and in Stewart (1978) show that volcanism has hugged the Colorado Plateau since the Mesozoic. Perhaps repeated vertical movement over tens of millions of years has periodically reopened paths up bounding faults for magma migration.

Volcanism along the northern margins of the Plateau is expressed as a series of small

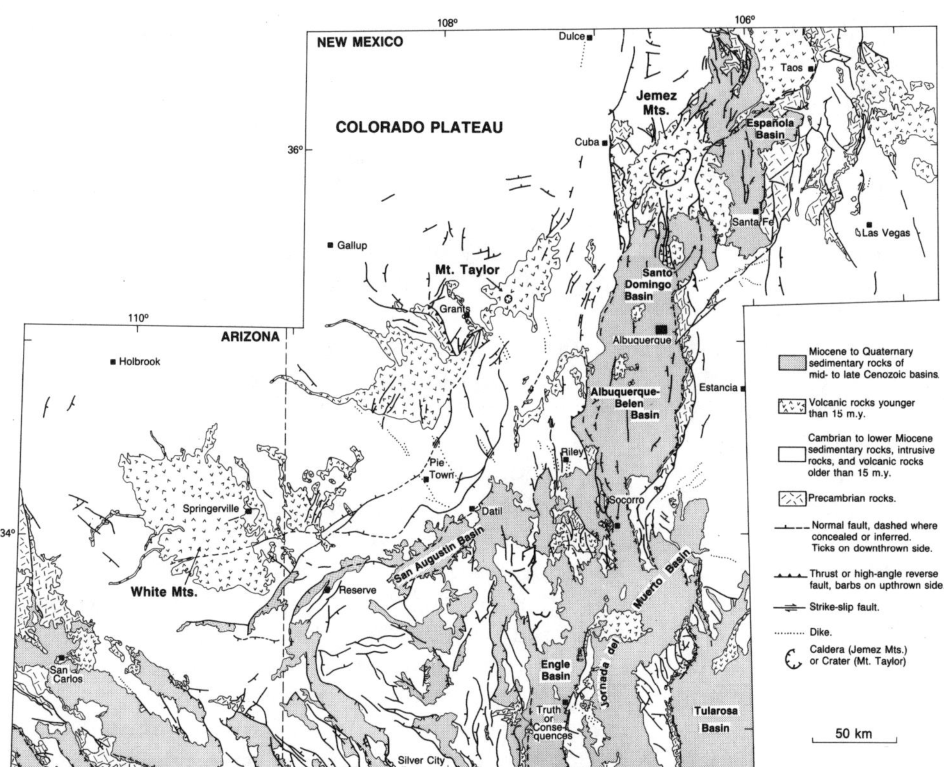

WESTERN USA FIGURE 3: Tectonic map of the central Rio Grande rift and southeastern Colorado Plateau, showing northeast-trending alignment of late Cenozoic volcanic centers (Jemez Lineament). After Baldridge et al. (1983).

monogenetic fields, with examples in northwest Colorado being remarkably isolated, small, and little known. The Dotsero flows are reported to be ~4,100 yr old (Simkin et al., 1981) – why did one of the younger eruptions in North America occur in Colorado, the least volcanically active state in the American west? In Arizona and New Mexico the Colorado Plateau-bounding volcanism consists of much larger mafic volcanic fields with silicic centers such as San Francisco Peak, Mount Baldy (Springerville field), and Mount Taylor. Some of these fields may owe their large size to the intersection of Colorado Plateau faults with the Jemez Lineament.

Rio Grande Rift–Jemez Zone

Volcanism in New Mexico is controlled by two major tectonic features (Fig. 3). The largest volcanic fields, including basalt to rhyolite central volcanoes and basaltic fields erupted from monogenetic cones and low shields, lie along a northeast-trending line, the Jemez lineament or zone. In contrast, nearly all of the isolated small fields of dominantly basaltic flows and cinder cones occur along the north–south-striking Rio Grande Rift. The largest central volcanic complex in New Mexico (the massive Jemez volcanic field) and the most voluminous basalt field (Taos Plateau volcanic field) lie at the intersection of these two tectonic features.

The Jemez Lineament, defined mainly as a broadly linear, northeast-trending array of late Cenozoic volcanic centers, extends 600 km from the Raton–Clayton volcanic field in northeastern New Mexico to the Springerville and White Mountains volcanic fields in east-central Arizona (see, e.g., Mayo, 1958). The lineament separates the axial grabens of the Rio Grande rift and a broad transition zone along the southeastern margin of the Colorado Plateau from a less deformed Plateau "core" to the northwest. However, it is not demarked by faults or fracture zones, except in the Rio Grande rift, and does not correspond to any single, simple structure in the upper crust (Baldridge et al., 1983). In the rift, the lineament corresponds to a series of faults which, in part, transfer crustal extension from the San Luis graben, north of the lineament, to the Espanola graben to the south (Aldrich, 1986). The Jemez lineament very likely corresponds to a major boundary or zone of weakness in the deep lithosphere.

Late Cenozoic volcanism along the lineament began earliest in the Jemez volcanic field (>13 Ma; Gardner et al., 1986) and in the White Mountains volcanic field (~8 Ma). Mount Taylor and most of the basaltic fields formed after 5 Ma. Basaltic volcanism occurred as recently as ~1,000 years ago near Grants, New Mexico (Crumpler, 1982). No systematic progression of ages, either in the inception of volcanism or in the beginning of silicic activity, occurs along the lineament, ruling out its origin as a hot spot trace (cf. Lipman, 1980).

The Rio Grande rift, a major Cenozoic continental rift, consists of a north-northeast-trending series of complex, asymmetrical grabens extending >1,000 km from central Colorado through New Mexico to west Texas and Chihuahua, Mexico (Olsen et al., 1987). The rift, closely related to the Basin and Range province, formed in response to extensional forces affecting the entire southwestern USA (Eaton, 1979). North of 34°N the rift is situated between the Colorado Plateau and the Great Plains (part of the North American craton), and is a well-defined physiographic and tectonic feature. In contrast, the southern rift, where the earliest and greatest extension occurred (Seager et al., 1984), is physiographically indistinguishable from the adjacent Basin and Range province. The rift is located at the crest of a broad, bilaterally symmetrical uplift extending for ~2,000 km across the southwestern and south-central USA (Eaton, 1987). The crust beneath the rift is thinned, ranging from 33 km in the north to 30 km in the south, compared to 45 km beneath the adjacent Colorado Plateau and 50 km beneath the Great Plains (Sinno et al., 1986).

Volcanism in the Rio Grande rift is minor compared to the great volumes of magma in parts of the East African rift (e.g., the Ethiopian and Kenyan portions), but comparable to that of the Rhine graben.

The amount and extent of magmatic rocks within or at the base of the crust is currently unknown, partially due to conflicting data. Lower crustal P-wave velocities (6.4-6.8 km/s), derived from wide-angle seismic refraction studies, do not suggest any large quantities of mafic rocks at depth beneath the rift axis. Yet heat flow and regional elevation data suggest that large quantities of magma may have been emplaced into or at the base of the crust beneath the entire region. Geophysical data, including S-wave shadowing and COCORP reflection profiling, suggest that a sill-like body of magma exists in the crust at a depth of 15-20 km near Socorro. No volcanic rocks that can be related to the magma chamber are present at the surface.

Compositions of basalt in the northern and central Rio Grande rift include a range of alkalic and tholeiitic basalt types. In the southern rift, basalt is dominantly alkalic. Recent detailed stratigraphic, compositional, and isotopic studies of basalt from the rift area show late Miocene to Holocene basaltic volcanism occurred in two distinct pulses, separated by a regionally significant lull in volcanism (e.g., Lucero volcanic field). Tholeiitic basalt is associated dominantly or entirely with the younger cycle. Neodymium and strontium isotope ratios in alkalic basalt correlate strongly with geographic location (Perry et al., 1988). Basalt erupted in areas of little crustal extension, such as the Great Plains and the northern rift, has relatively "enriched" isotopic signatures which are correlated with lithospheric mantle. Basalt from areas having undergone greater crustal extension, such as the southern rift and Basin and Range province, has relatively "depleted" isotopic signatures, indicative of asthenospheric mantle. The greater asthenospheric component is a consequence of thinned lithosphere beneath these areas.

Eastern California

A long zone of scattered volcanic centers extends from the Mono Lake area to the Salton Sea in eastern California and adjacent Nevada. Basaltic monogenetic volcanic fields are common along this diffuse zone, but it also contains various rhyolitic features including the Long Valley caldera and nearby domes. Excepting the Long Valley complex, there is a general trend for volcanism to be initiated earlier in the north and extend later (and have smaller volumes) in the south; a third of the fields also started or resumed activity at ~4.5 Ma, evidence of regional stresses controlling volcanism.

This volcanic province includes three separate volcanic loci of Smith and Luedke (1984): the California zone, Western Cordillera rift, and the Salton Sea rift. Smith and Luedke note that many of the northern volcanoes in this area occur in north–south rifts such as the Tahoe, West Walker–Carson, Inyo, and Mono rifts and the Owens Valley graben. The tectonic stress causing the east–west extension, and hence north to south rifting, is probably related to the westward encroachment of the Basin and Range, although extension in the Basin and Range itself is to the northwest.

Why did Long Valley form in this region? The caldera is clearly an anomaly, with eruption products orders of magnitude more voluminous than from any other volcanic center in California. Hill et al. (1985) propose that activity at Long Valley began with a major offset in the fault system bounding the eastern Sierra Nevada occurring in response to westward migration of Basin and Range extension.

References

Field trip guides exist for various areas of volcanism in the western USA, but most are gray literature and thus difficult to find. Along with the extensive new guides from the Geological Society of America and the International Association of Volcanology and Chemistry of the Earth's Interior, mentioned in the Introduction, two other books are very useful for visits to the Cascades. Stephen Harris's (1988) *Fire Mountains of the West: The Cascade and Mono Lake Volcanoes*, is well illustrated and full of historical details that usually don't appear in more technical publications. *Northwest Volcanoes: A Roadside Geologic Guide* by Lanny Ream (1983) has a very practical short text with pictures and detailed information for reaching geologically interesting outcrops and viewing points.

Armstrong, R. L., Leeman, W. P., and Malde, H. E., 1975, K-Ar dating, Quaternary and Neogene volcanic rocks of the Snake River Plain, Idaho: *Am. J. Sci. 275*, 225-251.

Atwater, T., 1970, Implications of plate tectonics for the Cenozoic tectonic evolution of western North America: *Geol. Soc. Am. Bull. 81*, 3513-3536.

Baldridge, W. S., Bartov, Y., and Kron, A., 1983, Geologic map of the Rio Grande rift and southeastern Colorado Plateau, New Mexico and Arizona (1:500,000): supplement to Riecker, R. E. (1979).

Braile, L. W., et al., 1982, The Yellowstone–Snake River Plain seismic profiling experiment: Crustal structure of the eastern Snake River Plain: *Jour. Geophys. Res. 87*, 2597-2609.

Brott, C. A., Blackwell, D. D., and Mitchell, J. C., 1978, Tectonic implications of the heat flow of the western Snake River Plain, Idaho: *Geol. Soc. Am. Bull. 89*, 1697-1707.

Christiansen, R.)L., 1984, Yellowstone magmatic evolution: Its bearing on understanding large-volume explosive volcanism: in *Explosive Volcanism: Inception, Evolution, Hazards*, Natl. Acad. Press, Washington, DC, 84-95.

Christiansen, R. L., and Lipman, P. W., 1972, Cenozoic volcanism and plate tectonic evolution of the western United States II. Late Cenozoic: *in* A discussion on volcanism and the structure of the Earth; *Phil. Trans. R. Soc. London A217*, 249-284.

Clowes, R. M., et al., Lithoprobe-southern Vancouver Island: Cenozoic subduction complex imaged by deep seismic reflections: *Can. J. Earth Sci. 24*, 31-51.

Crumpler, L. S., 1982, Volcanism in the Mount Taylor region: *N. M. Geol. Soc. Guidebook 33*, 291-298.

Eaton, G. P., 1979, A plate tectonic model for late Cenozoic crustal spreading in the western United States: in Riecker, R. E., 7-32.

Eaton, G. P., 1982, The basin and range province: Origin and tectonic significance: *Ann. Rev. Earth Sci. 10*, 409-440.

Eaton, G. P., 1987, Topography and origin of the southern Rocky Mountains and Alvarado Ridge: *in* Coward, M. P., Dewey, J. F., and Hancock, P. L. (eds.), Continental extensional tectonics, *Geol. Soc. (Lond.) Spec. Pub. 28*, 355-369.

Eaton, G. P., et al., 1978, Regional gravity and tectonic patterns: their relation to late Cenozoic epeirogeny and lateral spreading in the western Cordillera: *in* Smith, R. B., and Eaton, G. P. (eds.), Cenozoic Tectonics and Regional Geophysics of the Western Cordillera; *Geol. Soc. Am. Mem. 152*, 51-92.

EMSLAB Group, 1988, The EMSLAB electromagnetic sounding experiment: *EOS Trans. Am. Geophys. Un. 69*, 89-99.

Gardner, J. N., Goff, F., Garcia, S., and Hagan, R. C., 1986, Stratigraphic relations and lithologic variations in the Jemez volcanic field, New Mexico: *J. Geophys. Res. 91*, 1763-1778.

Gilluly, J., 1965, Volcanism, tectonism, and plutonism in the western United States: *Geol. Soc. Am. Spec. Pap. 80*, 69 pp.

Greeley, R., 1982, The Snake River Plain, Idaho: Representative of a new catagory of volcanism: *J. Geophys. Res. 87*, 2705-2712.

Guffanti, M., and Weaver, C. S., 1988, Distribution of late Cenozoic volcanic vents in the Cascade Range: Volcanic arc segmentation and regional tectonic considerations: *J. Geophys. Res. 93*, 6513-6529.

Harris, S. L., 1988, *Fire Mountains of the West: The*

Cascade and Mono Lake Volcanoes, Mountain Press, Missoula, MT, 379 pp.

Hearn, B. C., Donnelly-Nolan, J. M., and Goff, F. E., 1981, The Clear Lake volcanics: Tectonic setting and magma sources: *in* Research in the Geysers–Clear Lake Geothermal Area, Northern California, McLaughlin, R. J., and Donnelly-Nolan, J. M. (eds.), *USGS Prof. Pap. 1141*, 25-46.

Hill, D. P., Bailey, R. A., and Ryall, A. S., 1985, Active tectonic and magmatic processes beneath Long Valley caldera, eastern California: An overview: *J. Geophys. Res. 90*, 11,111-11,120.

Hughes, J. M., Stoiber, R. E., and Carr, M. J., 1980, Segmentation of the Cascade volcanic chain: *Geology 8*, 15-17.

Hyndman, R. D., Riddihough, R. P., and Herzer, R., 1979, The Nootka fault – a new plate boundary off western Canada: *Geophys. J. R. Astron. Soc. 58*, 667-683.

Lipman, P. W., 1980, Cenozoic volcanism in the western United States: Implications for continental tectonics: in *Studies in Geophysics: Continental Tectonics*, Natl. Acad. Sci., Washington, DC, 161-174.

Lipman, P. W., Prostka, H. J., and Christiansen, R. L., 1971, Evolving subduction zones in the western United States, as interpreted from igneous rocks: *Science 174*, 821-825.

Lucchitta, I., 1979, Late Cenozoic uplift of the southwestern Colorado Plateau and adjacent lower Colorado region: *Tectonophysics 61*, 63-95.

Luedke, R. G., and Smith, R. L., 1981, Map showing distribution, composition and age of late Cenozoic volcanic centers in California and Nevada: *USGS Misc. Invest. Map 1091-C.*

Luedke, R. G., and Smith, R. L., 1982, Map showing distribution, composition and age of late Cenozoic volcanic centers in Oregon and Washington: *USGS Misc. Invest. Map 1091-D.*

Mayo, E. B., 1958, Lineament tectonics and some ore districts of the Southwest: *Am. Inst. Mining Engin. Trans. 10*, 1169-1175.

McKee, E. H., and Noble, D. C., 1986, Tectonic and magmatic development of the Great Basin of western United States during Late Cenozoic time: *Mod. Geol. 10*, 39-49.

Olson, K. H., Baldridge, W. S., and Callender, J. F., 1987, Rio Grande rift: An overview: *Tectonophysics 143*, 119-139.

Perry, F. V., Baldridge, W. S., and DePaolo, D. J., 1988, Chemical and isotopic evidence for lithospheric thinning beneath the Rio Grande rift: *Nature 332*, 432-434.

Prowell, D. C., 1974, Geology of selected Tertiary volcanics in the central Coast Range Mountains of California and their bearing on the Calaveras and Hayward fault problems: Ph.D. Thesis, Univ. Calif., Santa Cruz, 182 pp.

Ream, L. R., 1983, *Northwest Volcanoes: A Roadside Geologic Guide* : B. J. Books, Renton, WA, 124 pp.

Riddihough, R., 1984, Recent movements of the Juan de Fuca plate system: *J. Geophys. Res. 89*, 6980-6994.

Riecker, R. E. (ed.), 1979, *Rio Grande rift: Tectonics and magmatism*, Am. Geophys. Un. Spec. Pub., Washington, DC, 438 pp.

Seager, W. R., Shafiqullah, M., Hawley, J. W., and Marvin, R. F., 1984, New K-Ar dates from basalts and the evolution of the southern Rio Grande rift: *Geol. Soc. Am. Bull. 95*, 87-99.

Simkin, T., et al., 1981,*Volcanoes of the World*: Hutchinson Ross, Stroudsburg, PA, 232 pp.

Sinno, Y. A., et al., 1986, Crustal structure of the southern Rio Grande rift determined from seismic refraction profiling: *J. Geophys. Res. 91*, 6143-6156.

Smith, R. L., and Luedke, R. G., 1984, Potentially active volcanic lineaments and loci in western coterminous United States: *in Explosive Volcanism: Inception, Evolution, Hazards*, Natl. Acad. Press, Washington, DC, 47-66.

Sonder, L. J., England, P. C., Wernicke, B. P., and Christiansen, R. L., 1987, *in* Continental Extensional Tectonics, *Geol. Soc. Spec. Publ. No. 28*, 187-201.

Sparlin, M. A., Braile, L. W., and Smith, R. B., 1982, Crustal structure of the eastern Snake River Plain determined from ray trace modeling of seismic refractor data: *J. Geophys. Res. 87*, 2619-2633.

Stewart, J. H., 1978, Basin-range structure in western North America: A review: *Geol. Soc. Am. Mem. 152*, 1-31.

Thompson, G. A., and Zoback, M. L., 1979, Regional geophysics of the Colorado Plateau: *Tectonophysics 61*, 149-181.

Weaver, C. S., and Baker, G. E., 1988, Geometry of the Juan de Fuca plate beneath Washington – Evidence from seismicity and the 1949 South Puget Sound earthquake: *Bull. Seis. Soc. Am. 78*, 264-275.

VOLCANOES OF WASHINGTON

Baker, 155
Glacier Peak, 156
Rainier, 158
Goat Rocks, 160
St. Helens, 161
Adams, 164
Simcoe, 165
Indian Heaven, 166
Marble Mountain–Trout Creek Hill Zone, 167

BAKER
Washington

Type: *Stratovolcano*
Lat/Long: *48.79°N, 121.82°W*
Elevation: *3,285 m*
Eruption History:
Cone construction: Pleistocene
Eruptions: 6,750, 6,525 yr BC, 1,100 AD
9 possible eruptions in 19th century
Intense fumarole activity in 1975
Composition: *Andesite with minor basalt*

Mount Baker is the northernmost and most isolated of the Cascade volcanoes in the USA. The andesitic cone rises nearly 2 km above the older metamorphic and sedimentary rocks at its base and it is almost completely covered by glaciers – hence its original Nooksack Indian name "White Steep Mountain."

Knowledge of Mount Baker is limited to those parts of its history not covered by glaciers or later volcanic rocks. Thus, the different eruptive centers that have been recognized do not necessarily represent the most significant phases of its activity. The oldest (0.4-Ma) known flows are from the **Black Buttes,** two strongly eroded remnants of a predecessor stratovolcano ~3 km west of Mount Baker. Although most of these rocks are andesitic lava flows, like nearly all of the exposed parts of Mount Baker, some of these ancient rocks are hypersthene basalt. The earliest eruptions from the present location of Mount Baker produced widespread, fluid lava flows which traveled down canyons previously eroded in the older basement rocks. The existence of large glacially carved cirques on the north slopes of the volcano demonstrate that its construction was largely completed by the time of the last major Pleistocene glaciation, some time between 25,000 and 10,000 yr ago.

From deposits mapped around the volcano it has been recognized that during the last 10,000 yr there were 1 pyroclastic flow, at least 4 small tephra units, 2 or more lava flows, and at least 8 mudflows. A 760-m-wide, 100-m-high cinder cone and its 11-km-long lava flow at **Schreibers Meadow** is one of the more recent satellitic cones on Mount Baker. Mudflows remain the most likely hazard from Mount Baker.

It is unclear how all of the distally mapped

BAKER: Vertical aerial view from US National High Altitude Photography program.

eruption products relate to activity at Baker's summit, which is totally covered by snow and ice. However, **Sherman Crater**, a 450-m-wide vent, 350 m lower and 800 m south of the summit, may have formed in the 18th or early 19th century, based on Indian traditions of drastic changes near the summit area of the volcano.

There are various accounts of activity at Mount Baker in the mid-1800s. An 1843 eruption resulted in a major fish kill in the Baker River, a large forest fire, and a dusting of volcanic ash over the adjacent countryside. Further eruptions occurred in the 1850s, and the first expedition to Sherman Crater in 1868 reported active fumarole fields. Steam activity continued at Sherman Crater and at the Dorr fumarole field on Baker's north flank until the 1940s and 1950s, by which time steaming was uncommon. After resumption of mild activity in the 1960s, a major episode of steam activity persisted at Sherman crater from March 1975 to early 1976. A large jet shot pressurized steam to 760 m, new fumaroles were active, crevasses developed in the ice concentric to the crater walls, a 70-m-wide plug of ice collapsed to form a warm water lake, and minor amounts of non-juvenile tephra were spread around the crater area.

How to get there: *Mount Baker is ~50 km east of Bellingham, WA. The best view of the mountain is from the Glacier Creek Road off of Highway 542. A 10-km hike, taking off from Dead Horse Road (No. 3907) affords closer views of Baker's north side. On the south side of the mountain Forest Service Road 372, taking off from Baker Lake Road, ends near the Schreibers Meadow cinder cone.*

References

Easterbrook, D. J., 1975, Mount Baker eruptions: *Geology 3*, 679-682.

Hyde, J. H., and Crandell, D. R., 1978, Postglacial volcanic deposits at Mount Baker, Washington, and potential hazards from future eruptions: *USGS Prof. Pap. 1022-C*, 17 pp.

Charles A. Wood

GLACIER PEAK
Washington

Type: *Dacite dome cluster*
Lat/Long: *48.12°N, 121.12°W*
Elevation: *3,213 m*
Eruption History:
Cone construction: Pleistocene
Pyroclastic eruptions: 11,500-11,000 yr BP, 5,500-5,000 yr BP, 1,800 yr BP, 1,100-1,000 yr BP, 18th century (?)
Composition: *Andesite to dacite*

Glacier Peak is a small Cascade Range stratovolcano. Although its summit reaches >3,000 m above surrounding valleys, the main cone of Glacier Peak is perched on a high ridge, and the volcanic pile is no more than 500-1,000 m thick. More than a dozen glaciers occur on the flanks of the volcano, and unconsolidated pyroclastic deposits >12,000 yr old have been largely removed by glaciation. Lava flows locally cap ridges to the northeast of the volcano, indicating topographic reversal, and glacial and fluvial downcutting of >2,000 m has occurred since the earliest cone-building eruptions. While small basaltic flows and cones are found at several points around the flanks of Glacier Peak, the main edifice is largely dacite and andesite. Lava flows extend no more than a few kilometers from the summit.

Glacier Peak is probably best known as the source of voluminous tephra eruptions dated to

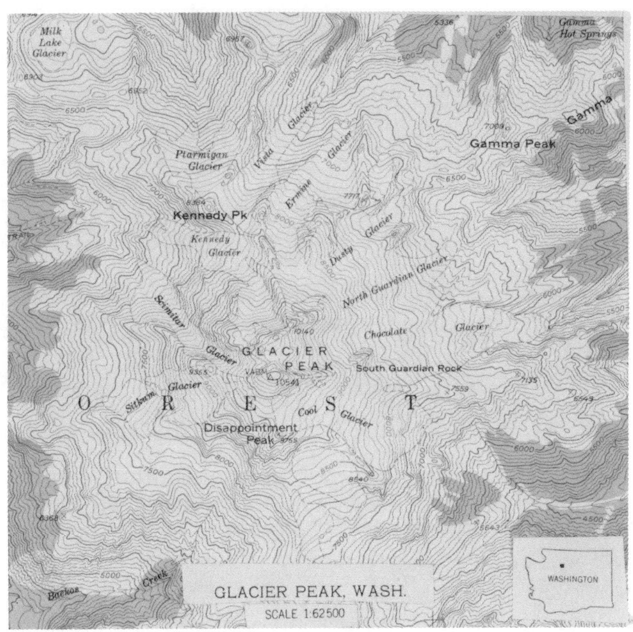

GLACIER PEAK: USGS topographic map of central region of Glacier Peak.

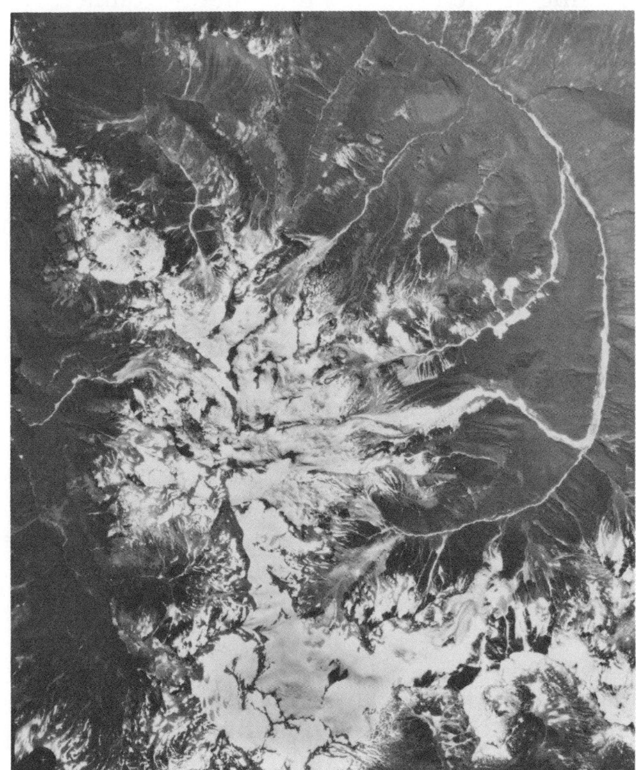

GLACIER PEAK: Vertical aerial view from US National High Altitude Photography program.

GLACIER PEAK: Oblique aerial view, August 1972; USGS photograph.

11,250 yr BP. Two tephra layers produced at this time have been identified as far as 800-1,000 km to the east, and are widely used by geologists, anthropologists, and paleoecologists to date late Pleistocene sediments. Also at this time, an extensive valley fill of pumiceous lahars and alluvium was deposited downriver to the west, blocking valleys and affecting drainages as far as 80 km from the volcano.

After these major eruptions, Glacier Peak apparently was dormant for 6,000 yr. The volcano reawoke 5,500-5,100 yr BP, and intermittent eruptions of pyroclastic flows and tephra have occurred since that time. Perhaps the most dramatic geologic features at Glacier Peak are enormous and relatively undissected late Pleistocene and Holocene pyroclastic fans which almost completely fill valleys on the eastern and western flanks of the volcano.

Indian legends and a thin tephra fall preserved east of the volcano may record a recent eruption in the 18th century, although no eruptive activity has occurred during at least the last 150 yr. Three hot springs surround the volcano, and warm ground and snow-free areas occur near fresh-appearing dacite domes which form subsidiary summits both north and south of the ice-covered main summit.

How to get there: *Glacier Peak is located in the Glacier Peak Wilderness of the North Cascade Range, ~150 km northeast of Seattle. Graded* *dirt roads from Darrington, Washington, approach to within 15-25 km of the volcano on the west, and connect with a network of hiking and climbing trails.*

References

Beget, J., 1982, Recent volcanic activity at Glacier Peak: *Science 215*, 1389-1390.

Beget, J., 1984, Tephrochronology of late Wisconsin Deglaciation and Holocene glacier fluctuations near Glacier Peak, North Cascade Range, Washington: *Quat. Res. 21*, 304-316.

Jim Beget

RAINIER
Washington

Type: *Stratovolcano*
Lat/Long: *46.85°N, 121.75°W*
Elevation: *4,392 m*
Relief: *>2,200 m*
Volcano Basal Diameter:
7 km at 2,100 m elevation
17 km at 1,800 m elevation
Eruptive History:
>0.7 Ma for main cone
6.5–4.0 ka
2.2 ka for major Holocene activity
Composition: *Basaltic andesite to andesite*

Mount Rainier, the highest and third most voluminous volcano in the Cascade Range, is potentially the most dangerous volcano in the range because of the large population living around its lowland drainages. These areas are at risk because of the mountain's great relief and the huge area and volume of ice and snow on the cone (92×10^6 m^2 and 4.4×10^9 m^3, respectively) that could generate lahars during eruptions. In addition, large ($>2 \times 10^8$ m^3) sector collapses of clay-rich, hydrothermally altered debris from the cone have occurred at least 3 times in the last 6,000 yr (Osceola, Round Pass, and Electron mudflows).

Relatively little is known of the eruptive history, composition, and age of Rainier compared to other Cascade Range peaks; fewer than two dozen chemical analyses of its products have been published. Most of its Holocene history has been assembled from work on fragmental deposits, chiefly tephras and lahars.

RAINIER: View of the east face of Mount Rainier showing Emmons Glacier and its well-defined lateral and terminal moraines. Photograph by L. Topinka, USGS.

Mount Rainier rests on Tertiary volcanic rocks which were gently folded along northwest trends and then intruded by granodiorite and quartz monzonite of the Tatoosh Pluton (17.5-14.1 Ma). Petrologic changes in clastic deposits east of the volcano indicate post-Tatoosh unroofing at the present site of Mount Rainier. An apparent 11-m.y. hiatus in volcanic activity and plutonism preceded the first evidence of a proto–Mount Rainier. The Lily Formation (2.9-0.84 Ma), a thick sequence of volcaniclastic debris west of the mountain, is the earliest Rainier deposit.

Early lava flows of present Mount Rainier formed a small shield on a dissected surface of the Tertiary basement which has as much as 700 m of relief. The present edifice is dominantly lava flows and breccias, 90% of which are composed of a petrographically uniform two-pyroxene andesite; the few chemical analyses available are of medium-K silicic andesite, with three analyses marginally dacitic. Mafic olivine-phyric basaltic andesite was erupted during the late Pleistocene from two satellite cones on the northwest flank of the mountain, **Echo Rock** and **Observation Rock.** Approximately 270 km^3 of lava have been erupted from Mount Rainier in the past 1 m.y.

A thick pumice layer northeast, east, and southeast of the volcano is interpreted to have erupted from Mount Rainier between 70 ka and 30 ka. Estimates based on limited outcrops suggest this unnamed layer is an order of magnitude larger in volume than any of Rainier's Holocene tephra layers.

Holocene explosive eruptions at Rainier produced 11 tephra beds totaling over 0.5 km^3. Eight layers (30-40% of the tephra volume) were erupted between 6,500 and 4,000 ^{14}C yr BP. Some of these tephra layers (e.g., S and F) are rich in lithic components and are thought to be the result of phreatic or phreatomagmatic eruptions. Layer F is unique among the post-glacial tephras because of its large percentage of clay minerals. Montmorillonite, with some illite and kaolinite, were formed before deposition, and thus were probably emplaced by a violent phreatic or phreatomagmatic event that penetrated an area of hydrothermal alteration. Layer F is similar in age and clay content to the Osceola Mudflow (discussed below) but is not found on top of the Osceola. The two deposits thus seem correlative.

Layer C, which is ~60% of the post-glacial tephras by volume, is also the most widespread, covering much of the eastern half of Mount Rainier National Park with 2-30 cm of lapilli, blocks, and bombs. It is also the coarsest of the Rainier tephras; 25-30 cm bombs can be found 8 km to the east of the summit. Vertical and horizontal changes in scoria, lithic fragments and pumice, and lateral variations of layer C indicate it was deposited by more than one event, or in a single, extended eruption during a change in wind direction. Isopachs and isopleths for this layer indicate an origin at the summit of Mount Rainier; however, it does not occur on snow-free parts of **Columbia Crest** cone. Apparently this 250-m-high summit cone is younger than the 2,200-yr age of Layer C. The age and lithologic similarity of this tephra layer to a block-and-ash flow in the South Puyallup River west of Mount Rainier, and a granular, monolithologic lahar found in the lower Puyallup valley and in the Puget Lowland, suggest the three occurred during the same episode of volcanic activity.

Post-glacial deposits at Mount Rainier are dominated by lahars; over 60 have been identified. Although relations between Holocene tephra and flowage deposits remain speculative, at least some lahars were probably eruption induced, most notably the Paradise lahar and the Osceola Mudflow. The Osceola Mudflow, which has been dated at 5,040 ^{14}C yr BP, had a volume $>10^9$ m^3 and a profound geomorphic effect on the Puget Sound shoreline, over 100 km from the mountain. As interpreted from well logs (neglecting minor relative sea-level changes), syn- and post-Osceola sedimentation has pushed the shoreline seaward 25 and 50 km, respectively, in two Puget Sound embayments and added ~460 km^2 of new land surface. Wood from buried trees in the Round Pass Mudflow has been dated at 2,600 ^{14}C yr BP. This clay-rich diamict is characterized by great thickness (locally >250 m), limited downvalley extent, hummocky terrain, and megaclasts of lithologically homogeneous material. It probably began as a debris avalanche of hydrothermally altered material from high on the western slopes, and most of it was deposited in the upper 20 km of the Puyallup River valley.

Another clay-rich lahar, the Electron Mudflow, has been dated at 530 ^{14}C yr BP. This lahar, which evidently began as a failure of part of the western edifice, has not been correlated with any eruptive activity at Mount Rainier and may have occurred without precursory eruptive phenomena. The Electron Mudflow apparently was very fluid and underwent minimal downstream attenuation of discharge. This is demonstrated ~36 km west of Mount Rainier where the Electron was still ~60 m deep as it exited the Cascade mountain front and flowed onto the Puget Lowland. Future lahars such as this pose the greatest risk to populated areas near Mount Rainier, particularly on downstream floodplains of the Nisqually and Puyallup River valleys and to sections of the White River valley upstream from Mud Mountain Dam.

Several of Mount Rainier's 26 named glaciers have been the focus of classic studies of neoglacial moraine development and glacial dynamics. Glacial-outburst floods from South Tahoma Glacier, typically between June and November on warm afternoons or after heavy rain, have repeatedly scoured Tahoma Creek during the late 1960s and during the middle and late 1980s. Similar events have occurred in many other drainages, most notably Kautz Creek and Nisqually River during historical time.

Studies of the Rainier hydrothermal system indicate that a narrow, central thermal system perpetuates snow-free areas at the summit craters and forms caves in the summit icecap. Heat flux is substantial in comparison to other volcanoes in the Cascade Range.

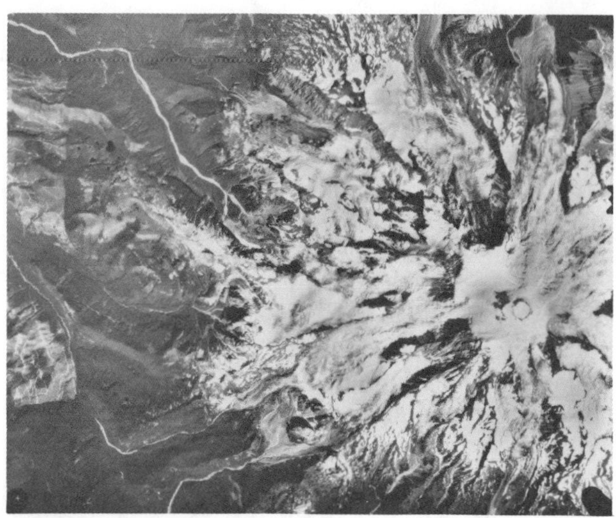

RAINIER: Vertical aerial view from US National High Altitude Photography program, showing Columbia Crest summit crater.

An average of ~30 earthquakes occur under Mount Rainier per year, making it the most seismically active volcano in the Cascade Range after Mount St. Helens. Trilateration and tilt networks established on the volcano in 1982 have shown no untoward displacements.

How to get there: *Mount Rainier National Park is located in west-central Washington, ~108 km south-southeast of Seattle and 158 km north-northeast of Portland, Oregon. The south entrance is accessible year round from US Highway 12 and state Highways 7 and 706, while the northeast entrance is accessible on a seasonal basis via state Highway 410.*

Tephra layers from Mount Rainier

Layer	Age (BP)	Main materials	Volume ($10^6 m^3$)
X	±150	Pumice	1
C	2,200	Pumice, scoria, lithic fragments	300
B	>4,000	Scoria, lithic fragments	5
H	<5,000	Pumice, lithic fragments	1
F	5,000	Lithic fragments, pumice, crystals, clay	25
S	5,200	Lithic fragments	20
N	5,500	Lithic fragments, pumice	2
D	6,000	Scoria, lithic fragments	75
L	6,400	Pumice	50
A	6,500	Pumice, lithic fragments	5
R	>8,750	Pumice, lithic fragments	25

From Mullineaux (1974). Age of layer X determined from tree rings; all others in ^{14}C years.

References

Frank, D. G., 1985, Hydrothermal processes at Mount Rainier, Ph.D. Thesis., Univ. Wash., 195 pp.

Mullineaux, D. R., 1974, Pumice and other pyroclastic deposits in Mount Rainier National Park, Washington: *USGS Bull. 1326*, 83 pp.

Patrick Pringle

GOAT ROCKS
Washington

Type: *Stratovolcano*
Lat/Long: *46.50°N, 121.45°W*
Elevation: *1,340 to 2,494 m*
Relief: *1,135 m*
Basal Diameter: *~11 km*
Eruptive History: *3 to <0.73 Ma; possibly renewed activity in late Pleistocene*
Composition: *High-K_2O pyroxene and minor hornblende andesite*

The Goat Rocks volcano is a deeply eroded, glaciated volcanic center in an area of widespread Pliocene and Pleistocene volcanism along the Cascade crest in southern Washington. Volcanism began ~3.2 Ma with eruption of at least 650 m of high-silica rhyolite tuff (perhaps a caldera fill), which is exposed on the east flank of the subsequent Goat Rocks volcano. The silicic volcanism ended ~3 Ma, and olivine basalt was locally erupted onto the rhyolitic rocks. Soon thereafter lava flows of high-K_2O andesite, dominantly pyroxene phyric but including flows with significant hornblende, formed the Goat Rocks volcano. The volcano was probably built between ~2.5 and 0.5 Ma. Some large volume lava flows moved many kilometers down paleovalleys away from the volcano. The most notable such flow is the 1.0 Ma Tieton Andesite, which advanced ~80 km eastward down the ancestral Tieton and Naches rivers, and is the longest known andesite flow on Earth. Thick sections of flows with radial dips of 10-20° surround the hydrothermally altered core of the volcano; the flows are cut by numerous dikes that define several sectors of a radial swarm. The Cispus Pass pluton occupies the southern part of the altered core and is questionably as young as 1 Ma.

GOAT ROCKS: *Looking southwest across Ives Peak area in core of Goats Rock volcano. USGS photograph by A. Post in 1962. Northwest base of Mount Adams in upper left corner.*

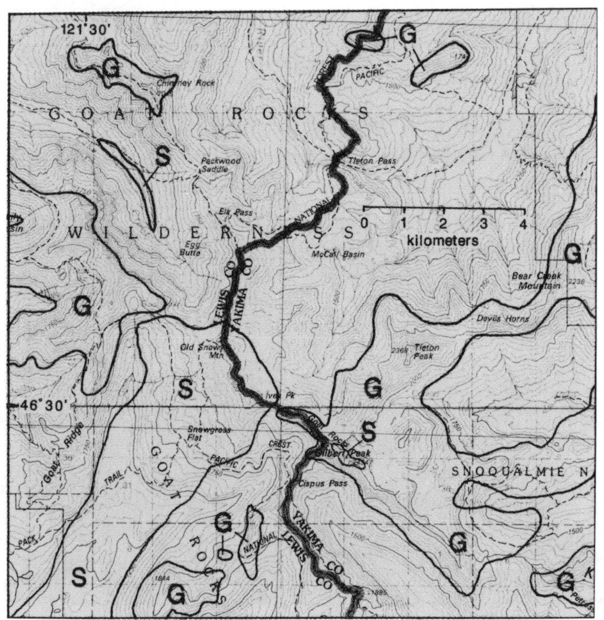

GOAT ROCKS: Map of core area of Goat Rocks volcano. G: pyroxene andesite of Goat Rocks volcano; S: hornblende andesite of Old Snowy Mountain.

Following extensive erosion, lava flows of hornblende andesite erupted onto the dissected flanks of Goat Rocks volcano. Gilbert Peak (2,494 m), the highest point in the area, is capped with hornblende andesite. Old Snowy Mountain, along the Cascade crest, erupted flows of hornblende andesite that poured westward into the glaciated Cispus River valley. These young flows themselves were glaciated and so are middle to late Pleistocene, not Holocene, in age. Whether the hornblende andesite records rejuvenation of the Goat Rocks volcano or independent volcanism is not clear.

Coeval and younger volcanism occurred north and south of Goat Rocks volcano. Eruption of >200 olivine basalt and basaltic andesite lava flows formed the 700-m-high **Hogback** Mountain shield volcano just south of White Pass during the late Pliocene and early Pleistocene. These lava flows intertongue with those from Goat Rocks volcano. At least five late Pleistocene andesitic volcanoes formed on the Tumac Plateau north of Goat Rocks. Young olivine basalt erupted from several vents between the Goat Rocks area and the north base of Mount Adams, 25 km to the south.

How to get there: *The Goat Rocks volcano is in southern Washington, 70 km west of Yakima and 15 km south of White Pass. Access is by foot along the Pacific Crest trail system from White Pass or several feeder trails east and west of the crest.*

References

Clayton, G. A., 1983, Geology of the White Pass area, south-central Cascade Range, Washington: M.S. Thesis, Univ. Wash., Seattle, 212 pp.

Swanson, D. A., and Clayton, G. A., 1983, Generalized geologic map of the Goat Rocks wilderness and roadless areas (6036, Parts A,C, and D), Lewis and Yakima Counties, Washington: *USGS Open-file Rpt. 83-357*, 10 pp.

Donald A. Swanson

ST. HELENS
Washington

Type: *Stratovolcano*
Lat/Long: *46.20°N, 122.19°W*
Elevation: *1,000 to 2,500 m (2,950 before 18 May, 1980)*
Relief: *1,550 m (1,950 before 18 May, 1980)*
Volcano Diameter: *8.5 km*
Eruptive History: *Intermittently since 50-40 ka, with apparent breaks from 36-21 ka,*

18-13 ka, and 10-4 ka; 10 eruptions in 1800s
Avalanche deposit eruption in 1980
Composition: *Basalt to rhyodacite, chiefly andesite and dacite*

Mount St. Helens is young. Its oldest known products were erupted 40-50 ka, and the symmetric cone that partly collapsed in 1980 was built almost entirely within the past 2,200 yr. Since its birth it has produced more than 60 individual tephra layers, several tens of volcanically induced debris flows, at least 6 of which entered the Columbia River 100 km downstream, and the equivalent of 60 km^3 of dacitic lava. It has been the most active volcano in the Cascades during the Holocene, and for that reason its eruption in 1980 came as no surprise. Mount St. Helens has been studied intensively, and its eruptive history is known with greater clarity than that of any other Cascade volcano.

Past activity of the volcano has been divided into 4 eruptive stages, each lasting 2,000 yr or more. The latest, the Spirit Lake eruptive stage, contains 9 eruptive periods lasting decades to centuries.

Before ~2,200 yr BP, only dacite, silicic andesite, and rare rhyodacite are known to have erupted, mostly as domes, pyroclastic flows, tephra, and lahars. A major change occurred thereafter, at the onset of the Castle Creek eruptive period. Numerous mafic lava flows, most of which are formally classed as borderline trachybasalt and trachybasaltic andesite but are more often termed olivine basalt and two-pyroxene basaltic andesite, were erupted from all sides of the volcano, especially the southwest and north. The Cave Basalt issued from an unknown vent probably near the southwest base of the present cone ~1,700 ^{14}C yr BP; it contains 3.4-km-long Ape Cave, the longest known uncollapsed segment of a lava-tube, as part of its 8.3-km-long tube system. The period began, was dominated by, and ended with basalt, basaltic andesite, and lesser andesite, but at least two dacitic tephras and pyroclastic flows and possibly a dome were also formed.

Sugar Bowl dome on the north flank of the volcano just east of the mouth of the present crater was emplaced ~1,200 ^{14}C yr BP. Two lateral blasts, the larger of which threw lithic debris 10 km northeast of the vent, probably accompanied dome growth. **East Dome,** at the east base of the volcano, is chemically similar to Sugar Bowl; it has not been dated, but bracketing radiocarbon ages allow it to be of Sugar

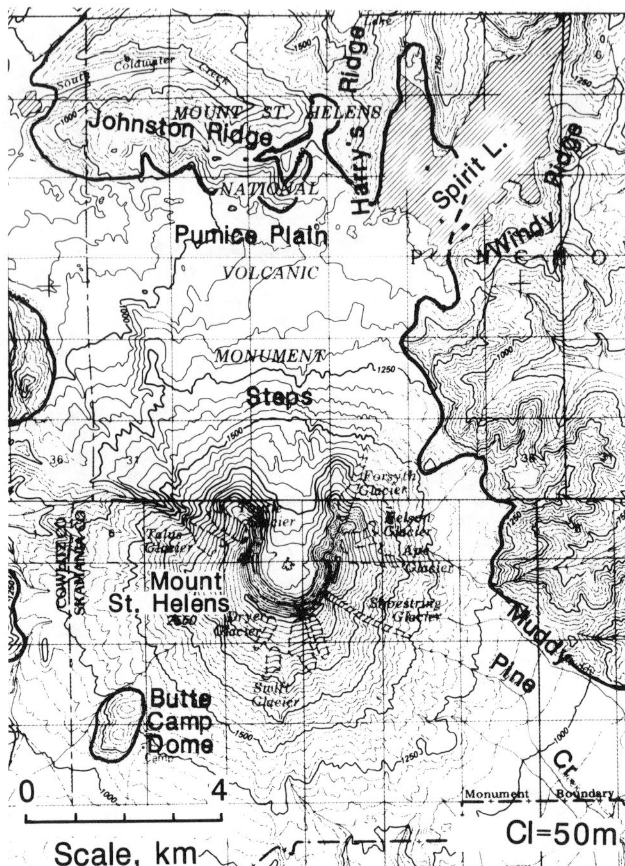

ST. HELENS: Map with area of 1980 flowage deposits outlined. Air-fall tephra and blast deposits extend beyond map boundary.

ST. HELENS: Looking into the amphitheater and the dome from Harrys Ridge, 1988. Photograph by L. Topinka, USGS.

ST. HELENS: Vertical aerial view from US National High Altitude Photography program, August 6, 1981.

Bowl age. The rhyodacitic compositions of the domes are among the most silicic yet found at Mount St. Helens.

The Kalama eruptive period began in the winter or early spring of 1479-1480 A.D., as deduced from dendrochronologic dating of the dacitic Wn tephra, the most voluminous tephra from Mount St. Helens since the Y tephras ~4,000 yr BP. Another widespread tephra fall, We, took place in the winter or early spring of 1481-1482. Episodic activity thereafter, possibly continuing into the 19th century, produced voluminous silicic andesite flows that together with the Castle Creek flows constructed the symmetric cone well known before 1980. The Kalama activity also included the emplacement of a dacite dome at the summit and the eruption of numerous lahars and pumiceous and lithic pyroclastic flows.

The Goat Rocks eruptive period began in 1800 A.D. with the eruption of dacitic tephra T and ended in 1857. The "floating island" silicic andesite flow was erupted before 1838, and an explosion sent lithic ash 100 km downwind in 1842. The **Goat Rocks** dome was extruded on the northwest flank of the volcano 600-700 m below the summit within several years after the 1842 explosion, possibly during or before 1847, when Paul Kane painted a famous canvas of its growth. Contemporary accounts indicate activity several times during the 1840s and 1850s but are not specific and in part contradictory. The last significant activity before 1980 was "dense smoke and fire" in 1857, although minor, unconfirmed eruptions were reported in 1898, 1903, and 1921.

The top of the cone was destroyed on 18 May 1980, when a 2.7-km^3 landslide suddenly removed pressure on a growing cryptodome, leading to a massive debris avalanche, lateral blast, plinian column, and dacitic pyroclastic flows and tephra. The blast (at times supersonic) devastated ~600 km^2. The debris avalanche buried the upper 24 km of the North Fork Toutle valley to a depth of 50 m with material from the collapsed volcano. Lahars, formed mostly from dewatering of the debris avalanche, closed the shipping channel of the Columbia River. The upper 400 m of the cone were removed, leaving a 600-m-deep, 2-km-wide crater exposing rocks as old as ~4 ka. Five other explosions occurred in 1980, and thereafter a dacitic dome grew in the crater, reaching ~260 m high and nearly 1 km wide in October 1986 (the latest activity as of Nov. 1988).

How to get there: *Mount St. Helens is in the Cascade Range of southwest Washington ~75 km northeast of Portland, Oregon. A network of paved Forest Service roads (90, 25, 26, and 99) accessed from Highways 12 and 503 leads to the Windy Ridge scenic area northeast of the volcano. Gravel roads from near Cougar go to the south side of the mountain. A paved state highway is being constructed to link I-5 in Castle Rock with Johnston Ridge 8 km northwest of the crater. Logistical information can be obtained from the Mount St. Helens National Volcanic Monument headquarters in Amboy, Washington, or the visitor center on Highway 504 east of Castle Rock.*

References

Lipman, P. W., and Mullineaux, D. R., 1981, The 1980 eruptions of Mount St. Helens, Wash., *USGS Prof. Pap. 1250*, 844 pp.

Swanson, D. A., et al., 1987, Growth of the lava dome at Mount St. Helens, Washington (USA), 1981-83, *in* Fink, J. (ed.), The emplacement of silicic domes and lava flows, *Geol. Soc. Am. Spec. Pap. 212*, 1-16.

Donald A. Swanson

ADAMS
Washington

Type: *Compound stratovolcano with ~60 peripheral vents*
Lat/Long: *46.2°N, 121.5°W*
Elevation: *3,742 m*
Height: *~2,700 m*
Cone Dimensions: *30 × 25 km*
Volcanic Field Dimensions: *55 × 30 km*
Eruptive History:
0.52–0.45 Ma: Oldest andesite cone
0.25, 0.15, 0.18–0.10 Ma: Basaltic shields
<0.02 Ma: Present andesite cone
11 Holocene vents; no historic activity
Composition: *Andesite, basalt, dacite to rhyodacite*

Mount Adams stands astride the Cascade Crest some 50 km due east of Mount St. Helens. The towering stratovolcano is marked by a dozen glaciers, most of which are fed radially from its summit icecap. In the High Cascades, Mount Adams is second in eruptive volume only to Mount Shasta, and it far surpasses its loftier neighbor Mount Rainier (which is perched on a pedestal of Miocene granodiorite). Adams's main cone exceeds 200 km^3, and at least half as much more was eroded during late Pleistocene time from earlier high-standing components of the compound edifice; peripheral basalt adds another 70 km^3 or so.

Nearly all the high cone above 2,300 m in elevation was constructed during latest Pleistocene time, probably between 20 and 10 ka, explaining the abundance of late-glacial till and the scarcity of older till. Products of this eruptive episode range from 54% to 62% SiO_2 on the main cone. Contemporaneous and younger peripheral vents yielded lavas and scoria in the range 48-57% SiO_2, and along with Mount Adams, they define a recently active north–south eruptive alignment 40 km long and only 5 km wide. If we include vents as old as 0.3 Ma, this zone lengthens to 55 km, trending N10°W, attesting to modest east–west extension in the upper crust. Compositionally, basalt within this zone is extremely varied from quartz- to nepheline-normative and from 0.16 to 1.60 wt% K_2O (at 48-49% SiO_2). Andesite also reflects this heterogeneous ancestry, ranging from 0.9 to 2.1 wt % K_2O at 57.5% SiO_2.

ADAMS: South slope with Adams Glacier to right and remnants of cinder cones. USGS photograph by A. Post.

Beneath the young volcano eroded stumps of at least two andesite-dacite edifices are as old as 0.5 Ma, representing several eruptive episodes and a spectrum of rock compositions extending continuously from 52 to 69% SiO_2. Coalescing shields of basalt (48-53% SiO_2) peripheral to the stratovolcano complex interleave with the andesite-dacite lavas erupted focally, demonstrating long-persistent contemporaneity and an absence of systematic petrochemical evolution. Mount Adams is the most potassic, petrochemically most "intracontinental," stratovolcano in the Cascades. Its products contain 2.0-2.6 wt % K_2O at 60% SiO_2 (and 4% at 69% SiO_2), which is double that of Mount St. Helens.

There have been no recorded eruptions of Mount Adams, and, of the 11 Holocene vents, none is known certainly to have erupted products younger than 3,500 yr. Seven of the Holocene eruptions took place at flank vents 2,000-2,500 m in elevation and produced a wide variety of compositions (49-61% SiO_2); the other four vents are peripheral to the main cone at 1,100-1,600 m and 48-54% SiO_2.

Dacitic lavas and a few block-and-ash flows (63-69% SiO_2) erupted several times from the focal area and rarely on the periphery, but none is younger than 0.1 Ma. A single rhyodacite

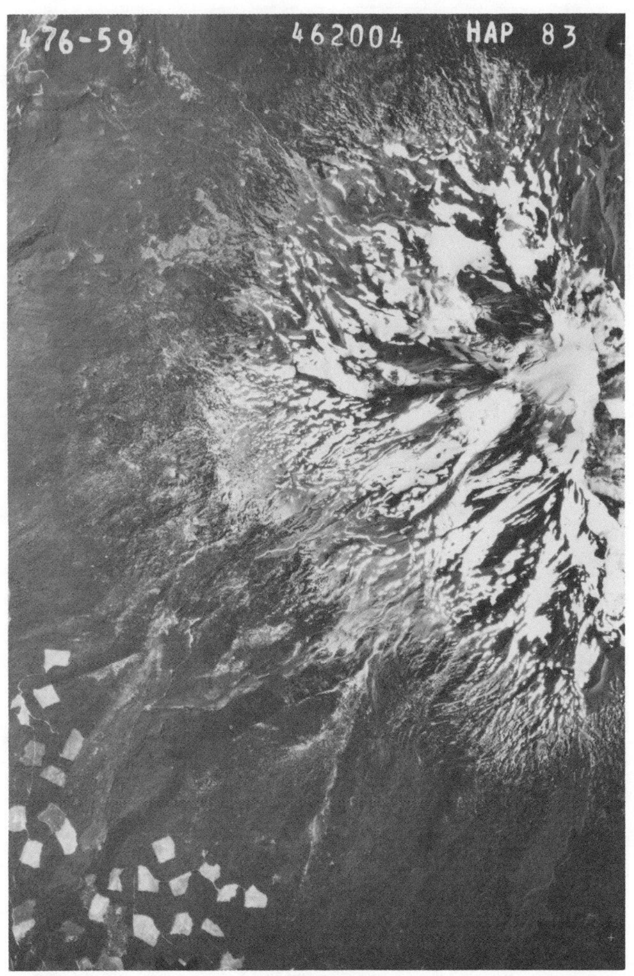

ADAMS: Vertical aerial view from US National High Altitude Photography program.

(72% SiO_2) is undated but also old. The antiquity of all known silicic units, in conjunction with the heterogeneous mafic compositions of the late Pleistocene summit cone and of Holocene lavas erupted around that cone, make it appear unlikely that Mount Adams is now underlain by an upper crustal magma reservoir.

Weak H_2S-bearing fumaroles still rise from crevasses in the summit icecap. Subjected to this solfataric flux, the breccia-and-scoria core of the stratovolcano has suffered severe acid-sulfate leaching and deposition of alunite, kaolinite, silica, gypsum, sulfur, and iron oxides. Where exposed in glacial headwalls, the 4-km^2 rotten core is a persistent source of avalanches and debris flows; the longest traveled 40 km after breaking loose ~5,000 yr ago, creating the southwest notch and shelf for the perched White Salmon Glacier.

How to get there: *Fifty km north of the Columbia River, Mount Adams is reached most rapidly from Trout Lake, which is two hours drive from Portland, Oregon, by paved road. US Forest Service roads from Trout Lake, Glenwood or Randle, Washington, lead toward the volcano.*

References

Hildreth, W., and Fierstein, J., 1985, Mount Adams: Eruptive history of an andesite-dacite stratovolcano at the focus of a fundamentally basaltic volcanic field: *USGS Open-file Rpt. 85-521*, 44-50.

Hildreth, W., and Fierstein, J., 1989, Geologic map of the Mount Adams volcanic field: in prep.

Wes Hildreth

SIMCOE
Washington

Type: *Monogenetic volcano field*
Lat/Long: *46.1°N, 120.9°W*
Eruptive History: *4.5-0.5? Ma*
Composition: *Olivine basalt, andesite, and rhyolite*

East and southeast of Mount Adams, extending south to the town of Goldendale, is a broad expanse of folded and faulted basalt that is little known. A north–south line of ~24 cinder cones crosses the center of the field, and south of the 46th parallel another dozen cones extend the line to the southeast. The oldest units are thick flows of rhyolite in the eastern Simcoe Mountains. These are contemporaneous with pyroxene-olivine basalt (POB) and olivine basalt (OB) of the Simcoe Mountains. The POB forms short, 6-10-m-thick flows, whereas the OBs are 1-6-m-thick flows extruded from small shields and cinder cones. A few of the above units are dated at 4.5 to 0.9 Ma. East of Mount Adams are a shield volcano and the lava flows forming **Lincoln Plateau**; all are olivine basalt probably younger than 0.9 Ma. Perhaps the youngest unit of this volcanic field is the dacite that forms **Signal Peak**, east of Mount Adams, that is believed to be 1.0-0.5 Ma.

How to get there: *US Highway 97 passes east of Mount Simcoe and associated ski facilities. Part of the Simcoe volcanic field is on the Yakima Indian Reservation.*

References

Hammond, P. E., 1980, *Reconnaissance Geologic Map and Cross Sections of the southern Washington Cascade Range*: Dept. Earth Sci., Portland State Univ.

Sheppard, R. A., 1967, Geology of the Simcoe Mountains volcanic area, Washington: *Wash. Div. Mines Geol., Geol. Map GM-3*.

Charles A. Wood

INDIAN HEAVEN
Washington

Type: *Polygenetic volcanic field*
Lat/Long: *46.00°N, 121.75°W*
Elevation: *720 to 1,806 m*
Volcano Area: *600 km²*
Eruptive History: *730 ka to 8,150 yr BP; no historic activity*
Composition: *Olivine tholeiite, calc-alkaline basalt and andesite*

The Indian Heaven volcanic field is midway between Mount St. Helens and Mount Adams; its principal feature is a 30 km long, N10°E-trending linear zone of coalescing, polygenetic shield volcanoes, cinder cones, and flows, with a volume of 100 km³. The shield volcanoes, which form the backbone of the volcanic field, are from north to south: **Sawtooth** Mountain (1,632 m in height), **Bird** Mountain (1,739 m), **Lemei Rock** (1,806 m), **East Crater** (1,614 m), **Gifford Peak** (1,636), **Berry** Mountain (1,523 m), and **Red** Mountain (1,513 m). Most of these cones are deeply glaciated, but a few retain characteristic shield morphology.

Of the 48 eruptive units in the field, monogenetic cinder cones account for 67%, polygenetic shields compose 29%, and monogenetic tuyas and mobergs form 4%. Cinder cones have basal diameters ranging from 305 to 1,600 m, and volumes from 0.9 to 185 m³. Shield volcanoes range from 1.3 to 9.5 km in basal width, with volumes of 60 m³ to 11.8 km³. Mobergs at **Crazy Hills** grew within glacial ice, but never reached the surface. They are made of pillow lavas and palagonitized hyaloclastitic breccia. The mobergs cover 21 km² and have a volume of 1.7 km³. **Lone Butte** tuya is an isolated volcano that grew through a glacier and rose above its surface. Its bottom part is similar to the mobergs but its upper part is made of surge deposits, air-fall scoria, and subaerial lava flows. The tuya has 1,027 m of relief, covers >2.5 km², and has a volume of 0.3 km³. About half of the volume has been removed by more recent glaciation.

INDIAN HEAVEN: Map of volcanic field showing topography and main vents.

Basalt to mafic andesite lava flows range from 0.4 to 24 m in thickness, whereas andesite flows are up to 90 m thick. Individual flows extend up to 46 km in length, have areas to 116 km², and volumes to 1.2 km³. Most flows <150 ka contain extensive lava tubes, making the Indian Heaven Volcanic Field an important speleological area.

Lava flow units are separated stratigraphically into two main groups. The older group has been extensively eroded during the Hayden Creek Glaciation, ~250 to 150 ka. A younger group ranges in age between Hayden Creek Glaciation and Evans Creek Glaciation (~25 to 15 ka). The youngest unit is **Big Lava Bed**, dated by radiocarbon at 8,150 yr BP (uncorrected). The oldest lavas are believed to be <0.73 Ma due to their relative freshness and normal magnetic polarity. If the entire field formed since 0.73 Ma, the average eruption rate would be 375 m³/d. If the last eruption was Big Lava Bed (volume 0.88 km³) the field is overdue for another eruption.

How to get there: *The center of the field lies ~60 km east of Vancouver, and ~35 km north of the*

INDIAN HEAVEN: View northward along the Indian Heaven chain of volcanoes from the summit of Red Mountain. From left to right the volcanoes are Berry Mountain, Gifford Peak, East Crater, and Lemei Rock. Mount Adams lies on the right.

Columbia River, in the Gifford Pinchot National Forest. The field is accessible from the south from Washington Highway 14 at Carson via US Forest Service roads 30 and 65; from the west via USFS 30 from the Lewis River; and from the east from Trout Lake via USFS 24. A network of logging roads and trails, connecting with the Pacific Crest trail no. 2000, provides access to most areas.

References

Hammond, P. E., 1987, Lone Butte and Crazy Hills: Subglacial volcanic complexes, Cascade Range, Washington: in Hill, M.L. (ed.), *Geol. Soc. Am. Centennial Field Guide – Cordilleran Section*, 339-344.

Walsh, T. J., Korosec, M. A., Phillips, W. M., Logan, R. L., and Schasse, H. W., 1987, Geologic Map of Washington – Southwest Quadrant: *Wash. Div. Geol. Earth Res. Geol. Map GM-34..*

Paul E. Hammond

MARBLE MOUNTAIN–TROUT CREEK HILL ZONE
Washington

Type: *Series of cones and shields*
Lat/Long: *45.9°N, 122.0°W*
Elevation: *150 to 1,378 m*
Eruptive History: *All activity <0.7 Ma; most eruptions >0.16 Ma, last three eruptions between 8,060 and 7,700 yr BP*
Composition: *Mostly calc-alkaline, medium-K basalt and andesites*

The Marble Mountain–Trout Creek Hill (MMTC) volcanic zone is located immediately south of Mount St. Helens, Washington, along the southwestern projection of the St. Helens seismic zone. The northern end of the zone includes the basaltic lava flows erupted from the flanks of St. Helens (Cave Basalt) and the prominent 1,255-m elevation Marble Mountain shield volcano at the south base of St. Helens. From here the 20-km-wide linear zone extends 65 km to the Columbia River. It includes Trout Creek Hill volcano, 893 m elevation, along the Wind River. The zone is underlain by deformed and altered volcanic strata and small hypabyssal intrusions, both of Oligocene to Miocene age.

In the MMTC zone the volcanic centers are chiefly monogenetic scoria cones with 3 to 4 extensive lava flows. At least 22 centers are recognized; many are unmapped topographic features. A few volcanoes are described to characterize the volcanic zone. All volcanoes except those <12,000 yr have been glaciated. Although all lava flows show normal magnetic polarity, the oldest dated flows are the deeply glaciated basalt of Soda Peaks at 0.36 Ma.

The largest volcano in the zone is the **Marble** Mountain shield volcano, 1,255 m elevation. It rises 350 m, covers ~32 km^2, and has a volume of ~1.8 km^3. Three scoria cones aligned north–south form the summit of the volcano. The last eruption (0.16 Ma) was from an andesitic cone on its south flank.

Bare Mountain and West Crater are two unusual centers in the volcanic zone. The highest part of **Bare** Mountain, 1,329 m elevation, consists of Tertiary intrusive andesite. Judging by the glaciated surface and deep incision of its lava flows, at more than 0.16 Ma, the center initially erupted a series of extensive flows into Siouxon Creek. At 7,700 yr BP the center again erupted – the last event in the zone – destroying the cone and creating a phreatic crater 400 m in diameter and 275 m deep. Lithic fragments of the Tertiary bedrock and abundant andesite scoria, >10 m in depth, blanket an oval-shaped

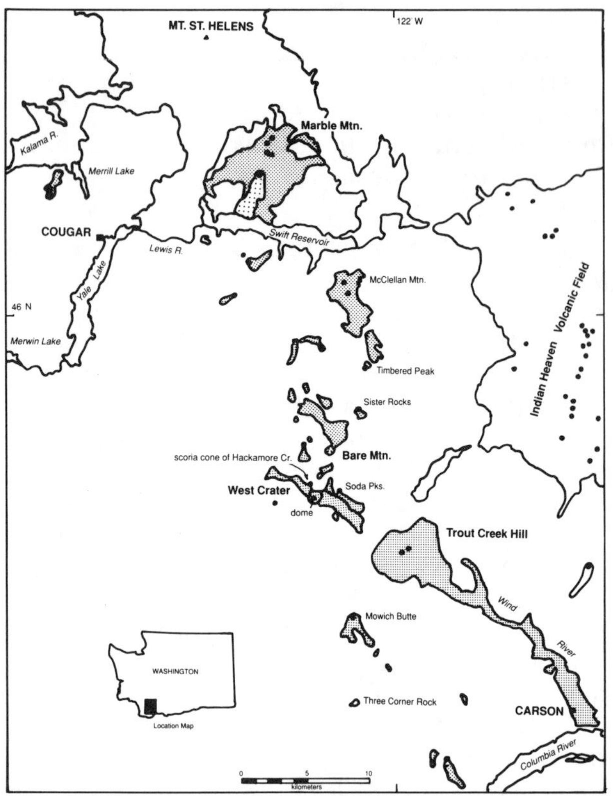

MARBLE MOUNTAIN–TROUT CREEK HILL: Map showing distribution of eruptive centers and lava flows forming the volcanic zone, and its geographic relationship to St. Helens and the Indian Heaven volcanic field. Dots are eruptive centers; cross-hatching shows andesite flow of south flank of Marble Mountain.

area ~6.4 km^2. No lava flows were erupted during this last event. The crater and its flanks are well exposed.

West Crater, 1,282 m in elevation, is an andesitic dome with two large adjacent lava flows. The dome is an inconspicuous volcanic feature enclosed on its south side by a higher ridge of Tertiary volcanic and intrusive rocks. It erupted 8,060 yr BP in the floor of a cirque carved in those rocks. The dome is 290 m high and 965 m in diameter, about the same size as the present dome in the crater of St. Helens. The dome contains a summit crater 200 m wide and 24 m deep. A final lobe of domal lava sits in the bottom of the crater.

The west lava flow of West Crater appears to overlap the low snout of a scoriaceous andesite flow from the small cone of Hackamore Creek. The cone erupted a black tephra covering ~1.9 km^2. Carbon fragments from the base of the deposit yield a radiocarbon date of 8,000 yr BP.

The closeness in age and in location of Bare Mountain, West Crater, and Hackamore Creek scoria cone suggest that all eruptions were related to the same magmatic event and controlled by the same structure(s).

How to get there: *The various volcanoes in this zone are readily accessible via a number of Forest Service roads in the Gifford Pinchot National Forest and the Mount St. Helens National Volcanic Monument. The volcanoes on the south side of St. Helens are reached via roads Washington 503 and Forest Service 90, ~50 km east of Woodland, which is on Interstate 5. Volcanoes just south of Swift Reservoir are on private land but are accessible via logging roads going south from either end of the reservoir; travel here is best on weekends to avoid traffic. The middle parts of the zone are accessible from Carson, which is 80 km east of Portland. Go north on the Wind River highway, taking Forest Service road 54 to reach Trout Creek Hill, West Crater, and Bare Mountain. Forest Service road 64, further to the north, goes to Silver Rocks and other nearby volcanoes. The southernmost volcanoes can be reached via Washington Dept. of Natural Resources roads leaving from Washington Highway 14 along the Columbia River west of Cougar.*

References

Hammond, P. E., and Korosec, M. A., 1983, Geochemical analyses, age dates and flow-volume estimates for Quaternary volcanic rocks, southern Cascade Mountains, Washington: *Wash. Div. Geol. Earth Res. Open-file Report 83-13*, 36 pp.

Polivka, D.R., 1984, Quaternary volcanology of the West Crater–Soda Peaks area, southern Washington Cascade Range: M.S. Thesis, Portland State University, Portland, Oregon, 78 pp.

Paul Hammond

VOLCANOES OF OREGON

Western Cascades, 169
Boring Lava, 170
High Cascades: Columbia River to Mount Hood, 172
Hood, 173
High Cascades: East of Mount Hood to Clear Lake, 175
High Cascades: Clear Lake to Olallie Butte, 176
Jefferson, 177
High Cascades: South of Mount Jefferson to Santiam Pass, 178
Sand Mountain, 180
Washington, 181
Belknap, 182
Broken Top, 183
Three Sisters, 184
Bachelor, 185
High Cascades: South of Three Sisters to Willamette Pass, 187
High Cascades: Willamette Pass to Windigo Pass, 189
High Cascades: Windingo Pass to Diamond Lake, 190
Thielsen, 191
Crater Lake, 193
Pelican Butte, 195
Sprague River Valley, 196
McLoughlin, 197
Brown Mountain, 199
Newberry, 200
Devils Garden, 203
Fort Rock Basin, 203
Bald Mountain, 205
Silicic domes of south-central Oregon, 206
Iron Mountain, 207
Silver Creek, 207
Diamond Craters, 208
Saddle Butte, 209
Jordan Craters, 210
Jackies Butte, 211

WESTERN CASCADES
Oregon

Type: *Many shield volcanoes, cinder cones, and lava flows*
Lat/Long: *43-45°N, 122-123°W*
Elevation: *100-2,000 m*
Eruptive History: *5–0.5 Ma*
Composition: *Basalt to dacite*

The Cascade Range in Oregon is customarily divided into two physiographic subprovinces: Western Cascades and High Cascades. The High Cascades subprovince is built of rocks mainly younger than 3.5 Ma and is the modern Cascade Range volcanic arc. In contrast, the Western Cascades encompass a deeply eroded pile of chiefly Oligocene to Pliocene volcanic and volcaniclastic rocks. Vents younger than 5 Ma in the Western Cascades mostly are small or intermediate-sized cinder cones and shields that erupted basalt and basaltic andesite lava along the eastern margin of the subprovince; many are poorly exposed, and their extent is generally too small to map separately on the accompanying location map. Listed here (from north to south) are the few Pliocene and Quaternary volcanoes in the Western Cascades (exclusive of the Portland area).

Battle Ax Mountain is a 1-2-Ma shield volcano that surmounts a high ridge north of Detroit, Oregon. Battle Ax erupted chiefly andesite lava, though its flows range from basaltic andesite to dacite.

Snow Peak, a Pliocene volcanic center 40 km east of Albany, remains unstudied. Two K-Ar ages indicate its age is ~3 Ma.

Harter Mountain, 5 km northwest of Iron Mountain, is a relatively unknown Quaternary basalt or basaltic andesite shield. No published K-Ar ages or chemical analyses exist for Harter.

Several Pliocene basaltic andesite shields form a ridge just west of the High Cascades, southwest of Mount Jefferson. The best known volcano in this group is **Iron** Mountain, which towers above US Highway 20 near Tombstone Pass. The cliffs of Iron Mountain expose bedded cinders intruded by dikes and sills. The summit lookout tower can be reached by an excellent trail that is well known for its diverse botany, including plants characteristic of both the Western Cascades and High Cascades. Other

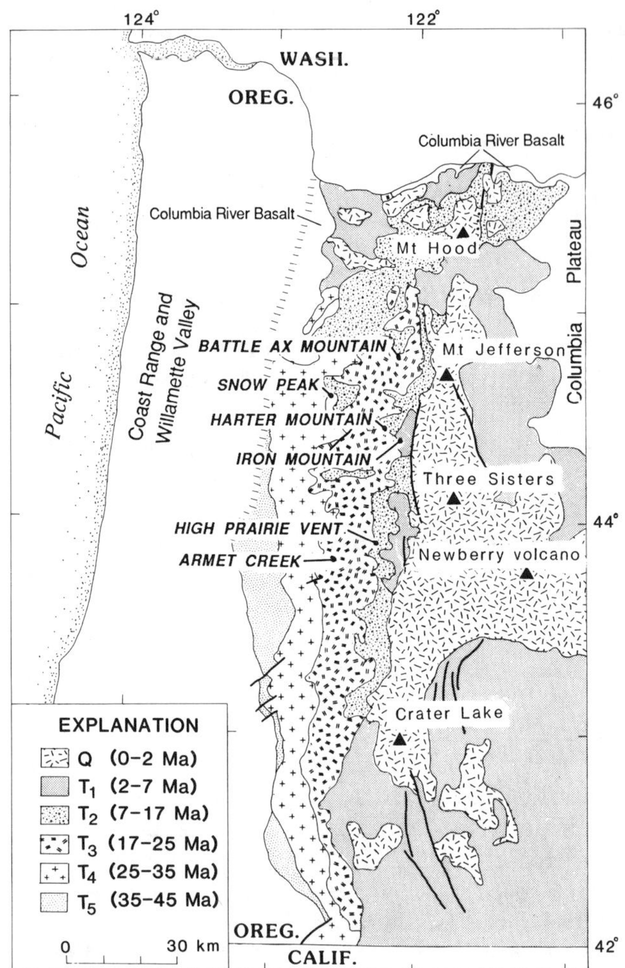

WESTERN CASCADES: Map showing generalized distribution of rocks by age in Cascade Range and locations of Quaternary and Pliocene volcanoes in Western Cascades subprovince.

vents are located near Crescent Mountain, Three Pyramids, and Bachelor Mountain. K-Ar ages range from 6 to 4 Ma for the lava from these volcanoes.

High Prairie is a sequence of low-potassium basalts ~1.98 Ma that forms benches along the North and Middle Forks of the Willamette River near Oakridge. The High Prairie vent is buried beneath outwash deposits.

A small basaltic andesite lava flow and cinder cone in **Armet Creek** are probably the most southerly Quaternary volcanic rocks erupted in the Western Cascades. The Armet Creek flow, exposed on the north side of Lookout Point Reservoir and visible from Oregon Highway 58, is ~0.56 Ma.

How to get there: *Several state highways traverse the Western Cascades. US Highway 20 passes near Harter and Iron Mountains; State Highway 58 near Armet Creek and High Prairie. Maps for the Mount Hood and Willamette National Forests (US Forest Service) are helpful when attempting to negotiate the maze of back roads.*

References

Walker, G. W., and Duncan, R. A., 1988, Geologic map of the Salem 1° × 2° sheet, Oregon: *USGS Misc. Invest. Map I-1893*, scale 1:250,000.

Woller, N. M., and Priest, G. R., 1983, Geology of the Lookout Point area, Lane County, Oregon, *in* Priest, G. R., and Vogt, B. F. (eds.), Geology and geothermal resources of the central Oregon Cascade Range: *Ore. Dept. Geol. Min. Indust. Spec. Pap.15*, 49-56.

David R. Sherrod

BORING LAVA
Oregon

Type: *Monogenetic volcano field*
Lat/Long: *45.30 °N, 122.50°W*
Elevation: *91–1,236 m ; mostly 200–300 m*
Eruptive History: *3 K-Ar ages: 2.56, 1.53, 1.3 Ma; no historic activity*
Composition: *High alumina olivine basalt*

Metropolitan Portland, Oregon, like Auckland, New Zealand, includes most of a Plio-Pleistocene volcanic field. The Boring Lava includes at least 32 and possibly 50 cinder cones and small shield volcanoes lying within a radius of 21 km of Kelly Butte, which is 100 km west of Mount Hood and the High Cascade axis. Only the Clear Lake volcanics in California lie as far west in the coterminous United States. Unlike Clear Lake, Boring Lava vents have been inactive for at least 0.3 Ma.

The three dated samples show reversed remanent magnetism, but since tens of other determinations of Boring Lava have about equal normal and reversed magnetic polarities, the volcanoes were probably active from at least 2.7 to perhaps <0.5 Ma.

Northwest of the town of Boring, 20 eruptive centers are concentrated within ~100 km^2. Vents in the east part of this cluster average <2.6 km in diameter and 333 m in height above their

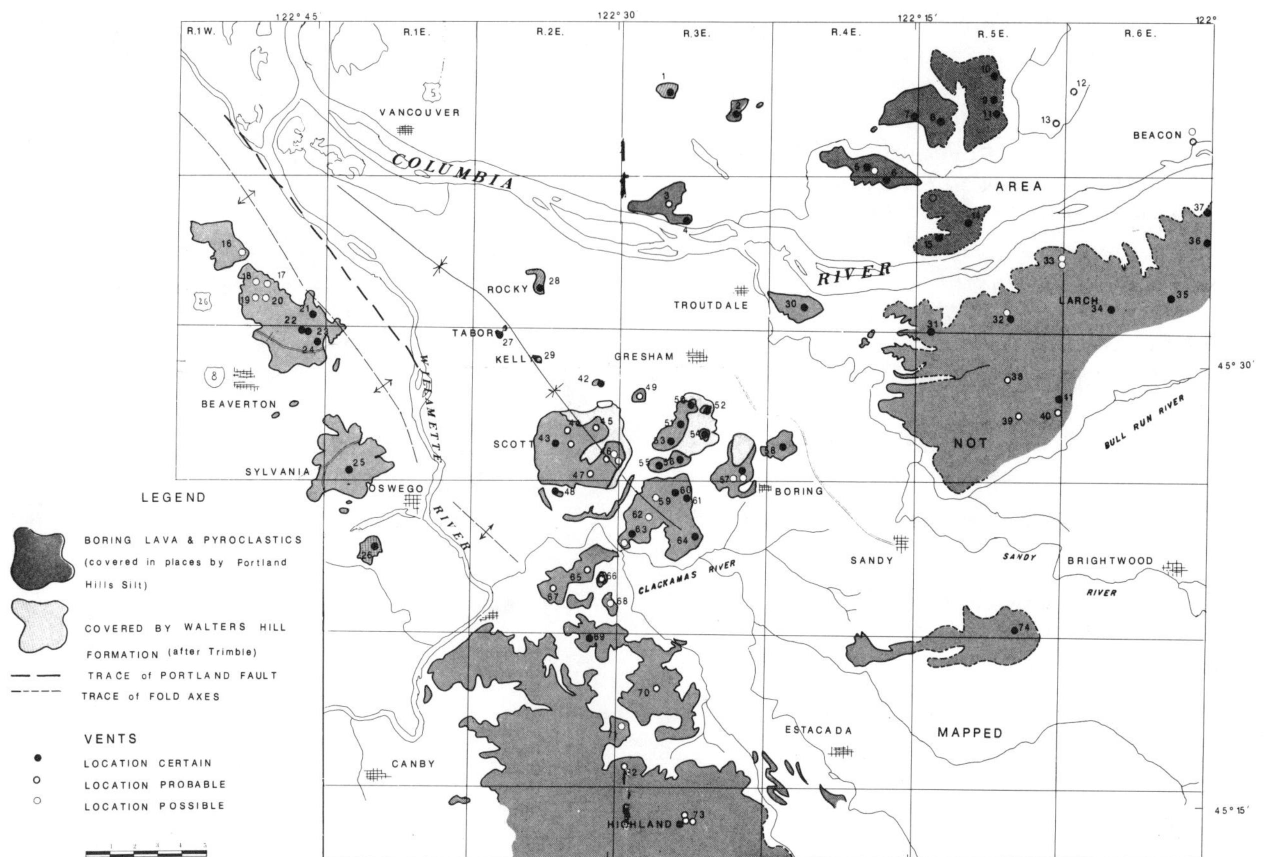

BORING LAVA: Locations of vents and distribution of Boring Lava, from Allen (1975). Volcanoes referred to in text: Kelly Butte (#29), Highland Butte (#73), Rocky Butte (#28), Larch Mountain (#34), Bobs Hill (#9), Mount Tabor (#27).

bases. Lava from **Highland Butte** and **Larch** Mountain shield volcanoes form gently sloping plains covering many tens of square kilometers. Well logs indicate that in most places except near vents, Boring Lava is between 30 and 60 m thick.

Partial summit craters remain only at **Bobs Hill**, 33 km northeast of Portland, and at a low cone enclosing a lake north of Battleground, Washington, 33 km north of Portland. Most other volcanoes still have a low cone shape and are mantled with loess above 122 m elevation. Below this they were scoured by the cataclysmic Bretz floods from Glacial Lake Missoula ~13,000-15,000 yr ago.

Boring lava is characteristically a light-gray phyric olivine basalt. A specimen from Rocky Butte is predominantly labradorite, with phenocrysts of olivine, mostly altered to iddingsite. The volcanoes locally contain scoria, cinders, tuff, tuff breccia, and ash. Weathering

BORING LAVA: Looking S88°E toward Mount Hood, with at least 8 Boring volcanoes (Mount Tabor on left) visible above downtown Portland. Oregon State Highway Dept. Photograph #6112.

may extend to depths of 8 m or more, the upper 2-5 m commonly being a red clayey soil.

How to get there: *The best and most accessible exposure is the cross-section of the cinder cone in the amphitheater in Mount* **Tabor** *Park at about NE 64th St. between Hawthorne and Stark avenues. Numerous quartzite-pebble xenoliths from the underlying Mio-Pliocene Troutdale gravels which make up the bulk of Mount Tabor have been found in the cinders here. The best view of the volcanic field is from the summit in Rocky Butte Park, at about NE 96th St., and Shaver Ave. where massive cliffs of flood-scoured lava form the NE face of the Butte.*

References

Allen, J. E., 1975, Volcanoes of the Portland area, Oregon: *Ore-Bin (Ore. Dept. Geol. Min. Indust.) 37*, 145-157.

Trimble, D. E., 1963, Geology of Portland, Oregon and adjacent areas: *USGS Bull. 1119*, 119 pp.

John E. Allen

HIGH CASCADES

Oregon

COLUMBIA RIVER TO MOUNT HOOD

Type: *Shield volcanoes and cinder cones*
Lat/Long: *45.50°N, 121.66°W*
Elevation: *60-1,500 m*
Eruptive History: *Sporadically throughout Quaternary*
Composition: *Basalt to andesite*

The most notable volcanic landforms in this part of the Cascade Range are Mount **Defiance** (andesite, 57-59% SiO_2) and **Lost Lake Butte** (54-55% SiO_2). Their morphology suggests they are younger than 0.25 Ma, perhaps substantially younger.

In the northern Hood River valley, **Booth Hill** and **Lenz Butte** form cinder cones that erupted olivine basalt(?). They are purportedly of Quaternary age. Probably at about the same time, a small unnamed shield volcano near **Long Prairie** was active on the Hood River escarpment. The Holocene **Parkdale** flow in the southern Hood River Valley is a basaltic andesite lava erupted 6,890 ^{14}C yr ago from a cinder cone at its southern margin.

Several basaltic andesite cinder cones erupted lava in the **Red Hill** area north of Mount Hood. These vents are undated, but their morphology and normal polarity suggest they are younger than ~0.73 Ma. Fifteen kilometers to the east in the Dog River area, 3 aligned cinder cones erupted basaltic andesite lava that flowed at least 15 km northeast down **Mill Creek**. The undated Mill Creek lava is <0.73 Ma, on the basis of its position near the floor of Mill Creek and its normal polarity magnetization.

How to get there: *Paved roads diverging from Oregon Highway 35 get to within 1-2 km of these vents. Starting points are the towns of Hood River (I-84 along Columbia River) and Government Camp (south side of Mount Hood).*

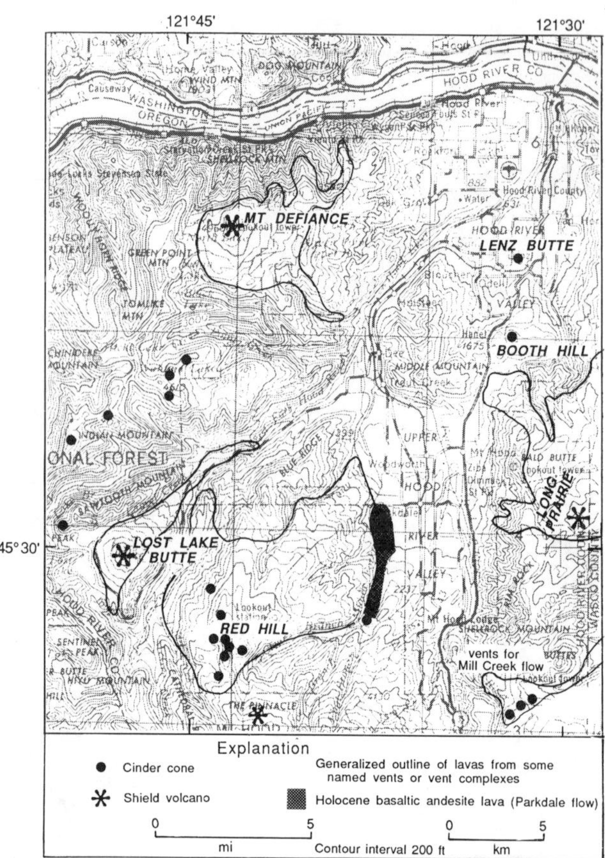

COLUMBIA RIVER TO MOUNT HOOD: Volcano map for Cascade Range from Columbia River to Mount Hood, northern Oregon.

Reference

Wise, W. S., 1969, Geology and petrology of the Mt. Hood area: a study of High Cascade volcanism: *Geol. Soc. Am. Bull. 80*, 969-1006.

David R. Sherrod

HOOD: View north, showing Holocene block and ash deposits (Timberline, Zigzag, and Old Maid eruptive episodes) that form debris fan on south flank. Crater Rock nestled beneath summit at apex of fan.

HOOD
Oregon

Type: *Stratovolcano*
Lat/Long: *45.37°N, 121.70°W*
Elevation: *1,250-3,426 m*
Eruptive History: *Quaternary*
Composition: *Andesite and dacite*

For the general public, Mount Hood is perhaps the most accessible and preeminent of Oregon's volcanoes, located only 75 km east-southeast of Portland, Oregon. It is the highest peak in the state (3,426 m) and one of the most often climbed peaks in the Pacific Northwest. In summer, Mount Hood's timberline wilderness is a pastoral garden for backpackers. In winter and spring the volcano's slopes host several downhill ski runs and cross-country tracks.

Mount Hood is a Quaternary stratovolcano composed of lava flows, domes, and volcaniclastic deposits. The bulk of the volcano is andesite that ranges from 57-62% SiO_2. Dacite (62-63% SiO_2) is limited to the products of the last 15,000 yr. Four eruptions during that time have spilled pyroclastic block-and-ash flows and lahars into the four river systems (Sandy, Salmon, Hood, and White Rivers) that drain the mountain and its 12 glaciers. Thus, the volcano poses a direct volcanic hazard to communities near the volcano and in downstream reaches.

The oldest rocks exposed in the Mount Hood area are flows of the middle Miocene Columbia River Basalt Group, a sequence of tholeiitic lava (erupted in eastern Oregon and Washington) that flooded through low points in the ancestral Cascade Range ~17-12 Ma. During and after these eruptions, Columbia River Basalt was folded into a series of northeast-trending anticlines and synclines with structural relief of at least 600 m. Folding diminished as middle and upper Miocene andesite lava and breccia accumulated in the synclines, probably culminating by ~10 Ma. The Laurel Hill stock and related plutons of hypabyssal quartz diorite were emplaced ~8-9 Ma.

Laurel Hill plutonic rocks underlie Government Camp and Multorpor Ski area at ~1,250 m elevation on the south flank of Mount Hood. Pre-Hood bedrock is exposed at ~1,250-1,500 m elevation throughout the area; therefore Mount Hood has constructional relief of only ~2,000 m. The volume of Hood's cone is ~30-60 km^3, depending on the amount of relief on the pre-Hood surface. (Most of a cone's volume is in its lower parts, so the large range in the volume estimate arises from the uncertainty in guessing the amount of pre-Hood topographic relief.)

A pre-Mount Hood volcano, the **Sandy Glacier** volcano (basaltic andesite and andesite), is exposed on the west side of Mount Hood at the 1,650-m elevation and is partly covered by Mount Hood rocks. The age of Sandy Glacier volcano is somewhat controversial because of two disparate K-Ar ages: ~3.2 and 1.3 Ma. There is no field evidence to favor one age over the other. Basaltic andesite was erupted from vents at **The Pinnacle** and **Cloud Cap** on the north side of Mount Hood since 0.73 Ma; Cloud Cap lava has a K-Ar age of ~0.5-0.6 Ma. These eruptive products are partly covered by and probably unrelated to the Mount Hood volcano.

Mount Hood itself is built of normally polarized andesite and is younger than 0.73 Ma. Potassium-argon ages from three main-cone lava flows are chiefly ~0.4-0.6 Ma, but rocks this young can be difficult to date. The Pleistocene volcanic history of Mount Hood is poorly understood because there has been no detailed geologic mapping of the main cone.

The summit area of Mount Hood comprises several andesite or dacite domes. Solfataric alteration has weakened these rocks and made them susceptible to potentially catastrophic

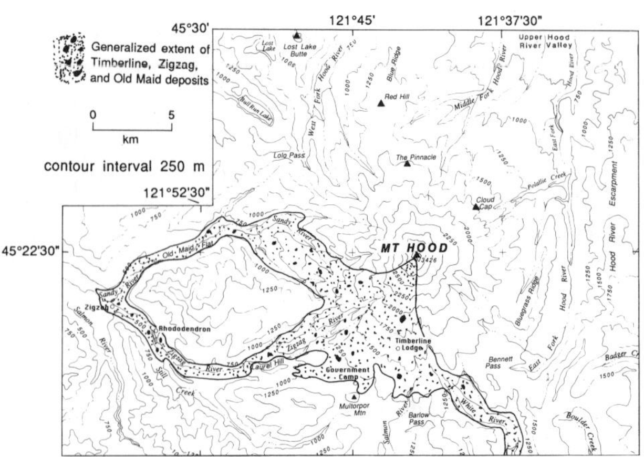

HOOD: Volcano map for Mount Hood, Cascade Range, northern Oregon.

slope failure. Pleistocene laharic deposits that are rich in altered Mount Hood andesite underlie terraces in the lower Hood River valley (35 km north of the volcano) and are present on both sides of the Columbia River near the town of Hood River, ~40 km north of the volcano. The lahar presumably began as a debris avalanche that incorporated large masses of preexisting rock on the flank of the volcano, inundated the Hood River Valley, and temporarily filled the Columbia River to a depth of 30 m. A single radiocarbon date from wood in the lahar is >38,000 yr (beyond the limit of ^{14}C dating when the sample was run). Old lahars are also exposed 35 km west of Mount Hood in the Sandy River drainage. Weathering horizons as much as 7 m thick indicate that the deposits west of Mount Hood are many tens of thousands of years old.

Volcanic hazard studies have illuminated Mount Hood's history of the last 15,000 yr, in which four eruptive periods are recognized: Polallie (15,000-12,000 yr ago), Timberline (1,800-1,400 yr ago), Zigzag (600-400 yr ago), and Old Maid (250-180 yr ago).

Deposits of Polallie age, which were formed chiefly by pyroclastic flows and debris flows, occur on all sides of Mount Hood, thus indicating a vent at or near the summit. Many Polallie deposits are restricted to ridge tops and valley sides, positions that were probably determined by the thickness of glacier ice in a given valley. Deposits along ridge tops accumulated when ice filled the adjacent valley, whereas deposits on valley sides were formed adjacent to glaciers after the glaciers had shrunk in size. Valley-floor deposits were laid down beyond the ends of the

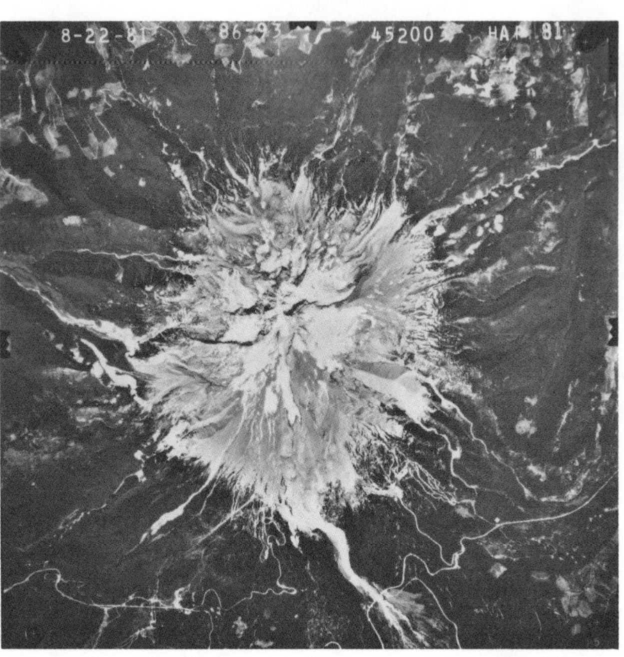

HOOD: Vertical air photograph from US National High Altitude Photography Program.

valley glaciers during or following both of the earlier stages. There are no radiocarbon ages from Polallie deposits. The 15,000-12,000-yr age is estimated from the degree of weathering and the inferred relation of Polallie deposits to glaciers on Mount Hood.

The Timberline eruptive period, which occurred 1,800-1,400 yr ago, produced between 0.7 and 1.1 km^3 of pyroclastic flows and debris flows. The vents for Timberline and all subsequent eruptions were high on the southwest flank, perhaps at the vent now filled by the Crater Rock dome. Deposits of Timberline age are mostly restricted to the southwest flank of the volcano, where they form a broad volcaniclastic fan across a 45° sector. Some Timberline debris flows, however, extend down the Sandy River for 80 km to its mouth at the Columbia River near Troutdale, Oregon.

The Zigzag eruptive period occurred 600-400 yr ago. Deposits of this age are of limited volume and have been found only in the valleys of the Sandy and Zigzag Rivers.

The Old Maid eruptive period occurred 250-180 yr ago. Rocks and unconsolidated deposits include a dome in Mount Hood's crater (Crater Rock), a pyroclastic flow, and many debris flows in the White and Sandy River valleys. Crater

Rock is a dacite dome ~300-400 m across its base and ~170 m high on its south side. Old Maid deposits represent ~0.15 km^3 of dacitic magma.

Old Maid eruptions (1760-1810 A.D.) occurred during a time when white men were exploring the Pacific Northwest. The Lewis and Clark expedition of 1804-1805 saw the Sandy River within a few years of the main Old Maid events, while the river was still clogged with eruptive debris. Their journal notations describe the Sandy as a shallow river choked with sediment and possessing a quicksand bottom. In contrast, the modern river is a deep, narrow, boulder-armored channel.

How to get there: *US Highway 26 crosses the south flank of Mount Hood, and Oregon Highway 35 meets it along the east side. Numerous paved or graded roads provide further access. A hiking trail encircles the volcano, much of which is protected within the Mount Hood Wilderness, part of the Mount Hood National Forest.*

References

Cameron, K. A., and Pringle, P. T., 1987, A detailed chronology of the most recent major eruptive period at Mount Hood, Oregon: *Geol. Soc. Am. Bull. 99*, 845-851.

Crandell, D. R., 1980, Recent eruptive history of Mount Hood, Oregon, and potential hazards from future eruptions: *USGS Bull. 1492*, 81 pp.

David R. Sherrod

HIGH CASCADES
Oregon

EAST OF MOUNT HOOD TO CLEAR LAKE

Type: *Many domes, cinder cones, and a shield volcano*
Lat/Long: *45.25°N, 121.50°W*
Elevation: *1,100-2,000 m*
Eruptive History: *Pliocene and Pleistocene*
Composition: *Basaltic andesite, andesite, and dacite*

The area includes chiefly andesite to dacite domes and lava flows that are Pliocene and early Pleistocene(?) in age. **Frog Lake Buttes** near Wapinitia Pass on US Highway 26 are a cluster of dacite domes. They are reversely polarized (thus older than 0.73 Ma) but undated by isotopic methods. A nearby dacite dome 1 km west of **Blue Box Pass** is probably the same age.

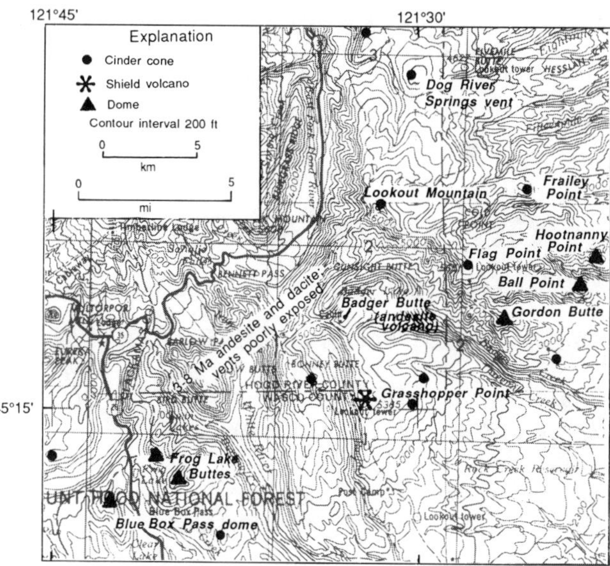

EAST HOOD TO CLEAR LAKE: Volcano map for Cascade Range from east of Mount Hood to Clear Lake, northern Oregon.

Badger Butte is a poorly exposed andesite volcano. **Gordon Butte**, **Ball Point**, and **Hootnanny Point** are previously unrecognized upper Pliocene(?) dacite domes that are similar in form, composition, and, presumably, age. These domes and Badger Butte are probably about the same age as 3-Ma andesite near Lookout Mountain.

Basaltic andesite was erupted from a few centers throughout this area. The largest mafic volcano is **Grasshopper Point**, an undated shield volcano of reversely polarized basaltic andesite lava flows. Cinder cones (now deeply eroded) dot the east side of the Grasshopper Point shield. Other cinder vents are exposed at **Flag Point**, **Frailey Point,** and near **Dog River Springs**. There are no isotopic ages from any of these centers.

How to get there: *US Highway 26 crosses the crest of the Cascade Range at Blue Box Pass. The area is enclosed by State Highway 35 and Forest Roads 42, 43, 44, and 27. Numerous graded roads provide access from these main roads. A map of the Mount Hood National Forest is helpful.*

References

Waters, A. C., 1968, Reconnaissance geologic map of the Dufur quadrangle, Hood River, Sherman, and Wasco Counties, Oregon: *USGS Misc. Invest. Map I-556.*

Wise, W. S., 1969, Geology and petrology of the Mt. Hood area: a study of High Cascade volcanism: *Geol. Soc. Am. Bull.*, *80*, 969-1006.

David R. Sherrod

HIGH CASCADES
Oregon

CLEAR LAKE TO OLALLIE BUTTE

Type: *Many shield volcanoes and cinder cones*
Lat/Long: *45.00°N, 121.75°W*
Elevation: *1,000-2,200 m*
Eruptive History: *Pleistocene*
Composition: *Basalt and basaltic andesite*

The crest of the Cascade Range from Clear Lake to Olallie Butte is built predominantly of Pleistocene basaltic andesite shield volcanoes, many of which are reversely polarized. The Cascade Range crest in this area has a lower average elevation than elsewhere in Oregon; consequently, the shields are not as deeply eroded by glacial processes. Indeed, much of this area has such thick soil cover that outcrops are hard to find and the magnetic polarity can be determined only locally.

Clear Lake Butte (reversely polarized, older than 0.73 Ma) and **Summit Butte** (normally polarized, probably younger than 0.73 Ma) are broad, low shields with slopes that average ~5°-7°. Their lavas, though unanalyzed, are petrographically similar to other basaltic andesite erupted in the area. Mount **Wilson,** a reversely polarized basaltic andesite shield (older than 0.73 Ma) was built atop the 2-3 Ma dacite dome of **Beaver Butte**. Consequently, its summit stands ~300 m higher than surrounding vents and has been glaciated several times. Two subsidiary basaltic andesite volcanoes west and north of Mount Wilson, **Wests Butte** and **North Wilson**, respectively, are probably normally polarized and younger than 0.73 Ma, on the basis of aeromagnetic anomaly maps.

The Pinhead Buttes comprise three mafic volcanoes that range widely in age. The youngest, **North Pinhead**, is a cinder cone with an uneroded summit crater. Its basalt lava flows are glaciated, however, indicating an age probably just over 11,000 yr BP. **South Pinhead** is a substantially older but undated basaltic andesite shield

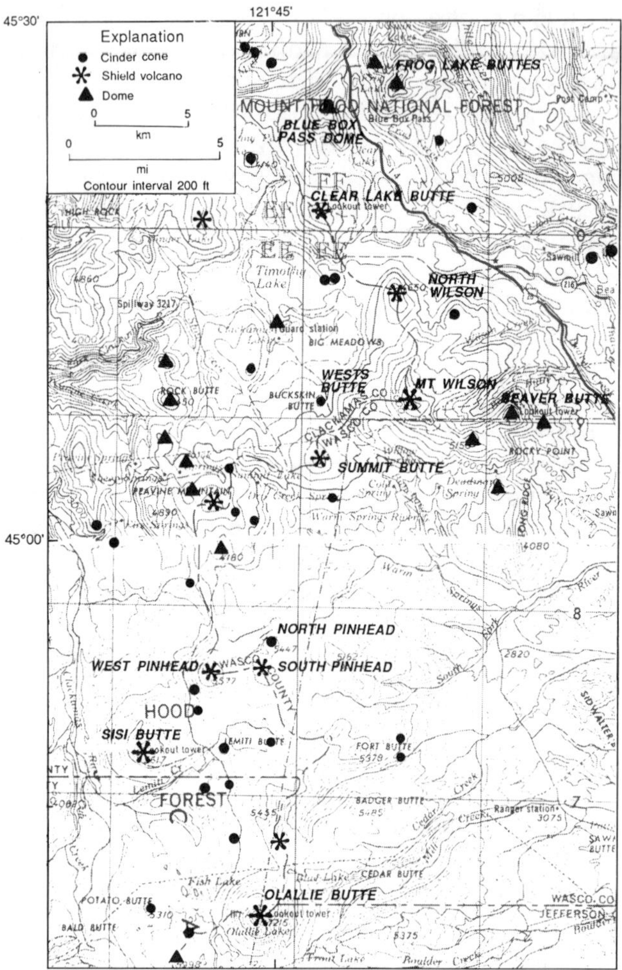

CLEAR LAKE TO OLALLIE BUTTE: Volcano map for Cascade Range from Clear Lake to Olallie Butte, northern Oregon.

volcano; there are no outcrops on South Pinhead and its magnetic polarity on aeromagnetic maps is ambiguous. **West Pinhead** is a deeply eroded, normally polarized basaltic andesite shield volcano that is older than North Pinhead and presumably younger than 0.73 Ma.

Sisi Butte is a prominent, normally polarized basaltic andesite shield volcano near the headwaters of the Clackamas River. The highest point on Sisi Butte, now marked by a US Forest Service fire lookout, is 200-300 m west of the central vent. Erosion has stripped away the pyroclastic cone and denuded a shallow, conduit-filling intrusion, so that the surrounding apron of lava now stands slightly higher than the volcano's eruptive center. Sisi Butte's morphology is similar to Mount Wilson in that glaciers have carved the northeast flank and left moraines strung out to the north and east. Sisi Butte is undated by isotopic methods; however, its erosional state, normal polarity magnetization, and geomorphic similarity to the reversely polarized shield of Mount Wilson suggest an age of 0.50-0.73 Ma.

Olallie Butte is a young analog of Sisi Butte: similar in composition (basaltic andesite) but substantially less eroded. Nevertheless, glaciers have exposed its central plug and carved broad U-shaped troughs on its northeast flank. The extent of Olallie Butte's glaciation indicates an age greater than 25,000 yr. Conceivably it is as old as 70,000-100,000 yr, a conclusion reached by comparing its morphology with Mount Jefferson, a composite volcano ~15 km to the south.

How to get there: *From US Highway 26, paved Forest Road 42 runs south along the Cascade Range crest to the Clackamas River, where it meets Forest Road 46. A map of the Mount Hood National Forest is helpful.*

References

Sherrod, D. R., and Conrey, R. M., 1988, Geologic setting of the Breitenbush–Austin Hot Springs area, Cascade Range, north-central Oregon, *in* Sherrod, D. R. (ed.), Geology and geothermal resources of the Breitenbush–Austin Hot Springs area, Clackamas and Marion Counties, Oregon: *Ore. Dept. Geol. Min. Indust. Open-file Rpt. O-88-5*, 14 pp.

White, C. M., 1980, Geology of the Breitenbush Hot Springs Quadrangle: *Ore. Dept. Geol. Min. Indust. Spec. Pap. 9*, 26 pp.

David R. Sherrod

JEFFERSON
Oregon

Type: *Stratovolcano*
Lat/Long: *44.66°N, 121.80°W*
Elevation: *3,199 m*
Eruptive History
Two periods of activity:
290 ka or less: 20 km³ of mainly andesite
~70 ka: 4 km³ of hornblende dacite
Composition: *Andesite to dacite*

Mount Jefferson is one of the major late Quaternary stratovolcanoes of the High Cascade Range. It consists of ~25 km³ of lavas erupted in two distinct episodes. The first episode constructed an andesite-dacite volcano perhaps slightly higher than the present Mount Jefferson. The remnant of this structure is composed of ~200 thin (3-10-m) flows of andesite and one thick (80-300-m) flow of dacite (the prominent pink mass in the core of the mountain when viewed from the west). Radial dikes are common, but pyroclastic material is very scarce except in the innermost core of the volcano. A

JEFFERSON: Aerial oblique view from the south. Photograph by D. Sherrod.

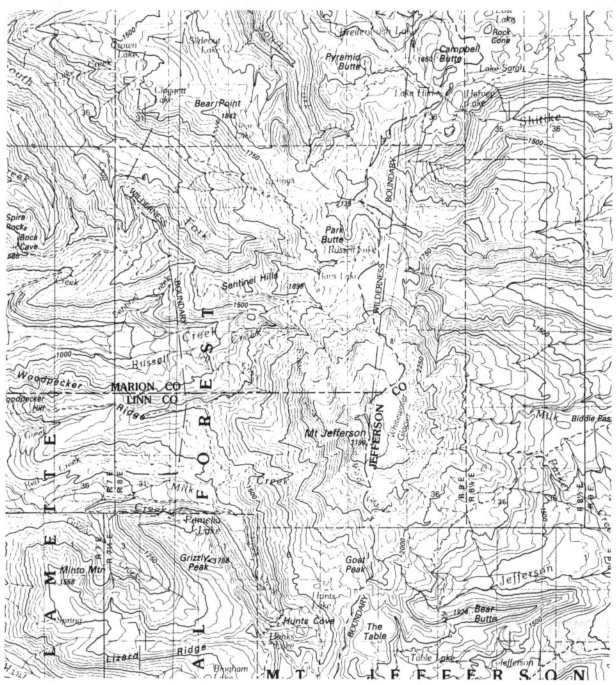

JEFFERSON: Topographic map.

JEFFERSON: Vertical air photograph from US National High Altitude Photography Program.

small plug is exposed in the core ~150 m below the present summit. Glacial erosion (~150 ka?) scoured canyons up to 300 m deep in the andesite cone before the succeeding dacite volcano was built atop it. Several hornblende dacite domes were erupted during this latter episode, along with a series of flows from a vent coincident with the present summit of the mountain. Pyroclastic eruptions, presumably associated with dome building, resulted in an extensive tephra deposit east of the mountain. Ash flows or hot avalanches due to dome collapse traveled at least 15 km down valleys both east and west of the volcano. One such deposit may be seen in road cuts along Highway 22 near Whitewater Creek. A deeply weathered exposure of this deposit is overlain by a biotite-bearing tephra, presumed to be Mount St. Helens set C.

The younger Jefferson dacite volcano was probably intraglacial. One piece of evidence for this is the fingerlike distribution of its lavas. Multiple flows piled up in single fingers, rather than spreading out into adjacent topographic lows. Other evidence includes "perched" flows on the sides of canyon walls, hundreds of meters above canyon floors, as well as extremely glassy and wildly fan-jointed preserved flow margins. All of these features are easily explicable if major ice streams filled the canyons, and a modest ice sheet encased the volcano during its eruptions.

Perhaps the most notable feature of the area around Mount Jefferson is the lack of mafic rocks. The basaltic andesite and basalt so typical of the High Cascade Range in Oregon are nearly absent from an area of ~200 km^2 which extends from Mount Jefferson northward some 15 km. Mount Jefferson is built on an older field of andesite and dacite volcanoes which extends back to at least 1.5 Ma, and perhaps to 2.5 Ma.

How to get there: *Mount Jefferson is located in the Mount Jefferson Wilderness area and the Warm Springs Indian Reservation, ~115 km southeast of Portland, Oregon, and 80 km northwest of Bend, Oregon. Highway 22 east of Salem, Oregon, provides access to Forest Service roads and trails which lead into the wilderness area.*

References

Conrey, R. M., 1986, Criteria for recognition of former ice-magma contacts in the Mt. Jefferson area, central Oregon High Cascade Range (abst.): *EOS, Trans. Am. Geophys. Un. 67*, 1251.

Sutton, K., 1974, Geology of Mt. Jefferson: M.S. Thesis, Univ. of Oregon, Eugene.

Richard M. Conrey

HIGH CASCADES
Oregon

SOUTH OF MOUNT JEFFERSON TO SANTIAM PASS

Type: *Several shield volcanoes, many cinder cones, a few domes*
Lat/Long: *44.50°N, 121.75°W*
Elevation: *1,200-2,390 m*
Eruptive History: *Pliocene(?) and Quaternary*
Composition: *Basalt to andesite*

The High Cascades in this area is a broad ridge built up by several shield volcanoes and numerous cinder cones. Most summits mark either relatively young vents or deeply eroded vent complexes. Reversely polarized basaltic andesite, basalt, and andesite lava older than 0.73 Ma are exposed in the walls of U-shaped canyons, but most of the area is mantled by nor-

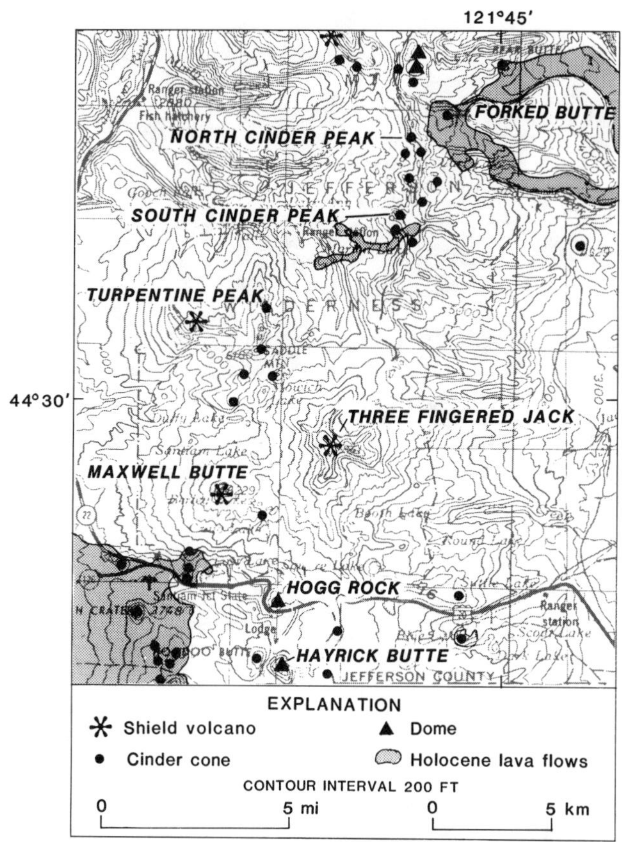

SOUTH OF JEFFERSON TO SANTIAM PASS: Volcano map for Cascade Range from south of Three Sisters to Willamette Pass, central Oregon: volcanoes near range crest.

mally polarized rocks younger than 0.73 Ma. Most of the area is within the Mount Jefferson Wilderness.

Three Fingered Jack (2,390 m) is the most distinctive volcano in this part of the range. This deeply glaciated basaltic andesite shield volcano has ~800 m of relief and is centered on a pyroclastic cone that underlies the summit of the mountain. The cone lacks a high-level conduit-filling plug, however, unlike other shield volcanoes such as nearby Mount Washington south of Santiam Pass. Three Fingered Jack is undated by radiometric methods, but its age probably lies between 0.50 and 0.25 Ma, as inferred from its erosional state compared to other shield volcanoes in the High Cascades.

Turpentine Peak forms another eroded, normally polarized basaltic andesite shield volcano of uncertain age, ~8 km northwest of Three Fingered Jack. Relatively uneroded **Maxwell Butte** basaltic andesite shield lies ~5 km southwest of Three Fingered Jack. Maxwell Butte is presumably younger than 0.25 Ma but remains undated.

Several cinder vents form a glaciated vent complex in the **North Cinder Peak** area. The individual vents are masses of basaltic andesite cinders, lava, and dikes or small plugs; many are visible along the Pacific Crest Trail. There are probably other similar vents along the Cascade Range crest between Three Fingered Jack and the North Cinder Peak area, but geologic mapping is sketchy there. **South Cinder Peak** is much younger than the North Cinder Peak complex, and probably <25,000 yr old. Glacial erosion has exposed its conduit on the north side, but the cone has probably not been glaciated more than once.

Holocene basalt and basaltic andesite lavas were erupted from at least three cinder cones north of Three Fingered Jack. These eruptions are younger than ~6,845 ^{14}C yr, which is the age of the widespread Mazama ash bed that underlies them, but none of the eruptions in the Mount Jefferson Wilderness has been dated directly. Using sedimentation rates determined from dated peat in nearby lakes, W. E. Scott assigned an age of 6,400-6,500 ^{14}C yr for eruptions from **Forked Butte**.

Hogg Rock and **Hayrick Butte** are two flat-topped andesite domes near Santiam Pass. The flat top is characteristic of tuyas, which are volcanic landforms associated with subglacial eruptions. Subsequent glaciation has stripped away other evidence, such as glassy margins or unusual jointing patterns.

How to get there: *Most of the area is in the Mount Jefferson Wilderness, and is traversed by the Pacific Crest Trail. Oregon Highway 22 and US Highway 20 cross the Cascade Range crest at Santiam Pass, and paved or graveled logging roads provide access around the other sides of the wilderness area. Maps of the Deschutes or Willamette National Forest are helpful aids in planning any trip away from highways in the area.*

References

Black, G. L, Woller, N. M., and Ferns, M. L., Geologic map of the Crescent Mountain area, Linn County, Oregon: *Ore. Dept. Geol. Min. Indust. Geol. Map Series GMS-47.*

Scott, W. E., 1977, Quaternary glaciation and volcanism, Metolius River area, Oregon: *Geol. Soc. Am. Bull. 88*, 113-124.

David R. Sherrod

SAND MOUNTAIN
Oregon

Type: *Linear chains of mafic cones.*
Lat/Long: *44.38°N., 121.93°W*
Elevation: *1,665 m*
Eruptive History: *Two principal episodes, ~3,800 and 3,000 yr BP*
Composition: *Subalkaline basalt and basaltic andesite*

The Sand Mountain chains of 23 cinder cones and associated lava fields cover 76 km^2 on the western margin of the central High Cascades of Oregon. Two north–south alignments of 42 distinct vents intersect beneath the largest cone (Sand Mountain, 250 m high), suggesting that a complex system of dikes and related conduits exists at depth. Early eruptions from a central group of cones produced a northern field of basaltic lavas and blocked drainages to form ephemeral lakes (Lava Lake, Fish Lake), 3,850 ^{14}C yr BP. Additional basaltic cones and lavas extended this eruptive episode along a north-northeast alignment and produced an ash-fall deposit recognized 12 km to the east. Minimum age of this ash is 3,440 ^{14}C yr BP. The last eruptive episode occurred 2,990 ^{14}C yr BP along the north-northwest alignment. Several large cones were formed, and basaltic andesite lavas spread into the path of an ancestral McKenzie River, establishing Sahalie Falls and Koosah Falls. The river was blocked, and a standing forest was drowned; preserved stumps are still rooted in the floor of Clear Lake and yield an age of 2,950 ^{14}C yr BP. Although a few cinder cones were breached by flows, most lavas issued from marginal vents up to 500 m away from the cones. Where early basalts spread westward down gentle slopes, a relatively smooth, ropy-to-blocky flow surface developed in association with lava tubes and tree molds. Arcuate pressure ridges mantled by scoria and blocks appeared on later basaltic andesite lavas, especially where they were confined in channels. The development of

SAND MOUNTAIN: Oblique aerial photo of Sand Mountain volcanic field, viewed south; North and South Sand Mountain cones, right center.

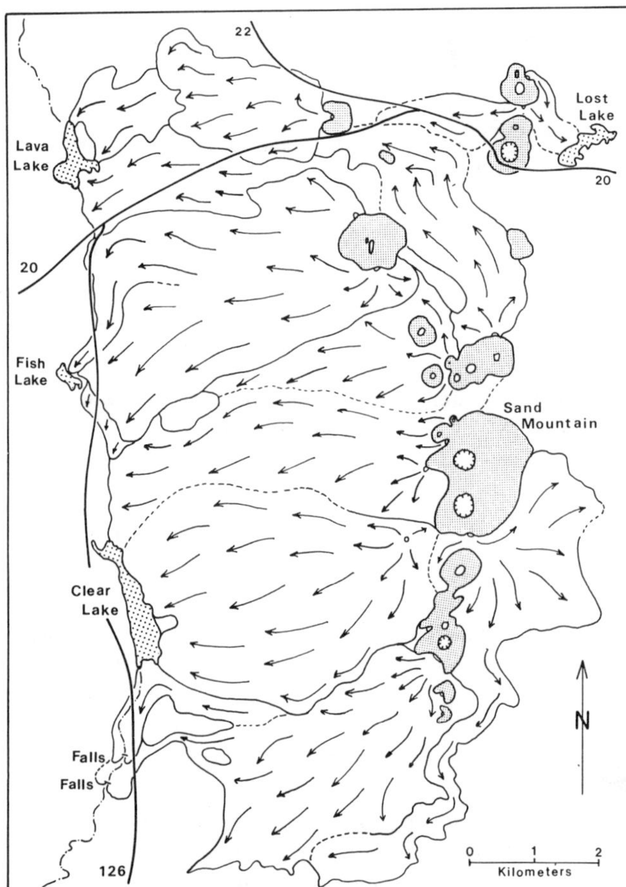

SAND MOUNTAIN: Map of Sand Mountain volcanic field, central High Cascades, Oregon.

the Sand Mountain field followed a tectonomagmatic pattern common in many Cascade volcanic systems. Faulting associated with High Cascade graben subsidence, ~5 Ma, produced elongate north–south zones of fractured rock, now buried under Pleistocene lava. Sand Mountain magmas were injected into these fractured zones, probably as dikes, and followed separate conduits to isolated vents on the surface. Parental basaltic magmas followed a north-northeast fracture system and were succeeded 800 yr later by more evolved basaltic andesite magmas along a north-northwest fracture system.

How to get there: *US 126 (east from Eugene) follows the west edge of the Sand Mountain lava field while US 20 (east from Albany) and US 22 (east from Salem) cross over the northern part. Access to Sand Mountain cones (summer months only) is via dirt roads from Big Lake near US 20 at Santiam Pass.*

References

Taylor, E. M., 1981, Central High Cascade roadside geology - Bend, Sisters, McKenzie Pass, and Santiam Pass, *in* Johnston, D. A., and Donnelly-Nolan, J. (eds.), Guides to some volcanic terranes in Washington, Idaho, Oregon, and northern California: *USGS Circ. 838*, 55-84.

Smith, G. A., Snee, L. W., and Taylor, E. M., 1987, Stratigraphic, sedimentologic, and petrologic record of late Miocene subsidence of the central Oregon High Cascades: *Geology 15*, 389-392.

Edward M. Taylor

WASHINGTON
Oregon

Type: *Mafic shield volcano*
Lat/Long: *44.33°N, 121.84°W*
Elevation: *2,376 m*
Eruptive History: *Late Pleistocene shield surmounted by composite summit cone of cinders, lavas, and intrusives*
Composition: *Subalkaline basalt and basaltic andesite*

The High Cascade Range adjacent to Mount Washington is covered by lavas of high-alumina diktyaxitic basalt. Part of this lava field was produced from scattered cinder cones, now represented by glacially scoured, inconspicuous buttes along the Cascade crest. Most of the basalt emerged from vents now buried beneath Mount Washington and poured down the east and west Cascade slopes. To the east, these basaltic lavas are exposed along the perimeter of

WASHINGTON: View to north with Washington (mid-foreground), Three Fingered Jack (middle ground), and Jefferson in hazy distance. Unforested lava flow in near foreground is Holocene basaltic andesite at McKenzie Pass. Photograph by D. Sherrod.

Mount Washington in the canyons of Cache Creek and Dry Creek, south to Dugout Mountain. Westward, basalt occurs as glaciated outcrops from Patjens Lakes to the McKenzie River, where they form intracanyon benches, 16 km from Mount Washington.

Subsequent eruptions of relatively uniform basaltic andesite lavas produced a shield volcano, 5 km in diameter, surmounted by a summit cone that probably reached an elevation of 2,600 m, ~1,200 m above the pre-existing basalt field. Mafic ash accumulated on the flanks of the shield and has been preserved as thick sections of palagonitic tuff on the southwest and northeast sides of the summit cone. The volcano was intruded by a micronorite plug which now forms the central pinnacle, 0.4 km in diameter. Dikes are exposed in many parts of the summit cone; their predominant orientation is north–south. A swarm of narrow dikes extends northward from the summit plug.

Although no isotopic ages are available, all of the Mount Washington lavas and the underlying basalt appear to be of normal paleomagnetic polarity; the age of Mount Washington is probably no more than a few 100,000 yr, similar to that of other central High Cascade stratovolcanoes. During the late Pleistocene, cirques were excavated into the flanks of the summit cone by valley glaciers which extended more than 12 km east and west. There is no evidence of recent reactivation of Mount Washington volcanism, but a series of aligned small basaltic andesite spatter cones erupted on the northeast flank 1,330 ^{14}C yr BP.

How to get there: *Access to the Mount Washington Wilderness is restricted to foot trails. The west and southwest sides of the mountain are crossed by the Skyline Trail, 5 km from a trailhead at Big Lake, near US 20. The best long-distance viewpoints on main highways are from US 20 near Blue Lake and from Oregon 242 at Windy Point.*

References

Taylor, E. M., 1981, Central High Cascade roadside geology – Bend, Sisters, McKenzie Pass, and Santiam Pass, in Johnston, D. A., and Donnelly-Nolan, J. (eds.), Guides to some volcanic terranes in Washington, Idaho, Oregon, and northern California: *USGS Circ. 838*, 55-84.

Taylor, E. M., Causey, J. D., and MacLeod, N. S., 1983, Geology and mineral resource potential map of the Mount Washington Wilderness, Deschutes, Lane, and Linn Counties, Oregon: *USGS Open-file Rpt. 83-0662.*

Edward M. Taylor

BELKNAP
Oregon

Type: *Shield volcano*
Lat/Long: *44.28°N, 121.54°W*
Elevation: *2,096 m*
Eruptive History: *Three principal episodes between 3,000 and 1,500 yr BP*
Composition: *Subalkaline basalt and basaltic andesite*

The Belknap shield volcano and its distal lava tongues cover 98 km^2 of the crest of the central High Cascades in Oregon. Prior to 2,900 yr BP, the first eruptive phase distributed basaltic cinders and ash over a broad area to the northeast and southeast, while basaltic lavas moved 10 km eastward from a growing shield. A second phase, 2,883 ^{14}C yr BP, produced an adventive shield of basaltic andesite on the east flank, known as "Little Belknap." The third phase was responsible for the bulk of modern Belknap volcano. It was constructed by effusion of basaltic andesite lavas from the central vent (Belknap Crater), 1,495 ^{14}C yr BP, and from a vent 2 km to the south (South Belknap cone), 1,775 ^{14}C yr BP. The final eruptions from the northeast base of Belknap Crater sent lavas

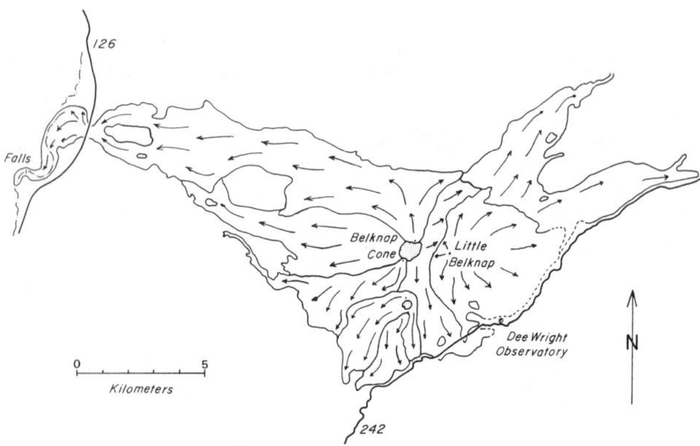

BELKNAP: Map of Belknap volcanic field, central High Cascades, Oregon.

BELKNAP: Central vent and associated lava flows. Photograph by L. Topinka, USGS.

15 km westward into the valley of the McKenzie River.

Belknap volcano is a well preserved Holocene example of the type of volcanic process responsible for construction of the Pleistocene High Cascade platform. Eruption of mafic lava and ash from a single vent area produced a broad shield with a core of cinders. Belknap is intermediate in scale between diminutive cinder cones with small lava flows such as **Twin Craters** or **Yapoah Cone** (south of Belknap) and large composite cones on a shield base which may reach elevations of 2,500 m, such as Mount Washington or Three Fingered Jack (north of Belknap).

How to get there: *From Eugene follow US 126 eastward to junction with Ore. 242, then on 242 to McKenzie Pass. From Sisters follow 242 westward to McKenzie Pass. Highways are paved but*

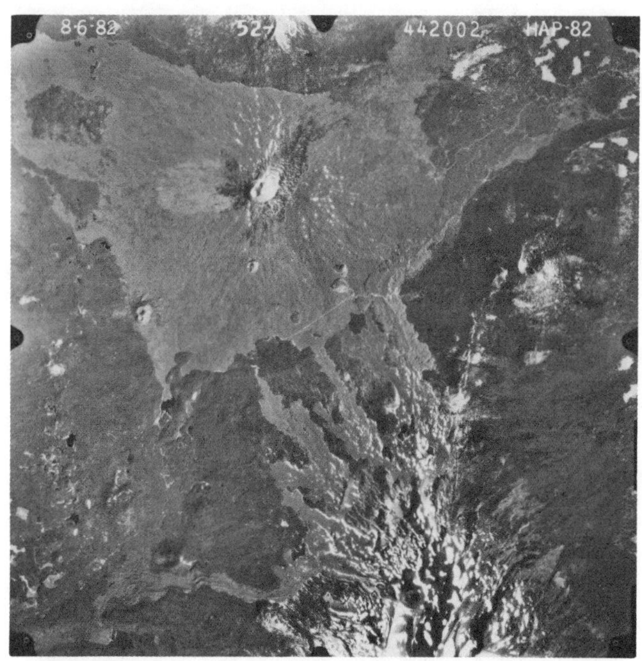

BELKNAP: Vertical air photographic from US National High Altitude Photography Program showing Belknap and flows from North Sister (bottom).

closed by snow during late fall, winter, and spring.

References

Taylor, E. M., 1987, Late High Cascade volcanism from the summit of McKenzie Pass, Oregon: Pleistocene composite cones on platform of shield volcanoes; Holocene eruptive centers and lava fields: *in* Hill, M. L. (ed.), *Centennial Field Guide Volume 1, Cordilleran section*, Geol. Soc. Am. 311-312.

Taylor, E. M., 1965, Recent volcanism between Three Fingered Jack and North Sister, Oregon Cascade Range: *Ore. Dept. Geol. Min. Indust., Misc. Pap. No. 10*, 121-147.

Edward M. Taylor

BROKEN TOP
Oregon

Type: *Mixed mafic-silicic stratovolcano.*
Lat/Long: *44.08°N, 121.70°W*
Elevation: *2,800 m*
Eruptive History: *Pleistocene*
Composition: *Dominant basaltic andesite with interbedded calc-alkaline andesite, dacite, and rhyodacite*

BROKEN TOP: Northwest side of Broken Top volcano. Extensive glacial erosion has exposed interbedded agglomerate and lavas, cut by sills, dikes, and a large plug.

Broken Top is a complex stratovolcano magnificently exposed by glacial erosion. Pleistocene eruptions of basaltic andesite lava produced a broad shield with a core of oxidized agglomerate invaded by dikes and sills. Subordinate silicic magmas were erupted intermittently; andesite, dacite, and rhyodacite lavas, intrusives, and pyroclastic flow deposits are associated with the predominant mafic lavas from the lower flanks to the summit of the volcano. The central crater of Broken Top was enlarged to a diameter of 0.8 km, probably by subsidence. The resulting depression was filled by thick flows of basaltic andesite and eventually the summit cone was buried beneath a shroud of thin, vesicular lavas. After the central conduit had congealed to a plug of micronorite, the core of the volcano was subjected to hydrothermal alteration. Glacial cirques have been carved into three sides of the mountain, revealing internal structure. Holocene eruptive activity on the flanks has produced basaltic cones, flows, and ash deposits interbedded with Neoglacial moraines and outwash.

How to get there: *Broken Top volcano is within Three Sisters Wilderness; vehicles are not permitted. For scenic views, follow Cascade Lakes Highway west from Bend. For close access to south slopes, follow secondary road north from Todd Lake, then west to trailheads at Crater Creek. For access to north slopes, follow Three Creek Lake Highway south from Sisters to trailheads north of Tam McArthur Rim.*

Reference

Taylor, E. M., 1978, Field geology of S.W. Broken Top quadrangle, Oregon: *Ore. Dept. Geol. Min. Indust., Spec. Pap. 2*, 1-50.

Edward M. Taylor

THREE SISTERS
Oregon

Type: *Mafic and silicic stratovolcano cluster*
Lat/Long: *44.17°N., 121.77°W. (N. Sister), 44.10°N. (M. Sister), 121.75°W. (S. Sister)*
Elevations: *3,075 m (N. Sister), 3,063 m (M. Sister), 3,158 m (S. Sister)*
Eruptive History *Late Pleistocene to 1,900 yr BP*
Composition:
Calc-alkaline basaltic andesite (N. Sister)
Interbedded calc-alkaline basalt, basaltic andesite, andesite, dacite, and rhyodacite (M. Sister)
Calc-alkaline basaltic andesite, andesite, dacite, and rhyodacite (S. Sister)

North Sister is an 8-km-wide shield volcano composed of basaltic andesite, built upon dissected remnants of an earlier basaltic shield of similar size (**Little Brother**), ~200,000 yr ago. The summit cone contains thin lavas interbedded with red and black cinders and ash, invaded by swarms of dikes, sills, and a central plug. Palagonitic ash is abundant on the northeast flank, probably because of more extensive ice cover during eruptions. Flat-lying lavas near the summit reflect a high level topographic irregularity, such as a large crater, which has not survived. Some late stage lavas reveal a progressive variation toward iron enrichment in hornblende-bearing andesites. Pleistocene and Holocene glacial erosion has excavated large cirques in the flanks of North Sister, revealing a complex, composite structure.

Middle Sister was constructed after North Sister; however, early Middle Sister lavas display iron enrichment trends similar to late North Sister lavas. Subsequent development of Middle Sister included lavas, ejecta, and intrusives of andesite, dacite, and rhyodacite dominated by an open-textured basalt porphyry which was extruded from fissures on the north,

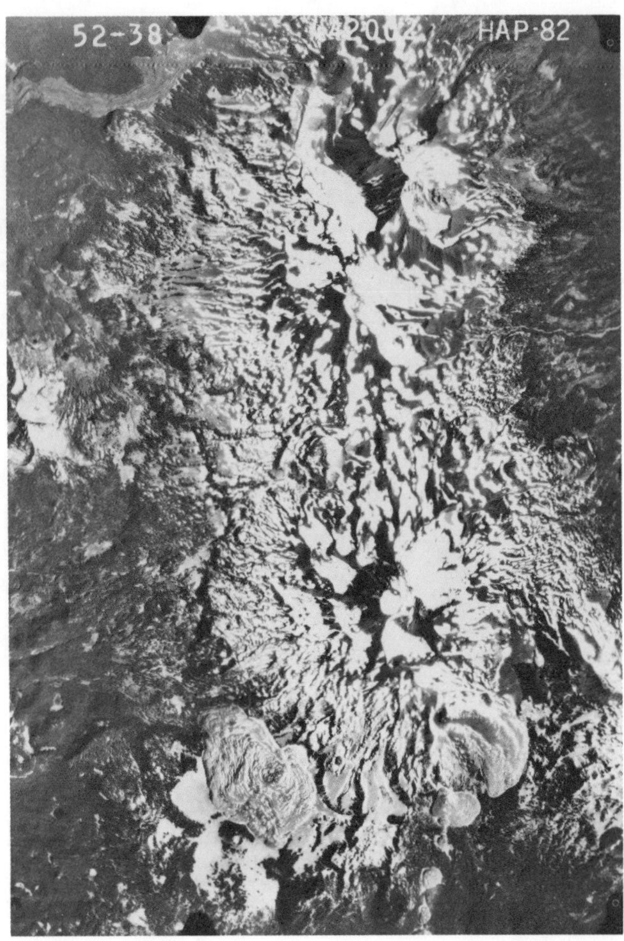

THREE SISTERS: Vertical air photograph from US National High Altitude Photography Program showing North, Middle, and South Sisters from top to bottom.

THREE SISTERS: East side of North Sister volcano, underlain by bedded palagonitic tuffs and dikes of basaltic andesite.

THREE SISTERS: South Sister and Newberry Lava Flow from vent on stratovolcano's flank, with Middle and North Sisters in the distance. Photograph by L. Topinka, USGS.

west, south, and southeast flanks. Andesitic lavas spread 7 km from the base of the cone. Dacitic and rhyodacitic lavas were largely restricted to thick flows and domes of more limited extent. The latest eruptions produced long, narrow tongues of dacitic lava from vents high on the north and south sides. Late Pleistocene and Holocene glacial erosion exposed part of a dacitic central plug.

Although early South Sister lavas are interbedded with Middle Sister lavas and were subjected to Pleistocene glaciation, South Sister is the youngest volcano of the Three Sisters group. The modern summit crater was formed by latest Pleistocene eruptions of now oxidized basaltic andesite; red lavas, cinders, and ash cover the crater rim. Earlier eruptions contributed andesite, dacite, and rhyodacite to the main cone and to flank eruptions up to 7 km away. Rhyodacitic eruptions of lavas, cinder cones, domes, and pumiceous tephra continued into Holocene time, as young as 1,900 yr BP. Holocene glaciation excavated high-level cirques but did not remove the summit crater.

How to get there: *Pacific Crest Trail crosses the west base of the Three Sisters from a northern trailhead at McKenzie Pass (Oregon 242) and a southern trailhead at Devils Lake (Cascade Lakes Highway).*

References

Scott, W. E., 1987, Holocene rhyodacite eruptions on the flanks of South Sister volcano, Oregon: *Geol. Soc. Am. Spec. Pap. 212*, 35-53.

Taylor, E. M., MacLeod, N. S., Sherrod, D. R., and Walker, G. W., 1987, Geologic map of the Three Sisters Wilderness, Deschutes, Lane, and Linn Counties, Oregon: *USGS Map MF-1952.*

Edward M. Taylor

BACHELOR
Oregon

Type: *Monogenetic volcano field*
Lat/Long: *43.78-44.02°N, 121.58-121.75°W (field)*
Elevation: *1,299 to 2,764 m*
Eruptive History: *Several episodes of activity during latest Pleistocene and early Holo cene, between ~18,000 and 7,000 yr BP*
Composition: *Basalt and basaltic andesite*

The Mount Bachelor volcanic chain is located in the eastern part of the central High Cascades of Oregon, south-southeast of the clustered composite volcanoes of the Three Sisters and Broken Top. The 25-km-long chain is composed of numerous scoria cones with related lava flows and three broad shield volcanoes, the northernmost of which is capped by the steep-sided summit cone of Mount Bachelor (formerly called Bachelor Butte). Together these features cover ~250 km^2 and constitute a total magma volume of 30-50 km^3. Individual volcanoes within the chain are typical of the thousands of monogenetic volcanoes that form the High Cascade landscape between major composite volcanoes.

Eruptions of the Mount Bachelor volcanic chain were dominantly effusive, but also included minor explosive activity that built cones of agglutinate spatter, bombs, and scoria. The eruptions can be divided into four episodes. The oldest episode occurred ~18,000-15,000 yr ago during glacier retreat. Eruptions at vents just west of Sparks Lake in the northwest part of the chain interacted with glacier ice and meltwater to form hyaloclastite; a thick sequence of lava flows was impounded on three sides by glacier ice and now forms a steep-sided plateau. However, most activity during the oldest episode was focused in the central part of the chain and built the shield volcano capped by **Sheridan** Mountain.

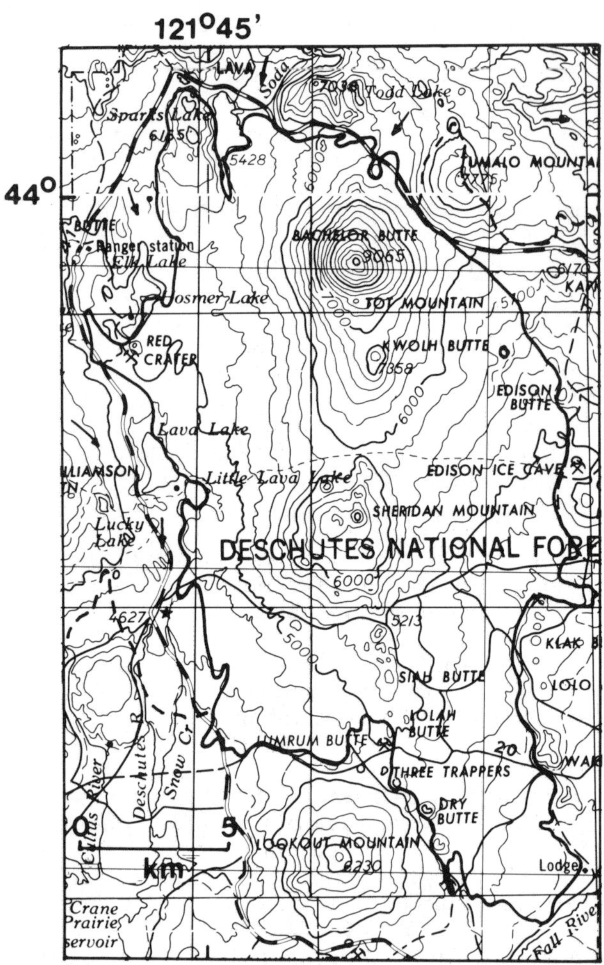

BACHELOR: Topographic map of Mount Bachelor volcanic chain showing extent of lava flows (solid line) and limit (dashed line) and ice-flow directions (arrows) of late Pleistocene glaciers. Base from Bend and Crescent 1° × 2° quadrangles, US Geological Survey.

During the second episode, eruptive activity shifted southward to a chain of scoria cones and lava flows that extends from near Sheridan Mountain to the south end of the chain. Eruptions during the third episode of activity built the shield volcano of **Kwolh Butte**, the shield that underlies Mount Bachelor, and, last, the summit cone of Mount Bachelor. By ~12,000 yr ago, the age of the oldest glacial moraines recognized on Mount Bachelor, the cone had nearly attained its present size. The moraines are overlain by the youngest lava flows of the third episode, which were erupted from vents on the north flank.

The final eruptive episode of the chain occurred ~8,000 to 10,000 yr ago and produced a scoria cone and lava flows on the lower north flank of Mount Bachelor. All activity ceased before 6,845 yr BP, as tephra of that age from the climactic eruption of Mount Mazama is found on all products of the Bachelor chain.

Petrologic data indicate that the lavas of the

BACHELOR: View of Mount Bachelor and Sparks Lake from the west. Cone of hyaloclastite and scoria and ice-contact lava flows of oldest eruptive episode (right foreground). Forested lava flows in center of view were erupted from cone visible on lower north flank during last eruptive episode of chain. Upper Holocene rhyodacite lava flow of South Sister (left foreground). Photograph by L. Topinka, US Geological Survey, Cascades Volcano Observatory.

BACHELOR: North side of Mount Bachelor showing cluster of lava domes, small pyroclastic cones, and plugs on the summit and 12,000-yr-old end moraines (A). Lava flows (B) that partly bury moraines and lava flows that issued from parasitic vents (C) postdate moraines. Prominent moraines (D) date from past 4,000 yr.

chain were derived from three parental magmas, each of which fed one to several magma bodies. Individual bodies underwent differing degrees of crystal fractionation, mixing, and assimilation before eruption, accounting for the variations seen in mineralogic and chemical compositions of lavas derived from the same parent. The linear distribution of vents, the linear structures of individual vents, and the essential synchrony of eruptions along segments of the chain indicate that magma rose to the surface along north–south zones of crustal weakness. Many of the conduits were probably fed by dikes.

The summit of Mount Bachelor consists of a northwest-trending cluster of small pyroclastic cones, shallow collapse craters, low blocky domes, and, in the cirque on the north side, several lava plugs. The steep (25°) upper flanks are covered with lava flows that display well preserved levees and flow fronts; many emerge from lava tubes. Several small parasitic shields on the south and north flanks impart a north–south elongation to the cone. Owing to a lack of deep exposures the internal structure of the cone is little known. Deposits of scoria and scoriaceous ash found as far as 10 km from the summit suggest that the lava-clad cone has a core that probably contains considerable pyroclastic debris. On the basis of similar rock composition and size, the internal structure of Mount Bachelor is probably comparable to that of older eroded cones in the High Cascades, such as Mount Thielsen, North Sister, and Mount Washington. The upper parts of these volcanoes are composed of alternating beds of pyroclastic material and agglutinate spatter, as well as lava flows, riddled with dikes and plugs.

No evidence of thermal activity exists on the volcano. Deep wells that form in the snowpack have been described as fumaroles; however, these features are formed by air moving in and out of the porous edifice with changes in atmospheric pressure.

How to get there: *The Mount Bachelor volcanic chain is located in the Deschutes National Forest, 20 km WSW of Bend in central Oregon. The north flank of Mount Bachelor is readily accessible during all seasons from the Cascade Lakes Highway (US Forest Service and Deschutes County Road 46). Several paved and graded gravel roads head west from US 97 and lead to other parts of the chain. Mount Bachelor Ski Area operates chairlifts to the summit during the summer as well as the ski season (weather permitting).*

Reference

Scott, W. E., and Gardner, C. A., in press, Geologic map of the Mount Bachelor volcanic chain and surrounding area, Cascade Range, Oregon: *USGS Misc. Invest. Series Map I-1967.*

William E. Scott

HIGH CASCADES
Oregon

SOUTH OF THREE SISTERS TO WILLAMETTE PASS (HIGHWAY 58)

Type: *Many shield volcanoes and cinder cones; a few domes*
Lat/Long: *43.75°N, 122.00°W*
Elevation: *1,500-2,400 m*
Eruptive History: *Pliocene(?) and Quaternary*
Composition: *Basalt to rhyodacite*

In contrast to the compositional diversity that characterizes volcanic rocks in the Three Sisters area, the High Cascades from Three Sisters to Crater Lake consist almost entirely of basaltic andesite and fewer basalt shield volcanoes and cinder cones that erupted during the past 2 or 3 million years. The crest in most places is built by normally polarized small-volume shield volcanoes or large cinder cones such as **Elk** Mountain, **Williamson** Mountain, and **Sixbit Point**, all younger than 0.73 Ma. Older rocks are exposed at **Little Packsaddle** and **Packsaddle** Mountains, two ridges of reversely polarized basaltic andesite that form part of a 0.73-Ma shield volcano. **Irish** Mountain is an elongate, normally polarized basaltic andesite shield. Its age is 0.25-0.73 Ma, estimated by comparing its state of erosion with other dated basaltic andesite shields such as **Cupit Mary** Mountain (K-Ar ages 0.30 and 0.68). Numerous basaltic andesite volcanoes fall into this general age range: **Charlton Butte**, **Taylor Butte**, **Fuji** Mountain, and the unnamed hill used by the **Willamette Pass** Ski Area for their downhill runs. In contrast, **Maiden Peak** and **The Twins**

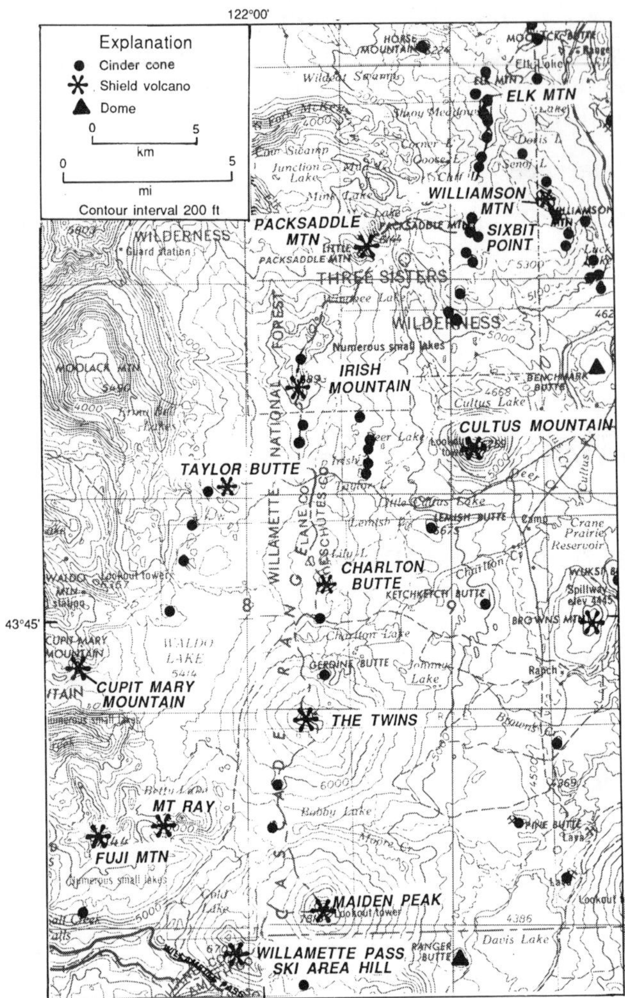

SOUTH OF THREE SISTERS TO WILLAMETTE PASS: Volcano map for Cascade Range from south of Three Sisters to Willamette Pass, central Oregon: centered at 122°W.

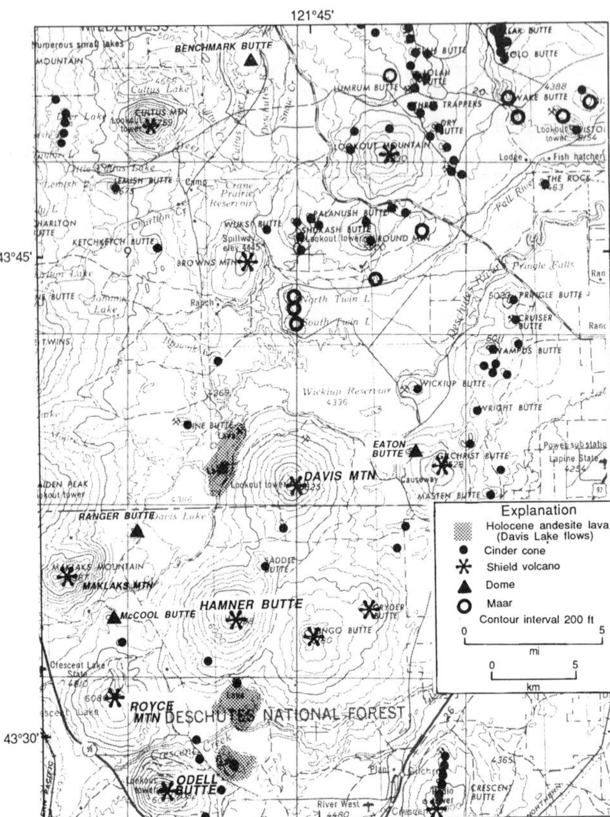

SOUTH OF THREE SISTERS TO WILLAMETTE PASS: Volcano map for Cascade Range from south of Three Sisters to Willamette Pass, central Oregon: volcanoes east of range crest; centered at 121°45'W.

are two basaltic andesite shields that are probably younger than ~0.25 Ma. These two volcanoes have substantial parts of their summit cones preserved. Mount **Ray**, a reversely polarized shield (older than 0.73 Ma), sits just west of the crest at the south end of Waldo Lake, ~8 km southwest of The Twins.

Several basaltic andesite shield volcanoes have erupted 15-20 km east of the crest. At least three of these – **Royce** Mountain and **Hamner** and **Odell** buttes – are reversely polarized and therefore >0.73 Ma. They are not deeply glaciated, however, owing to their position east of the range crest. The adjacent shield volcanoes of **Davis** and **Maklaks** Mountains are normally polarized and probably younger than 0.73 Ma.

The **Davis Lake** flows comprise three small Holocene cinder cones and their andesite lava, which lie nestled at the feet of Odell, Hamner, and Davis Mountains. The most northerly of these flows forms the natural dam to Davis Lake; the middle flow has a ^{14}C age of 4,740 yr. The three flows are compositionally similar and probably were erupted nearly simultaneously. They record the only known Holocene eruptions in the Cascade Range between latitudes 43-44° N (exclusive of Newberry Volcano).

Dacitic to rhyolitic vents are rare in this part of the High Cascades. The three domes of **McCool**, **Ranger**, and **Eaton** buttes are the only rhyodacite or rhyolite vents in the area. Eaton Butte has a K-Ar plagioclase age of 3.68 ± 3.3 Ma. The large analytical error results from the low K_2O content of the plagioclase and the very low radiogenic argon (1.1%) obtained during analysis. Skeptically, these domes should be considered undated. **Benchmark Butte**, an isolated, flat-topped dacite lava, is normally polarized and probably of late Pleistocene age. The flat top is characteristic of lava erupted beneath glacial ice, but erosion has destroyed any other evidence that indicates the lava erupted during a glacial advance.

How to get there: *Perhaps Oregon's most scenic volcano highway is Century Drive, passing through this part of the Cascade Range as it loops from Bend to Wickiup Junction (both towns on US Highway 97, between Cascade Range crest and Newberry volcano).*

References

Taylor, E. M., MacLeod, N. S., Sherrod, D. R., and Walker, G. W., 1987, Geologic map of the Three Sisters Wilderness, Deschutes, Lane, and Linn Counties, Oregon: *USGS Misc. Field Studies Map MF-1952.*

Williams, Howel, 1957, A geologic map of the Bend quadrangle, Oregon, and a reconnaissance geologic map of the central portion of the High Cascade Mountains: *Ore. Dept. Geol. Min. Indust.*, scales 1:125,000 and 1:250,000.

David R. Sherrod

HIGH CASCADES
Oregon

WILLAMETTE PASS TO WINDIGO PASS

Type: *Many shield volcanoes and cinder cones*
Lat/Long: *43.50°N, 122.00°W*
Elevation: *1,000-2,700 m*
Eruptive History: *Pleistocene*
Composition: *Basalt and basaltic andesite*

Diamond Peak, the dominant landform in the Willamette Pass area, is a basaltic andesite shield ~15 km^3 in volume. Like other shields in the area, it has a central pyroclastic cone (locally palagonitized but mostly fresh basaltic andesite cinders and glassy scoria) that is surrounded and surmounted by lava flows. Volcaniclastic rocks such as lahars and pyroclastic flows are unknown. Diamond Peak began erupting from a vent near its northern summit. A second vent later opened near the southern summit, piggybacking its lava and tephra over the previously erupted volcanic rocks. This vent migration likely involved only a small interval of time. Diamond Peak is probably <100,000 yr old, but is certainly older than the last glaciation, which ended ~11,000 yr ago.

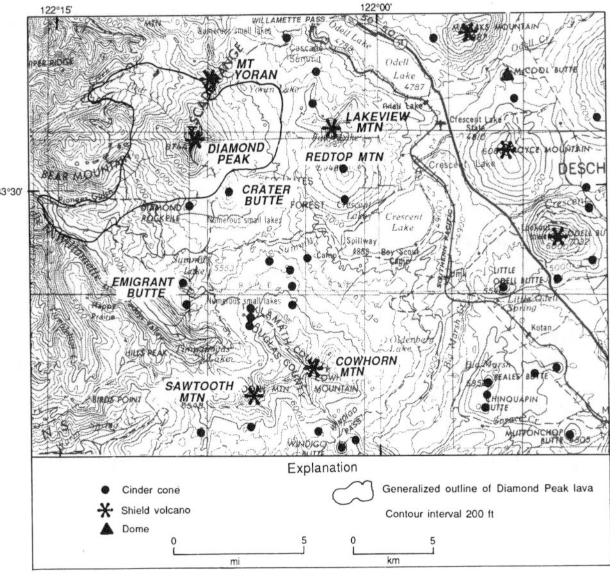

WILLAMETTE PASS TO WINDIGO PASS: Volcano map.

At Diamond Peak the eruptions probably became slightly more siliceous with time, though always in the range of olivine-bearing basaltic andesite. Early erupted flows (53-55% SiO_2) filled the valley of Pioneer Gulch on the southwest side of the volcano. Most lava in the shield contains 55-58% SiO_2. Other nearby volcanoes that are probably similar in age to Diamond Peak include **Crater Butte** (basaltic andesite) and **Redtop** Mountain (basalt), situated 4 km southeast and 10 km to the east, respectively. **Emigrant Butte** (basaltic andesite), located ~9 km south of Diamond Peak, is more deeply gutted by glaciation.

Substantially older than Diamond Peak is Mount **Yoran**, whose summit is a deeply eroded neck poking out from Diamond Peak's north slope. A sample of Mount Yoran's basaltic andesite lava, which is normally polarized, has K-Ar whole-rock ages of 0.52, 0.50, and 0.33; a weighted average of these ages is ~0.40 Ma. **Lakeview** Mountain, another basaltic andesite shield, probably erupted some time in the interval 0.25-0.73 Ma.

South of Summit Lake, lava from **Cowhorn** and **Sawtooth** Mountain shields interfinger along the range crest. These normally polarized central vents and surrounding cinder cones are chiefly basaltic andesite, but include lesser basalt and minor andesite. Both shields have been so deeply gutted by glaciation that their central conduit-filling plugs form the actual summits. In each case, the surrounding pyroclastic rocks are preserved only in the gulleys that separate the plug from the flanking aprons of lava flows. Sawtooth Mountain lava has a K-Ar whole rock age of 0.56 and 0.52 Ma.

How to get there: *Oregon Highway 58 crosses the Cascade Range crest at Willamette Pass; a narrow graded road crosses near Emigrant Butte. The Pacific Crest Trail passes from Willamette Pass to Diamond Peak and south to Cowhorn Mountain. Maps of the Deschutes and Willamette National Forests are excellent guides to trails and roads.*

References

Sherrod, D. R., Moyle, P. R., Rumsey, C. M., and MacLeod, N. S., 1983, Geology and mineral resource potential map of the Diamond Peak Wilderness, Lane and Klamath Counties, Oregon: *USGS Open-file Report 83-661*, 20 pp.

Williams, Howel, 1957, A geologic map of the Bend quadrangle, Oregon, and a reconnaissance geologic map of the central portion of the High Cascade Mountains: *Ore. Dept. Geol. Min. Indust.*, scales 1:125,000 and 1:250,000.

David R. Sherrod

HIGH CASCADES
Oregon

WINDIGO PASS TO DIAMOND LAKE

Type: *Many shield volcanoes and cinder cones; one eroded composite volcano*
Lat/Long: *43.25°N, 122.00°W*
Elevation: *1,500-2,800 m*
Eruptive History: *Pleistocene*
Composition: *Basalt to rhyolite*

Windigo Pass is a minor crossing in the Cascade Range, traversed by a graveled US Forest Service road. To the south, several cinder vents have erupted basaltic andesite and fewer basalt lava flows in the last 100,000 yr, an age estimated from the youthfulness of the cones and preservation of flow morphology. **Red Cinder Butte** and **Tenas Peak** are two such cones at higher elevations. The cinder cones **Kelsay Point**, **Thirsty Point**, and **Cinnamon Butte**, which erupted at lower elevations, still have well-preserved summit craters, and conceivably may be Holocene in age. Flows from Cinnamon Butte ramped through gaps in moraines deposited during late Pleistocene glaciation. All eruptions in this area are older than Mazama ash, however. Mazama ash is a widespread silicic tephra erupted ~6,845 yr BP during climactic eruptions of Mount Mazama (Crater Lake National Park, 30 km south of Cinnamon Butte).

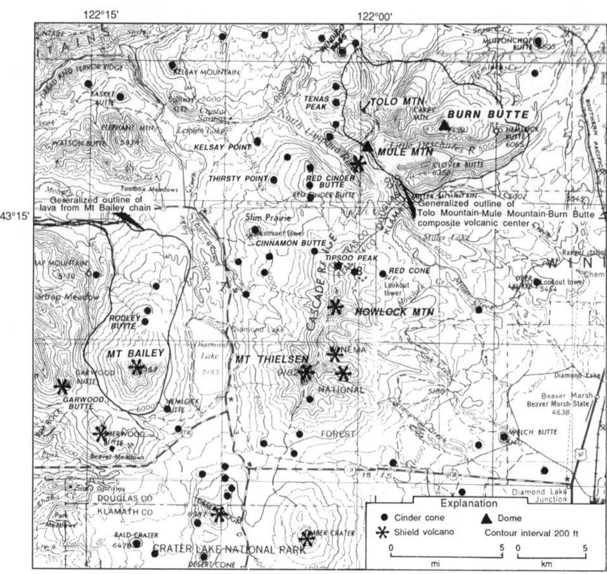

WINDIGO PASS TO DIAMOND LAKE: Volcano map.

Intermediate and silicic volcanism was centered at Tolo Mountain–Mule Mountain–Burn Butte, 5 km southeast of Windigo Pass. The morphology of vents has been largely destroyed by glaciation, however, and the area is buried by ~4 m of Mazama ash. **Burn Butte** is a rhyolite dome with a K-Ar plagioclase age of 2.52 Ma. A large analytical error, owing to low K_2O (0.385%) and low radiogenic argon (4.6% of total argon) leaves the absolute age ambiguous. A K-Ar whole rock age of 0.64 Ma (1.8% K_2O and 12% radiogenic argon) was obtained from a thick dacite dome or flow at nearby **Mule** Mountain. **Tolo** Mountain is almost entirely coarse near-vent andesitic tuff breccia and lava flows. The Tolo Mountain–Mule Mountain–Burn Butte area is interpreted as the eroded stump of a normally polarized, Pleistocene composite volcanic center, probably active ~0.6-0.7 Ma.

From Mule Mountain southward the crest of

the Cascade Range is a broad ridge surmounted by the summit cones of several basaltic andesite shield volcanoes, including **Howlock** Mountain, **Red Cone**, and Mount Thielsen, and several other nameless volcanic necks. Howlock is as deeply eroded as Sawtooth Mountain (K-Ar age ~0.5 Ma), suggesting an age of ~0.5 Ma for Howlock. **Tipsoo Peak** is a basaltic cinder cone that erupted lava flows, probably within the last 100,000 yr.

Mount **Bailey** is the southernmost volcano in a north–south-trending volcanic chain 10 km long that rises west of Diamond Lake. Bailey is about the same age as **Diamond Peak**, 43 km north: <100,000 yr but >11,000 yr old, on the basis of glacial evidence and morphologic comparisons with dated volcanoes. Like Diamond Peak, Bailey consists of a tephra cone surrounded by basaltic andesite lava. Bailey is slightly smaller (8-9 km^3) than Diamond Peak, and minor andesite erupted from the summit cone in its late stages, whereas Diamond Peak eruptions were never more siliceous than basaltic andesite.

The Mount Bailey chain includes **Rodley Butte** and other cinder cones to the north, all of which are similar in age (based on morphology) and magmatically related (on the basis of mineralogy, chemistry, and close spatial association of the vents). Volcanism along the Bailey chain migrated spatially from north to south while it evolved chemically. Basaltic andesite (55% SiO_2) was erupted from the vents north of Rodley Butte, whereas basaltic andesite and finally andesite (58-59% SiO_2) were erupted from the Mount Bailey volcano. Thus, the Mount Bailey chain may be broadly analogous in its temporal and chemical evolution to the more voluminous (30-40 km^3) Mount Bachelor volcanic chain (100 km northeast in the High Cascades), the age, timing, and composition of which are more thoroughly documented elsewhere in this book.

How to get there: *The area is north of Crater Lake National Park. Several paved highways enter the Diamond Lake area. The Pacific Crest Trail follows along the west flank of Mount Thielsen and along the crest to Tolo Mountain and Windigo Pass. A map of the Umpqua National Forest is helpful for locating other road and trail access.*

References

Sherrod, D. R., Benham, J. R., and MacLeod, N. S., 1983, Geology and mineral resource potential map of the Windigo–Thielsen Roadless area, Douglas and Klamath Counties, Oregon: *USGS Open-file Rpt. 83-660,* 22 pp.

Williams, Howel, 1957, A geologic map of the Bend quadrangle, Oregon, and a reconnaissance geologic map of the central portion of the High Cascade Mountains: *Ore. Dept. Geol. Min. Indust.,* scales 1:125,000 and 1:250,000.

David R. Sherrod

THIELSEN
Oregon

Type: *Shield volcano*
Lat/Long: *43.15°N, 122°.07°W*
Elevation: *2,800 m*
Eruptive History: *0.3 Ma*
Composition: *Basaltic andesite*

Mount Thielsen is a normally polarized shield volcano comprising ~8 km^3 of basaltic andesite built atop a broad pedestal (24 km^3) of older lava. Thielsen is remarkable even at a distance for its colorfully interbedded pyroclastic rocks that dip away from the jagged spire of the central plug, often called the "lightning rod of the Cascades." The most spectacular views are on the north and east sides (accessible only by foot or horseback) where now-vanished glaciers have carved precipitous cirque walls that reveal the construction. Thielsen's age is ~0.29 Ma (whole-rock K-Ar), and its geomorphology is a reference point for assigning Cascade Range volcanoes to the age divisions 0-0.25 Ma (younger than Thielsen) or 0.25-0.73 Ma (older than Thielsen). Very little of Thielsen's underpinnings are exposed because Holocene Mazama ash, which erupted from vents at Crater Lake National Park (20 km south), forms a shroud 4-20 m thick in the Thielsen area.

Mount Thielsen is similar to many of the basaltic andesite shields that form the bulk of the High Cascades in Oregon. It consists of a central pyroclastic cone built of scoriaceous to pumiceous cindery tuff and coarse breccia. Variations in grain size define the bedding, which is made more spectacular by the alteration of glassy tephra to colorful palagonite. The beds

THIELSEN: National High Altitude Photography Program vertical view.

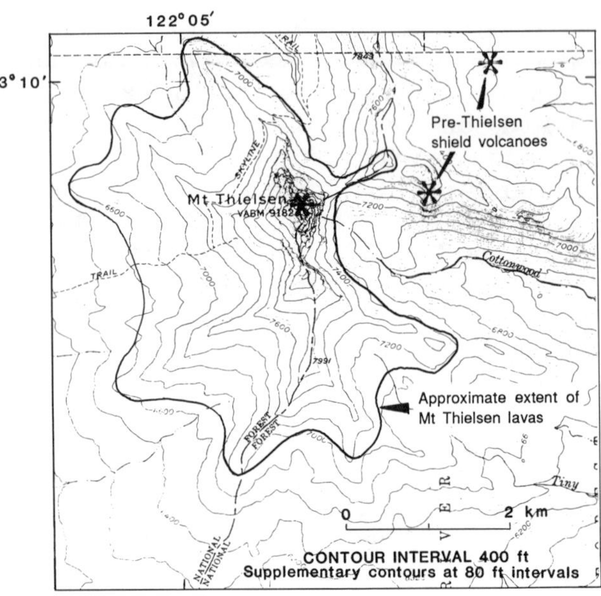

THIELSEN: Volcano map of Mount Thielsen, Cascade Range, central Oregon.

mainly dip 10-40° away from the central conduit-filling plug, although locally these beds have been steepened and even overturned during the plug's intrusion, a feature unreported from other shields in the Cascade Range. Dikes and sills lace the cone.

The lava of Mount Thielsen forms stacks of gently dipping flows and breccia as much as 100 m thick. Single lava flows are as thin as a few centimeters near their vents but thicken to more than 10 m downslope. Fountaining from dike-fed eruptions around the edge of the cone generated coalescing spatter, which formed many of the flows. In the eastern cirque wall, lava is preserved draining back toward the dikes. Lava moving downslope from these satellitic vents probably rafted away parts of the cone.

This sequence of events differs somewhat from that suggested by Howel Williams, who first described Mount Thielsen in detail. According to Williams, after the lava shield was built, a tremendous explosion cored out the center to create a summit crater; the pyroclastic cone was built within the crater. This scenario was based largely on his interpretation of the steep narrow zone where the pyroclastic cone abuts against lava flows. By my interpretation, the cone and lava ramparts were built simultaneously. Their steep contact is not the product of crater-forming blasts, but results from the fountaining of lava at dike-fed vents around the margin of the pyroclastic cone. Evidence against a crater-blasting explosion includes the interfingering of lava and tephra, and the lack of coarse breccia (for example, talus breccia or slump blocks) near the contact of the cone and flanking lava flows.

Thielsen's summit spire is a thick sheet of two-pyroxene basaltic andesite 500 m across at the lowest exposures but dividing upward. The conduit-filling magma congealed without multiple intrusion.

Thielsen has been glaciated many times since its eruptions ~300,000 yr ago. A talus rampart in the north cirque still protects a very small permanent snowfield ("Oregon's southernmost glacier") shown as Lathrop Glacier on the new 7.5-minute quadrangle map.

How to get there: *Mount Thielsen is located near Diamond Lake and north of Crater Lake National Park, in the Umpqua National Forest. There are a store, post office, and campgrounds at Diamond Lake. Trails from Oregon Highway 138 lead up the south and west sides of Thielsen.*

Reference

Williams, Howel, 1933, Mount Thielsen, a dissected Cascade volcano: *Univ. Cal. Pubs., Bull. Dept. Geol. Sci.* 23, 195-214.

David R. Sherrod

CRATER LAKE
Oregon

Type: *Collapse caldera, stratovolcanoes, monogenetic volcanoes*
Lat/Long: *42.94°N, 122.10°W*
Elevation: *2,470 m caldera rim, caldera 1,175 m deep*
Eruptive History:
400 ka to 50 ka: andesite/dacite stratovolcanoes
30 ka to 7 ka: rhyodacite pumice & lava
6845 ± 50 yr BP: zoned (rhyodacite, andesite, cumulates) caldera-forming eruption
Postcaldera andesite and rhyodacite
Composition: *Basalt to rhyodacite*

Crater Lake caldera formed by collapse during the catastrophic eruption of ~50 km^3 of magma, 6,845 ^{14}C yr BP. The 8 × 10 km caldera lies in the remains of Mount **Mazama**, a Pleistocene stratovolcano cluster covering 400 km^2 in the southern Oregon Cascades. Prior to its climactic eruption, Mount Mazama's summit had an elevation between 3,300 m and 3,700 m. Its southern and southeastern flanks were deeply incised by glacial valleys, now beheaded, that form U-shaped notches in the caldera wall.

Mazama was one of the major volcanoes of the High Cascades and is the largest edifice between Mount Shasta and the Three Sisters volcanoes. Around Mazama are monogenetic cinder cones, lava fields, and small shield volcanoes that produced calc-alkaline basalt and andesite, primitive tholeiite, and rare shoshonitic andesite. These range in age from ~600 ka to perhaps 40 ka, and are similar to monogenetic volcanoes up and down the High Cascades.

Beneath the eastern half of Mount Mazama and extending to the southeast is an extensive field of rhyodacite flows and domes, apparently 700-600 ka old, covering at least 350 km^2. Hornblende dacite underlies the rhyodacite in the southeastern part of the field. Generally north–south trending normal faults cut monogenetic vent lavas and rhyodacite, with most displacement down to the east. The same faults cut older Mazama flows, but displacements are not evident in the younger lavas and no tectonic faults have been detected in the caldera walls.

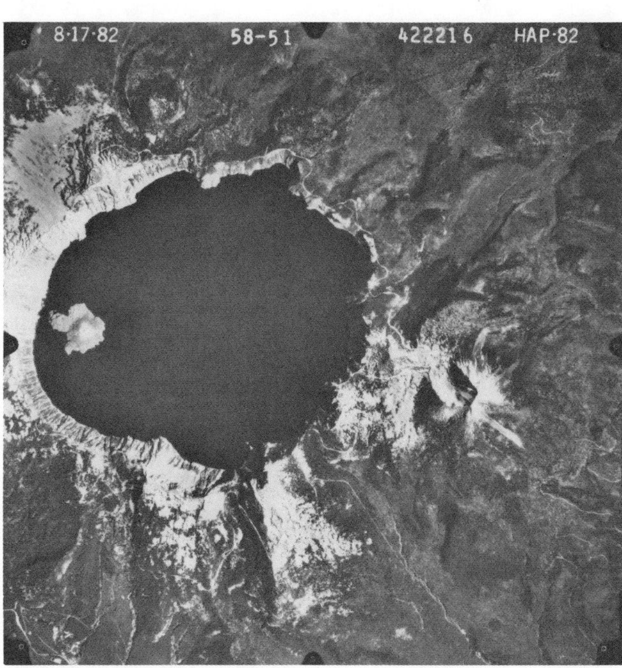

CRATER LAKE: National High Altitude Photography Program vertical view.

CRATER LAKE: View from the rim of Wizard Island. Photograph by P. Mouginis-Mark.

Individual stratovolcanoes and shields that make up Mount Mazama become younger in a west-northwest sense. The oldest Mazama lavas dated are flows near lake level at Phantom Ship and the lavas of Mount **Scott** (~400 ka). The youngest stratovolcano is **Hillman Peak** (~70 ka). Local andesite flows on the north rim are 50-40 ka old. Shields are composed of sheet flows of agglutinated mafic andesite that are typically ~5 m thick and form bands on the caldera walls. More viscous magma of andesite and dacite flows resulted in thicker flow units, up to 30 m, with larger proportions of flow-top rubble to dense lava. Even so, most of these also appear to be composed of agglutinated bombs. Undercooled inclusions of crystal-poor andesite magma are common in many andesite and dacite flows. Such flows make up Mount Scott (2,721 m), east of the caldera rim, and **Phantom Cone** and other centers evident in the southern caldera walls. Some andesite lava flowed into glacial valleys forming thick intracanyon flows, such as at Sentinel Rock. At many places in the caldera walls and on the flanks of Mazama exposures of small glassy columns and piles of monolithologic glassy breccia provide evidence of lava/ice interaction. Many flows bury glaciated lava surfaces.

Explosive silicic eruptions occurred at several vents around 70 ka. The most impressive was at Pumice Castle on the east wall, where layers of a dacitic Plinian fall deposit become densely welded near their vent. The deposit is non-welded to the south and in poor exposures on the north caldera wall from Cleetwood Cove to Steel Bay. Other welded dacite pumice fall deposits occur on the north side of Cloudcap and in the wall beneath the east flank of Llao Rock. Dacitic pyroclastic flow deposits are present below Llao Rock and in the southwest wall. The Watchman flow is also dacite, as are monolithologic breccias and lithic-pyroclastic-flow deposits in the head of Munson Valley. Dacite forms the silicic endmember in the basalt-andesite-dacite mingled lavas of **Williams Crater** (formerly **Forgotten Crater**). There is no evidence of andesitic or dacitic volcanism between 40 ka and the climactic eruption.

Rhyodacitic magma erupted as pumice and lava flows between 30 and 25 ka at **Grouse Hill**, **Steel Bay**, and **Redcloud Cliff.** All are hornblende phyric, chemically evolved rhyodacite that apparently leaked from the growing climactic magma chamber. **Sharp Peak** and adjacent rhyodacite domes form a northeast-trending linear array that probably vented in the latest Pleistocene or early Holocene. They are compositionally identical with climactic rhyodacite (70% SiO_2). The Llao Rock flow and preceding Plinian fall deposit were erupted 7,015 yr BP, and are zoned from 72 to 70% SiO_2. The Cleetwood flow and pumice were erupted 100-200 yr later, because the flow was still hot when the caldera collapsed during the climactic eruption at 6,845 yr BP. Cleetwood and climactic rhyodacites are compositionally homogeneous and identical. This rhyodacite makes up ~90% of the volume of climactic ejecta.

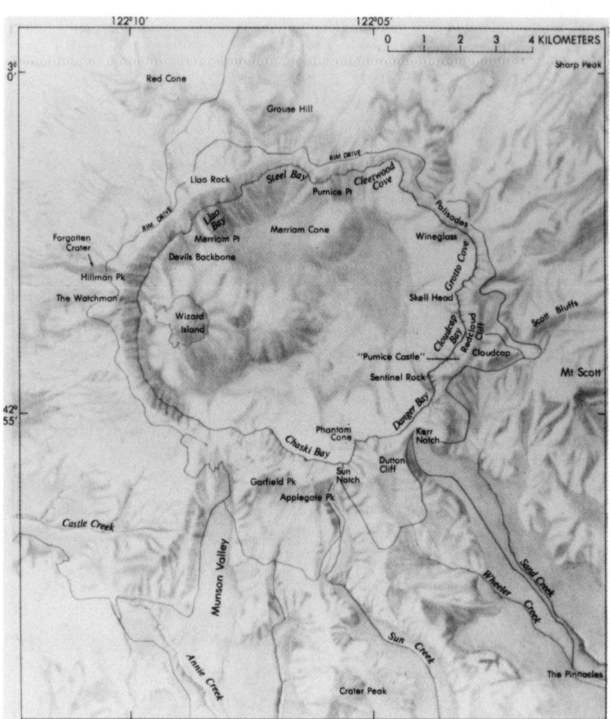

CRATER LAKE: Physiography of Crater Lake caldera and vicinity, Oregon (Bacon, 1983).

The climactic eruption began as a plinian column producing a widespread fall deposit mainly to the north-northeast. This column eventually collapsed and generated at least four flow units which cooled together to form the Wineglass Welded Tuff. The Wineglass is present only on the north and east flanks of Mazama, and thus the vent for this valley-hugging ignimbrite was north of the summit. Eruption of the Wineglass ceased with onset of caldera collapse when highly mobile pyroclastic flows, derived from many vents, traveled radially down the flanks of the mountain to fill all the major drainages. Near the caldera and on all high ground these ring-vent-phase deposits consist of lithic breccia, while pumiceous ignimbrite occupies valleys. This deposit is spectacularly compositionally zoned: homogeneous rhyodacite (70% SiO_2) followed by andesite (<61% SiO_2) and mafic cumulate scoria (to as low as ~48% SiO_2). Scoriae show a dramatic contrast in LILE contents, which can be divided into high-Sr and low-Sr groups. High-Sr magmas were ejected first. These scoria types indicate a complex history of growth and replenishment of the zoned system, with repeated influxes of, and crystallization of, andesitic and basaltic parental magmas. Preclimactic rhyodacite lava flows contain first low-Sr (Grouse Hill, etc.), then high-Sr andesitic magmatic inclusions (Sharp Peak, Llao Rock, Cleetwood) that record changes in parent magma with time and place.

Timing of caldera collapse is well documented, because the interior of the Cleetwood flow was sufficiently plastic at the time of the climactic eruption that remobilized lava flowed northeast in response to seismicity and also down the caldera wall at Cleetwood Cove. Scarps on the brittle flow surface expose Plinian fall deposit and thin Wineglass Welded Tuff, but the proximal lithic breccia of the ring-vent-phase ignimbrite is banked against them, indicating disruption of the Cleetwood flow surface after emplacement and welding of <2 m of Wineglass but before deposition of the last lithic breccia. Thick, densely welded Wineglass in paleovalleys at the caldera rim sagged towards the caldera, showing that collapse took place before complete solidification of this ignimbrite.

Post-caldera volcanic landforms are present beneath the lake surface and poke through to form Wizard Island. The central platform, **Merriam Cone**, and Wizard Island are all andesite evidently erupted within a few hundred years of caldera collapse. The small rhyodacite dome 30 m below lake level 1 km east of Wizard Island is the youngest feature.

How to get there: *Crater Lake National Park is accessible from the south and north via Oregon Highways 62 and 138, respectively. A paved road runs around the caldera rim. Access to the lake is limited to the trail at Cleetwood Cove, where tour boats provide a good close-up view of the caldera walls.*

References

Bacon, C. R., 1983, Eruptive history of Mount Mazama and Crater Lake caldera, Cascade Range, USA.: *J. Volc. Geotherm. Res. 18*, 57-115.

Bacon, C. R., 1987, Mount Mazama and Crater Lake caldera, Oregon. *Geol. Soc. Am. Centennial Field Guide 1*, 301-306.

Charles R. Bacon

PELICAN BUTTE
Oregon

Type: *Steep-sided shield*
Lat/Long: *42.51°N. 122.06°W*
Elevation: *1,280 to 2,449 m*
Eruptive History: *Definitely <1.2 Ma, probably all <0.7 Ma; no eruptions since last glaciation*
Composition: *Glassy phenocryst-poor platy andesite*

Pelican Butte is a normally polarized, steep-sided andesite shield built on faulted Pliocene and early Pleistocene basaltic andesite. Pleistocene glaciation carved a steep canyon and broad cirque in the northeast flank of the volcano, lowered the summit some tens of meters, and exposed a lava-filled intrusive conduit. However, the volcano's original shape is largely preserved. Pelican Butte (20 km^3) is volumetrically one of the larger Quaternary volcanoes between Crater Lake and Mount Shasta; it is larger by one-third than the nearby more scenic Mount McLoughlin. Two analyzed samples contain 58 and 60% SiO_2.

Two types of andesite make up most of the volcano. Thick platy flows are most common low on the mountain. These glassy flows traveled far down the slopes of Pelican Butte following preexisting channels. Phenocryst minerals are plagioclase, augite, hypersthene, and locally olivine; all are inconspicuous and collectively make up no more than a few percent of the lava.

The second type of andesite forms thin flows that have scoriaceous and blocky tops. Lava of this type is common near the summit of Pelican Butte. It is typically fine-grained and vesicular. Small snow-white plagioclase phenocrysts of sodic labradorite are characteristic. Phenocrysts of olivine, augite, and hypersthene are small and are difficult to identify in hand specimen; together they average between 5 and 10%.

Volcanic breccia is poorly exposed on the slopes of the mountain, although exposures near the summit and in the deep canyon on the northeast flank suggest that a significant part of the volcano consists of volcanic breccia sandwiched between lava flows. Tuff breccia and lapilli tuff exposed near the summit are part of a summit cone built during a late pyroclastic phase and later largely covered by lava flows. Erosion has removed most of the covering flows and much of the summit cone as well.

The age of the olivine andesite of Pelican Butte is constrained by stratigraphic relations, a potassium-argon age determination, magnetic data, and geomorphic development. A few flows on the west side of Pelican Butte abut a fault scarp but are not offset by the fault. The fault cuts basaltic andesite shield volcanoes dated at 1.17 Ma. Therefore, the maximum age of fault movement and the Pelican Butte andesites is ~1.2 Ma. Pelican Butte is no younger than 12 ka (age of the last glaciation) and probably no younger than 60 ka on the basis of correlation of glacial deposits in southern Oregon with those in the Sierra Nevada of California.

How to get there: *Pelican Butte's summit is high, the mountain is detached from the axis of the Cascades, and it formed on top of a system of down-to-the-east normal faults with large displacements that delineate the east side of the Cascade Range at this latitude. For these reasons, views from its summit are impressive. It offers a 180° panorama of Cascade Peaks from just south of Crater Lake past Mount McLoughlin and onto the volcanoes in the Mountain Lakes Wilderness. A reasonably well-maintained gravel road, open during snow-free summer months, extends to the summit. The road turns north off Highway 140 ~6 km east of the Lake of the Woods highway maintenance station. Although the last few miles are steep and narrow, most vehicles with high clearance should be able to make the trip. No water is available on the upper parts of the volcano.*

References

Blakely, R. J., 1986, Maps showing aeromagnetic data and interpretation of the Medford 1° by 2° quadrangle, Oregon–California: *USGS Misc. Field Studies Map MF-1383-B*,.

Smith, J. G., in press, Geologic map of the Pelican Butte quadrangle, Klamath County, Oregon: *USGS Quad. Map GQ-1653*.

James G. Smith

SPRAGUE RIVER VALLEY
Oregon

Type: *Maars, cinder cones, shield volcanoes, domes*
Lat/Long: *42.50°N, 121.50°W*
Elevation: *1,300-2,100 m*
Eruptive History: *Intermittently from 5-1.5 Ma*
Composition: *Basalt to dacite*

The Sprague River Valley is east of the Cascade Range, ~70 southeast of Crater Lake National Park, and 35 km northeast of Klamath Falls. The valley is notable for its maars (more than 23), many of them forming individual tuff cones or chains of cones. Palagonite tuff from some of the vents has coalesced to form extensive masses, but the original eruptive centers can be recognized on air photos by the radially dipping beds around each vent. The age of the maar field in the Sprague River Valley is bracketed by ages from overlying and underlying units; it is >3.43-Ma basalt from Calimus Butte and <5-6 Ma rhyolite domes near Beatty (2 km east of map). Although undated, the dacite or rhyolite domes of **Bug** and **Council** buttes are likely similar in age to the domes near Beatty.

The maar field coincides with a large area of Pliocene sandstone and diatomite, which indicates an extensive system of lakes. Thus, the abundance of maars probably results from interaction during the Pliocene of mafic magma with groundwater at shallow depths.

Subsequent volcanism resulted chiefly in lava flows and cinder cones or small shield volcanoes. **Fuego Mountain** and **Calimus Butte** are typical in size for shields in this area. Fuego is steeper, probably because its cone includes more viscous basaltic andesite and some andesite, whereas the Calimus shield is built of basalt and basaltic andesite. Both shields have been deformed by basin-range style normal faults. Indeed, the alignment of some chains of vents throughout the area probably results from magma rising along fractures and faults related to extensional tectonism.

The two youngest volcanoes are **Saddle** Mountain and **Taylor Butte**. Saddle Mountain is a steep-sided basaltic andesite shield. Taylor Butte is a cinder cone that erupted the basalt of

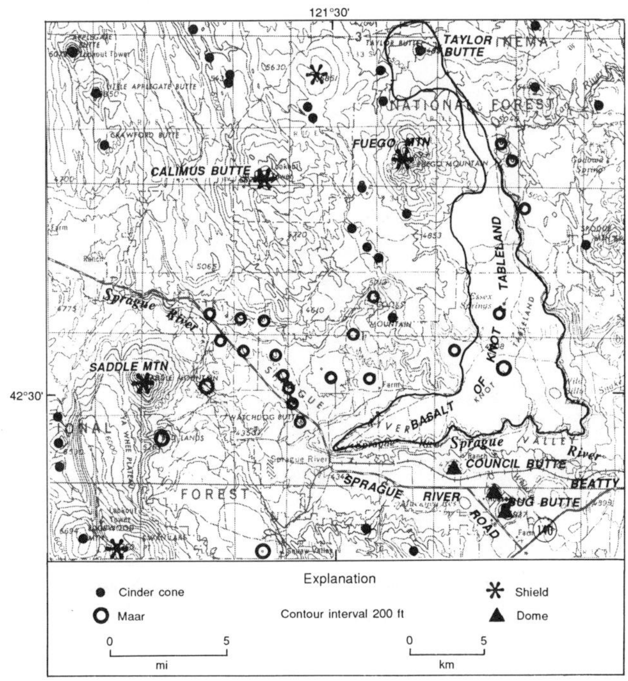

SPRAGUE RIVER VALLEY: Basin and Range near Sprague River valley, south-central Oregon.

Knot Tableland, an ancestral intracanyon lava flow. The Sprague River has entrenched itself south of Knot Tableland, creating inverted topography shown by the basalt of Knot Tableland. So far, efforts to date the basalt of Knot Tableland have been thwarted by its very low potassium content. The basalt is probably ~2.0 Ma, however, on the basis of its similarity with dated, topographically inverted lava flows elsewhere in eastern Oregon.

How to get there: *Oregon Highway 140 approaches the Sprague River Valley from Lakeview or Klamath Falls, Oregon. The Sprague River Road is a little-traveled but scenic highway following the Sprague River from Highway 140 west to Chiloquin on US Highway 97. Land along the Sprague River is privately owned; permission to trespass is necessary. The rest of the area is in the Winema National Forest.*

Reference

Peterson, N. V., and McIntyre, J. R., 1970, The reconnaissance geology and mineral resources of eastern Klamath County and western Lake County, Oregon: *Ore. Dept. Geol. Mine. Indust. Bull. 66*, 70 pp.

David R. Sherrod

McLOUGHLIN
Oregon

Type: *Steep-sided lava cone or shield*
Lat/Long: *42.44°N., 122.31°W.*
Elevation: *1,525 to 2,894 m*
Eruptive History: *Normally polarized, <700 ka; most <200 ka*
Composition: *Calc-alkaline basaltic andesite and andesite*

Mount McLoughlin (Mount **Pit** or **Pitt**) rises 1,200 m as a steep-sided, dominantly basaltic andesite lava cone above the low Pliocene and Pleistocene basaltic andesite shields on which it is built. McLoughlin is easily recognized from as far away as Medicine Lake in California, along I-5 between Yreka, California, and Medford, Oregon, or around the rim of Crater Lake. Although it is the tallest volcano between Shasta and Crater Lake, McLoughlin, with a volume of only 13 km^3, is dwarfed by the bulk of Shasta (350 km^3) and Mazama (130 km^3).

When viewed from the south or southeast it appears as a seemingly perfectly symmetrical Fuji-like volcano. However, when seen from the east, along the shores of Klamath Lake, or from the north along Crater Lake's rim, it is apparent that a major part of the mountain is missing. Late Pleistocene glaciers have carved away the entire northeast side of the mountain, lowering the summit about a hundred meters, excavated the large bowl-like cirque, and exposed the congealed lava that fills two small central conduits. Steeply dipping layers of pyroclastic breccia and tuff and numerous interlayered lava flows are exposed in the walls of the cirque. An explosive origin is sometimes ascribed to this cirque; however, there is no evidence of deposits that would have resulted from such an explosion. Glacial striae and other glacial features are common in the cirque, and glacial deposits such as moraines and till are present at the mouth of the cirque and around the north base of the mountain. Finally, the composition of McLoughlin's lava is much more mafic than that of other volcanoes at which explosive events of the required size have occurred.

Mount McLoughlin is a young volcano. A pronounced magnetic high centered just east of McLoughlin's main vent is interpreted as indi-

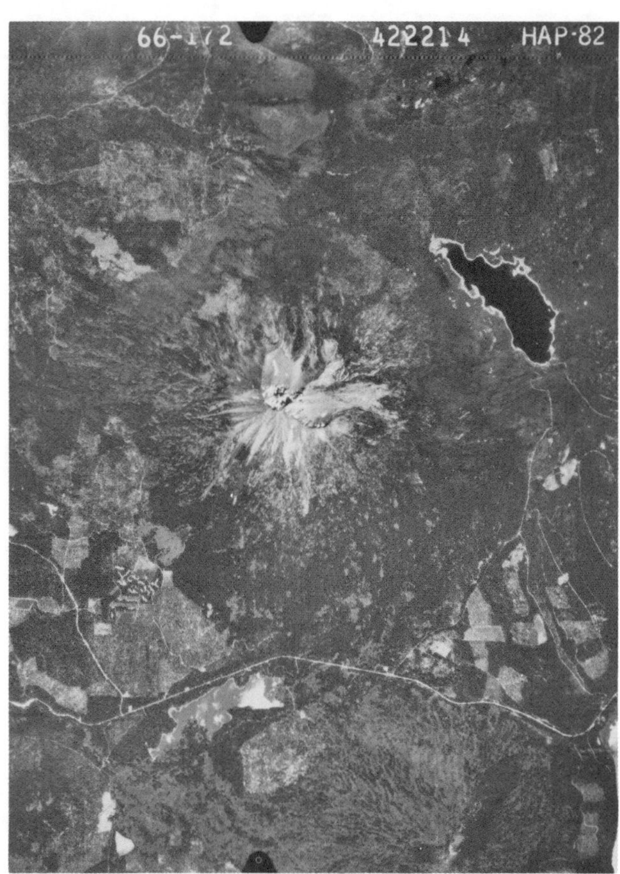

MCLOUGHLIN: National High Altitude Program vertical air photograph of McLoughlin (middle) and Brown Mountain (bottom).

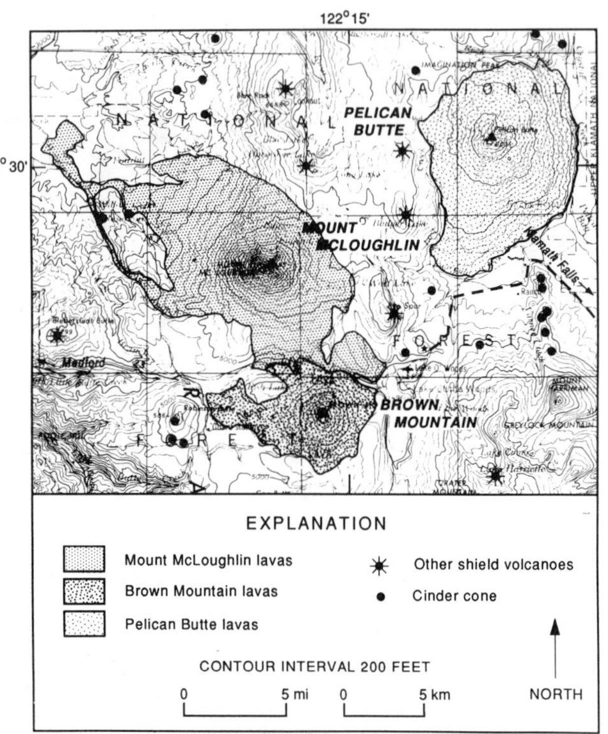

MCLOUGHLIN: Volcano map showing extent of lavas from Mount McLoughlin, Brown Mountain, and Pelican Butte and location of other small shields and cinder cones.

cating that most of the main cone is normally polarized and thus less than ~700 ka. The well-preserved shape of the mountain's west and south flanks, the lack of soil development on many flows, and preservation of primary flow features suggests that the bulk of the main cone is no older than 200 ka, with much of it probably younger. The main cone was essentially complete before the last major Pleistocene glaciation. Many flank flows are younger than the main cone; some may be as young as 20 to 30 ka.

Early thick flows of light-colored pyroxene-bearing lava extend east beyond the mountain's base for at least 10 km. They crop out along Highway 140 and are well exposed in the upper one-third of the steep grade on Highway 140 ~3 km east of the junction with the road to Lake of the Woods. Extensive black clinkery flank flows of olivine-bearing basaltic andesite and andesite poured from the northwest side of the volcano, filled the old valley of Fourbit Creek for ~6 km beyond the mountain's base, and stopped at the site of the present-day Big Butte Springs. These large-volume springs, whose catchment area is the whole northwest slope of Mount McLoughlin, gush from the end of the flows and are the domestic water source for Medford and other towns in the Bear Creek Valley.

Another clinkery black basaltic andesite flank flow emerges from the volcano above the 1,500-m level, fills a stream valley for ~11 km, and descends to an elevation of 900 m. Despite its great length and blocky character this remarkable flow is <160 m wide for more than 4 km of its length.

Detailed study of the main edifice in the early 1970s divided the eruptive history into 3 stages. Chemical analyses indicate that the composition of the rocks ranges from ~53 to ~57% SiO_2. Thus, although Mount McLoughlin is commonly called an andesitic volcano, as inferred from its steep-sided form, it is composed mostly of basaltic andesite.

How to get there: *Access to Mount McLoughlin and its close neighbors, Brown Mountain and*

Pelican Butte, is remarkably easy via Oregon Highway 140 between Medford and Klamath Falls. The thick conifer forests around the bases of these mountains with their many campgrounds are well known to southern Oregon residents but often overlooked by others in a hurry to get to better known Crater Lake, 55 km to the north. Recreational activities include hiking and fishing in the summer and snow sports in the winter. Parts of Mount McLoughlin and Pelican Butte are in the Sky Lakes Roadless Area. Mount McLoughlin can be climbed during mid to late summer after snow has melted from the trail. To reach the trail head, turn north from Highway 140, east of the Cascade Crest, onto the Four Mile Lake road and proceed ~4 km. A sign marks the trail's start. Carry water as no streams cross the trail. The 10-km-long, moderately steep trail is marked above tree line by red circles painted on rocks. Take care, as these inconspicuous trail markers are easily missed, especially during descent.

References

Maynard, L. C., 1974, Geology of Mount McLoughlin: M.S. Thesis, Univ. Oregon, Eugene, 139 pp.

Smith, J. G., 1983, Geologic map of the Sky Lakes Roadless Area and Mountain Lakes Wilderness, Jackson and Klamath Counties, Oregon: *USGS Misc. Field Studies Map MF-1507A.*

James G. Smith

BROWN MOUNTAIN
Oregon

Type: *Shield surmounted by cinder cone*
Lat/Long: *42.36°N, 122.27°W*
Elevation: *1,500 to 2,228 m*
Eruptive History: *Probably between 60 and 12 ka*
Composition: *Olivine-bearing basaltic andesite and andesite*

Brown Mountain is a small (5 km^3), youthful-looking shield topped by a cinder cone whose central depression is still 15 m deep. Much of the mountain is bare, unweathered, dark-colored, block-lava and clinkery aa flows. The flows are mostly olivine-bearing basaltic andesite and andesite in composition.

BROWN MOUNTAIN: Radar image of McLoughlin (center with pointed shadow), Brown Mountain (bottom) and other volcanic hills.

At first glance Brown Mountain looks no older than a few thousand years. However, a small glacial valley carved into the northeast flank and a cirque gouged out of the summit cinder cone belie its youthful appearance. The deposits left behind have features typical of glacial deposits from ~13 ka. Evidence of the next older glaciation is missing, and the age of Brown Mountain can be bracketed between ~60 and 12 ka.

How to get there: *A climb to the summit of Brown Mountain is mostly a scramble over fresh talus as there is no maintained trail to its summit. Because its summit is lower than that of nearby peaks, views from Brown Mountain are not as spectacular. However, it offers a close view of the south flank of Mount McLoughlin. Some scrambling can be saved by starting from the highest logging spur southeast of Fish Lake. Carry water; there is none on the mountain.*

Reference

Carver, G. A., 1972, Glacial geology of the Mountain Lakes Wilderness and adjacent parts of the Cascade Range, Oregon, Ph.D. Thesis, Univ. Wash., Seattle, 76 pp.

James G. Smith

NEWBERRY
Oregon

Type: *Shield*
Lat/Long: *43.68°N, 121.25°W*
Elevation: *2,434 m*
Relief: *~1,100 m*
Volcano/Caldera Diameter: *~40/5 × 7 km*
Eruptive History: *0.5 Ma to ~1,300 yr BP*
Composition: *Basalt to rhyolite*

Newberry volcano, one of the largest Quaternary volcanoes in the coterminous United States, lies 60 km east of the crest of the Cascade Range in central Oregon. It offers quite a wide variety of volcanic landforms and deposits. During its 0.5-m.y. history of magmatic activity, including caldera collapse and subsequent caldera-filling volcanism, Newberry has erupted often, including several times in the Holocene. The chemical composition of all volcanic products is largely bimodal with basaltic andesite and rhyodacite dominating, but a full range from basalt to rhyolite is represented.

Newberry has the shape of an elongate shield, 60 km north–south and 30 km east–west. The elevation of the topographic base of the volcano is ~1,350 m and its highest point, Paulina Peak, is 2,434 m. Although only ~1,100 m high, the volcano covers an area of 1,600 km^2 and has a very large volume of 450 km^3. The erupted volume may be substantially greater than this, because lava flows and alluvium from Newberry extend well beyond the base of the volcano. No glacial erosion or deposits are known.

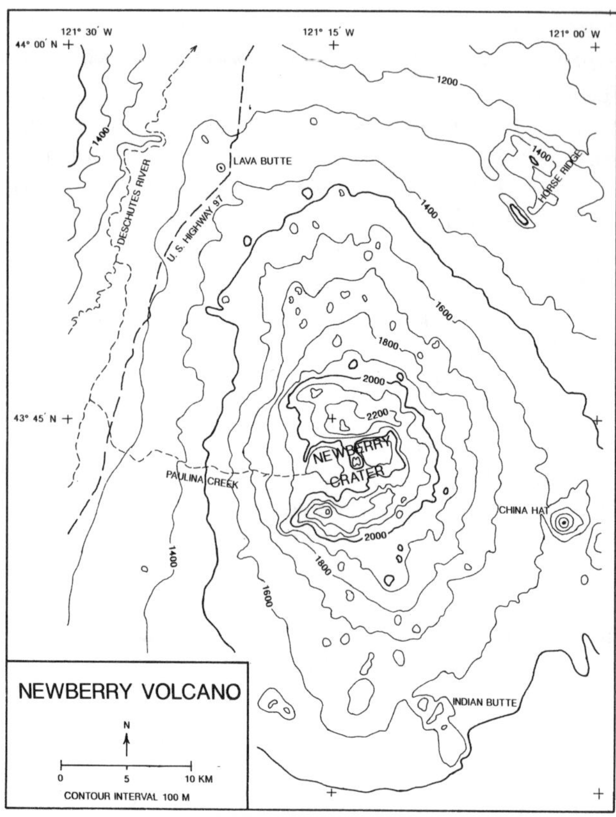

NEWBERRY: Topographic map.

Newberry volcano is surrounded by basalt flows, ash-flow tuffs, vent deposits, and volcaniclastic rocks of late Miocene, Pliocene, and Quaternary age. The volcano lies at the northwest end of a sequence of rhyolite domes and caldera-related ash-flow tuffs that show a well-defined northwest age progression. The sequence includes 10-m.y.-old rhyolite ~250 km east of Newberry. The rhyolitic rocks become younger at shorter distances, culminating in rocks <1 m.y. old at Newberry.

Newberry lies in an extensional tectonic setting marked by numerous Quaternary and Pliocene normal faults. Three major fault zones converge at Newberry: the Brothers (NW), Sisters (NNW), and Walker Rim (NE) fault zones. Faults cut only the oldest Newberry lavas. However, aligned cinder cones and fissure vent deposits on the north and south flank of Newberry trend mostly parallel to the Sisters and Walker Rim fault zones. Strata beneath Newberry are probably much more faulted than the volcano itself.

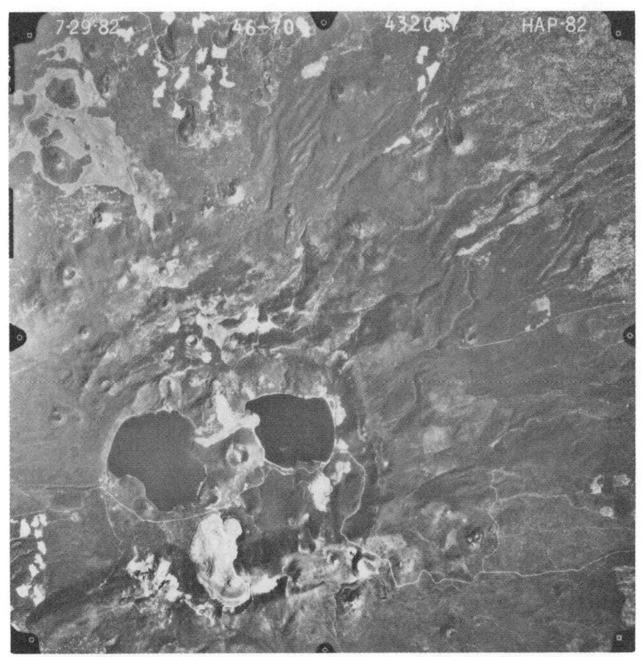

NEWBERRY: National High Altitude Photography Program vertical aerial view of Newberry caldera and northern flanks.

The outer flanks of Newberry typically slope 1-2° to the north and south and 2-3° to the east and west. The slopes abruptly steepen to 8° on the upper flanks. This striking slope change may represent (1) the slopes of an older, steeper volcano, (2) a circular fault line scarp along which the central part of the volcano has been raised pistonlike, or (3) a greater abundance of stubby silicic flows and domes and thick tephra near the center of the volcano.

The east and west flanks of Newberry volcano consist almost entirely of ash-flow and air-fall tuffs, alluvial deposits, and a few basaltic flows. The north and south flanks, on the other hand, consist mostly of basaltic andesite lava flows and numerous cinder cones.

Tuffs, alluvium, and lava flows of the east and west flanks are the oldest strata exposed and also the most eroded; subparallel ridges and valleys locally up to 30 m deep characterize these flanks. Two ash-flow tuffs, the Tepee Draw (rhyodacite) and "black lapilli" (basaltic andesite), are volumetrically significant; each is several tens of cubic kilometers in volume, and their eruption probably produced caldera collapse. The ages of all of the flank tuffs are probably in the range of 0.3-0.5 Ma.

Lava flows of the north and south flanks are interbedded with and locally underlie tuff and alluvium on the east and west flanks. Newberry lavas that extend north several kilometers beyond the edifice of the volcano contain many well-preserved lava tubes.

Ages of Newberry lavas are little known; few have been dated by K-Ar methods, and only the youngest can be dated by ^{14}C. Many youthful-looking flows are probably only 10,000-20,000 yr old, although only Holocene flows younger than Mazama ash (6,845 yr) have been dated by ^{14}C. Attempts to delineate groups of progressively older lava flows according to their relative youthfulness of flow surface features proved unsuccessful during mapping of the volcano.

Approximately 400 basaltic cinder cones and fissure vents occur on the flanks of Newberry. The vents are most abundant on the north and south flanks. Cinder cones are usually a few tens of meters high and a few hundred meters across, with the largest being 140 m high and 1 km wide. Cinder cones and fissure vent deposits on the upper flanks commonly contain inclusions of intermediate and silicic volcanic rocks. Many cinder cones and fissure vents are aligned. On the south flank, most vent alignments trend north-northeast, parallel to faults of the Walker Rim zone. On the north flank the vents are aligned north-northwest, parallel to faults of the Sisters fault zone. Many cones and fissures on the north flank as far as 1-4 km from the caldera rim form curved arrays parallel to the caldera walls.

Dacite, rhyodacite, and rhyolite domes and flows occur at many localities on the middle and upper flanks of the volcano. The largest dome forms **Paulina Peak**, the highest point on the volcano. This dome stands more than 500 m above the surrounding terrain, extends southwest 5 km from the caldera rim, and covers an area of ~7 km^2. Numerous smaller domes are ~50-100 m high and 100-1,500 m across. Flows are generally thinner but extend as far as 4 km.

The summit caldera of Newberry is ~7 km wide east–west and 5 km north–south. The west wall of the caldera is only a few tens of meters high, whereas elsewhere the walls are 200-500 m high. The caldera was initially deeper by at least 500 m but has been filled during late Pleistocene and Holocene time by pyroclastic rocks, flows, domes, and sedimentary rocks.

The surficial Pleistocene rocks in the cal-

NEWBERRY: Big Obsidian Flow, the south wall of Newberry caldera, and cinder cones and rhyolitic domes on the south flank of the volcano. Photograph by D. Sherrod.

dera include rhyodacite domes, a large obsidian flow, a basaltic andesite flow, palagonite tuff cones, and fluvial and lacustrine deposits.

Six Holocene eruptive episodes, four rhyolitic and two basaltic, have been recognized at Newberry; rhyolitic eruptions occurred mostly in the east half of the caldera, and basaltic eruptions on the flanks: (1) An estimated 8,000-10,000 yr ago, an obsidian dome and related obsidian flow erupted in the southeast part of the caldera. These have been almost completely buried by Mazama ash and younger Newberry tephra. (2) At nearly the same time, mafic cinders, scoria, spatter, and lava flows erupted from a fissure on the east rim of the caldera. (3) Several rhyolitic eruptions occurred in the caldera ~6,200 yr ago. They produced a widespread phreatomagmatic pumiceous tephra deposit, obsidian flows, large and small pumice cones, and a pumice ring. (4) About 6,100 yr ago, basaltic andesite lava and cinder cones erupted from extensive fissure vents on the northwest, north, and south flanks of Newberry. Spatter and cinders also erupted from a fissure on the north caldera wall. The lava flows range up to 9 km long and are more voluminous at lower elevations. **Lava Butte** (a cinder cone with a road to its top) and accompanying lava flow are strikingly located alongside US Highway 97. (5) About 3,500 yr ago, obsidian flows and associated pumice deposits in the caldera erupted from fissures. (6) About 1,300 yr ago, a 3-part sequence of rhyolitic air-fall tephra, ash-flow tephra, and an obsidian flow erupted from a common vent at the base of the south caldera wall. The initial Plinian eruption produced the Newberry pumice fall deposit which blanketed the east flank of the volcano and areas to the east. This pumice deposit thins to 25 cm at a distance of 60 km east of the caldera. The Paulina Lake ash flow followed the Plinian phase and spread from near the south caldera wall to Paulina Lake. The final phase of the eruption produced the Big Obsidian Flow, which covers an area of 20 km^2.

Geologic evidence suggests that a silicic magma chamber has been present beneath Newberry throughout the Holocene and probably still exists. Basaltic magma, represented in episodes 1 and 4, probably ascended from the mantle or lower crust, underplating the chamber, heating it, and episodically triggering rhyolitic eruptions.

The presence of a long history of silicic volcanism, Holocene volcanism, hot springs, an inferred silicic magma chamber, fault zones, caldera ring fractures, and steep geothermal gradients suggests a high potential for developable geothermal energy for electricity generation. Geothermal gradients observed in several drill holes suggest promising temperatures at depths of 3,000 m or less. The USGS Newberry 2 drill hole in the south-central part of the caldera had a temperature of 265° C at its base at 932 m.

How to get there: *Access to the volcano is generally excellent, with the most direct route being along Highway 97 going south from Bend. A system of roads for harvesting timber and for recreation covers all but the highest flanks. A road through the caldera and one to the top of Paulina Peak offer access to the top of the volcano. Nearly all the land of Newberry volcano is managed by the Deschutes National Forest.*

References

Fitterman, D. V., 1988, Overview of the structure and geothermal potential of Newberry Volcano, Oregon: *J. Geophy. Res. 93*, 10,059-10,066.

MacLeod, N. S., and D. R. Sherrod, 1988, Geologic evidence for a magma chamber beneath Newberry Volcano, Oregon: *J. Geophy. Res. 93*, 10,067-10,079.

Lawrence A. Chitwood

DEVILS GARDEN

Oregon

Type: *Lava flow field*
Lat/Long: *43.50°N, 120.90°W*
Elevation: *1,525 m*
Eruptive History: *10,000 to 50,000 yr BP*
Composition: *Basalt or basaltic andesite*

For the student of lava flow and vent features, the Devils Garden lava field is rich in excellent examples. An area of 117 km^2 is covered by multiple flows of fresh, inflated, pahoehoe lava that erupted from fissure vents in the northeast part of the Devils Garden. Several rounded hills and higher areas of older rocks are now kipukas completely surrounded by the black, basaltic lava. Near the main vent, much of the lava flowed through a narrow, open gutter and formed a large, sinuous, well-developed lava tube, Derrick Cave.

The main vent from which all the lava issued is surrounded by a low spatter rampart, but other vents along the fissure to the south are marked by well-preserved spatter cones, two of which, "**The Blowouts**," are exceptionally large. The spatter cones range 1-30 m in height, and 2-150 m in diameter.

The Devils Garden lavas display a wide range of pahoehoe features and excellent examples of inflated lava. The landforms in the lava field have resulted mostly from inflation, or swelling, of lobes and sheets of initially thin lava flows. These highly fluid lavas erupted onto a landscape of low gradient (2° average) and spread widely. Initial thicknesses of lobes and sheets typically increased from ~0.5 m to 5 m; that is, the lava swelled up 10 times its original thickness.

Distinctive morphologic forms developed during inflation. Tension cracks opened in brittle crust during swelling to form elongated trenches. Sheets of inflating lava developed into pressure plateaus characterized by broad, horizontal, elevated surfaces with steeply sloping sides. Plateau pits formed in the pressure plateaus. The bottoms of the pits were not flooded by lava and so consequently did not inflate. Long, narrow lobes of lava inflated to form pressure ridges with sides that tilted outward. One or more tension cracks may have opened up lengthwise along the top. Perhaps the most common features are cracked tumuli. These short pressure ridges, usually oval in plan view, represent swollen wide spots ("aneurysms") in otherwise narrow flow lobes. During inflation at concave flow edges, liftup caves opened under rising blocks of brittle crust. The small slabs, blocks, and columns of crust broken by contractive cooling were compressed at each concave edge to form more or less rigid, monolithic blocks that rose as units. In circular plateau pits where the edge of the lava is continuously concave, these caves may completely encircle the pits.

The age of the Devils Garden lava field is unknown but probably falls between 10,000 and 50,000 yr. Air-fall ash from Mount Mazama's cataclysmic eruption 6,845 ^{14}C yr ago fills cracks and depressions in the lava.

How to get there: *A forest road leading southwest from Brothers, Oregon (on Highway 20, southeast of Bend) leads to the Devils Garden area.*

References

Chitwood, L. A., 1987, Origin and morphology of inflated lava (abstract): *EOS, Trans. Am. Geophys. Un., 68,* 1545.

Peterson, N. V., 1965, Hole-in-the-Ground, Fort Rock, Devils Garden area field trip *in* Peterson, N. V., and E. A. Groh, (eds.), Lunar geological field conference guide book: *Ore. Dept. Geol. Min. Indust. Bull. 57,* 19-28.

Lawrence A. Chitwood

FORT ROCK BASIN

Oregon

Type: *Maar field*
Lat/Long: *43.30°N, 121.00°W*
Elevation: *1,320-1,716 m*
Eruptive History: *Pliocene and Pleistocene*
Composition: *Basaltic glass*

Nearly 40 maars, tuff rings, and tuff cones of Pliocene and Pleistocene age occur in the Fort Rock Basin of south-central Oregon. Most are significantly eroded, allowing excellent exposures of their lithology, bedding, and sedimentary structures; a few retain much of their original morphology.

The Fort Rock Basin is dry, internally drained, and largely filled with lacustrine sedi-

FORT ROCK BASIN: Fort Rock tuff cone. Photograph by D. Sherrod.

FORT ROCK BASIN: Hole-in-the-Ground maar, Fort Rock basin. Photograph by D. Sherrod.

ments which accumulated during the episodic existence of pluvial Fort Rock Lake. This area lies within the extensional environment of the Basin and Range Province and is characterized by numerous normal faults of Pliocene and Pleistocene age that cut volcanic rocks of similar age.

Maar volcanoes are low volcanic cones with broad, bowl-shaped craters. Three general kinds are well-represented in the Fort Rock Basin: *maar*, with a crater floor below original ground level, such as Hole-in-the-Ground; *tuff ring*, with a crater floor at or above original ground level, such as Fort Rock; and *tuff cone*, which is a tall tuff ring, such as Table Rock.

The maar volcanoes of the Fort Rock Basin are the result of the explosive interaction of rising basaltic magma and abundant surface or groundwater. Beyond the basin where surface or shallow groundwater was not available, eruptions produced cinder cones and lava flows. But where water was present, deposits ranging from explosion breccia to thinly bedded layers and massive beds can be seen.

Fort Rock is an isolated tuff ring with spectacular, wave-cut cliffs and terraces. The wave-cut remnant is ~1,400 m in diameter and 60 m high, and the present crater floor is 6 to 12 m above the floor of the lake basin. The south rim has been breached by waves of the former Fort Rock Lake providing easy access to the crater. The best developed wave-cut terrace is 20 m above the floor of the basin.

Orange-brown lapilli tuff in beds 1 cm to 1 m thick can be traced from within the crater to the outer flanks. Graded beds with accretionary lapilli are common. Inward-dipping beds are parallel to the crater walls and suggest that the crater is funnel shaped. The innermost beds dip inward at angles of 20° to 70°. On the west side is a distinct angular unconformity where the deposits, truncated by slumping into the crater, are plastered with younger beds. These younger beds are part of a continuous pyroclastic sequence on the outer flanks.

Hole-in-the-Ground is a nearly circular maar with a floor 150 m below and a rim 35 to 65 m above original ground level. Its diameter from rim to rim is 1,600 m. The volume of the crater below the original surface is only 60% of the volume of the ejecta. Only 10% of the ejecta is juvenile basaltic material. Most of the ejected material is fine grained, but some of the blocks of older rocks reach dimensions of 8 m. The largest blocks were hurled distances of up to 3.7 km from the center of the crater. Accretionary lapilli, impact sags, and vesiculated tuffs are well developed.

Basaltic magma came into contact with abundant groundwater at a depth of 300 to 500 m. Repeated slumping and subsidence along a ring fault led to intermittent closures of the vent, changes in the supply of groundwater, and repeated accumulations of pressure. Four major explosive events resulted from pressures of over 500 bars in the orifice of the vent. Ejection velocities during these periods reached 200 m/s.

The **Table Rock** maar complex consists of Table Rock (a tuff cone), 2 large tuff rings, and 6 smaller tuff rings and eroded vents. The complex forms a north-northwest-elongated oval 5.6 × 8.8 km. The highest point is the top of Table Rock, 395 m above the basin floor. The complex overlies a 220-m-thick section of lake sediments, interbedded tuffs, and sands and gravel.

These tuff rings consist of approximately 95% basaltic glass shards. Eruptions in ancient Fort Rock Lake produced underwater debris flows that deposited massive tuff breccia now exposed at the bases of some of these tuff rings. When the rings breached the lake surface, much of the ash was deposited from air fall and base surges to form thin beds showing bomb sags. Beds deposited by base surges formed radial dunes and antidunes, abundant cross bedding, and ash plastered onto slopes of 45 to 90°. Slump structures produced overturned anticlines and convoluted bedding.

Table Rock is a symmetrical tuff cone ~1,500 m in diameter at its base, tapering to a diameter of ~360 m. The cone is capped with a basalt lava lake now partially eroded. Dikes extend north-northwest and south-southeast of the crater lake, parallel to the long axis of the tuff ring complex. The lava lake and dikes suggest that before eruptions stopped, access of ground or surface water to the magma ceased.

How to get there: *The western edge of the maar field is readily accessible from Highway 31, which skirts Moffitt Butt and Big Hole, as well as other features. Roads leading east from Highway 31 pass near both Fort Rock and Table Rock.*

References

Heiken, G. H., 1971, Tuff rings: examples from the Fort Rock–Christmas Lake Valley basin, south-central Oregon: *J. Geophys. Res. 76*, 5615-5626.

Heiken, G. H., 1981, A field trip to the maar volcanoes of the Fort Rock–Christmas Valley basin, Oregon *in* Johnson, D. A., and Donnelly-Nolan, J. M., (eds.), Guides to some volcanic terranes in Washington, Idaho, Oregon, and Northern California: *USGS Circ. 838*, 119-140.

Lawrence A. Chitwood

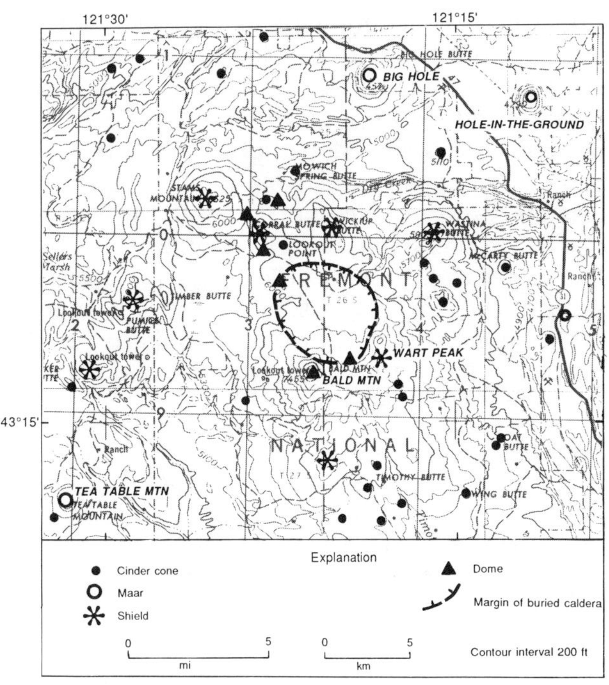

BALD MOUNTAIN: Volcano map for Basin and Range near Bald Mountain caldera, central Oregon.

BALD MOUNTAIN
Oregon

Type: *Caldera, domes, shield volcanoes, and cinder cones*
Lat/Long: *43.25°N, 121.25°W*
Elevation: *1,400-2,300 m*
Eruptive History: *Late Miocene and Pliocene*
Composition: *Bimodal (basalt-basaltic andesite and rhyodacite-rhyolite)*

Bald Mountain caldera is an informally named caldera 45 km south of Newberry volcano that has been buried by basalt and basaltic andesite. The caldera is inferred from the distribution of silicic domes and the semicircular ridge formed by Bald Mountain (silicic dome) and **Wart Peak** (basaltic shield volcano). The Peyerl Tuff, a sequence of biotite-bearing ash flows exposed near Oregon Highway 31, probably issued from vents near Bald Mountain 3.4-4.6 Ma. The caldera margin is ~5 km in diameter, which makes Bald Mountain similar in size but substantially older than Newberry volcano's caldera.

Several shield volcanoes and cinder cones have erupted basalt and basaltic andesite before and after the eruption of ash flows from the Bald Mountain caldera. Where the water table was shallow, as near geographic basins, mafic vents formed maars (e.g., **Big Hole**, **Hole-in-the-Ground**, **Tea Table** Mountain). The entire area forms a broad forested upland underlain by a bimodal suite of volcanic rocks at the western edge of the Basin and Range province.

How to get there: *Oregon Highway 31 passes near many of the features discussed here. A map of the Fremont National Forest is helpful for deciphering the network of graded roads in the Bald Mountain area.*

Reference

Peterson, N. V., and McIntyre, J. R., 1970, The reconnaissance geology and mineral resources of eastern Klamath County and western Lake County, Oregon: *Ore. Dept. Geol. Min. Indust. Bull. 66*, 70 pp.

David R. Sherrod

SILICIC DOMES OF SOUTH-CENTRAL OREGON

Type: *Domes and related flows and breccia*
Lat/Long: *43°-44° N., 120°-122° W.*
Elevation: *Regional, 1,500 to 2,440 m*
Eruptive History: *Age progression, 5 to <0.5 Ma*
Composition: *Rhyodacite to alkali rhyolite*

Silicic domes and related flows and breccia of south-central Oregon <5 m.y. old represent the western–and younger–part of a well-defined age progression in silicic volcanism that extends over a time interval of ~11 Ma to <0.5 Ma and a distance of 250 km. The silicic domes are part of a bimodal (basalt-rhyolite) assemblage; dating of the associated basalt flows and vents is currently inadequate to establish clearly whether they also are arranged, as suspected, in a similar progression. Interspersed among the larger, dated domes are a number of smaller rhyolitic masses, as well as numerous basaltic cinder and lava cones, small shield volcanoes, and extensive flow fields. The underlying cause of the age progression is unknown, but may relate to progressive westward fracturing of the crust through crustal plate interactions.

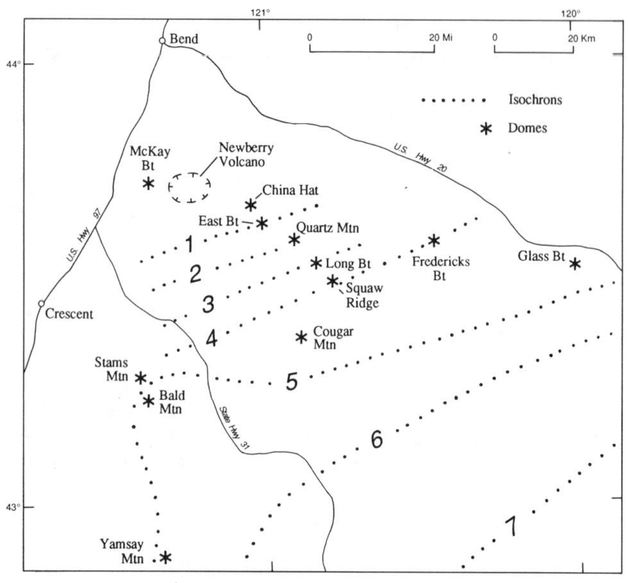

SILICIC DOMES: Age progression of silicic domes of south-central Oregon.

Domes within the younger part of the age progression include: **Stams** Mountain (5 Ma) and **Bald** Mountain (4.5-5 Ma), both associated with the Bald Mountain caldera, possibly **Yamsay** Mountain (dated at ~5 Ma), **Glass Butte** (5 Ma), **Fredericks Butte** and **Cougar** Mountain (4.4 Ma), **Squaw Ridge** (3.7 Ma), **Long Butte** (2.4 Ma), **Quartz** Mountain (1 Ma), **East Butte** (0.85 Ma), **China Hat** (0.8 Ma), Newberry Volcano (3 Ma-1,400 yr), **McKay Butte** (0.58-0.60 Ma).

Most of the domes are exogenous and partly to largely buried in their own flow-jointed and brecciated detritus. The domes vary in morphology with age; the younger masses are generally steep-sided and up to 450 m high, the related flows manifested by ramping and pressure ridges; the older masses eroded to gentler, rounded hills mostly 60 to 150 m high. Selvedges several meters thick of sheared, brecciated, and hydrated glass, commonly with residual eyes of unaltered glass (apache tears) occur on the peripheries of domes; flows of multicolored, locally spherulitic obsidian are associated with some domes.

How to get there: *The domes discussed here are within an area of more than 10,000 km^2, lying to the east and south of Newberry volcano, a prominent edifice located 37 km south of Bend, Oregon. US Highway 20, east of Bend, extends along the northern border of the area, and State Highway 31 traverses the area from US Highway*

97 southeastward toward Silver Lake and Lakeview, Oregon. Nearly all of the domes are accessible by gravel or poorly maintained dirt roads that lead from these highways.

References

MacLeod, N. S., Walker, G. W., and McKee, E. H., 1976, Geothermal significance of eastward increase in age of upper Cenozoic rhyolitic domes in southeastern Oregon: United Nations Symposium on the Development and use of Geothermal Resources, 2nd, San Francisco, 1975, *Proc. !*, 465-474.

Walker, G. W., 1974, Some implications of late Cenozoic volcanism to geothermal potential in the High Lava Plains of south-central Oregon: *The Ore Bin (Ore. Geol.) 36*, 109-119.

George W. Walker

IRON MOUNTAIN
Oregon

Type: *Dome*
Lat/Long: *43.25°N, 119.46°W*
Elevation: *1,310 to 1,640 m*
Eruptive History: *2.7 Ma*
Composition: *Rhyolite*

Iron Mountain is a prominent, steep-sided physiographic feature of southeast Oregon that projects above a vast field of late Miocene rhyolitic ash-flow tuff and Miocene and younger tuffaceous sedimentary rocks and basalt flows. It is an exogenous dome ~300 m high and 1.5 km in diameter, partly mantled at the base with erosional debris, and was emplaced at the intersection of several normal faults with displacements of several tens of meters. The dome is composed of vitrophyric, porphyritic rhyolite; phenocrysts include plagioclase, biotite, and hornblende.

How to get there: *Iron Mountain is 48 km southwest of Burns, Harney County, Oregon, ~31 km south of US Highway 20 and 21 km southeast of US Highway 395. Several gravel or dirt roads lead south from US Highway 20, and several dirt roads go east from US Highway 395 to Iron Mountain. It also can be reached by driving south of Burns on State Highway 205 to the Narrows, between Harney and Malheur Lakes, and from there heading west on any of several different graded gravel or poor dirt roads that lead to areas west of Harney Lake.*

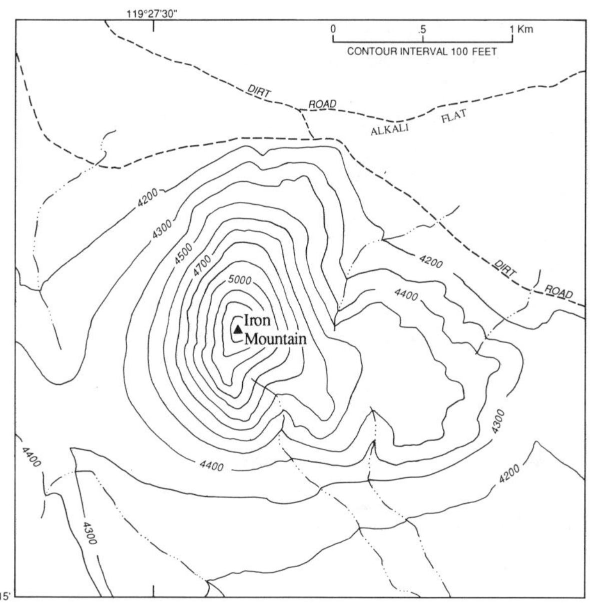

IRON MOUNTAIN: Topographic map.

References

Greene, R. C., Walker, G. W., and Corcoran, R. E., 1972, Geologic map of the Burns quadrangle, Oregon: USGS *Misc. Geol. Invest. Map I-680.*

Parker, D. J., 1974, Petrology of selected volcanic rocks of the Harney Basin, Oregon: Ph.D. Thesis, Ore. State Univ., Corvallis, 119 pp.

George W. Walker

SILVER CREEK
Oregon

Type: *Monogenetic volcanic field*
Lat/Long: *43.4°N, 119.5°W*
Elevation: *~1,375 m*
Eruptive History: *Unknown; <1 Ma?*
Composition: *Basalt?*

An unstudied basalt field occurs along the Brothers Fault Zone between the large patch of basalt east-southeast of Newberry Caldera and Diamond Craters. Because Silver Creek cuts through these flows, that name is applied here to the volcanic field. According to large-scale geologic maps there are ~24 cinder cones, 5 maars, and 3 or more small shield volcanoes within the ~1,600 km^2 lava field. A northwest–southeast trend for lines of vents and for the field itself suggest that faulting controls the distribution of vents. Four K-Ar ages from the eastern part of the field lie between 3-2 Ma.

How to get there: *The Silver Creek basalts are traversed by Oregon State Highway 20, with the village of Riley near the center of the field.*

Reference

Walker, G. W., 1977, Geologic map of Oregon east of the 121st meridian: *USGS Misc. Invest. Series Map I-902.*

Charles A. Wood

DIAMOND CRATERS
Oregon

Type: *Monogenetic volcanic field*
Lat/Long: *43.1°N, 118.7°W*
Elevation: *1,280-1,450 m*
Eruptive History: *3 main phases <0.06 Ma*
Composition: *Olivine tholeiite to transitional basalt*

The Diamond Craters volcanic field is located on the Oregon Plateau at the extreme southern edge of the Harney Basin and lies near the middle of a series of young basalt fields extending from Bend, Oregon, southeast to the Jordan Craters area. The entire field is Quaternary in age and is underlain by a variety of Pleistocene and Pliocene deposits including stratified tuffaceous silt and sandstones, basaltic tephra, and tholeiite basalt flows characteristic of the Harney and Danforth formations. The Diamond Craters field is comprised of ~60 km^2 of olivine tholeiite to transitional basalt lava flows and pyroclastic ejecta, including numerous cones and craters (explosion and collapse), a central vent complex, and numerous fissure and graben systems.

Three principal eruption phases account for the varied deposits and morphology of the Diamond Crater area. Initial activity was probably fissure-fed, leading to the formation of a large roughly circular pahoehoe basalt field. Phase 2 commenced with a doming of the basalt field followed by phreatomagmatic and strombolian activity from numerous individual vents. This pyroclastic activity resulted in deposition of basaltic lapilli, scoria and cored bombs. Near the end of phase 2, activity concentrated around a central vent complex, initially as highly explosive pulses (ash "halo" deposit) and finally as less violent pyroclastic eruptions from as many as 20 craters and cones. Minor effusion of fluid lava around the margins of the central vent complex marks the end of this eruption phase. Phase 3 saw additional doming followed by collapse and formation of a northwest–southeast-trending graben, outflow of fluid lava within many of the older craters, and formation of open fissures reaching a maximum dimension of ~5 m × 15 m. Minor sporadic pyroclastic erup-

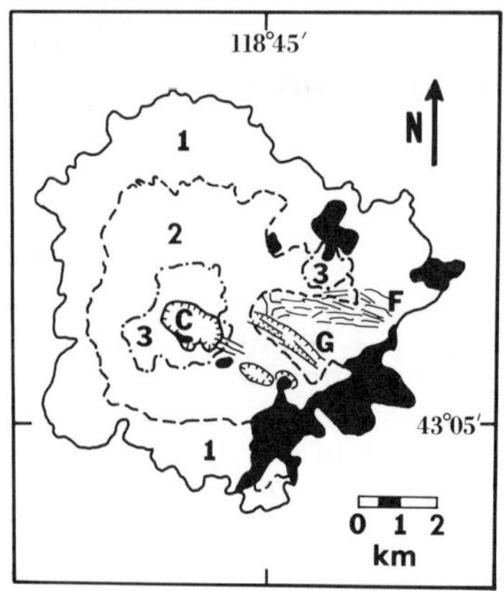

DIAMOND CRATERS: Location and extent of Diamond Craters volcanic field. Numbers represent extents of (1) early basalt flows and (2) early and (3) late phase 2 pyroclastic deposits. The black regions are phase 3 lava flows. Other major features are the central vent complex (C), the northwest–southeast-trending graben (G), and the fissure swarm (F). Modified from references.

DIAMOND CRATERS: Central vent complex.

tions followed the third phase resulting in formation of additional scoria cones.

In addition to the abundant explosion and pit craters, excellent pahoehoe flow structures and ubiquitous cored bombs, the Diamond Craters field exhibits some rather unusual features. The central vent complex with its 20 or so individual cones and craters illustrates the complex nature of both the structural and magmatic evolution of the field. This is further evidenced by the formation of the graben, and by the development of a large fissure system, both near the eastern edge of the field. In addition, early Diamond Craters basalt flows altered the drainage patterns of at least two rivers flowing northward from the Steens Mountain area of south-central Oregon.

How to get there: *The Diamond Craters field is located at the southern edge of a broad alluvial plain ~95 km south-southeast of Burns, Oregon. Access is best achieved by following Oregon State Highway 78 south from Burns or north from Burns Junction to New Princeton, Oregon. From there, well-marked, all-weather roads head southwest for ~32 km leading directly to the eastern and southeastern portions of the field. Many of the most interesting features are visible along a dirt road that passes east-west across the southern portion of the field.*

References

Peterson, N. V., and Groh, E. A., 1964, Diamond Craters, Oregon: *Ore Bin (Ore. Geol.) 26*, 17-34.

Russell, J. K., and Nicholls, J., 1987, Early crystallization history of alkali olivine basalts, Diamond Craters, Oregon: *Geochim. Cosmochim. Acta 51*, 143-154.

William K. Hart

SADDLE BUTTE
Oregon

Type: *Monogenetic volcanic field*
Lat/Long: *43.0°N, 117.8°W*
Elevation: *1,125-1,525 m*
Eruptive History: *Unknown; 1? to <0.01? Ma*
Composition: *High-alumina basalt*

Latest Pliocene(?) and Pleistocene basalt of the Saddle Butte area is separated from rocks of similar age from the greater Jordan Craters volcanic field to the east only by the Owyhee River; presumably the two volcanic fields were once one. What sets them apart, however, is the occurrence of two unconnected, young (<0.01-Ma) patches of basalt, the Saddle Butte and Jordan Craters volcanic fields, *senso stricto*. The Pleistocene field of Saddle Butte covers ~1,100 km^2, with the superposed, younger basalt having an area of ~240 km^2. The younger basalt appears to have no central vents, and the older basalt unit (a single age of 0.43 Ma) has only a few small cinder cones and small shields, some at the northeast edge of the Sheepshead Mountains. This volcanism is similar to the "plains style" of the Snake River Plain.

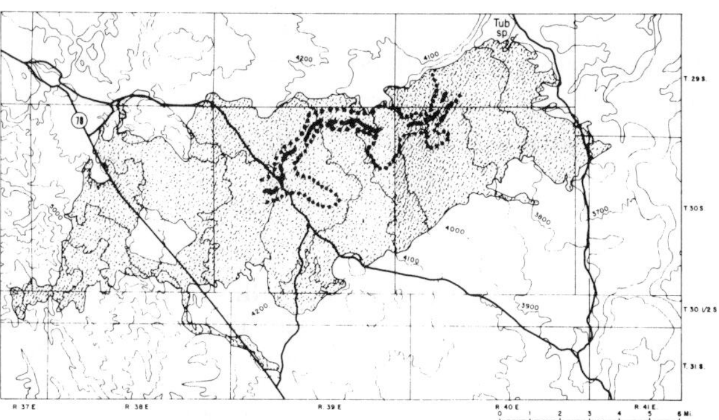

SADDLE BUTTE: Outline map of the extent of flows, from Ciesiel and Wagner, 1969. Dashed line indicates most recent flow, and solid line is trace of a lava tube. Scale bar shows 6 miles (10 km).

Only one aspect of the geology of this area seems to have been studied. Lava caves or tubes occur in abundance in these shallow-dipping flows. Forty-Mile Cave (so-called in press reports; actual length of a series of interconnected caves is ~13.5 km, an exaggeration factor of 4.7) is the most impressive, being 1.1 km long and up to 18 m wide and 14 m high. The smooth-walled tube lining is still preserved as are high-lava marks. Lava tubes such as this form by the draining of a liquid core after the surrounding lava has cooled sufficiently to remain rigid. Lava tubes are very common in basaltic lava flows; ~60% of Kilauea's flows in Hawaii were emplaced by tubes.

How to get there: *The Saddle Butte lava field is about midway between the villages of Crane and Jordan Valley, southeast of Burns, Oregon. State Highway 78 crosses the western end of the flows* *and smaller roads lead to the lava tubes. Lava tubes are dangerous because of the sharpness of stoped blocks fallen from their roofs, and because they may collapse at any time.*

References

Ciesiel, R. F., and Wagner, N. S., 1969, Lava-tube caves in the Saddle Butte area of Malheur county, Oregon: *Ore Bin (Ore. Geol.) 31*, 153-171.

Sherrod, D. R., et al., 1988, Mineral resources of the Sheepshead Mountains, Wildcat Canyon, and Table Mountain Wilderness Study areas, Malheur and Harney Counties, Oregon: *USGS Bull. 1739-A*, 31 pp.

Charles A. Wood

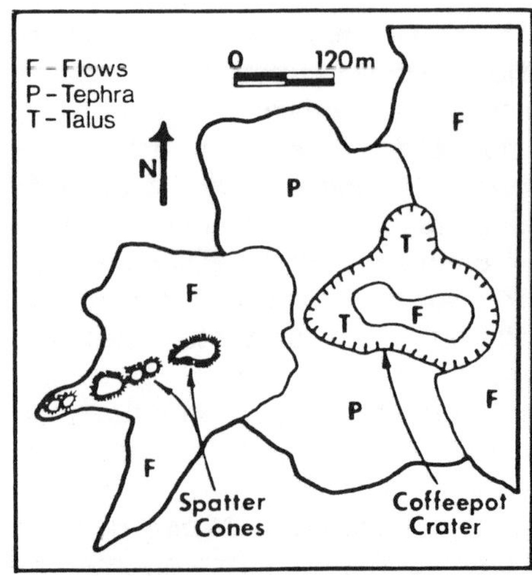

JORDAN CRATERS: Sketch map of the Coffeepot Crater area, modified after Otto and Hutchinson.

JORDAN CRATERS
Oregon

Type: *Monogenetic volcanic field*
Lat/Long: *43.1°N, 117.4°W*
Elevation: *1,200-1,400 m*
Eruptive History: *2 main phases <0.03 Ma*
Composition: *Transitional to alkali basalt*

The Jordan Craters volcanic field is located on the Owyhee–Oregon Plateau at the southeastern end of a series of young basalt fields extending from near Bend, Oregon, through Diamond Craters in south-central Oregon. The Jordan Craters field, as defined here, is limited to the northernmost and youngest portion of a larger (250-km^2) Quaternary alkaline basalt field that has three major north–south aligned vents following surficial and inferred expressions of regional Basin and Range faulting. Fluid pahoehoe basalt flows emanated from each of these sources, with pyroclastic activity confined to small scattered cones and to a larger crater–cone complex (Coffeepot Crater) at the northernmost edge of the Jordan Craters portion of the field. This entire Quaternary alkaline basalt field is part of a larger field (nearly 800 km^2) that includes Pleistocene and Pliocene olivine tholeiite to transitional basalt flows and vents. The older basalt is underlain by, and interfingered with, complex sequences of Pliocene tuffaceous sediments, Miocene basalt, silicic flows and ash-flow tuffs, and tuffaceous sediments.

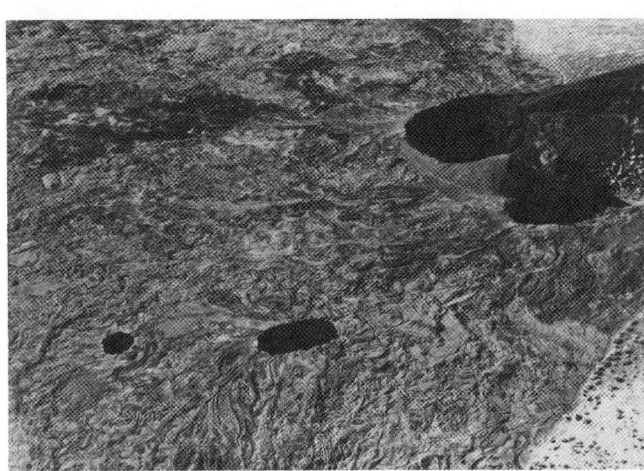

JORDAN CRATERS: Oblique aerial view of Jordan Craters vent (Coffepot Crater) and pit craters and pahoehoe flow features. Photograph by Oregon Dept. of Geology and Natural Industries.

Jordan Craters is highlighted because of its well-preserved vents and striking flow features. The area is dominated by a 75-km^2 (>1.5 km^3) olivine basalt lava-flow field which originated from **Coffeepot** Crater, the main eruptive center. Near vent, the flow surfaces are highly vesiculated shelly pahoehoe, and grade to massive, tube-fed ropy pahoehoe in the distal regions. Coffeepot Crater is a heart-shaped tephra cone constructed of numerous overlapping lobes of alternating densely to weakly welded scoriaceous lapilli and bombs. The walls of the crater show good evidence for a fluctuating lava pond which appears to have broken through and rafted away portions of the northeastern and southeastern crater walls. Backflow of the lava pond into the conduit is indicated by pahoehoe crust on the present crater floor. Additional material was vented from a series of S45°W trending spatter cones. A two stage eruptive history is suggested: (1) initial strombolian tephra cone and spatter cone-building eruptions, and (2) minor fire-fountaining, creation of a lava pond, voluminous outflows of lava, and small-volume phreatomagmatic activity.

The Jordan Craters lava flow field is mineralogically and chemically homogeneous, whereas proximal tephra and flow deposits exhibit significant chemical heterogeneity which correlates with the eruptive history. Also noteworthy are the many excellent examples of basalt flow and eruption features, including ropy and shelly pahoehoe surfaces, lava channels, lava blisters, pressure ridges and squeeze-ups, pit craters. lava-tube skylights, vertically striated internal crater walls, and re-fused cobbles of rhyolitic country rock. Additionally, the southeasterly flowing lava altered ancestral drainage patterns, giving rise to a natural dam and the formation of two small lakes (Upper and Lower Cow Lakes).

How to get there: *The Jordan Craters field is located on a high plateau ~200 km southwest of Boise, Idaho. Access to Coffeepot Crater is best achieved by following US 95 north out of Jordan Valley, Oregon, for ~12.5 km to the intersection with a dirt road heading west. This road is passable in the summer but is not well marked. Ignoring all of the numerous ranch roads, remain on the main road for ~42 km, at which point a left turn down a narrow dirt road leads to the base of the crater. The crater and lava flow field are visible to the left of the main road starting at ~km 30. The southern edge of the lava flow field is reached by traveling west out of Jordan Valley, Oregon, for ~8 km to a dirt road that leads to Cow Lakes (follow the signs).*

References

Hart, W. K., and Mertzman, S. A., 1983, Late Cenozoic volcanic stratigraphy of the Jordan Valley area, southeastern Oregon: *Ore. Geol. 45*, 15-19.

Otto, B. R., and Hutchison, 1977, The geology of Jordan Craters, Malhuer County, Oregon: *Ore Bin (Ore. Geol.) 39*, 125-140.

William K. Hart

JACKIES BUTTE
Oregon

Type: *Monogenetic volcanic field*
Lat/Long: *42.6°N, 117.6°W*
Elevation: *~1,525 m*
Eruptive History: *Unknown; <1 Ma?*
Composition: *Basalt*

Jackies Butte (sometimes called **Bowden Crater**) is a small lava field with two small cinder cones and two small shields. Lavas, presumably basalts, covering ~325 km^2, comprise this virtually unstudied volcanic field south of Saddle Butte and Jordan Craters.

How to get there: *Jackies Butte is on the east side of Highway 95, ~30 km south of Burns Junction, Oregon, heading toward the Nevada border.*

Reference

Walker, G. W., 1977, Geologic map of Oregon east of the 121st meridian: *USGS Misc. Invest. Series Map I-902.*

Charles A. Wood

VOLCANOES OF CALIFORNIA

Medicine Lake, 212
Shasta, 214
Lassen, 216
Pre-Lassen centers, 219
Lassen: other cones, 221
Eagle Lake, 222
Tuscan Formation, 224
Sutter Buttes, 225
Clear Lake, 226
Sonoma, 228
Aurora–Bodie, 229
Adobe Hills, 230
Mono Craters, 231
Inyo Craters, 233
Long Valley, 234
Big Pine, 236
Ubehebe, 237
Saline Range, 238
Coso, 239
Cima, 240
Black Mountain, 241
Pisgah, 242
Amboy, 243
Salton Buttes, 245

MEDICINE LAKE
California

Type: *Shield with caldera*
Lat/Long: *41.6°N, 121.6°W*
Elevation/Height: *2,412 m/1,200 m*
Volcano Diameter: *50 km*
Caldera Diameter: *7 × 12 km*
Eruptive History: *Oldest flow: undated (1 Ma?)*
Youngest eruption: Glass Mountain rhyolite and dacite flow: 1,000 yr?
Composition: *Basalt to rhyolite*

Medicine Lake volcano is a shield volcano of basaltic through rhyolitic composition that lies east of the main axis of the Cascade Range, ~50 km east-northeast of Mount Shasta. Its estimated volume (600 km^3) is larger than that of Shasta, which is the largest of the Cascade stratovolcanoes. The small lake from which Medicine Lake volcano derives its name lies within the 7 × 12-km central caldera. The volcano is thought to be younger than ~1 Ma, and consists of calc-alkaline and tholeiitic lavas. The most recent eruption occurred ~1,000 yr ago when rhyolite and dacite erupted at Glass Mountain and associated vents near the caldera's eastern rim. No field evidence has been found to substantiate a report of an eruption in 1910.

Primitive high-alumina tholeiitic basalt has erupted throughout the history of Medicine Lake volcano and is common around the lower flanks. The largest eruptions are basaltic, perhaps 5 times larger than the largest known rhyolitic lava flow, Glass Mountain (1 km^3). Basalt and andesite also appear to be more common than rhyolite, with dacite being relatively rare. Vents for both mafic and silicic flows typically form linear arrays, most trending within a few degrees of north–south, also the most common direction for faults. Medicine Lake volcano lies in a strongly east–west extensional tectonic environment behind the main Cascade arc. The most easterly alignment of vents is N55°E, not very different from the east-northeast regional alignment of vents that form a highland between Mount Shasta and Medicine Lake volcano. Medicine Lake volcano is very similar to Newberry volcano, also located east of the main arc.

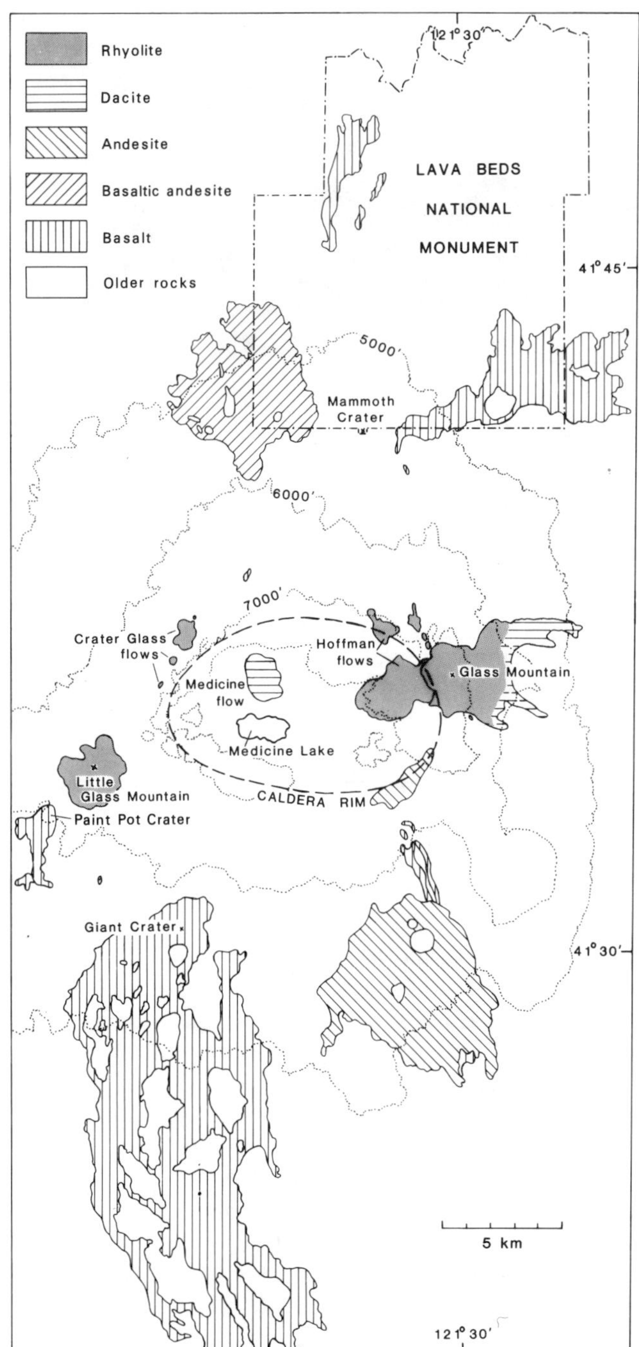

MEDICINE LAKE: Geologic sketch map showing Holocene lavas. Dashed line indicates approximate position of caldera rim. From Donnelly-Nolan, 1988.

Estimates of the size of a possible silicic magma chamber beneath Medicine Lake volcano differ, but geophysical studies constrain it to a very small volume. The magmatic system probably consists of numerous small bodies of

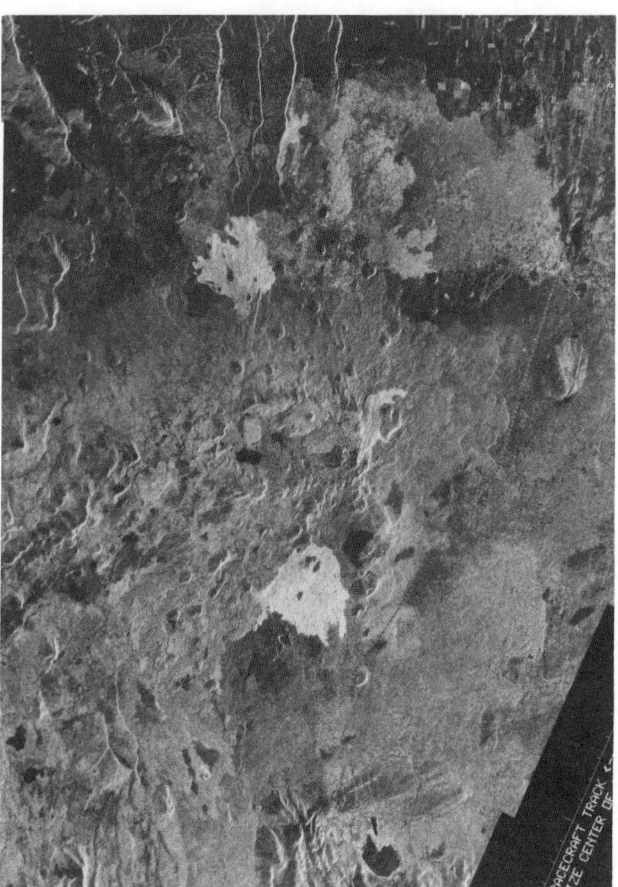

MEDICINE LAKE: Mosaic of Seasat radar images of Medicine Lake area. Black rectangle in middle is Medicine Lake itself, and the two most conspicuous bright areas (lower right and upper left) are young basaltic andesite lava flows shown on sketch map. Lava Beds National Monument is at the top of the image. In radar images smooth surfaces – such as water – reflect little energy and are dark, but rough surfaces – e.g., lava flows – are bright. Image provided by Ron Blom of Jet Propulsion Laboratory.

differentiated magma together with a plexus of mafic dikes that represents periodic influx of the primitive basalt that provides the ultimate heat source for the volcano. Geologic mapping of the volcano's many units suggests eruptions usually are relatively small and occur frequently during episodes of activity. Only one ash-flow tuff is known, an andesite tuff of late Pleistocene age, which does not appear to be related to caldera formation. Instead, the caldera probably formed by repeated subsidence over a long period as fluid lavas from summit vents flowed mostly outward down the flanks of the volcano.

MEDICINE LAKE: Profile view from the northeast of the shield of Medicine Lake volcano. Width of view ~30 km.

Holocene lava flows range in composition from high-alumina basalt to andesite to dacite and rhyolite. Some of these are compositionally zoned, including **Glass** Mountain, which first erupted dacite containing mafic magmatic inclusions and last erupted rhyolite without inclusions. Lava compositions can be explained by different petrologic processes or combinations of processes including fractional crystallization, contamination with fractionation, and magma mixing. Granitic inclusions are present in many of the flows, providing evidence of crustal contamination. The inclusions may be from geophysically inferred granitic bedrock akin to that of the Sierra Nevada batholith exposed further south.

There are no hot springs and very few cold springs in this dry area in the rain shadow east of Mount Shasta. One fumarolic area is present at the Hot Spot near Glass Mountain.

Basalt flows on the north and south flanks of the volcano contain lava tubes that have collapsed in numerous places. These collapses give access to the lava-tube caves that are a prime attraction for visitors to Lava Beds National Monument, located on the north flank of Medicine Lake volcano.

How to get there: *Medicine Lake volcano is ~55 km northeast of Mount Shasta, California, and ~70 km south of Klamath Falls, Oregon. Dirt roads approach and cross the shield volcano from the north via Lava Beds National Monument, and from the west via a road from McCloud, California, which is just east of Interstate Highway 5. There is a fishing camp at Medicine Lake itself. Glass Mountain is readily accessible via dirt roads down the east flank of the shield.*

References

Anderson, C. A., 1941, Volcanoes of the Medicine Lake Highland, California: Univ. Calif. Pubs., *Bull. Dept. Geol. Sci. 25*, #7, 347-422.

Donnelly-Nolan, J. M., 1988, A magmatic model of Medicine Lake volcano, California: *J. Geophys. Res. 93*, 4412-4420.

Julie Donnelly-Nolan

SHASTA
California

Type: *Stratovolcano*
Lat/Long: *41.40°N, 122.18°W*
Elevation: *4,317 m*
Eruptive History:
Initiation of activity: 0.59 Ma
Cone collapse and avalanche: 0.3 Ma
Sargents Ridge Cone: <0.25 Ma
Misery Hill Cone: <0.13 Ma
Shastina Cone: ~9,500 yr BP
Hotlum Cone: <9,500 yr BP
10 or more additional Holocene eruptions
Composition: *Silicic andesite to dacite*

Mount Shasta, a compound stratovolcano rising 3,500 m above its base to an elevation of 4,317 m, dominates the landscape of northern California. The largest stratovolcano of the Cascade chain at ~350 km^3, it compares in volume to such well known massive stratovolcanoes as Fuji-san (Japan) and Cotopaxi (Ecuador). Mount Shasta hosts five glaciers, including the Whitney Glacier, the largest in California.

SHASTA: Oblique aerial view of Shasta (left) and Shastina (right), seen from the northwest. Hotlum cone is at the summit of Shasta and the Bolam and Whitney glaciers are on the left and right sides of Shasta, respectively. USGS photograph by R. Krimmel.

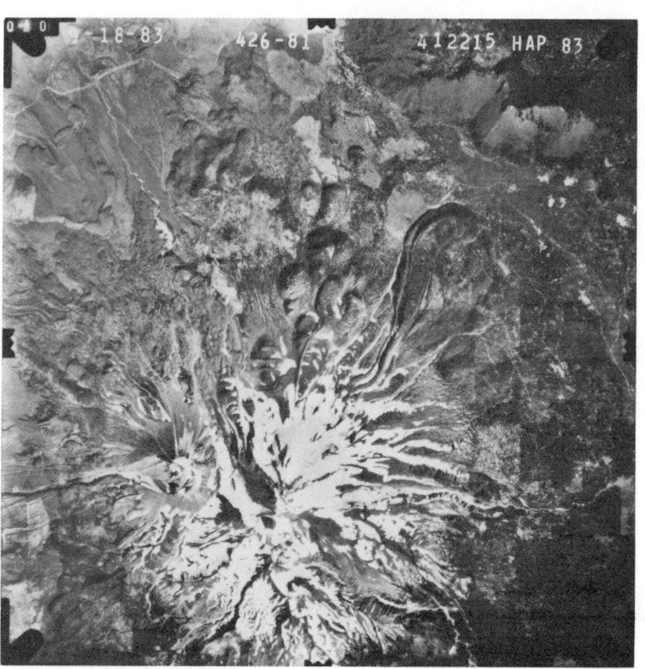

SHASTA: Vertical aerial view, north up, from National High Altitude Photography Program. Shastina is left of the center of Shasta.

Its morphology, structure, and stratigraphy show that the evolution of Mount Shasta was complex and discontinuous. The flanks of the stratovolcano grade summitward from broad aprons of andesitic debris flows to composite sections of interstratified andesitic and dacitic lavas and fragmental deposits – including both debris flows and pyroclastic flows; the higher edifice comprises mainly lavas. Few permanent streams drain the volcano. The south side is deeply scarred by the valley of Mud Creek and by several smaller glaciated valleys, but the north and northeast sides show little erosion. **Shastina** is a large subsidiary cone that rises to 3,758 m on the west flank of the compound volcano.

Four major cone-building episodes built most of the stratovolcano around separate central vents. The main bulk of the cones built in each of these episodes appears to have accumulated in a short time, lasting perhaps only a few hundred or a few thousand years, during which numerous lavas erupted, mainly from the central vent; the final major eruptions from each of the central craters produced dacitic domes and dense-fragment pyroclastic flows. After each episode of rapid cone building, the volcano underwent significant erosion while less frequent eruptions occurred, both from the central vent and from numerous flank vents. The flank eruptions typically produced cinder cones, small monogenetic lava cones, or domes, the latter commonly accompanied by pyroclastic flows. Pyroclastic flows are particularly conspicuous on the west flank of Shastina and its major flank vent, **Black Butte**.

The Mount Shasta magmatic system has evolved more or less continuously for at least 590,000 yr, but the ancestral cone was virtually destroyed by an enormous volcanic sector avalanche and landslide (~45 km^3) around 300,000 yr ago. Only a small remnant of this older edifice remains on the west side of the stratovolcano. Shasta Valley to the north is largely floored by the debris of the sector collapse, likely representing a considerable fraction of the volume of the ancestral cone. The **Sargents Ridge** cone, oldest of the four major edifices that formed the present compound volcano after the major sector collapse, is younger than ~250,000 yr, has undergone two major glaciations, and is exposed mainly on the south side of Mount Shasta. The next younger **Misery Hill** cone is younger than ~130,000 yr, has been sculpted in one major glaciation, and forms much of the upper part of the mountain. The two younger cones are Holocene. **Shastina**, west of the cluster of other central vents, was formed mainly between 9,700 and 9,400 yr; the **Hotlum** cone, which forms the summit and the north and northwest slopes of Shasta, may overlap Shastina in age, but most of the Hotlum cone is probably younger. Mount Shasta has continued to erupt at least once every 600-800 yr for the past 10,000 yr. Its most recent eruption probably was in 1786. Evidence for this eruption, recorded from sea by the explorer La Perouse, is somewhat ambiguous, but his description could only have referred to Mount Shasta. A small craterlike depression in the summit dome, containing several small groups of fumaroles and an acidic hot spring, might have formed during that eruption; lithic ash preserved on the slopes of the volcano and widely to the east yields charcoal dates of ~200 yr.

Although lavas have dominated most eruptions of Mount Shasta, several episodes of dacitic pumice eruption are also recognized. The most recent and widespread of these, 9,700 yr ago, produced the Red Banks (ledges of largely oxidized, partly agglutinated dacite and mixed dacite-andesite pumice on the upper slopes of the volcano) and more distal pumiceous fallout and pyroclastic-flow deposits. The Red Banks eruptive episode also produced associated debris flows on the lower flanks of Mount Shasta. Virtually all the lavas of the major cone-building episodes are two-pyroxene andesite; the flank eruptions typically produced either basalt to femic andesite or dacitic to rhyodacitic domes and pyroclastic flows. Probably ~90% of the composite volcano consists of silicic andesite to dacite in the range 58-64% SiO_2. The most abundant phenocrysts are plagioclase, usually large, partly resorbed, and conspicuously zoned. Hypersthene typically, but not invariably, exceeds augite. Quartz is rare except as partly resorbed xenocrysts; sanidine is lacking. The rocks characteristically contain conspicuous inclusions of fine-grained microdiorite, more mafic than the host, that appear to be fragments of quenched-liquid blobs. Although the dominant andesite probably erupted from relatively deep magmatic reservoirs, the somewhat abundant dacite, both lava and pumice, probably indicates intermittent development of high-level magma chambers in which more extensive differentiation occurred.

SHASTA: USGS topographic map; contour interval 200 ft. The hummocky area of Shasta Valley, northwest of Mount Shasta, is underlain by deposits from a large volcanic sector avalanche of ~300,000 yr ago.

How to get there: *Mount Shasta and Shastina dominate the lumber village of Weed on Highway I-5, the major north-south artery of northern California. The volcano can be circumnavigated via the Military Pass (dirt) road which runs from ~22 km north of Weed to McCloud.*

References

Crandell, D. R., 1989, Gigantic debris avalanche of Pleistocene age from ancestral Mount Shasta volcano, California, and debris-avalanche hazard zonation: *USGS Bull.*, in press.

Crandell, D. R., Miller, C. D., Glicken, H. X., Christiansen, R. L., and Newhall, C. G., 1984, Catastrophic debris avalanche from ancestral Mount Shasta volcano, California: *Geology 12*, 143-146.

Robert L. Christiansen

LASSEN
California

Type: *Stratovolcano, dacite dome field, and peripheral lava flow complexes*
Lat/Long: *40.50°N, 121.50°W*
Elevation: *~1,280 to 3,187 m*

Eruptive History: *0.6 Ma to 1917 A.D.*
Composition: *Basaltic andesite to rhyolite*

The Lassen volcanic center in Lassen Volcanic National Park consists of an andesitic stratovolcano, a dacite dome field, and peripheral small andesitic shield volcanoes. Lassen's long and complex eruptive history is a 600,000-yr-long record of volcanism associated with the generation, rise, emplacement, and evolution of a plutonic magma body in the crust.

The evolution of the Lassen volcanic center began with the construction of **Brokeoff** volcano, an andesitic stratovolcano. Glacial erosion, enhanced by hydrothermal alteration of permeable cone rocks, has resulted in deep erosion of Brokeoff volcano. The major erosional remnants, Brokeoff Mountain, Mount Diller, Mount Conard, and Diamond Peak, enclose a central depression that marks the position of Brokeoff volcano, which was ~3,350 m high, had a basal diameter ~12 km, and a volume of ~80 km^3. K-Ar ages of lavas from Brokeoff volcano range from 0.59-0.39 Ma. Thus, Brokeoff was active for ~200,000 yr. The core of Brokeoff volcano is composed of thin, glassy olivine and pyroxene andesite lava flows with abundant intercalated fragmental rocks erupted from a central vent. The flanks consist primarily of thicker, more crystalline lava flows with sparse fragmental rocks, probably erupted from vents on the flank of the volcano.

Coincident with the extinction of Brokeoff volcano was a major change in the character of volcanism at the Lassen volcanic center. Activity shifted to the north flank of Brokeoff volcano and became more silicic and episodic. Three sequences of silicic lavas and a group of hybrid lavas were erupted in the last 400,000 yr. A conspicuous feature of the lavas of the dacite dome field is the ubiquitous presence of blobs of quenched andesitic magma, formed by small amounts of andesite mixing into the silicic magma system.

The first expression of the silicic magma system was eruption of small rhyodacite lava flows and a rhyolite dome at 0.4 Ma. They were quickly followed by eruption of more than 50 km^3 of rhyolitic magma as air-fall tephra and ash flows. This eruption is thought to have produced a small caldera on the northern flank of

LASSEN: Northeast side of Lassen Peak, showing revegetation in the 1915 devastated area. Photograph by L. Topinka, USGS.

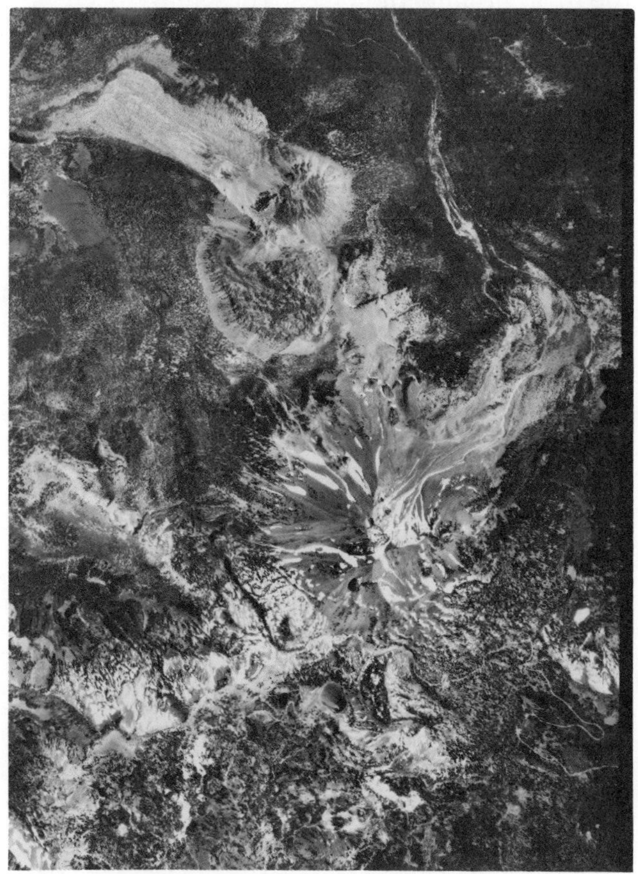

LASSEN: Vertical air view with Lassen Peak stratovolcano, Chaos Crags domes to north-northwest, and Chaos Jumbles avalanche deposit (bright lobe) to northwest. National High Altitude Photography Program photograph.

Brokeoff volcano that is now filled by two subsequent sequences of lava comprising the dacite dome field. The lavas of the early sequence are best preserved on Raker Peak and Mount Conard. The ash flows are preserved only to the west in the Manton area; the air-fall tephra is widely recognized in nearby states.

The second and third sequences of lavas at the Lassen volcanic center comprise the dacite dome field. The second sequence erupted from 0.25-0.20 Ma Porphyritic two-pyroxene-hornblende dacite magma produced a cluster of 12 lava domes and and associated thick flows with a total volume of ~15-25 km^3. The vents for these lavas were concentrated along the inferred edge of the caldera on the north flank of Brokeoff volcano. **Bumpass** Mountain, **Mount Helen**, **Ski Heil Peak**, and **Reading Peak** are the most prominent domes. These have been extensively glaciated so that no trace of their original pumiceous carapace is preserved and their massively jointed, devitrified interiors are exposed.

The third sequence consists of porphyritic hornblende-biotite rhyodacite erupted as lava domes, short thick lava flows, and pyroclastic flows erupted in at least 12 episodes during the past 100 ka. Rocks of this sequence form the northern and western portions of the dacite dome field, and their vents are concentrated in linear chains near the inferred western edge of the old caldera. The most prominent features of this sequence are **Eagle Peak** (55 ka), **Sunflower Flat** domes (35 ka), Lassen Peak (~25 ka), and Chaos Crags (1,050 yr BP). Variable amounts of the domes' primary surface morphology is preserved. The surface morphology of the Chaos Crags domes is nearly unmodified by erosion processes. The Chaos Crags domes are steep-sided with extremely rough terrain on the top. The talus cones and piles around their bases are primary crumble breccia. The Sunflower Flat domes are 35 ka old, but unglaciated. Thick forest hides their rugged topography. Many of the older domes have been glaciated, and primary loose talus has been removed.

A group of hybrid andesite lavas, called quartz basalt by previous geologists, are associated with the silicic lavas. They form lava-flow complexes with agglutinate cones marking their vents. They are characterized petrographically by the coexistence of magnesian olivine and quartz and are not basalt but rather are black, glassy olivine or augite andesite. These hybrid andesites were produced by thorough mixing of mafic magma injected into silicic magma. A volume of ~10 km^3 of hybrid andesite erupted around the margins of the dacite dome field in ten episodes over the last 300,000 yr. **Hat** Mountain, **Raker Peak**, and **Cinder Cone** are the most prominent features. The products of the 1915 eruption of Lassen Peak belong to this sequence.

Three episodes of volcanism have occurred at the Lassen volcanic center in the past 1,100 yr. These are the complex eruption at Chaos Crags, the eruptions at Cinder Cone, and the summit eruptions of Lassen Peak in 1914-1917.

The development of **Chaos Crags** forms a typical silicic volcanic cycle. The deposits originating from the Chaos Crags indicate a complex eruption 1,100-1,000 yr ago. Initial activity included formation of a tephra cone, emplacement of two pyroclastic flows, and growth of a dome that plugged the vent. After a quiet interval of ~70 yr, the dome was destroyed by a violent eruption that emplaced a pyroclastic flow and an air-fall tephra lobe. These deposits are best seen in the drainages of Manzanita and Lost Creeks within 10 km of the Chaos Crags. This violent eruption was followed by the growth of 5 domes, 3 of which had hot, dome-collapse avalanches. **Chaos Jumbles** formed 700 yr later when one of the domes collapsed in a series of 3 cold rockfall-avalanches.

The eruption of **Cinder Cone** is typical of mafic volcanism in the Lassen area. The Cinder Cone eruptive episode consists of 4 basaltic andesite lava flows, a complex vent cone, and an ash blanket covering 200-300 km^2. Two lava flows erupted before the ash blanket and two followed the ash blanket. All of these erupted over a short time interval ~425 ^{14}C yr ago.

The eruption of Lassen Peak in 1914-1917 included a diverse series of events that illustrate the destructive potential of snow- or ice-covered volcanoes. Mild vent-opening and crater-forming phreatic explosions persisted for a full year before the violent explosions of May 19 and 22, 1915. Lava first appeared in mid-May, 1915 as a small dome filling the crater. On the evening of May 19, an explosion destroyed the lava dome. Fragments of the still-hot lava were ejected onto the snow-covered upper slopes of Lassen Peak and generated an avalanche, debris flow, and flood that affected downslope valleys for at least 50 km. On May 22, a subplinian eruption deposited an air-fall tephra lobe. Partial collapse of the eruption column generated a pyroclastic flow that devastated the northern flank of the volcano. Subsequent melting of snow incorporated in the pyroclastic flow generated a fast-moving debris flow. Hot tephra that fell onto the snow-covered upper slopes of the volcano generated additional debris flows.

How to get there: *The Lassen volcanic center is within Lassen Volcanic National Park, ~65 km east of Redding, California, or 240 km west of Reno, Nevada. Access within the Park is by California Highway 89, which runs through the the volcanic center. The periphery of the volcanic center is accessible via a good network of USFS gravel roads. The National Park Service maintains many trails in the area. Lassen Peak may be climbed via an easy 4-km trail that starts from the Park road.*

References

Williams, Howel, 1932, Geology of Lassen Volcanic National Park, California: Univ. Calif. Pub. *Dept. Geol. Sci.* *21*,195-385.

Clynne, M. A., and Muffler, L. J. P., 1989, Lassen Volcanic National Park and Vicinity: *in* Muffler, L. J. P. (ed.), *IAVCEI Excursion 12B*, South Cascades Arc Volcanism, California and Southern Oregon.

Michael A. Clynne

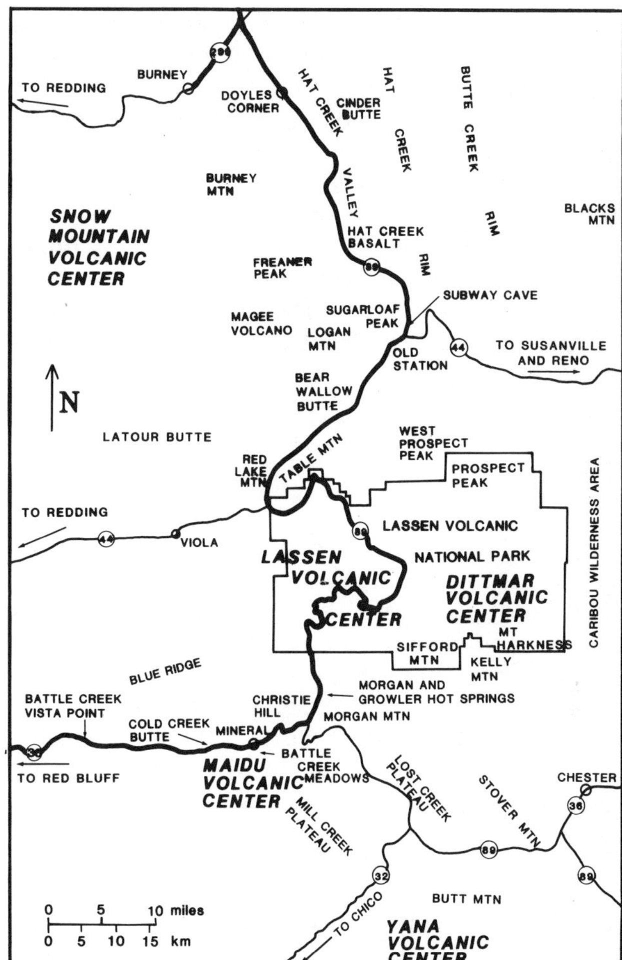

PRE-LASSEN CENTERS: Sketch map showing locations of Dittmar, Maidu, Yana, and Snow Mountain volcanic centers. Other cones in the Lassen area are also labeled.

PRE-LASSEN CENTERS
California

Type: *Stratovolcano complexes*
Lat/Long:
Snow Mountain: 40.75°N, 121.75°W
Dittmar: 40.45°N, 121.30°W
Maidu: 40. 30°N, 121.60°W
Yana: 40.15°N, 121.45°W
Elevation: *~2,100 to 2,400 m*
Eruptive History: *3 to 0.7 Ma*
Composition: *Basaltic andesite, andesite, rhyolite*

Five andesitic volcanic centers (composite cones flanked by silicic domes and flows) younger than ~3 Ma have been recognized in the Lassen Volcanic National Park area. Each had a similar history, consisting of three stages: (1) an initial cone-building period of mafic andesite and andesite lava flows and pyroclastics; (2) a later cone-building period characterized by thick andesite and silicic andesite lava flows; and (3) silicic volcanism flanking the main cone. The silicic magma chamber of the third stage provided a heat source for a hydrothermal system that developed within the core of the main cone. Alteration of permeable rocks of the cone facilitated glacial and fluvial erosion of the volcano's center. The result is a caldera-like feature in which a resistant rim of thick, late cone-building lava flows and flanking silicic rocks is selectively preserved surrounding a central depression, which replaces the altered and eroded core of the composite cone.

Four such volcanic complexes in the Lassen area, the Yana, Snow Mountain, Dittmar, and Maidu volcanic centers have reached this stage, and their hydrothermal systems are extinct. The youngest, the Lassen volcanic center, hosts active silicic volcanism and a well developed hydrothermal system, including the thermal features in Lassen Volcanic National Park and Mill Canyon.

YANA

The Yana volcanic center is 20 km southwest of Chester. Butt Mountain, Ruffa Ridge, and Humboldt Peak are the major remnants of a deeply eroded andesitic stratovolcano that was 15-20 km in diameter. Propylitized rocks of the core of the cone are exposed in the valley of Butt Creek. Basaltic andesite and andesite containing olivine and pyroxene are the principal rock types of the Yana cone. Block-and-ash flows are especially abundant. These fragmental deposits, were remobilized and redeposited as volcanic debris flows, forming a significant portion of the Tuscan Formation. The abundance of silicic rocks associated with the Yana volcanic center is unknown. The Yana volcanic center was active ~3 to 2 Ma ago.

SNOW MOUNTAIN

The Snow Mountain volcanic center is 16 km southwest of Burney. Snow Mountain, Green Mountain, and Clover Mountain are the largest remnants of a volcanic center that was ~10 km in diameter. The center is deeply eroded, and the hydrothermally altered core of the andesitic stratovolcano is exposed in the canyon walls of creeks draining the area. Rock types range from pyroxene andesite to hornblende-biotite rhyolite. The Snow Mountain volcanic center contributed fragmental material to the Tuscan Formation. Two K-Ar dates suggest that the Snow Mountain volcanic center was active from ~2 to 1 Ma.

DITTMAR

The Dittmar volcanic center is in the southeastern part of Lassen Volcanic National Park. It had a diameter of ~20 km and was centered in the upper Warner Creek Valley just north of Kelly Mountain. Saddle Mountain, Pilot Mountain, Kelly Mountain, and Mount Hoffman are the largest remnants of the deeply eroded andesitic stratovolcano. The cone was constructed of olivine and pyroxene andesite, but two-pyroxene silicic andesite is abundant in the upper parts. Silicic rocks are not as abundant as in the other volcanic centers. Dacite domes are located in the deep canyons of Hot Springs and Kings Creek, and small remnants of silicic rocks are found buried by rocks of the Lassen volcanic center on the central plateau of Lassen Volcanic National Park. Some of the silicic rocks in the area of Stover Mountain may be related to the Dittmar volcanic center. Only a single K-Ar age is available for the Dittmar volcanic center, 1.4 Ma for an andesite lava from the upper part of the Dittmar stratovolcano. Stratigraphy suggests that the Dittmar volcanic center was active from ~2.5 to 1 Ma.

MAIDU

The Maidu volcanic center is in the area around Battle Creek Meadows near the town of Mineral. It was at least 25 km in diameter. Hampton Butte and Turner Mountain are the largest remnants of the hydrothermally altered and deeply eroded andesitic stratovolcano, which is well exposed in the canyon walls of Mill Creek. Early workers suggested that the central depression (Battle Creek Meadows) was a caldera; however, no evidence has been found to support the hypothesis. Rock types range from olivine basaltic andesite to hornblende-biotite rhyolite. Spectacularly porphyritic pyroxene andesite with zoned augite phenocrysts up to 1 cm is common in the upper parts of the stratovolcano. Silicic lavas are abundant. The rhyolite of **Blue Ridge** is a 300-m-thick lava flow covering 80 km^2 on the west flank of the Maidu volcanic center, and is probably the largest rhyolite flow in the Cascade Range. Other large rhyolite flows occur on the east flank of the Maidu volcanic center (Lost Creek and Mill

Creek Plateaus). A dacite dome field occurs on the north flank of the Maidu volcanic center. **Red Rock** Mountain, **Rocky Peak**, **Christie Hill**, **Morgan** Mountain, and **Doe** Mountain are some of the more prominent domes. The Maidu volcanic center was active from ~2 Ma to 0.7 Ma.

How to get there: *Dittmar volcanic center is within Lassen Volcanic National Park and is generally accessible via park roads. The Maidu volcanic center is crossed by Highway 36, which goes from Red Bluff through Mineral, to Lassen National Park. Forest Service gravel roads provide access to the others of these older, forested volcanic remnants.*

Reference

Clynne, M. A., and Muffler, L. J. P., 1989, Lassen Volcanic National Park and Vicinity: *in* Muffler, L. J. P. (ed.), *IAVCEI Excursion 12B*, South Cascades Arc Volcanism, California and Southern Oregon.

Michael A. Clynne

LASSEN: OTHER CONES

California

Type: *Many individual shields, cones and domes*
Lat/Long: *40.35-40.85°N, 121.00-121.45°W*
Elevation: *~1,200 to ~2,500 m*
Eruptive History: *~2 Ma to ~15 ka*
Composition: *Basalt and andesite*

Hundreds of small shield volcanoes, lava cones, and cinder cones surround the volcanic centers in the Lassen area. Virtually every high point marks a young volcanic vent or older faulted and eroded edifice. Most are of basaltic to andesitic composition. These volcanoes range in age from Late Miocene to Recent.

Cold Creek Butte is a basaltic cinder cone just west of Battle Creek Meadows. A road metal quarry visible from California Highway 36 exposes the layers of ash that form the cinder cone. Battle Creek Meadows were formed when the lava flow from Cold Creek Butte blocked the channel of Battle Creek, ~20,000 yr ago.

LASSEN: OTHER CONES: Portion of a Skylab 3 photograph showing shield volcanoes north and east of Lassen (L). B = Burney Peak, F = Freaner Peak, C = Crater Peak, S = Sugarloaf Peak, and P = Prospect Peak. Photograph # SL3-25-053. Northeast up.

Mount **Harkness** is an andesitic shield volcano in the southeast corner of Lassen Volcanic National Park. It is actually a rather small volcano that is built on a remnant of the Dittmar volcanic center. A trail to the summit of Mount Harkness leads to a lookout that has a good view of the area. Mount Harkness has a cinder cone at its summit that may be post-glacial (<20,000 yr old).

Sifford Mountain is a basaltic to andesitic shield volcano at the southern boundary of Lassen Volcanic National Park. It has a long eruptive history, probably spanning the last 100 ka. Sifford Mountain is notable for its symmetrical shape and small cinder cone at its summit.

Latour Butte is an andesitic lava cone north of Viola. It was a relatively long-lived, moderately large volcano. Erosion has exposed several intrusive necks on its upper flanks. It is probably 1-2 Ma old.

Red Lake Mountain is a large cinder cone surrounded by lava flows erupted from its base. Its composition ranges from basalt to andesite. One of the later flows is exposed in road cuts on California Highway 44 ~2.5 km west of the junction with California 89. Red Lake Mountain lavas overlie a pyroclastic flow from **Eagle Peak** in Lassen Volcanic National Park that has been dated at 57 ka.

Table Mountain is an andesitic shield volcano just north of Lassen Volcanic National Park. Its morphology and composition are typical of many volcanoes in the area to the east. The entire edifice is composed of lava flows of a single type of black, glassy, two-pyroxene andesite. Pyroclastic deposits are lacking. Table Mountain probably formed at 1-2 Ma.

Approximately 9 km northwest of Red Lake Mountain, the highway skirts block lava flows of andesite from a cinder cone called **Bear Wallow Butte**, which is the southernmost of a chain of vents called the **Tumble Buttes**. These vents are among the youngest in the area, and are probably <20 ka.

Just east of the Tumble Buttes is an andesitic lava cone called **Logan** Mountain. Although its cone is well preserved, it is cut by two faults, indicating that it is an older volcano.

Prospect Peak and **West Prospect Peak** are two volcanoes along the northern boundary of Lassen Volcanic National Park. West Prospect Peak is a lava cone of basaltic andesite to andesite, whereas Prospect Peak is an andesitic shield volcano with a cinder cone at its summit. Prospect Peak is the younger volcano, although both are less than a few hundred thousand years old. There is a good gravel road to the summit of West Prospect Peak, where a lookout provides spectacular views of the surrounding region, especially the Hat Creek graben and volcanoes north and east of Lassen Volcanic National Park.

Along the highway for 25 km north of Old Station, the valley is filled with tholeiitic basalt that issued from a fissure just south of Old Station. This basalt flow has been called the Hat Creek flow or Hat Creek Basalt. Stratigraphic evidence suggests that the **Hat Creek Basalt** is ~15 ka. It was a very fluid lava that flowed in lava tubes. A row of spatter cones marks the location of the fissure. A trail departs from the Hat Creek Campground ~1.5 km southwest of Old Station and winds through the vent area. Just north of the intersection of California Highways 44 and 89 at Old Station is Subway Cave, a piece of a lava tube open to the public.

Just north of Old Station is **Sugarloaf**, an impressive andesitic lava cone. Despite its sparse tree cover, Sugarloaf is older than the Hat Creek flow and is probably 20-30 ka.

The **Magee** volcano in the Thousand Lakes Wilderness Area is a deeply eroded andesitic composite cone. Intrusive andesite is spectacularly exposed in the walls of glacial cirques below the rim of Fredonyer, Magee, and Crater Peaks. The Magee volcano is 1-2 Ma.

North of the Magee volcano are Freaner Peak and Burney Mountain. **Freaner Peak** is an olivine andesite shield volcano, and **Burney** Mountain is a large dacite dome cluster. Five separate domes are built one on top of another. The sparsely phyric, black, glassy pyroxene dacite lavas forming Burney Mountain probably erupted at unusually high temperature. Burney Mountain has been dated at 0.23 Ma.

Cinder Butte is a basaltic andesite shield volcano located at the north end of the Hat Creek Valley. Its lavas have extremely rough topography. A row of parasitic vents probably mark the location of the dike that fed the volcano. Cinder Butte is slightly older than the Hat Creek Basalt.

Blacks Mountain is 25 km east of Hat Creek Valley and is the most conspicuous of a number of andesitic shield volcanoes. These shield volcanoes have basal diameters up to 10 km and usually are composed of a single rock type. Blacks Mountain is 1-2 Ma.

How to get there: *These smaller volcanic cones are all in Lassen Volcanic National Park and immediate environs. As per the description above, National Park or Forest Service roads provide views or access to most of the cones.*

Reference

Clynne, M. A., and Muffler, L. J. P., 1989, Lassen Volcanic National Park and Vicinity: *in* Muffler, L. J. P. (ed.), *IAVCEI Excursion 12B*, South Cascades Arc Volcanism, California and Southern Oregon.

Michael A. Clynne

EAGLE LAKE
California

Type: *Polygenetic cones in volcano-tectonic depression*
Lat/Long: *40.61°N, 120.75°W*
Elevation: *1,555 to 2,342 m*
Volcano Field Size: *6 × 15 km*

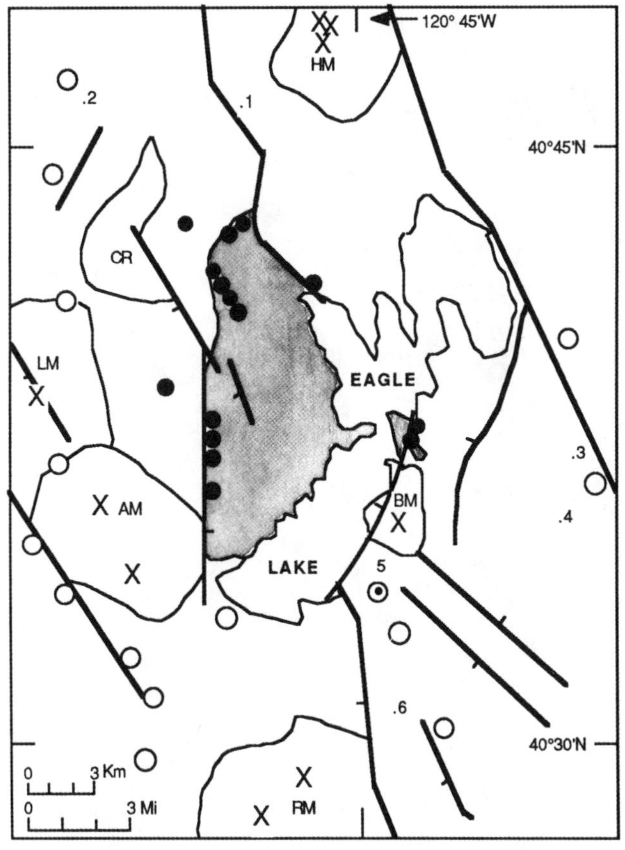

EAGLE LAKE: Map of Eagle Lake Volcanic Field. Shaded area: Eagle Lake basalt flows. Major eruptive centers: AM = ***Antelope*** *and* ***Fox*** *Mountain volcanoes, BM, Black Mountain; CR, Rhyolite domes of Champs Flat; HM, Heavey Mountain volcano; LM, Logan Mountain volcano; RM, Roop Mountain volcano. Filled/dark circles: cinder cones with basaltic flows, Late Quaternary.Open circles: plug and dike centers, Plio-Pleistocene, andesitic. X: central vents of major volcanoes. Plio-Pleistocene, basaltic and mafic andesite; Heavy line: fault, normal-slip with bar on downthrown side. Ages (Ma): BM: 0.17; CR: 3.8; HM: 2.1; RM: 3.9; 1: 6.9; 2: 4.3, 3: 2.5; 4: 3.1; 5: 4.6; 6: 2.3.*

Eruptive History: *Two main periods of activity:*
4.6 to 2.1 Ma
0.17 to ~0.05 Ma
Composition: *Tholeiitic olivine basalt, andesite, minor rhyolite*

The Eagle Lake volcanic field is located at the junction of three geological provinces: Sierra Nevada, Cascades, and Basin and Range. The field is obviously defined by late Pleistocene flows of olivine tholeiitic basalt covering 125 km^2 within the Eagle Lake volcano-tectonic depression. Fifteen cinder cone-and-flow vents, <100 m high, fed the flows during two eruptive periods, each very brief, probably between 50 and 100 ka. **Black** Mountain, a severely faulted mafic andesite volcano, 0.17 Ma old, represents a distinctly earlier eruptive period – probably the first associated with the formation of the volcano-tectonic depression. Vents align along normal-slip and right normal-slip faults that outline the Eagle Lake volcano-tectonic depression, a north-northeast pull-apart basin within the northwest-trending Walker Lane fault system. The cones and flows are essentially unaffected by erosion, but fault scarps and en echelon rifts locally offset the lava surfaces by several meters. The flows are uniform, highly fluid, diktytaxitic, high-alumina, low-potassium, tholeiitic olivine basalt. They represent the southernmost occurrence of widespread late Quaternary basaltic magmatism believed to be associated with backarc spreading in the northwestern Great Basin.

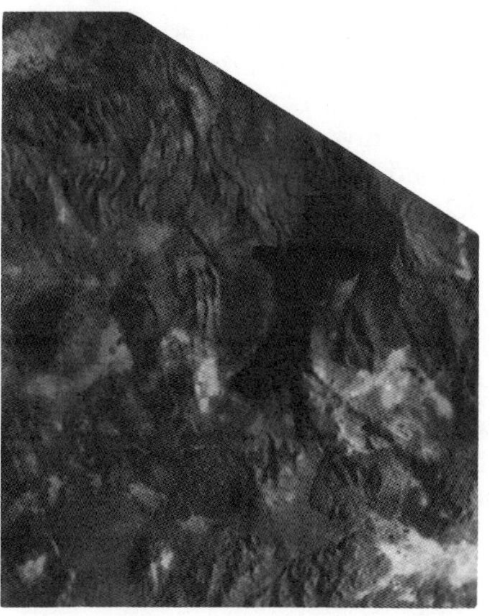

EAGLE LAKE: Skylab photograph showing Eagle Lake area. The lake itself is black. Skylab photograph SL3-25-054.

Several low composite volcanoes and small shields of Pliocene age occur just outside the late Pleistocene field. **Roop** Mountain volcano, 650 m high, is mainly composed of mafic andesite flows. **Antelope** Mountain volcano, 600 m high, consists of andesite flows with minor dacite flows and pyroclastic debris. **Logan** Mountain shield, 400 m high, is basaltic and mafic andesitic. **Heavey** Mountain shield, 250 m high, is also made of basalt and mafic andesite flows. Two-million-year old, multiple-vent structures are well preserved on Heavey Mountain. The only known silicic flows in the area include the deeply eroded flow domes of **Champs Flat** exposed in an area of 3 km^2.

The areas between these volcanoes are underlain by numerous coalescing and interfingering flows with local interbeds of tuff, mostly of Pliocene age. They are derived from many small local vents.

How to get there: *The Eagle Lake volcanic field is in Lassen County ~20 airline km northwest of Susanville. Drive State Route 36 ~4 km west of Susanville, then turn north on paved County Road A1 ~35 km to the southern end of the field at Eagle Lake. This road continues on to the northeast through the field for ~30 km to its junction with State Route 139.*

References

Grose, T. L. T., and McKee, E. H., 1986, Potassium-argon ages of Late Miocene to Late Quaternary volcanic rocks in the Susanville-Eagle Lake area, Lassen County, California: *Isochron West 45*, 5-11.

Grose, T. L. T., Bean, S. M., Roberts, C. T., Tuppan, E. J., and Youngkin, M. T., 1981, Geology of the southeastern Cascades, California: 3rd Tech. Ann. Rept., USGS Grant No.-08-0001-G-624, map scale 1:100,000.

T. L. T. Grose

TUSCAN FORMATION
California

Type: *Polygenetic lahar field*
Lat./Long.: *40.12°N, 121. 87°W*
Elevation: *100 to 2,000 m*
Eruptive History: *3.99 to ~2.0 Ma*
Composition: *Basaltic and basaltic andesite lahars; minor basalt flows, rhyolite tephra, debris flows, and volcanic alluvium*

The Tuscan Formation is composed primarily of basaltic and basaltic andesite lahars derived from Pliocene stratovolcanoes at the southern end of the Cascade Range in northeastern California. Volcanic debris flows from Pliocene Mount **Yana**, Pliocene and Pleistocene Mount **Maidu**, and probably other sources, extended to the Sacramento Valley forming a westward thinning, wedge-shaped lahar field that once covered an area of 5,180 km^2. Near the Sacramento River thin distal lahars and interbedded volcanic conglomerate and sandstone grade laterally into fine-grained alluvial deposits of the coeval **Tehama Formation**. Rhyolitic ash flow of the Nomlaki Tuff Member occurs near the base of both the Tuscan and Tehama Formations and records an explosive event dated at 3.4 Ma. The **Sifford** Ring has been proposed as the topographic expression of the caldera that formed during the eruption of the Nomlaki but little geologic evidence supports this conclusion. The Ishi Tuff Member of the Tuscan Formation records a smaller explosive event ~2.6 Ma.

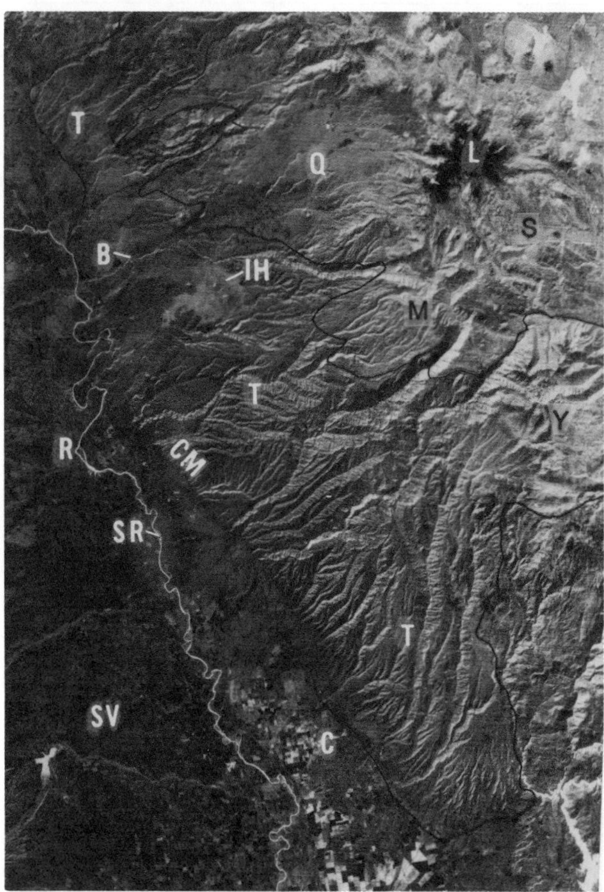

TUSCAN FORMATION: Landsat image of the Tuscan Formation (T) and overlying Quaternary volcanic rocks (Q): Mount Yana (Y), Mount Maidu (M), Sifford Ring (S), Lassen Peak (L), Inskip Hill (IH), Chico monocline (CM), Sacramento River (SR), Sacramento Valley (SV), Chico (C), Red Bluff (R), Battle Creek fault zone (B).

Development of the lahar field began around Mount Yana and progressed northward.

In the south, the lahars are as much as 520 m thick, overlie the Nomlaki Tuff Member, and were deposited between 3.4 and 2.4 Ma. Northeast of Red Bluff the lahars are generally <200 m thick, overlie the Ishi Tuff Member, and were deposited between 2.6 and 2.0 Ma. A distinctive avalanche-lateral blast deposit, which erupted ~2 Ma, occurs at the top of the northern lahar section and is characterized by a relatively undissected constructional surface. In contrast, the older lahars to the southeast are deeply dissected, partly because of their greater age and partly because they were uplifted and tilted to the west along a basement fault below the Chico monocline between 2.4 and 1.1 Ma. North of Red Bluff, the Tuscan Formation was folded about northeast-trending folds that structurally control the large meanders in the Sacramento River, and faulted by the northeast-trending Battle Creek fault zone between 1.0 and 0.5 Ma. To the east, the Tuscan Formation is overlain by lower to middle Pleistocene flows of basalt, andesite and rhyolite and rhyolite ash flow tuff. Late Pleistocene high alumina basalt flows and cinder cones, such as **Inskip Hill**, are the youngest volcanic features in the area.

How to get there: *The Tuscan Formation borders the northeast part of the Sacramento Valley ~170 km northwest of Sacramento, California. Highway 32 northeast from Chico, California, and Highway 36 northeast from Red Bluff, California, provide paved access to and excellent views of the Tuscan Formation.*

References

Harwood, D. S., and Heeley, E. J., 1987, Late Cenozoic tectonism of the Sacramento Valley: *USGS Prof. Pap. 1359*, 46 pp.

Lydon, P. A., 1968, Geology and lahars of the Tuscan Formation, northern California, *in* Coats, R. R., Hay, R. L., and Anderson, C. A. (eds.), Studies in volcanology, a memoir in honor of Howel Williams: *Geol. Soc. Am. Mem. 116*, 441-475.

David S. Harwood

SUTTER BUTTES
California

Type: *Dome field*
Lat/Long: *39.22°N, 121.80°W*
Volcano Diameter/Height: *16 km/0.6 km*

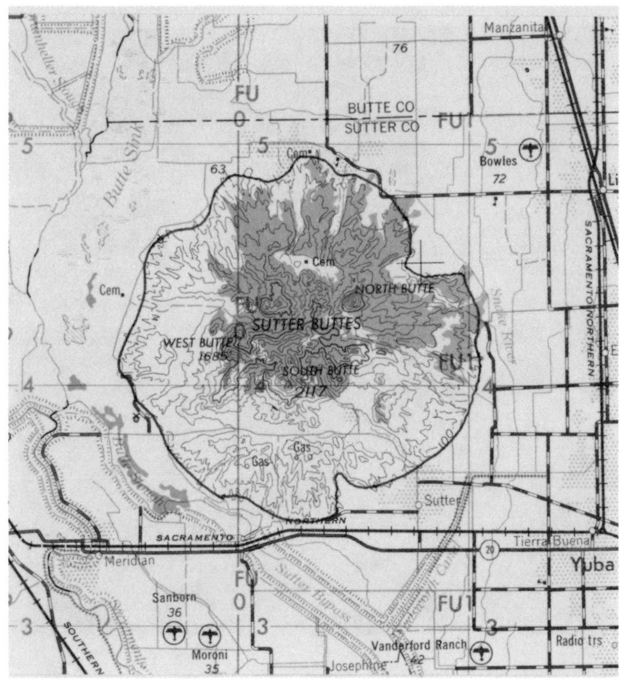

SUTTER BUTTES: USGS topographic map with buttes outlined.

Eruptive History: *2.5-1.5 Ma*
Composition: *Andesite, dacite, rhyolite*

Sutter Buttes is an anomalous volcanic landform rising starkly from the flat plain of the Sacramento Valley. As interpreted by Howel Williams, rising magmas uplifted Cretaceous and Tertiary sediments, which quickly eroded. Explosive eruptions, extending perhaps 0.5 m.y., accompanied the emplacement of viscous intrusions and extrusions at the center and periphery of the uplift. These pelean domes, which strongly uparched the intruded beds, are andesitic in the central core of Sutter Buttes, and are surrounded by a halo of dacitic to rhyolitic domes. There is no temporal succession of petrologic types, however. Geochemically, the andesites have much higher than normal values of K, Ba, and Sr.

The coalescing central domes are surrounded by gently dipping andesitic sediments interpreted by Williams as waterlaid volcaniclastics produced by reworking of air-fall material ejected from vents in the center of the Buttes. This interpretation may be correct, but "water-laid volcanic deposits" at many other volcanoes have often been reinterpreted as primary base surge deposits. Sutter Buttes needs re-examination.

SUTTER BUTTES: Vertical aerial view from National High Altitude Photography Program.

It is often suggested that Sutter Buttes constitutes the southernmost Cascade volcano, and the feature does occur along the continental extension of the Mendocino Fracture Zone in a region which might have once covered a subduction zone. However, there are great differences in age and morphology compared to the large and young Cascade stratovolcanoes. Additionally, the lack of continuing volcanism at Sutter Buttes confounds analogues with the conventional Cascade volcanoes.

How to get there: *Sutter Buttes is 90 km north of Sacramento and 45 km west of Marysville, California. The buttes are a privately owned nature preserve; permission is required for visits, but public roads pass close for visual inspection.*

Reference

Williams, H., and Curtis, G. H., 1977, Sutter Buttes of California: Univ. Calif. Pub. *Bull. Dept. Geol. Sci.* *116*, 56 pp.

Charles A. Wood

CLEAR LAKE
California

Type: *Polygenetic volcano field*
Lat/Long: *38.9°N, 122.7°W*

CLEAR LAKE: Aerial view looking west across the southeastern arm of Clear Lake toward the 1-km-high composite dacite volcano, Mount ***Konocti****, which is ~0.35 Ma. In the foreground, covered by brush and houses, is the 90-ka rhyolite of* ***Borox Lake****, the youngest silicic flow in the Clear Lake volcanic field.*

Elevation: *400 to 1,400 m*
Field Area: *500 km^2*
Eruptive History: *4 episodes from 2.1 Ma to 10? ka, with gaps of 0.15 to 0.2 Ma*
Composition: *Basalt through rhyolite; dacite dominant*

The Clear Lake volcanic field (late Pliocene to Holocene) lies in a tectonically active, complex geologic setting within the San Andreas transform fault system in the northern Coast Ranges of California. Clear Lake and the volcanic field are located within a fault-bounded, locally extensional basin. The lake is the largest natural freshwater lake entirely within California; it is probably volcano-tectonic in origin, but is not a caldera lake. The volcanic field is the northernmost of a series of young Cenozoic volcanic fields in the Coast Ranges. Within the field, eruptive loci have migrated northward through the last 2.1 Ma. Eruptive centers are numerous; a clear central focus of activity is lacking. Volcanism appears to be related to extension in a pull-apart basin within the San Andreas fault system and is not directly related to subduction, which ceased off the California coast at this latitude ~3 Ma.

The Clear Lake volcanics range from basalt through rhyolite in composition. Basalt is rare, and the dominant composition is dacite. Four eruptive episodes separated by time gaps of 0.15-

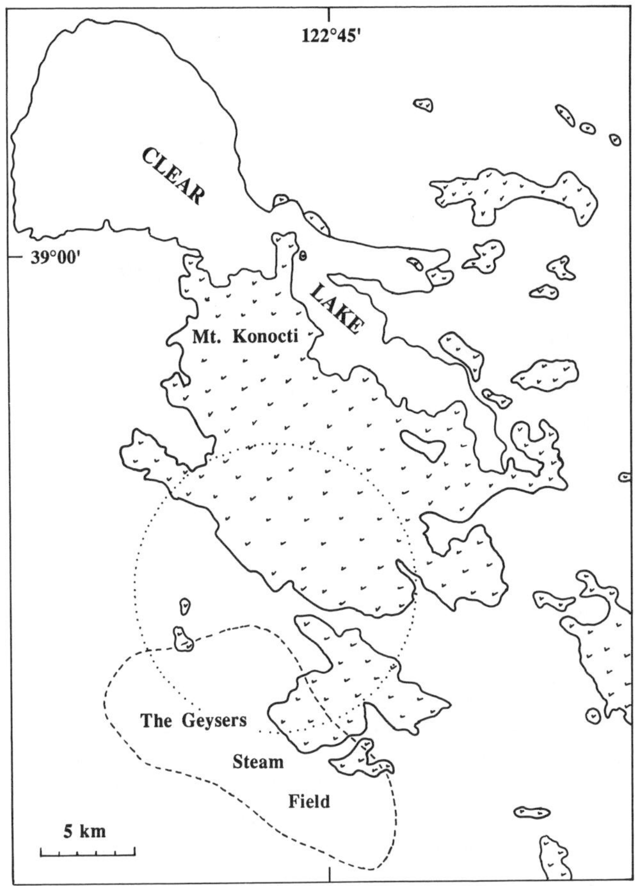

CLEAR LAKE: Map showing distribution of the upper Pliocene to Holocene Clear Lake volcanics (pattern), major faults, and associated thermal features and ore deposits. Dotted circle is vertical projection of inferred magma body.

0.2 Ma were each characterized by different compositional ranges. Total erupted volume probably exceeds 70 km^3. Volcanism has been largely non-explosive with only one major silicic air-fall tuff and no ash-flow tuffs. Numerous young locally vented deposits of palagonitic mafic tuff occur around the southeast shore of Clear Lake.

Gravity and teleseismic studies suggest that a large silicic magma chamber, ~14 km in diameter, lies 7 km and deeper beneath the volcanic field. This reservoir is thought to be the heat source for the Geysers geothermal field (on the southwest side of the volcanic field), which is the largest producing geothermal field in the world, with installed electrical generating capacity of ~2,000 megawatts in 1988, enough electricity for about two cities the size of San Francisco. Numerous thermal springs occur along

CLEAR LAKE: Air photo of Clear Lake. Top of photograph is oriented ~N40°W. Mount Konocti, largest edifice of the Clear Lake volcanic field, lies just south of the center, and the northern part of the volcanic field lies in the lower half of the photograph.

northwest to north-northwest-trending faults that are subparallel to the main San Andreas fault. Associated epithermal deposits of mercury and gold include the Sulphur Bank mine (still the site of active mercury deposition) and the McLaughlin mine (a major disseminated gold deposit in an outlier of the Clear Lake volcanics), and associated hot springs deposits.

Many Clear Lake lavas appear to be contaminated by crustal materials. Basaltic andesite and andesite often contain xenocrystic quartz up to several centimeters across and known locally as "Lake County Diamonds." The dacite and rhyolite typically contain mafic magmatic inclusions, most often andesitic in composition. The dacite in particular exhibits signs of magma mixing, often containing in a single hand specimen quartz, both sodic and calcic plagioclase, sanidine, biotite, hornblende, orthopyroxene, clinopyroxene, and forsteritic olivine. Strontium isotope data show that phenocrysts in rhyolite and dacite are commonly out of equilibrium with their host matrix. Many lavas also typically have high MgO contents, perhaps reflecting contamination by the abundant serpentinite underlying the area.

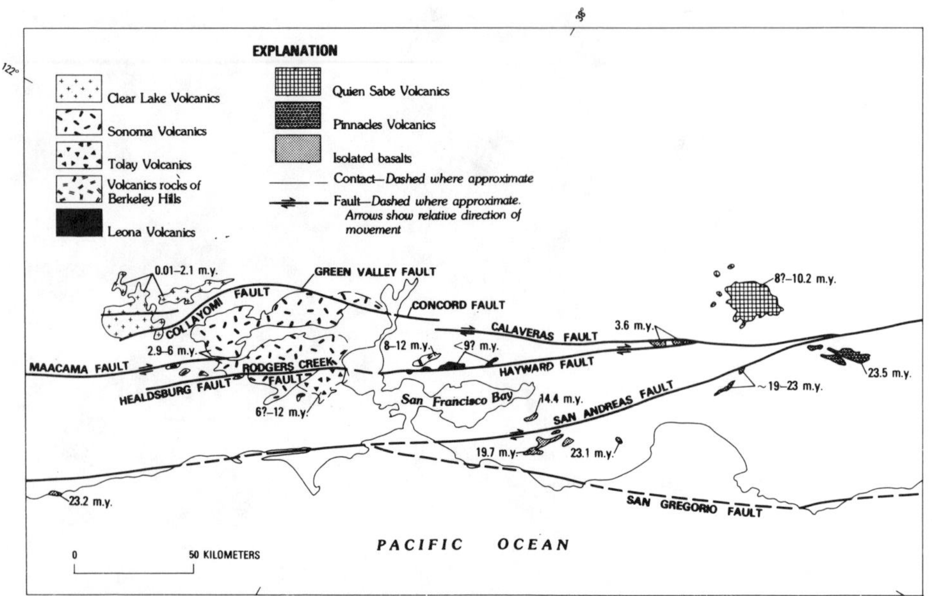

SONOMA: Map showing relationship of Sonoma volcanics to Clear Lake volcanics and other Cenozoic volcanic fields in the Central California Coast Ranges. From Hearn et al., 1981.

How to get there: *Clear Lake is a popular recreation center with resorts, camping, and boating facilities. State Highway 20 skirts the east side of the lake, and numerous smaller roads provide access to volcanic features.*

References

Donnelly-Nolan, J. M., Hearn, B. C., Jr., Curtis, G. H., and Drake, R. E., 1981, Geochronology and evolution of the Clear Lake Volcanics: *USGS Prof. Pap. 1141*, 47-60.

Hearn, B. C., Jr., Donnelly-Nolan, J. M., and Goff, F. E., 1981, The Clear Lake Volcanics: Tectonic setting and magma sources: *USGS Prof. Pap. 1141*, 25-46.

Julie M. Donnelly-Nolan

SONOMA
California

Type: *Lava and ash-flow field*
Lat/Long: *38.42°N, 122.48°W*
Field Area: *~250 km^2*
Eruptive History: *8.9 – 2.9 Ma*
Composition: *Andesite, dacite, and rhyolite*

The Sonoma volcanics lie amid a system of faults east of and paralleling the San Andreas. The field stretches ~90 km, overlapping the Clear Lake volcanics to the north-northwest in latitude and age. As in the Clear Lake field there is a northward younging of the volcanics, with the southern portion of the field being dominantly late Miocene andesite and the northern area being Pliocene dacite and rhyolite. Lava flows form three linear mountain ranges in the south: the Sonoma Mountains, and Mounts **Veeder** and **George**. In contrast, the younger, more northerly portion of the Sonoma volcanics is largely a pile of welded to partially welded ash flows, in some places capped by basaltic lava flows. There are no known caldera vents for the ash flows. Vents are generally rare in the entire field, except for silicic domes.

The Sonoma volcanics are both faulted and folded, with an estimated 85 km of right lateral displacement having occurred during the last 8 million years. Kenneth Fox suggests that the Sonoma volcanics, like the Berkeley Hills to the south and Clear Lake to the north, were not subduction driven, but rather represent magmas that welled up behind the trailing edge of the Juan de Fuca plate as it moved northward.

How to get there: *Many geologists have probably traveled through the Sonoma field unwittingly, for the volcanics occur above the distracting vineyards and wine-tasting rooms of the Napa and Sonoma valleys. Although the morphology of the volcanic rocks is muted by weathering, various roads provide good access.*

Reference

Fox, K. E., 1983, Tectonic setting of Late Miocene, Pliocene, and Pleistocene rocks in part of the Coast Ranges north of San Francisco, California: *USGS Prof. Pap. 1239*, 33 pp.

Charles A. Wood

AURORA-BODIE

California-Nevada

Type: *Monogenetic volcanic field*
Lat./Long.: *38.2°N, 118.8°W*
Eruptive History: *15 to 8 Ma, 4 Ma to Holocene*
Composition: *Basalt to andesite*

The Aurora-Bodie volcanic field occurs east-northeast of Mono Lake, between the Sierra Nevada and the Great Basin. Calc-alkaline andesite, dacite and trachyandesite lavas, breccias and ashflow tuffs, dated 15 to 8 Ma, underlie a tighter concentration of younger alkaline-calcic cinder cones and flows. The older volcanics cover ~80 km^2 with a volume of ~35 km^3.

Andesite domes and flows, 4.5 to 2 Ma in age, occur at **Cedar Hill** and other areas in the field, and Pleistocene to late Holocene basaltic rocks forming well-preserved cinder cones and flows cover ~100 km^2. The probable late Pleistocene age **Mud Springs** volcano consists of a steep-fronted bulbous flow surrounding a depressed vent area, and a 7-km-long ridged flow, together creating a remarkably distinctive landform. Although trees cover the flows, partially accounting for their dark color, the volcano is clearly one of the freshest in the Aurora–Bodie field.

Aurora Crater, ~12 km west of Mud Springs volcano, is a 1.8-km-wide breached crater of ~0.25 Ma totally surrounded by lava flows, with an estimated total volume of 2 km^3. Studies of basalt in this area suggest average denudation rates of ~0.1 cm/1,000 yr.

Many of the volcanoes are cut by faults, and Pleistocene basalt has been warped as well. The topography of the entire area has been softened by ash probably erupted by the younger Mono Craters to the southwest. Gold and silver found in quartz veins in the Miocene (but not younger) volcanic rocks were mined until ~1950, and only the ghost towns of Bodie, California, and Aurora, Nevada, remain.

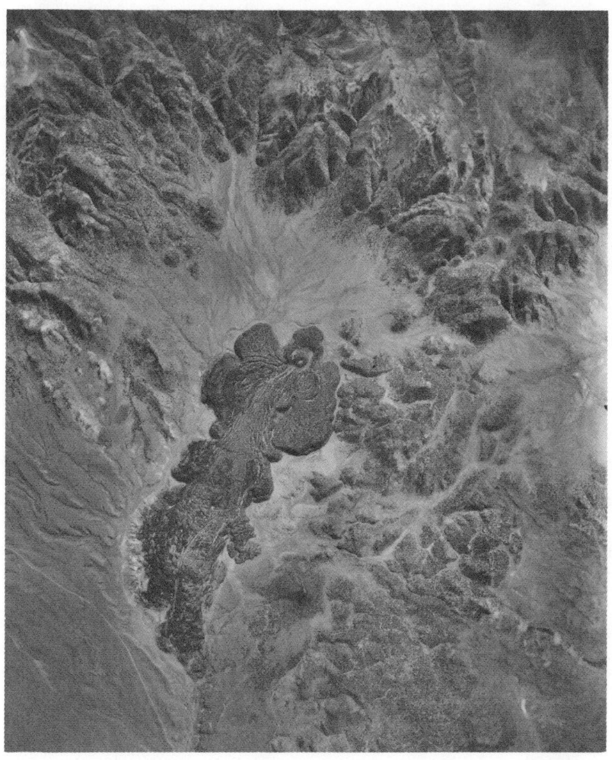

AURORA-BODIE: Mud Springs volcano, south up.

AURORA-BODIE: Aurora Crater; southeast up. USAF/USGS U-2 photograph 018V-047.

How to get there: *California Highway 167, which passes north of Mono Lake and leads to Nevada, is south of the Aurora–Bodie volcanic field; tertiary roads approach the main cones.*

References

Kleinhampl, F. J., et al., 1975, Aeromagnetic and limited gravity studies and generalized geology of the Bodie Hills region, Nevada and California: *USGS Bull. 1384*, 38 pp.

Chesterman, C. W., 1968, Volcanic geology of the Bodie Hills, Mono County, California: *in* Studies in Volcanology – A memoir in honor of Howel Williams, *Geol. Soc. Am. Mem. 116*, 45-68.

Charles A. Wood

ADOBE HILLS
California–Nevada

Type: *Lava flow field*
Lat/Long: *38.05°N, 118.55°W*
Length: *60 km*
Width: *10–25 km*
Elevation: *1,980 to 2,705 m*
Eruptive History: *4.5 to 2.6 Ma*
Composition: *Predominantly olivine basalt*

The Adobe Hills volcanic area is comprised of a sheetlike mass of overlapping olivine basalt flows that covers an irregularly shaped, generally east–west-trending area of ~800 km^2. This area, which extends eastward for ~60 km from the eastern shore of Mono Lake, is roughly bisected by the California–Nevada border. The irregular sheet of basalt flows locally attains thicknesses of nearly 200 m in its central part, where multiple flows overlap, but thins to individual flows, with thicknesses of less than 10 m, along its northern and southern margins. Its highly irregular shape and subregional extent suggest an origin by numerous eruptions from a number of widely spaced centers. Subdued conical piles of cinder and scoria are widely distributed and are particularly abundant in the Adobe Hills; however, many of the flows were probably produced by fissure eruptions. Limited observations indicate that individual flows are typically extensive and thin. These flows are rich in olivine and are typically microcrystalline with an intergranular texture of labradorite, augite, olivine, and magnesite.

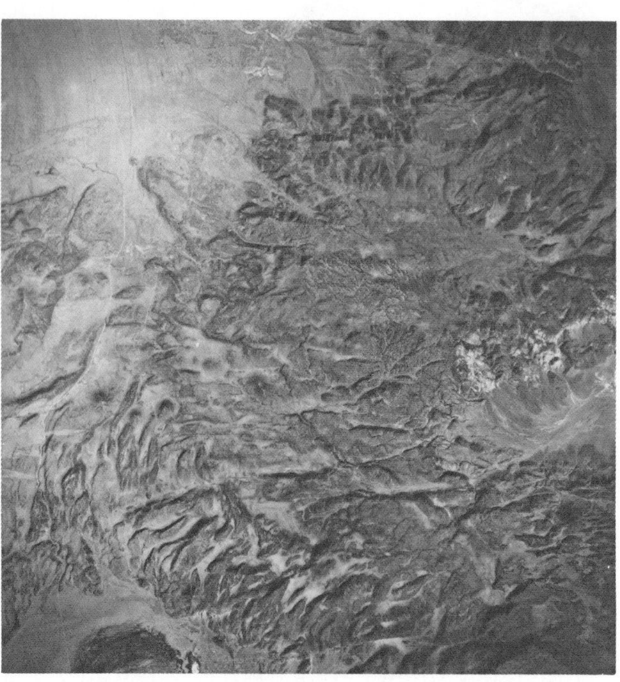

ADOBE HILLS: Vertical aerial view of the complexly faulted Pliocene basalt flows that form the western part of the Adobe Hills. USAF/USGS U-2 photograph 018V-050.

ADOBE HILLS: Oblique aerial view of the western part of the Adobe Hills. View south toward the White Mountains (background).

The lava flows of the Adobe Hills area have been pervasively deformed during late Pliocene and/or Quaternary time by many subparallel, northwest- to northeast-trending faults. The resultant topography is dominated by elongate, fault-bounded ridges and basins that largely obscure the original constructional form of the basalt sheet. A substantial part of this faulting may have been synvolcanic; locally, cinder cones are situated along faults that have displaced underlying flows.

How to get there: *The Adobe Hills area is ~20 km east of Lee Vining, California, the eastern gateway to Yosemite National Park. The area is flanked on the south by California State Highway 120 and US 6 and on the north by California State Highway 167 and Nevada State Highway 359. Access into the interior of the area, however, is limited to ungraded dirt roads and tracks that, for the most part, are passable only to vehicles designed for off-road travel.*

Reference

Gilbert, C. M., Christensen M. N., Al-Rawi, Yehya, and Lajoie, K. R., 1968, Structural and volcanic history of Mono Basin, California–Nevada: *Geol. Soc. Am. Mem.116*, 275-329.

John C. Dohrenwend

MONO CRATERS
California

Type: *Chain of silicic domes and craters*
Lat/Long: *37.88°N, 119.00°W*
Elevation: *2,325 m*
Eruptive History: *35 ka to 13th century A.D.; Mono Lake volcanism to 18th century A.D.*
Composition: *Rhyolite, basalt, and dacite for Mono Lake volcanism.*

The Mono Craters form a conspicuous 12-km-long arc of young silicic domes and explosion craters between Long Valley caldera and Mono Lake. The arc marks the eastern side of an ~18-km-wide ring-fracture system that has subsided at least 200 m since 0.7 Ma. Modeling of teleseismic data suggests that a magma chamber of 200-600 km^3 lies directly below the Mono Craters at a depth of 8-10 km.

Long Valley, the Mono arc, and the Mono Lake volcanics appear to represent three phases in the development of large mid-crustal magma chambers and associated calderas. Long Valley

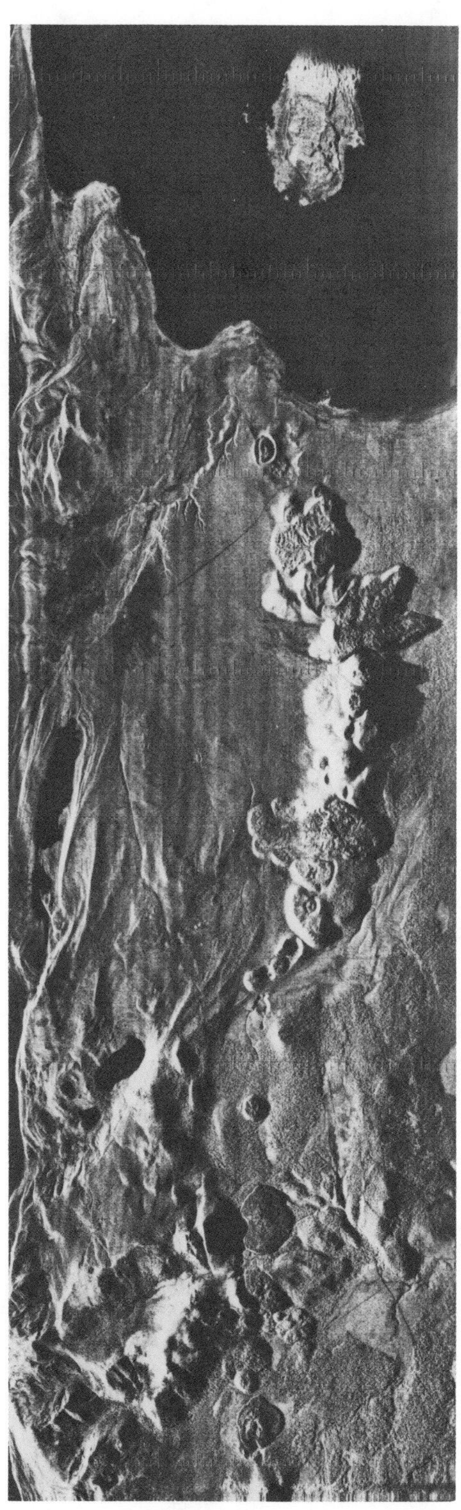

MONO CRATERS: Radar image of Mono Craters and Inyo Craters. Source of image unknown.

MONO CRATERS: Northwest Coulee (left) and Southern Coulee (right) as photographed from the west in red light of wavelengths 6400 to 7000 angstroms. Photograph from J. F. Cronin, 1967, Terrestrial Multispectral Photography: Air Force Cambridge Res. Lab. AFCRL-67-0076, 26 pp.

has reached the caldera stage, the Mono arc has accumulated enough crustal magma for downsagging to occur, and the Mono Lake volcanics are at the earliest stage of dike penetration of the crust to form isolated volcanoes.

The Mono Craters area includes, for purposes of this description, 3 large rhyolitic (obsidian) flows, 6 to 8 steep-sided rhyolite domes, a number of explosion craters, as well as the dacitic islands in Mono Lake, and Black Point, a basaltic tephra cone on the north shore of the lake. All of these were formed within the last 35,000 yr, with the majority of the Mono domes and flows being <10, 000 yr old. The volcanic islands in Mono Lake are <1,800 yr old, and dacitic lavas and ashes on the northern edge of Paohoa island are reported to be only ~200 yr old.

Panum Crater is the most spectacular of the Mono Craters, with a rim of ash and lapilli enclosing a dome of flow-banded rhyolite. Careful mapping has revealed throat-clearing breccias, pyroclastic flows, surge deposits, and rhyolite and obsidian domes. Panum Crater is ~0.6 km wide, with a rim height of 70 m.

The most recent eruptions at Mono Craters occurred in the northern portion of the field (including the Panum activity) between 1325 and 1365 A.D. The products include a series of locally extensive tephras with a total volume of ~0.2 km^3, and ~0.4 km^3 of lava domes and flows. These eruptions occurred within a year or so after the latest activity at the Inyo Crater chain to the south.

The Mono domes and flows are remarkably steep, blocky, and barren, reflecting both their glassy texture and youthful age. The domes are large – **Northern Coulee** is 0.4 km^3 – and often composed of multiple flows emanating from explosion craters or fissures. The alignment of vents for the most recent activity suggest that the domes, flows, and craters erupted in association with intrusion of a thin dike ~ 6 km long.

Black Point is a very unusual volcano, being a flat-topped pile of horizontally bedded basaltic cinders and tuff-breccia. The basaltic ash is largely palagonite – a brownish-green basaltic glass commonly found in maar rim deposits. It has been proposed that Black Point erupted under water during the late Pleistocene when Mono Lake was 72 m higher than the cone's summit. If true, the horizontal bedding would have formed by slow settling of ejecta around the vent.

How to get there: *The Mono Craters (and Inyo Craters to the south) are easily seen and explored from Highway 395, between Lee Vining and the junction with Mammoth Lake, California. A network of dirt roads provides easy access to most flows and craters, with Highway 120 closely approaching Panum Crater. The internal structure of Black Point is revealed in a quarry on its western flank.*

References

Achauer, U., Greene, L., Evans, J. R., and Iyer, H. M., 1986, Nature of the magma chamber underlying the Mono Craters area, eastern California, as determined from teleseismic travel time residuals: *J. Geophys. Res. 91*, 13,873-13,891.

Sieh, T., and Bursik, M., 1986, Most recent eruption of the Mono Craters, eastern central California: *J. Geophys. Res. 91*, 12,539-12,571.

Charles A. Wood

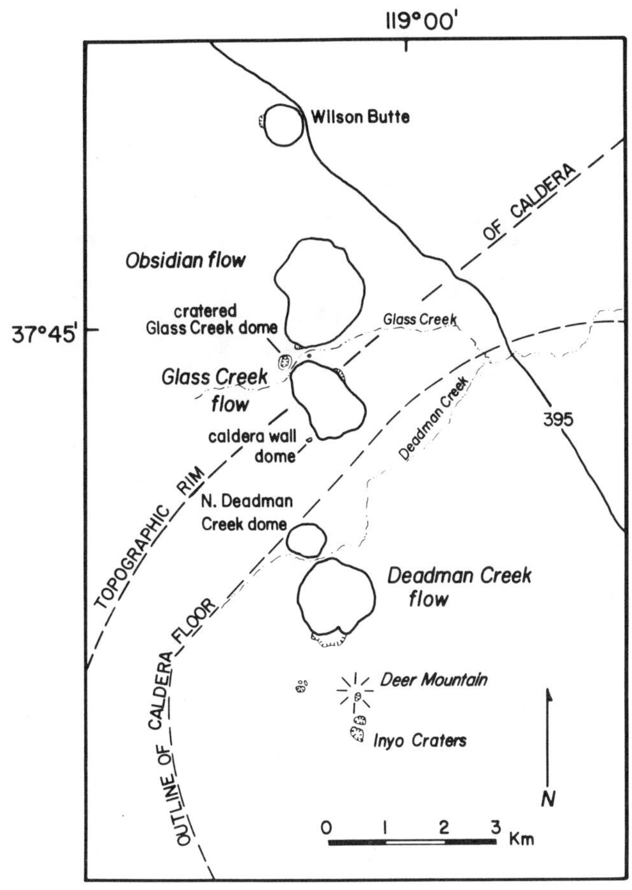

INYO CRATERS: Map showing locations of domes, flows, and phreatic craters (hachured) of the Inyo Crater chain. The rim and floor of Long Valley caldera are indicated. Map from Sampson, 1987, in GSA Spec. Pap. 212, 89-101.

*INYO CRATERS: Oblique aerial view northward across **Glass Creek** (foreground) and **Obsidian Domes,** 600-yr-old rhyolite domes. Obsidian Dome is 2 km in the north–south dimension. Photograph provided by J. Eichelberger.*

INYO CRATERS
California

Type: *Silicic dome and crater chain*
Lat/Long: *37.69°N, 119.02°W*
Elevation: *2,681 m*
Eruptive History: *6,000 yr BP to ~1350 AD*
Composition: *Rhyolite and rhyodacite*

The Inyo Crater chain of silicic domes, flows, tephra deposits, and explosion craters extends ~12 km along a north–south line from south of the Mono Craters to just inside the Long Valley caldera. Phreatic explosion pits in Mammoth Mountain extend the volcanic chain a few additional kilometers to the south. The vents for all these features were localized by north–south-trending faults and tension cracks which cut through Long Valley caldera. Though the erupted magma may somehow have been associated with the Long Valley caldera magma chamber, regional east–west extension appears to have controlled the locations of the vents.

Five large rhyolitic domes and lava flows, two or three 100-m-wide scale domes, and a few small phreatic explosion craters comprise the Inyo Craters chain. A total of ~0.8 km^3 of rhyolitic-rhyodacitic magma have been emplaced during the last 6,000 yr. Many of the Inyo domes and craters were formed ~600 yr ago when activity occurred along the entire volcanic chain. Drilling demonstrated that a 6-m-wide solid dike (at a depth of 600 m) fed magma to the northern domes, but under **South Inyo Crater**, at the south end of the 600-yr-old activity, the dike was a ~20-m-wide breccia zone. The proposed interpretation is that the southern dike rose into wet fill within Long Valley caldera, leading to formation of phreatic craters and quenching of the rising magma. Outside the caldera, in a drier environment, the dike rose to the surface yielding extrusive domes and lava flows.

Two texturally and chemically distinct rhyolitic lavas are mixed together in some of the flows – coarsely porphyritic rhyodacite and finely porphyritic obsidian – with a north-south decrease in the abundance of the rhyolitic obsidian.

How to get there: *The Inyo Craters and domes are in a more wooded area than the Mono Craters and are not as well seen from a distance. Nevertheless, numerous dirt roads branching from Highway 395 provide reasonable access. The volcanic line is entirely within the Inyo National Forest.*

References

Miller, C. D., 1985, Holocene eruptions at the Inyo volcanic chain, California: Implications for possible eruptions in Long Valley caldera: *Geology 13*, 14-17.

Eichelberger, J. C., et al., 1988, Structure and stratigraphy beneath a young phreatic vent: South Inyo crater, Long Valley caldera, California: *J. Geophys. Res. 93*, 13,208-13,220.

Charles A. Wood

LONG VALLEY
California

Type: *Silicic caldera with associated basaltic activity*
Lat/Long: *37.70°N, 118.87°W*
Elevation: *2,070 to 2,590 m*
Caldera Diameter: *17 × 32 km*
Eruptive History:
1.92 to 0.09 Ma: pre-caldera silicic domes of Glass Mountain
0.708 Ma: main eruption of Bishop Tuff; rhyolite and caldera formation
0.673 to 0.126 Ma: post-caldera flows, domes, and resurgence

*LONG VALLEY: South moat of Long Valley caldera, seen from the summit of **Mammoth** Mountain (a dome complex in the southwest corner of the caldera). The village of Mammoth Lakes sits in the left mid-ground, with the southern margin of the resurgent dome complex beyond it. The steep wall of the caldera is composed of various Paleozoic and Mesozoic metamorphic and igneous rocks. The region of seismic swarms and soil gas anomalies occupies the central portion of the photograph. The White Mountains, of the California–Nevada border, lie in the distance.*

650 yr BP: Inyo craters and domes
Composition: *Basalt to high-silica rhyolite*

The Long Valley caldera is one of only three large Quaternary calderas in the western USA that are young enough potentially to have residual magma bodies beneath them. That the caldera is still dangerous is indicated by a series of strong (four $M = 6$) earthquakes one week (May 25, 1980) after the catastrophic eruption of St. Helens. The caldera formed in an embayment of the Sierra Nevada Mountains, with its east side open toward the Basin and Range extension in Nevada.

The voluminous (500 km^3) eruption of the Bishop Tuff plinian pumice and ignimbrite at 0.7 Ma led to formation of the caldera, with resurgent dome eruptions beginning a few thousand years later. Activity continued over an extended period, to produce moat rhyolite and basalt. After a 0.12-Ma lull, more recent silicic eruptions have produced a linear chain of vents (the Inyo domes and the Mono domes and craters) to the north. Geophysical evidence suggests that a large body of magma remains beneath the western part of the caldera surface at

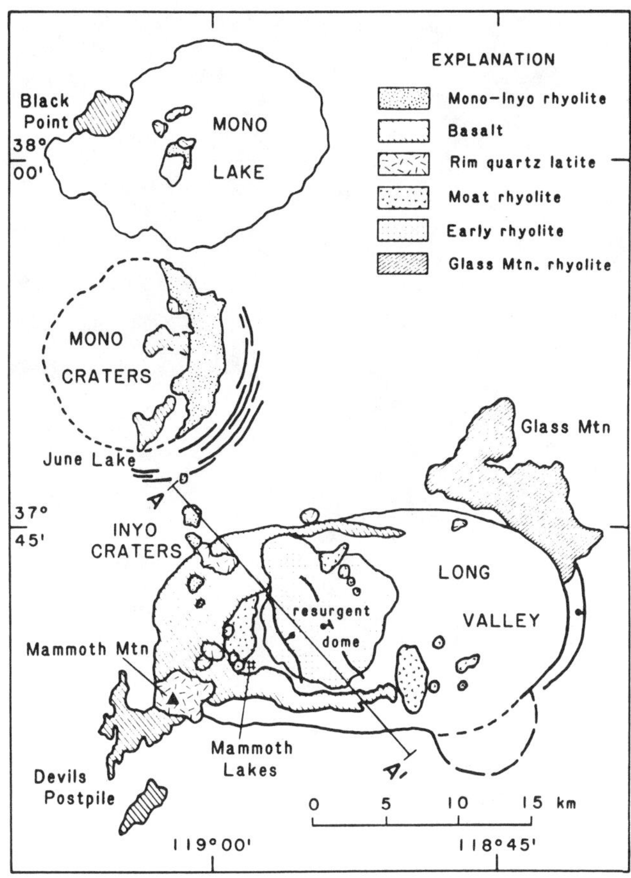

LONG VALLEY: Generalized map of Long Valley caldera and the Mono–Inyo Craters volcanic system, from Hill et al., 1985.

from 4.5 to 13 km. A major hydrothermal system, receiving inflow from the Sierra Nevada, produces numerous hot springs in the center of the caldera, which have been commercially exploited on a trial basis.

The post-1980 activity has consisted of earthquake swarms, most near the south margin of the caldera. The resurgent dome complex was uplifted by more than 44 cm between 1975 and 1985, and the south moat showed evidence of extension of several microradians per year during that same period. Changes in fumarole and soil gas chemistry have also been reported from a zone in the south moat over the site of many swarms. Deformation has been interpreted to indicate the injection of 0.1 to 0.2 km^3 of magma into the residual magma chamber, although it is now clear that the tectonic component to the activity was more important than previously recognized.

LONG VALLEY: Regional view, showing Mono Lake (top), Sierra Nevada Mountains (snow-covered on left), and Long Valley caldera (center, containing small lake). Space Shuttle photograph S17-45-73.

Earliest magmatism at Long Valley began with the extrusion of the **Glass Mountain** rhyolite, which now occupies the northeastern caldera rim. The Bishop Tuff is a crystal-rich high-silica rhyolite that contains up to 30% of phenocrysts of quartz, sanidine, plagioclase, biotite, and Fe-Ti oxides. It has been argued that the dominant differentiation mechanism in this unusual magma was diffusive, with SiO_2 varying only from 75.5 to 77.4%, while concentrations of some trace elements vary by factors of 2 to 3. Modern petrologic studies of the Bishop Tuff have fundamentally changed many notions concerning the origin of large silicic magma bodies, the mode of their emplacement into the upper crust, and their eruption.

Exposures of the entire Long Valley sequence are unusually good, and the area offers rich opportunities for field geology. The volcanic tableland to the southeast of the caldera (near the town of Bishop) is an excellent example of the original depositional form of a major outflow sheet of ignimbrite.

How to get there: *Long Valley caldera is in a reentrant into the eastern base of the Sierra Nevada range, ~265 km south of Reno, Nevada, and 110 km east of the entrance to Yosemite National Park. US 395 crosses the caldera from northwest to southeast, with the village of Mammoth Lakes occupying a position in the central western portion of the caldera. Access is facilitated throughout the region by the numerous logging roads.*

References

Hill, D. P., Bailey, R. A., and Ryall, A. S., 1985: Active tectonics and magmatic processes beneath Long Valley caldera, Eastern California: An overview: *J. Geophys. Res. 90*, 11,111-11,120.

Goldstein, N. E., and Stein, R. S., 1988, What's new at Long Valley: *J. Geophys. Res. 93*, 13,187-13,190.

Stanley N. Williams

BIG PINE
California

Type: *Bimodal volcano field*
Lat/Long: *37.05°N, 118.25°W*
Elevation: *1,300 to 1,950 m*
Eruptive History: *Sporadic since 1.2 Ma or earlier; most recent activity, latest Pleistocene*
Composition: *Basanite to olivine and silicic alkali basalt; rhyolite*

The Big Pine volcanic field is one of several late Cenozoic subalkaline-alkaline basalt fields in the southwestern USA. The field contains ≥40 vents and lies within a ~1000-km^2 region in Owens Valley, a 2–3-km-deep structural trough between the Sierra Nevada and Inyo–White ranges at the western margin of the Basin and Range province. Roughly 120 km^2 (~2.5 km^3) of lava is now exposed, and more is buried. The greatest volume was erupted from the northern margin of the field, near the boundary between the two abutting northwest–southeast grabens, now buried by alluvium, that comprise the deepest part of Owens Valley. Vents are generally localized along normal faults bounding the ranges. The largest cones contain ~0.05 to 0.15 km^3 of agglutinate and cinders and rise to heights of 220 to 260 m above their bases; the largest flows are 9 km in length and 3 km in width. The field contains a single rhyolite dome (1.0 Ma) consisting of ~0.05 km^3 of pumiceous rhyolite and obsidian.

BIG PINE: Landsat TM image of the Big Pine field, showing the dark basaltic flows and cones against the lighter-toned granitic alluvium of the Owens Valley. Image is ~35 km across; north is up.

The Big Pine field appears to have been episodically active throughout the entire Quaternary. The oldest dated lavas are 1.2-Ma ridge-capping flows and eroded cinder cones. Flows of this age underlie remnants of fan deposits, and cones are heavily eroded. Most of the exposed flows date from ~0.1 to 0.5 Ma and are lightly mantled by silt and partly alluviated; the flanks of cones are steep-sided and not gullied, and their bases are buried by alluvium. Erosion is less than for that coeval cones from Cima volcanic field, ~150 km to the south, perhaps because there was less influx of aeolian silt in Owens Valley. The youngest flows and cones, erupted only shortly before the end of the latest Pleistocene glaciation (~10-25 ka), are virtually unweathered and uneroded.

Dated lavas are too few to identify periods of exceptional activity. Two vent areas were active throughout the interval 1.2 to 0.1 Ma; others were also active repeatedly, but perhaps over a shorter interval.

Basalt of the Big Pine field is rich in granitic xenoliths derived from the upper-crustal plutons through which the lavas erupted. They also contain lherzolite and wehrlite xenoliths with metamorphic textures, evidently derived from the upper mantle.

Basalt of the Big Pine field shows no consistent trends in composition when viewed as a whole. However, successive flows from individual centers tend to become increasingly silicic with time. This trend is accompanied by modest changes in total alkalies and in the Fe/(Fe+Mg) ratio, but the sense of these changes is different for different vents. This is taken to reflect a number of processes affecting the evolving magma bodies differently.

The volcanic field has been interpreted as a product of magma plumes, rising rapidly from an upper mantle source. Initial eruptions consist of this material, poor in assimilated upper crustal rocks. Subsequent flows from the same vent are increasingly affected by differentiation of the source material and by assimilation of granitic country rocks.

Faulted and dated lava flows give estimates of vertical displacement rates for faults in Owens Valley ranging from ~0.3 to ~0.7 mm/yr over the past ~60 to 400 ka.

How to get there: *The Big Pine field is located in Owens Valley, between the towns of Big Pine and Independence. US 395, from Los Angeles, California, to Reno, Nevada, bisects the field. Many of the flows and cinder cones are accessible via paved and 4WD secondary roads. The Big Pine field is ~385 km north of Los Angeles, a 4-5-hour drive.*

References

Darrow, A. C., 1972. Origin of the basalts of the Big Pine Volcanic Field: M.S. Thesis, Univ. California, Santa Barbara, 61 pp.

Moore, J. G., and Dodge, F. C. W., 1980. Late Cenozoic Volcanic Rocks of the Southern Sierra Nevada, California: I. Geology and Petrology: *Geol Soc. Am. Bull. 91*, part II, 1995-2038.

Alan R. Gillespie

UBEHEBE
California

Type: *Maar volcanoes*
Lat/Long: *37.02°N, 117.45°W*
Elevation: *752 m*
Eruptive History: *Holocene?*
Composition: *Alkali basalt*

The Ubehebe Craters include over a dozen maar volcanoes formed during hydrovolcanic eruptions of alkali basalt through permeable fanglomerate deposits on the north side of Tin Mountain. The craters formed along a range-bounding fault that marks the western margin of Tin Mountain. The volcanic center is named after Ubehebe Crater, the largest tuff ring (0.8 km wide, 235 m deep) of the volcanic field. Initial eruptive activity at the center developed a scoria cone south of Ubehebe Crater. Subsequent activity was predominantly hydrovolcanic and produced two clusters of explosion craters and tuff rings west of, and south of, Ubehebe Crater. The southern cluster includes **Little Hebe Crater**, the second youngest crater in the field. The youngest eruptive events at the center were the episodic hydrovolcanic explosions that formed Ubehebe Crater. Ejecta from the crater covers all preexisting craters in the area. The deposits also overlie lake beds of Lake Rogers, ~4 km north of the vent. This stratigraphic position, and the lack of erosional modification of the pyroclastic surge apron, suggest the youngest activity was Holocene.

Tuff from Ubehebe Crater was deposited by air fall and pyroclastic surge processes. Cross-bedded sequences in surge deposits are well-exposed in the road cuts on the northeast side of Ubehebe Crater and in gully exposures southeast of Little Hebe. The scoria cone deposits of the earliest phase of volcanic activity at the center are exposed in cross section in the south and southwest wall of Ubehebe Crater.

How to get there: *The Ubehebe Craters are accessible via a paved road, west of Scotty's Castle, at the north end of the Death Valley National Monument. The road leads to an overlook at the west side of Ubehebe Crater.*

Reference

Crowe, B. M., and Fisher, R. V., 1973, Sedimentary structures in base-surge deposits with special reference to cross-bedding, Ubehebe Craters, Death Valley, California: *Geol. Soc. Am. Bull. 84*, 663-682.

Bruce M. Crowe

SALINE RANGE
California

SALINE RANGE: Aerial view of the pervasively faulted basalt carapace of the Saline Range. View to northeast across Saline Valley. USAF/USGS U-2 photograph 374-181.

Type: *Monogenetic volcanic field*
Lat/Long: *36.7°N, 117.8°W*
Elevation: *450 to 2,150 m*
Length: *25 km*
Width: *20 km*
Eruptive History: *4.6 Ma to 2.5 Ma*
Composition: *Alkali-olivine basalt to trachybasalt and trachyandesite*

Pliocene trachyandesite and basalt flows and agglomerate form a low broad caprock dome across ~330 km^2 of the Saline Range. More than 75 highly degraded cinder cones, many aligned along north-northeast-trending faults, mark the eruptive sites of these mafic lavas. In addition, relatively minor amounts of silicic to intermediate lava flows, ash-flow, air-fall, and water-reworked tuffs are interlayered with the mafic flows, and two small, fine-grained to pegmatitic trachyandesite plugs intrude the tuffs and lava flows. Ridge-capping lava flows in the adjacent Dry Mountains and Last Chance Range are interpreted as having been continuous with the Saline Range lavas. Some of the older flows are now perched as much as 1,400 m above the floors of the intervening basins, and the caprock-forming flows of the Saline Range have been intensively faulted along many closely spaced, northeast-trending faults. Locally, however, the youngest flows cascade down range flanks. These relations indicate that volcanism, and at least early basin-and-range faulting, were approximately contemporaneous in this area, and that Saline and Eureka Valleys are two of the youngest basins in the Basin and Range province.

How to get there: *The Saline Range separates Saline Valley (to the south) from Eureka Valley (to the north) and is located approximately midway between northern Death Valley and central Owens Valley in the southwest corner of the Great Basin. Access to Eureka and Saline valleys is afforded by a combination of paved secondary roads and graded dirt roads; however, the Saline Range itself is not accessible to motorized vehicles.*

References

Elliot, G. S., Wrucke, C. T., and Nedell, S. S., 1984, K-Ar ages of late Cenozoic volcanic rocks from the northern Death Valley region, California: *Isochron West 40*, 3-7.

Ross, D. C., 1970, Pegmatitic trachyandesite plugs and associated volcanic rocks in the Saline Range–Inyo Mountains region, California: *USGS Prof. Pap. 614-D*, 29 pp.

John C. Dohrenwend

COSO
California

Type: *Monogenetic volcanic field*
Lat/Long: *36.00°N, 117.75°W*
Elevation: *730 to 2,400 m*
Eruptive History: *Two principal periods:*
4 to 2.5 Ma: 31 km^3
1 to 0.04 Ma: 2.5 km^3
Composition:
Older period: Basalt to rhyolite
Younger period: Basalt and rhyolite

The Coso volcanic field is located at the west edge of the Basin and Range province. Initiation of volcanism at Coso preceded the onset of Basin and Range crustal extension there, as expressed by normal faulting. The earlier of the two principal periods of volcanism began with the emplacement of basalt flows over a surface of little relief. Then, during the ensuing period of ~1.5 million yr, eruptive activity included chemically more evolved rocks erupted upon a faulted terrain of substantial relief. Following a 1.5-million-yr hiatus with few eruptions, a bimodal field of basalt lava flows and rhyolite lava domes and flows developed on Basin and Range terrain of essentially the same form as today's landscape. Many of the young basalt flows are intracanyon, occupying parts of the presentday drainage system.

A plot of cumulative volume of the Pleistocene lavas versus age shows a time-predictive relation consistent with eruption being controlled by Basin and Range crustal extension. This intriguing model suggests that the next eruption of basalt is due now, whereas that of rhyolite is due in ~60,000 yr. Such forecasts are speculative, but abundant north-trending normal faults cutting materials as young as Holocene, in conjunction with historically high levels of seismicity, indicate ongoing east-west crustal extension within the volcanic field.

The Coso volcanic field is best known for its Pleistocene rhyolite. Thirty-eight rhyolite domes and flows form an elongate array atop a north-trending 8 × 20-km horst of Mesozoic bedrock. Nearly uneroded constructional forms are exhibited by most domes. Many are nested within tuff-ring craters, and a few filled and overrode their craters to feed flows a kilometer or two long. The two oldest domes contain several percent phenocrysts; the rest are essentially aphyric. Obsidian is exposed locally on most extrusions, and analyses of fresh glass indicate that all of the rhyolite is of the so-called high-silica variety; SiO_2 content is essentially constant at 77%. Other major-element constituents are nearly invariant. However, trace-element contents vary and help define 7 age groups, each of unique chemical composition.

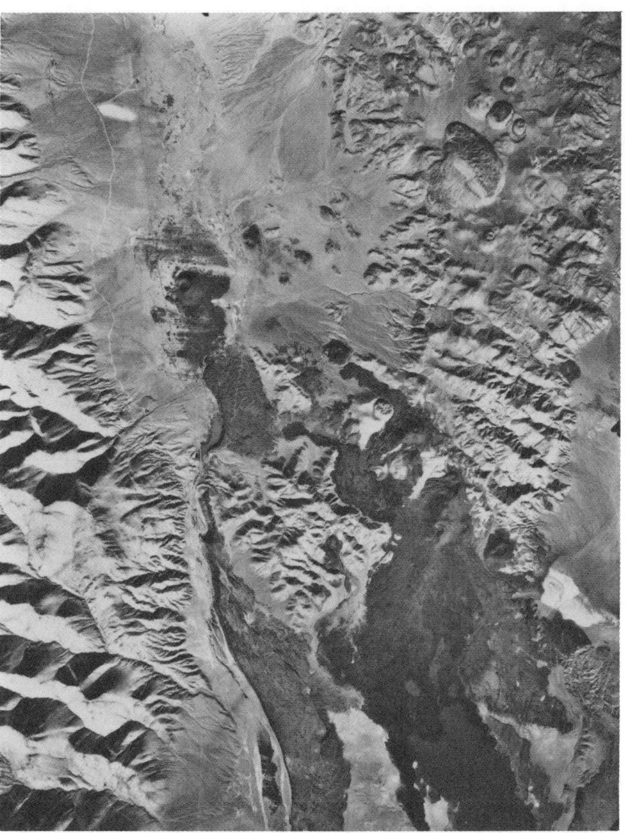

COSO: Vertical aerial photograph of the Pleistocene part of the Coso volcanic field and part of the Sierra Nevada, adjacent to the west. Highway 395 roughly follows the base of the Sierran escarpment. Basalt scoria cones and lava flows are dark and are concentrated along the south edge of a northeast-trending horst upon which rhyolite lava domes and flows occur.

The Coso volcanic field is also well known as a geothermal area. Fumaroles are present along faults bounding the rhyolite-capped horst and locally within the rhyolite field. A multidisciplinary program of geothermal assessment carried out in the 1970s defined a potential re-

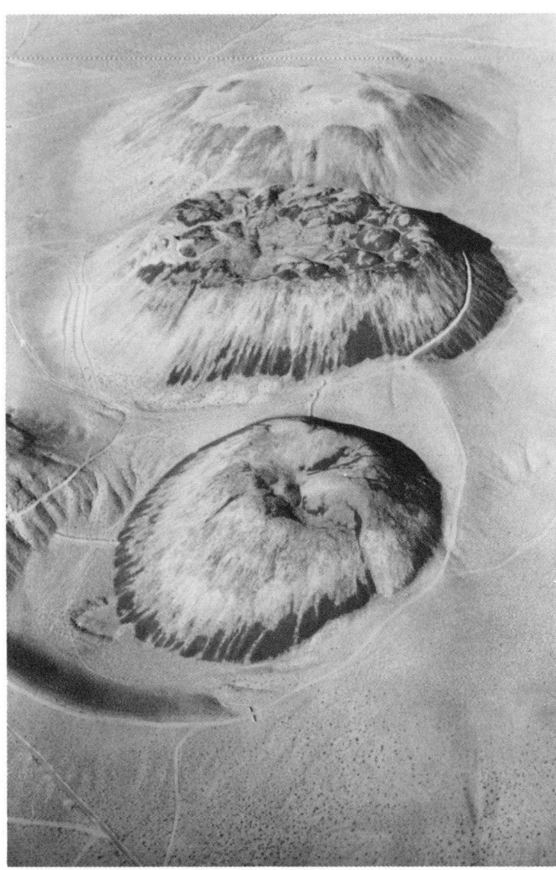

COSO: Pleistocene rhyolite domes of the Coso field, aligned northwest-southeast, one of the principal directions of faulting in the area. Dome in foreground partly fills a 1-km-wide tuff-ring.

source of 650 megawatts electric with a nominal life span of 30 yr. Judged by the youthfulness of the rhyolite lavas and by a zone of low seismic velocity crust roughly beneath the rhyolite, a magma body may be the source of thermal energy for the geothermal system. Commercial development beginning in the 1980s resulted in the startup of a geothermal steam-driven 3-MW electric power plant in 1987; considerable developmental drilling for additional power plants continues as of 1989.

How to get there: *The west side of the Coso volcanic field is crossed by Highway 395 at the village of Little Lake, ~34 km north of Inyokern, California. Most of the field, however, including all of the Pleistocene rhyolite, is a few to several kilometers to the east, within the China Lake Naval Weapons Center. Access is controlled by the Navy, and permission to visit must be arranged through officials at China Lake.*

References

Duffield, W. A., Bacon, C. R., and Dalrymple, G. B., 1980, Late Cenozoic volcanism, geochronology and structure of the Coso Range, Inyo County, California: *J. Geophys. Res. 85*, 2381-2404.

Duffield, W. A., and Roquemore, G. R., 1988, Late Cenozoic volcanism and tectonism in the Coso Range area, California: *in* Weide, D. L. (ed.), Geol. Soc. Am. Cordilleran Section, *1988 Field Trip Guidebook*, 159-176.

Wendell A. Duffield

CIMA
California

Type: *Monogenetic volcanic field*
Lat/Long: *35.25°N, 115.75°W*
Length: *35 km*
Width: *10-18 km*
Elevation: *650 to 1,509 m*
Eruptive History:
Three principal periods of activity
7.55 Ma to 6.47 Ma
5.12 Ma to 3.27 Ma
1.14 Ma to ~0.015 Ma
Composition: *Predominantly alkali-olivine basalt to hawaiite*

The Cima volcanic field is one of several late Tertiary and Quaternary, alkaline to subalkaline basalt fields in the southwestern United States. Its 52 vents and more than 65 flows collectively cover an area of ~150 km^2. More than 30 cones and associated flows of Pleistocene age are located in the southern part of the field. The mean heights of these Pleistocene cones range from 25 to 155 m, and mean diameters range from 200 to 920 m. The craters surmounting the younger cones are as deep as 45 m and as wide as 330 m. The Pleistocene flows, as much as 9.1 km long and 1.7 km wide, range from 2.5 m to as much as 30 m thick. Tephra rings and basal surge deposits are associated with at least 4 of the Pleistocene vents. Both the cones and flows display progressive morphologic changes with increasing age. The youngest cones are steep-sided and relatively uneroded, and the older

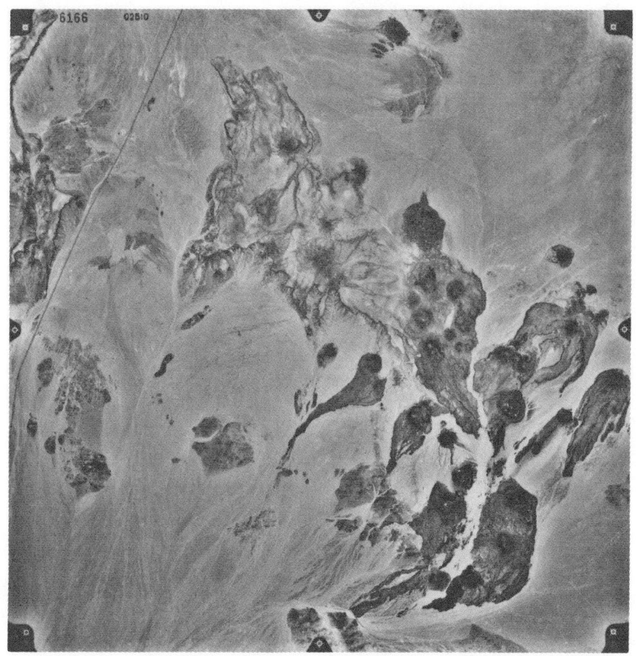

CIMA: Vertical aerial view of the Cima volcanic field (north toward upper right). Lighter, more northerly flows range in age from latest Miocene to middle Pliocene; darker, more southerly flows are Quaternary in age. USGS U-2 photograph 6166.

CIMA: 0.63-Ma cone and flow in the Cima volcanic field. The cone is a tephra ring of base surge deposits overlain by a carapace of air-fall cinders. Oblique aerial view to northeast.

cones are generally deeply dissected with exposed agglutinate layers and feeder dikes. The surfaces of the younger flows are highly irregular and little modified, whereas older flow surfaces have been partly to entirely buried beneath mantles of aeolian silt and fine sand and are relatively flat and featureless.

Among the more significant characteristics of the Cima field is the abundance of mafic and ultramafic xenoliths. These xenoliths consist of mantle spinel peridotite and websterite, olivine microgabbro, and olivine-free pyroxenite, gabbro, and microgabbro; partial melting phenomena are present in all of the medium- to coarse-grained xenoliths. These relations indicate a long history of melting, crystallization, and remelting in a progressively depressurized upper mantle.

Another noteworthy characteristic of the Cima field is the polygenetic character of individual cinder cones. At least four of the Pleistocene cones have erupted discontinuously over periods of as much as several hundred thousand years.

How to get there: *The Cima field is located in the eastern Mojave Desert, ~110 km southwest of Las Vegas, Nevada, and 105 km northeast of Barstow, California. The paved Kel–Baker Road from Baker, California, affords easy access to the field's southern margin and most of the other parts of the field are accessible via graded dirt roads.*

References

Turrin, B. D., Dohrenwend, J. C., Drake, R. E., and Curtis, G. H., 1985, Potassium-argon ages from the Cima volcanic field, eastern Mojave Desert, California: *Isochron West 44*, 9-16.

Wilshire, H. G., 1988, Geology of the Cima volcanic field, San Bernardino County, California: *in* Weide, D. L. (ed.), *Geol. Soc. Am. Cordilleran Sect. 1988 Field Trip Guidebook*, 210-213.

John C. Dohrenwend

BLACK MOUNTAIN
California

Type: *Lava-capped mesas*
Lat/Long: *35.15°N, 117.2°W*
Length: *16 km*
Width: *10 km*
Elevation: *670 to 1,200 m*
Eruptive History: *Mostly <2.55 Ma*
Composition: *Basalt*

The Black Mountain volcanic area consists of two adjacent, generally east- to northeast-

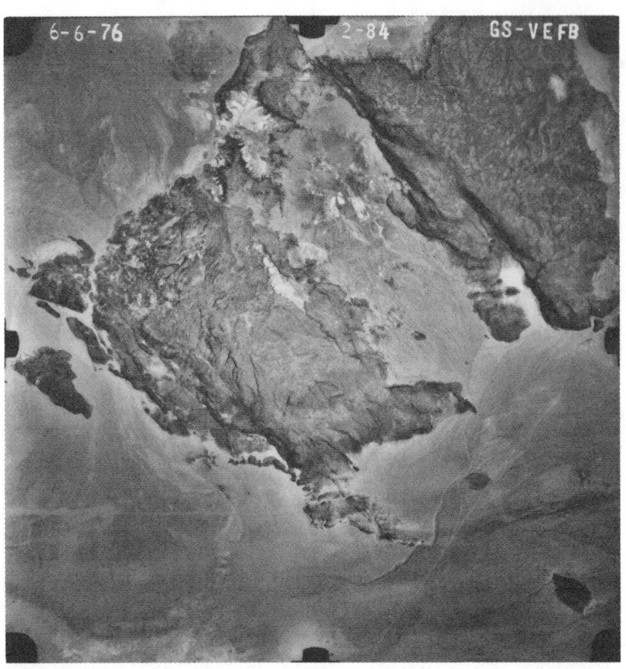

BLACK MOUNTAIN: Vertical aerial view of Black Mountain. Complex oblique slip faulting has caused extensive right-lateral offset and vertical uplift along the southwest flank of the mountain and a general eastward tilting of the late Pliocene and Quaternary(?) basalt flows that cap it. USGS aerial photograph GS-VEFB 2-84.

sloping mesas that are capped by an undetermined number of basalt flows. Each mesa is crudely rectangular and measures ~10 km long and 7 km wide. A K-Ar age of 2.55 Ma on one of the oldest flows, and a general absence of original constructional morphology on flow surfaces, indicate a latest Pliocene and early Quaternary(?) age for these flows. The volcanology and petrology of the flows have not been studied in any detail. However, the overall morphology of the Black Mountain area suggests extrusion via fissure eruptions.

The flows of the Black Mountain area have undergone a significant amount of latest Pliocene and/or Quaternary deformation. The two mesas are bounded along their southwest margins by the Blackwater and Harper faults, two of the young, northwest-trending, right-lateral fault systems that cut obliquely across the central Mojave Desert. Each mesa has been complexly and pervasively fractured by one of these northwest-trending faults. The capping flows have been uplifted from 80 to nearly 400 m, tilted to the east, and offset in a right-lateral sense by as much as 3 km.

How to get there: *Black Mountain is located ~30 km northwest of Barstow in the central Mojave Desert of California. The area can be reached from Barstow by traveling 15 km west on California State Highway 58 and then 25 km north on Hinkley Road (a paved and graded dirt road). From Hinkley Road, ungraded dirt roads and jeep trails afford access to the periphery of the mountain.*

Reference

Burke, D. B., Hillhouse, J. W., McKee, E. H., Miller, S. T., and Morton, J. L., 1982, Cenozoic rocks in the Barstow Basin area of southern California – stratigraphic relations, radiometric ages, and paleomagnetism: *USGS Bull. 1529-E*, 16 pp.

John C. Dohrenwend

PISGAH
California

Type: *Monogenetic volcano field*
Lat/Long: *34.75°N, 115.38°W*
Elevation: *550 to 774 m*
Eruptive History: *Recent*
Composition: *Olivine basalt*

Pisgah lava field in the Eastern Mojave Desert, California, is one of many small Cenozoic basaltic volcanoes within the Basin and Range physiographic province. Lava erupted onto alluvial fan and playa lake deposits from vents in the vicinity of Pisgah Crater. The lava field is a composite of numerous thin flows, and extends ~18 km west of the vent and 8 km to the southeast. It is 6.4 km across at its widest point and covers 100 km^2. Three eruption episodes can be recognized on the basis of variations in the size of plagioclase and olivine phenocrysts, which increase with decreasing age of eruption. The groundmass content for all three eruptive phases is similar, consisting of olivine, plagioclase, titanaugite, magnetite, and ilmenite.

Pisgah lava field is predominantly pahoehoe with the exception of the eastern Phase 2, intermediate-age lavas, which are aa. A dominant mesoscale texture, at spatial scales of tens of meters and with several meters of relief, is present within all three units in the form of tumuli, pressure ridges and pressure plateaus. Because of the preservation of fresh flow features across most of the field, activity at Pisgah is considered to be Recent.

Pisgah Crater is the youngest vent and is composed of cinders and small bombs interbedded with a few layers of agglutinate. The cinder cone is the most prominent feature on the field, being 98 m high and 488 m wide at its base. The cone is currently being mined for aggregate.

Alluvial and aeolian processes have modified the margins of the lava field. Aeolian deposits, derived from alluvium to the northwest and transported by the dominant winds in this area, are concentrated on the western side of the flow. Windblown material is deposited in low areas between tumuli and ridges and is locally ~2 m thick. Elsewhere, subsequent drainage has ponded against the lava and deposited alluvial sediments that bury the edge of the lava field.

Because of the variety of geologic features and volcanic surface textures present, Pisgah lava field has long been used as a test site for various remote sensing instruments and techniques.

How to get there: *The Pisgah lava field is ~65 km east of Barstow, California, and 185 km southwest of Las Vegas, Nevada. Access is provided by the National Trails Highway (old US Highway 66) which parallels Interstate 40 north of the lava flow. Exits from I-40 are at Ludlow and Hector. The southern part of the lava flow is on Twenty-Nine Palms Marine Corps Air-Ground Combat Center and is inaccessible.*

References

Dellwig, L. F., 1969, An evaluation of mutifrequency radar imagery of the Pisgah Crater area, California: *Mod. Geol. 1*, 65-73.

Wise, W. S., 1965, Geologic map of the Pisgah and Sunshine Cone lava fields: Earth Resources Survey Program *Tech. Lett. NASA-11*, 8 pp.

Eileen Theilig

AMBOY

California

Type: *Monogenetic volcanic field*
Lat/Long: *35.5°N, 115.8°W*
Elevation: *300 m*
Eruptive History: *Holocene ?*
Composition: *Feldspathoidal basalt*

The Amboy lava field covers ~70 km^2 and consists primarily of vesicular pahoehoe. The field is in an alluvial-filled valley between the Bullion Mountains to the southwest and the Bristol Mountains to the northeast. Within the valley, it lies between Bagdad Dry Lake to the west and Bristol Dry Lake to the east; both are playa lakes typical of the Mojave Desert.

Amboy Crater is a prominent, undissected cinder cone in the northeastern quadrant of the lava field. The volcano erupted along the northern border of Bristol Dry Lake and poured lava onto its surface, dividing it into the two present playas. The cone rises 75 m above the surrounding lava flows and is ~460 m in basal diameter. It is composed of a loose accumulation of volcanic ejecta with secondary amounts of agglutinated ejecta and flows. The ejecta include angular scoriaceous tephra plus ropy, ribbon- and almond-shaped bombs. Some lithic non-vesicular accessory basaltic ejecta are present, but included xenolithic fragments are absent.

Amboy Crater is not a single cone but is composed of at least four nearly coaxial nested cones. The outer slopes of the main cone are gullied by erosion. Within the main outer cone, there is a remnant of a second cone on the west side; both cones are breached on the west. In addition to the two main cones, there are two relatively undisturbed cone walls within the main crater. These innermost cones are composed almost entirely of angular scoriaceous cinders.

Most of the Amboy lava field is composed of undifferentiated flow units of relatively dense, "degassed" pahoehoe lavas that form a hummocky terrain. The surface relief on this unit ranges from 2 to 5 m. The flow is characterized by abundant tumuli and pressure ridges and, as is typical for this type of flow, a fractured surface. Lava tubes are not present in any of the flows, nor are blisters or shelly-type pahoehoe;

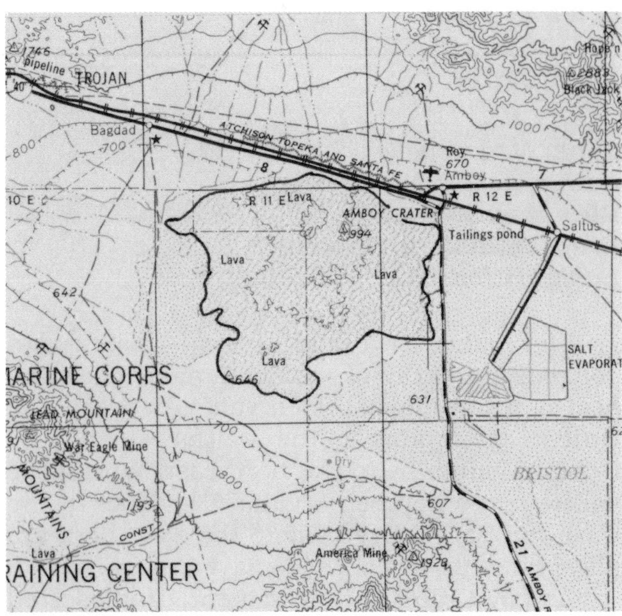

AMBOY: USGS topographic map with Amboy lava field outlined.

AMBOY: Aerial photograph showing part of the lava field. North is to the top; old US Highway 66 is in the upper right corner. A wind streak of dark cindery material extends to the southeast of the cone.

only a few lava channels are present. Low-lying areas on the flow are filled with windblown sediments which range from a few centimeters to >1 m thick. The thickness of aeolian deposits increases toward the southwest part of the field. Sand-blasting is prevalent over the entire flow, and wind-faceted pebbles of basalt are common.

The oldest flows occur in the eastern and southeastern part of the field. They are characterized by numerous collapse depressions up to ~10 m in diameter and several meters deep. Although the name would imply that "collapse" depressions formed by the collapse of a crust over fluid lava, at least some depressions may form as a result of inflation of an emplaced, but still plastic, crust by molten lava around a general void in the flow.

Two other types of flows occur in the field: platform units and vent lavas. Platform units are isolated areas of the flow that have relatively flat surfaces. These areas appear to represent stagnant parts of basalt flows that solidified in place with relatively little lateral movement after emplacement. The surfaces of the platform units consist of a layer of fist-sized cobbles and smaller fragments of basalt weathered in situ and windblown sediments. Vent lavas appear to be situated over some of the vents for the Amboy lava field. They are topographically raised areas ~10 m high and consist of relatively dense pahoehoe. The flows appear to have cooled in stagnant lava ponds which either drained back down the vent, causing slight crustal collapse on the surface, or broke out through subsequent flows around the edges of the pond. The surfaces are relatively flat and unfractured except around the edges of the pond. Individual collapse craters, which could represent drainback in the vents, occur in the middle of the ponds. Like the platform areas, the surfaces of the vent lavas have weathered basalt fragments and aeolian particles to a depth of several centimeters.

How to get there: *Amboy lava field is readily accessible west of Amboy, California, on Old US Highway 66, ~5 km northeast of Palm Springs.*

References

Greeley, R., and Iversen, J. D., 1978, Field guide to Amboy Lava Flow, San Bernardino County, California: *in* Greeley, R., Womer, M. B., Papson, R. P., and Spudis, P. D. (eds.), *Aeolian Features of Southern California: A Comparative Planetary Geology Guidebook*, 23-52, NASA, Washington, DC.

Parker, R. B., 1963, Recent volcanism at Amboy Crater, San Bernardino County, California: *Calif. Div. Mines Geol. Spec. Rept. 76,*

Ronald Greeley

SALTON BUTTES

California

Type: *Dome field*
Lat/Long: *33.20°N, 115.62°W*
Elevation: *-72 to -40 m*
Eruptive History: *Pleistocene*
Composition: *Rhyolite*

The Salton Buttes comprise five small rhyolite domes extruded onto Quaternary sediments of the Colorado River delta. **Rock Hill** and **Mullet Island** are simple domes; Mullet Island is notable for its symmetrical "onion-skin" pattern of foliation, attributed to endogenous growth. Obsidian Butte consists of a central dome surrounded on all sides by a single flow. **Red Island** is made up of two domes, each mantled by subaqueous pyroclastic deposits. Xenoliths of basalt, partly melted granite, deltaic sediments, and their hydrothermally metamorphosed equivalents are common in the rhyolites of **Obsidian Butte** and Red Island. All the domes exhibit wave-cut benches carved during various stands of pre-historic Lake Cabuilla, and have been partly buried by lacustrine and aeolian deposits.

The Salton Buttes lie within the Salton Sea Geothermal Field, where temperatures at 1.5 to 2.5 km reach 360°C, and sediments of the Colorado River delta are being metamorphosed to greenschist facies. Both rhyolitic and basaltic intrusive rocks have been encountered by geothermal wells. A 5 × 8-km magnetic high beneath the Salton Buttes appears to represent either a batholith or a large dike swarm at depth. The dome field, the intrusive rocks, and the geothermal system are all manifestations of a spreading center beneath the sediments of the Colorado River delta, as part of the leaky transform fault that is transitional from the Gulf of California to the San Andreas fault system.

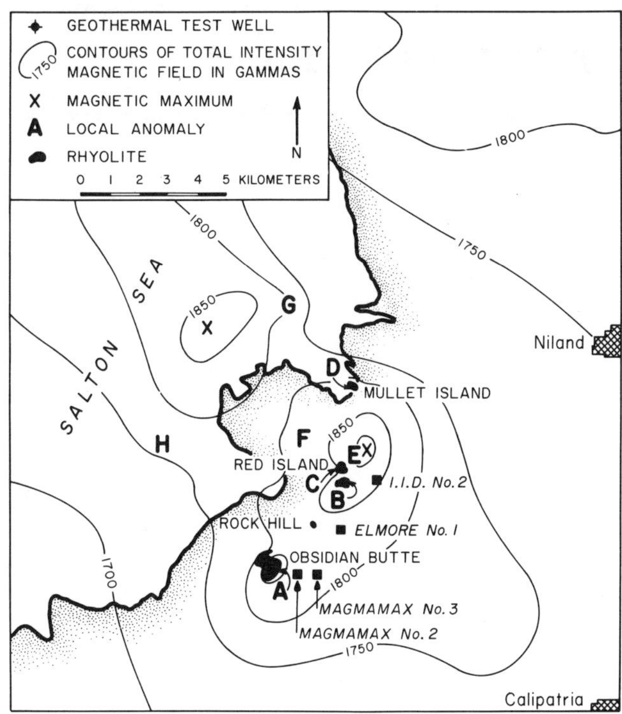

SALTON BUTTES: Sketch map of volcanic landforms south of Salton Sea.

The Salton Buttes are the youngest extrusions of a bimodal basalt-rhyolite system that probably existed throughout the Quaternary. Extrusions older than latest Pleistocene, however, are now buried by sediments of the Colorado River delta.

How to get there: *The Salton Buttes are at the southeast end of the Salton Sea, in the Imperial Valley of southern California, ~165 km ENE of San Diego. Obsidian Butte, Red Island, and Rock Hill are accessible by road; Mullet Island can be reached only by boat.*

References

Griscom, A., and Muffler, L. J. P., 1971, Aeromagnetic map and interpretation of the Salton Sea geothermal area, California: *USGS Geophys. Invest. Map GP-754*, scale 1:62,500.

Robinson, P. T., Elders, W. A., and Muffler, L. J. P., 1976, Quaternary volcanism in the Salton Sea geothermal field, Imperial Valley, California: *Geol. Soc. Am. Bull. 87*, 347-360.

L. J. Patrick Muffler

VOLCANOES OF IDAHO

Snake River Plain, 246
Craters of the Moon, 248
King's Bowl, 250
Menan Buttes, 251
Split Butte, 252
Wapi, 254

SNAKE RIVER PLAIN
Idaho

Type: *Volcanic plain*
Lat/Long: *42.7°N, 113.5°W (center)*
Elevation: *700 to 2,300 m*
Eruptive History: *~Mid-Miocene to ~2,000 yr BP; generally shifting from earliest in the west to recent activity on the Yellowstone Plateau*
Composition: *Bimodal rhyolite-basalt*

The Snake River Plain forms a broad arch across the southern part of Idaho extending 600 km eastward from the Oregon border to the Yellowstone Plateau. Its width ranges from 65 to 100 km. The western part is a complex graben bounded by a system of normal faults. Structure in the eastern part is less certain, but may involve both downwarping and faulting. To the north the plain cuts Mesozoic–Tertiary plutonic rocks and folded Paleozoic–Mesozoic rocks. To the south, the plain is bounded by basin-and-range fault-block mountains and Tertiary rhyolitic and basaltic rocks. There is no evidence that the Snake River Plain existed as a structural feature prior to the Miocene.

Eruptions of high-volume, bimodal rhyolite and basalt began in the middle Miocene in the southwest region, and activity shifted east-northeast where it is now focused on the Yellowstone Plateau. The present surface of the Snake River Plain is dominated by basaltic lava flows as recent as ~2 ka BP. Although a thin veneer of loess and windblown sand covers parts of the plain, most of the primary surface features are preserved.

The Snake River Plain represents a style of volcanism between flood basalt eruptions and Hawaiian volcanism. Like Hawaiian volcanism, plains volcanism involves multiple, thin (3-5-m) flow units erupted from central vents, and minor fountaining to produce cinder cones. However, like flood basalt eruptions, the vents are often aligned on rift zones, and some of the flows are fissure fed. The surface of flow accumulation is planar, because the vents are spread over a wide area, not focused in a central zone. Typical of plains volcanism, most flows on the Snake River Plain accumulate as (1) small, low shields, (2) fissure flows, and (3) or large tube-fed flows. All were probably emplaced relatively

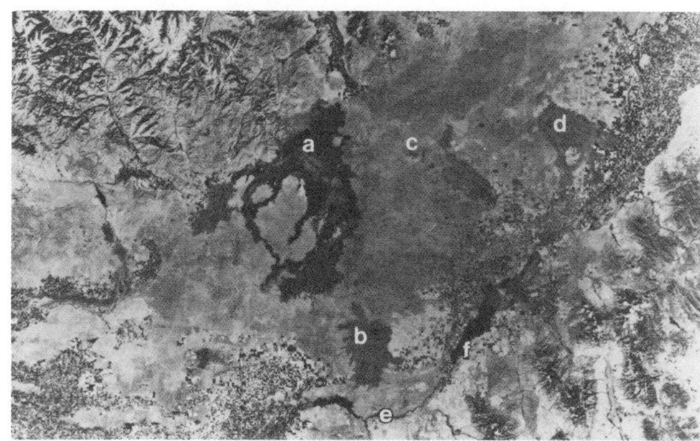

SNAKE RIVER PLAIN: Landsat images of the eastern Snake River Plain: (a) Craters of the Moon, (b) Wapi lava field, (c) Big Southern Butte, (d) ***Hells Half Acre*** *lava field, (e) Snake River Plain River, and (f) town of American Falls and reservoir to northeast.*

SNAKE RIVER PLAIN: Oblique aerial view northeastward across the Inferno Chasm rift zone showing various small vents (Inferno Chasm is out of view to the right).

slowly, often advancing only a few meters per hour, forming "toey" lava flows with hummocky surfaces of several meters relief. Pressure ridges and collapse craters are common.

Low shields are low-profile volcanoes having slopes of ~0.5° and diameters of up to 15 km. In the Snake River Plain the summits of low shields commonly steepen and have one or more craters. Low shields are often aligned on rift zones, with the shields along each rift being roughly of the same age. The shields (typified by the Wapi lava field) are composed of numerous thin flow units, some of which are fed by small lava tubes and channels. It appears that once the shields reach a maximum size, effusion ceases and/or shifts to other vent positions, leading to multiple, coalescing low shields.

Fissure flows are commonly associated with rift zones and consist of sheets of lava erupted from linear vents. Although fissure flows can cover dozens of square kilometers (as in the Craters of the Moon area), their areal extent and total volume are less than a hundredth or a thousandth as large as flood basalt eruptions. Moreover, eruption at point-sources along the fissure may produce localized spatter and cinder cones, both of which are rare or absent in flood basalt provinces.

Large lava tube flows play an important role in maintaining the relatively flat surface of basaltic plains by filling the areas between low shields. Lava tube systems longer than 20 km are common in the eastern Snake River Plain, where perhaps more than a fifth of the flows contain lava tubes and channels. This type of flow is long, narrow, somewhat sinuous, and emplaced by tubes typically >10 m across. Although the sources for the tubes are not always obvious, the flows mainly come from low shields and fissures.

Several types of craters occur in the Snake River Plain, including calderas, pit craters, maar craters, and small collapse depressions. Calderas are rare and tend to be small, except for the Island Park caldera on the northeastern boundary of the plain. Pit craters on the Snake River Plain are generally <1 km in diameter and are commonly found at the summits of low shields. Maar craters and tuff rings occur in several parts of the plain. The largest, Menan Buttes, may have formed by interaction of magma with surface water, possibly the ancestral Snake River. Others, such as **Sand Crater** and Split Butte, are young maar craters that apparently have involved groundwater. They consist of tephra deposits around central craters 500-

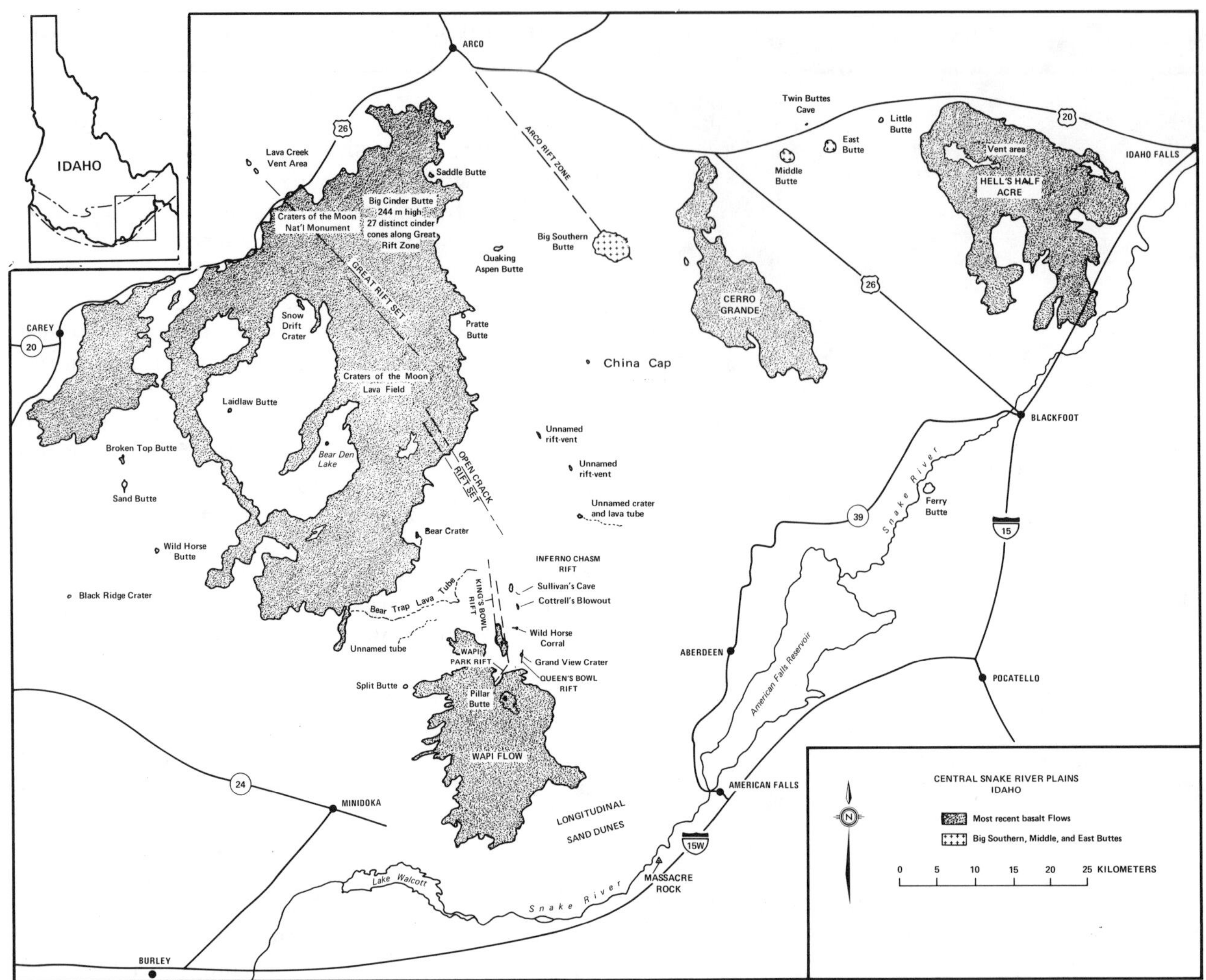

SNAKE RIVER PLAIN: Schematic map of most recent basalt flows (stipple) and silicic buttes (dot pattern) in the eastern Snake River Plain.

800 m in diameter that developed in lava flows. They appear to have formed by an early phreatomagmatic phase followed by an effusive phase that emplaced a lava lake, which then subsided to form the craters.

Rhyolitic hills occur in several places in the eastern Snake River Plain. One of the most prominent is **Big Southern Butte**. It rises 760 m above the plain, is 6.5 km across, and comprises 2 coalesced cumulo domes of 0.3 Ma rhyolite and an elevated section of older basalt flows.

How to get there: *The Snake River Plain is traversed by Interstate Highways 15, 86, and 84, plus a relatively good network of US, state, and county roads.*

References

Baker, V. R., Greeley, R., Komar, P. D., Swanson, D. A., and Waitt, R. B., Jr., 1987, Columbia and Snake River Plains: *in* Graf, W. L. (ed.), *Geomorphic Systems of North America*, Boulder, Colorado, *Geol. Soc. Am., Centennial Spec. Vol. 2*, 403-468.

Greeley, R., 1982, The Snake River Plain, Idaho: Representative of a new category of volcanism: *J. Geophys. Res. 87*, 2,705-2,712.

Ronald Greeley

CRATERS OF THE MOON

Idaho

Type: *Monogenetic volcanic field*
Lat/Long: *43.2°N, 113.5°W*

CRATERS OF THE MOON: Low oblique aerial view of Big Craters area. Parking area and trails (white lines) allow inspection of numerous spatter cones. The cones align along a fracture with some flows covering older cinder cones (left center). Inferno Cone (upper right) is composed of cinders and has no summit crater.

Elevation: *1,630 to 1,810 m*
Eruptive History: *8 eruptive periods from 15,000 to ~2,000 yr BP*
Composition: *Olivine basalt*

Craters of the Moon lava field lies along the northern border of the Snake River Plain, midway between Arco and Carey, Idaho. It consists of Holocene to Pleistocene lava flows, cinder cones, spatter cones, lava tubes, and other features typical of basaltic volcanism. Much of the field lies within the Crater of the Moon National Monument, administered by the National Park Service.

The lava field covers 1,600 km^2 and consists of more than 60 mappable flows, ~25 cinder cones, and 8 fissures/fissure systems. Detailed mapping and evaluation of dated flows have allowed USGS scientists to determine that the magma output rate was constant at ~1.5 $km^3/10^3$ yr during 15,000-7,000 yr BP, then increased to 2.8 $km^3/10^3$ yr from 7,000-2,000 yr BP.

The primary fissure system, called the Great Rift, passes through the Monument as a set of en echelon fissures which strikes N35°W and occurs in a zone up to 3 km wide. Many of the flows and cinder cones have been described in detail. Some of the more prominent and easily visited features are mentioned here.

Big Cinder Butte – the highest cone within the monument – has a total volume of 2×10^8 m^3 and rises 250 m above the plain. It has no summit crater. Many bombs are found on the cone, including cow dung, breadcrust, and spindle types. The rims of five or six older volcanoes are visible around its base.

Several flows issued from Big Cinder Butte. The youngest originated from a fissure on the north side of the cone and is a "Blue Dragon"-type pahoehoe flow. This type of flow is characterized by a shining iridescent surface.

Radiocarbon dates for many of the eruptions have been obtained by tunneling beneath the flows to obtain charcoalized rootlets from preflow soils.

Sunset and **Grassy** cones rise ~160 m above the surrounding plain. Three flows issue from Sunset cone but are largely blanketed with ash and soil. Two pahoehoe flows breached the crater in the northeast and in the west. On the northwest flank of the crater, an aa flow ~6 m thick issued from a small parasitic cone. Both cones represent some of the earliest eruptive activity in the Monument. Two craters on the north side of Grassy Cone erupted lava simultaneously to produce flows which merged and extended northward. A network of lava tubes, some as large as 5 m in diameter, developed within the flows.

The **North Crater** flow possesses features termed "monoliths" which appear to be fragments of cinder cones broken apart and rafted to their present positions by flowing lava. They range in length from ~0.5 to 200 m long; some are up to 25 m high. The massive Highway flow issued northward from the North Crater vent and was confined to the valley between Sunset and Grassy Cones. Its surface is braided with channels 3-6 m deep, and its flow margins range from 3 to 15 m high. Flow units include pahoehoe, aa, and block flows. A fault near the highway forms a distinct scarp 5-15 m high which cuts the Highway flow, exposing it in cross section. The fault scarp faces North Crater and must have formed soon after the emplacement of the Highway flow, because pahoehoe lava from within the flow is draped over the scarp. The fault may have resulted from post-eruption subsidence.

Big Craters is a series of spatter cones and open fissure flows. The slope of each cone is 50-60° near the summit and decreases to ~30° near the base. The upper slopes are angular and rough, but the lower slopes are smoother and have profiles similar to those of cinder cones. The cones range in height from 12 to 20 m and have chimneylike vent areas, most of which have collapsed, displaying a rubbly interior. Toward the northwest portion of the Big Crater area, however, two vents remain intact.

Several large pahoehoe and aa flows were erupted from these vents. The northeast flow is a short pahoehoe unit that issued from an open fissure north of the spatter cones. Many of the older flow units are exposed in the walls of the fissure. The northwest flow evolved from flat pahoehoe to slab lava and eventually to a mixture of aa and pahoehoe. Part of this flow extends northward, where it is overlain by the North Crater flow. The southwest flow is an extensive aa flow mixed with small pahoehoe units which contain numerous channels and overlie the northwest flow.

The area east of **Inferno Cone** consists of a series of lava "domes." Many of the flow units contain lava tubes, and individual flow units are generally "Blue Dragon" pahoehoe. The lava tubes in this area may be part of an extensive network in a single flow. Indian Tunnel is one of the largest tubes in the Monument, reaching widths of over 15 m. Crescentric rock heaps in this area were probably used to secure tepees of the Indians who frequented the caves. Numerous other tunnels and caves are common in this area, ranging from centimeters to meters in size.

How to get there: *Craters of the Moon is 29 km southwest of Arco, Idaho, on US Alternate Route 93 in Butte and Blaine counties. Paved roads are found within the National Monument (US Dept. of Interior Park Service) open to vehicular traffic; the southern part of the Monument is a wilderness area closed to vehicles.*

References

Kuntz, M. A., Champion, D. E., Spiker, E. C., and Lefebvre, R. H., 1986, Contrasting magma types and steady-state, volume-predictable, basaltic volcanism along the Great Rift, Idaho: *Geol. Soc. Am. Bull. 97*, 579-594.

Kuntz, M. A., Champion, D. E., Lefebvre, R. H., and Covington, H. R., 1988, Geologic map of the Craters of the Moon, Kings Bowl, and Wapi Lava Fields, and the Great Rift Volcanic Rift Zone, South-Central Idaho: *USGS Misc. Inv. Series Map I-1632.*

Ronald Greeley

KING'S BOWL
Idaho

Type: *Composite flow field*
Lat/Long: *42.95°N, 113.20°W*
Elevation: *1,500 m*
Eruptive History: *2,130 yr BP*
Composition: *Basalt*

The King's Bowl Field is small and covers <2.6 km^2. It is situated on the King's Bowl Rift Set, one of several such sets which collectively make up Idaho's Great Rift, a series of tension fractures that cross cut the eastern Snake River Plain. The King's Bowl field is a composite feature made up of flows from several point sources along the Rift as well as a larger, apparently dike-fed sheet flow, which for a time was held in a lava lake. These flows locally overlap, indicating that the eruptive sequence was complex and issued from different vents at different times.

In spite of its small size, the King's Bowl field shows some remarkably diverse and fresh features associated with the eruption, including squeezeups, spatter cones and basalt mounds. These features are associated with the sheet flow which was contained for a time as a lava lake. The outline of the lava lake is approximately defined by a series of basalt mounds interpreted as remnants of a levee system. The basalt mounds are prominent relief features clearly visible at the site and are frequently separated by outflow channels. At the north end of the field the mounds are traceable into a section of lava levee.

An area of ash dominates the east side of the rift at the King's Bowl. Ash is not abundant on the west side of the rift, suggesting a prevailing westerly wind during eruption.

KING's BOWL: Aerial view. Note the tension fractures that collectively make up the Great Rift. The large crater near the middle of the flow is King's Bowl. The light area on the east side of King's Bowl is ash generated during the phreatomagmatic stage of the eruption. The lavas stretch ~6 km. USGS Photograph CYN-7B-103.

King's Bowl itself is a large ovoid crater (85 m × 30 m × 30 m deep) apparently generated by an explosive event coupled with collapse. The explosion ejected blocks of basalt dominantly to the west. There is a noticeable size variation of these blocks, with the largest (≤1.5 m) closest to the King's Bowl. Squeezeups (bulbous masses of basalt) were extruded at the intersection of crustal plates of the lava lake or at points where the thin solid crust of the lake was broken. Many of these squeezeups are solid whereas others are hollow, indicating drainback. Some of the squeezeups are fractured or have impact holes in them. Some actually contain the impacting block.

How to get there: *From American Falls, Idaho, follow State Route 39 across American Falls Dam and north. Obvious signs indicate the route to Crystal Ice Cave (a US National Landmark), located on the Rift at the King's Bowl flow.*

References

Greeley, Ronald, Theilig, Eileen, and King, J. S., 1977, Guide to the Geology of King's Bowl Lava Field: *in* Greeley, R., and King, J. S. (eds.), *Volcanism of the Eastern Snake River Plain, A Comparative Planetology Guidebook*, NASA, 171-188.

King, John S., 1977 Crystal Ice Cave and King's Bowl Crater, Snake River, Plain, Idaho: *in* Greeley, R., and King, J. S. (eds.), *Volcanism of the Eastern Snake River Plain, A Comparative Planetology Guidebook*, NASA, 153-163.

John S. King

MENAN BUTTES
Idaho

Type: *Tuff cones*
Lat/Long: *43.77°N, 111.98°W*
Eruptive History: *Mid- to late-Pleistocene*
Composition: *Olivine tholeiitic basalt*

North and South Menan Buttes are the two most prominent phreatomagmatic cones of the Menan Complex, a group of six cones roughly aligned along a north-northwest trend, 16 km west-southwest of Rexburg, Idaho. North Menan Butte is the larger, standing 250 m above the surrounding Snake River Plain. It is elliptical in plan with axes 3.5 and 2.5 km in length. South Menan Butte measures 3 km × 2 km and has 145 m of relief. The buttes are asymmetrical with a greater accumulation of material on their northeast flanks, presumably due to strong southwest winds during eruption.

MENAN BUTTES: Aerial view of North and South Menan Buttes. The river is the Henry's Fork. Note the arrested development of a third cone (now partially covered by farm fields) between North and South Menan Buttes. USDA Photograph 40 16065 278-210.

The large size of these two buttes is atypical compared to other tuff cones on the Snake River Plain. In most Snake River tuff cones, early eruptive activity was phreatomagmatic, but with continuing activity the vent usually became sealed from the water source and basaltic flows resulted. There are no lava flows at Menan Buttes, indicating sustained access of water to the vents during the entire eruptive process.

Both North and South Menan Buttes show a noticeable change of slope in profile and are interpreted as composite structures consisting of tuff cones resting on tuff rings.

The eastern Snake River Plain is a thick accumulation of tholeiitic basalt. Major element analyses of the tuffs of the Menan Buttes show them to be very similar in composition to the olivine tholeiitic basalt of the eastern plain. Although isotopic dates are lacking, a Pleistocene age of the Menan Buttes is suggested by overlap relations with surrounding flows.

How to get there: *From Rexburg, follow State Route 33 west. Four kilometers west of the bridge across Henry's Fork, turn south onto an improved road for 3 km to North Menan Butte, which is on Bureau of Land Management land. South Menan Butte is on private property.*

References

Creighton, D. N., 1987, Menan Buttes, southeastern Idaho: *in Centennial Field Guide, V.2,* Decade of North American Geology Series, Geol. Soc. Am., 109-111

Creighton, D. N., 1982, The Geology of the Menan Complex, a Group of Phreatomagmatic Constructs in the Eastern Snake River Plain, Idaho: M.S. Thesis, State Univ of N.Y. at Buffalo.

John S. King

SPLIT BUTTE
Idaho

Type: *Tuff cone/pit crater*
Lat/Long: *43°N, 113.3°W*
Elevation: *1,410 m*
Eruptive History: *>2,270 yr BP*
Composition: *Olivine basalt*

Split Butte on the south-central Snake River Plain overlies basalt flows of the Snake River Group and was encroached from the southeast by a lobe of the Wapi lava flow, which has been dated at 2,270 yr BP. The butte consists of vitric ash forming a ring 600 m across. The ring is asymmetrical, having a greater accumulation on the east, the result of prevailing westerly winds during the eruption. Although the eroded ring stands 50 m above the surrounding plain, an original ash thickness of 80 m on the east is estimated from the dip of the beds and the ring diameter. A topographic notch or erosional "split" ~150 m wide occurs in the thick eastern ash accumulation.

Basal tephra layers tend to be massive, fine grained, poorly sorted, and extensively palagonitized. Above the basal zone lie thin planar beds of coarser and better sorted ash. Palagonitization and oxidation are less prevalent in these upper layers, which constitute the bulk of the

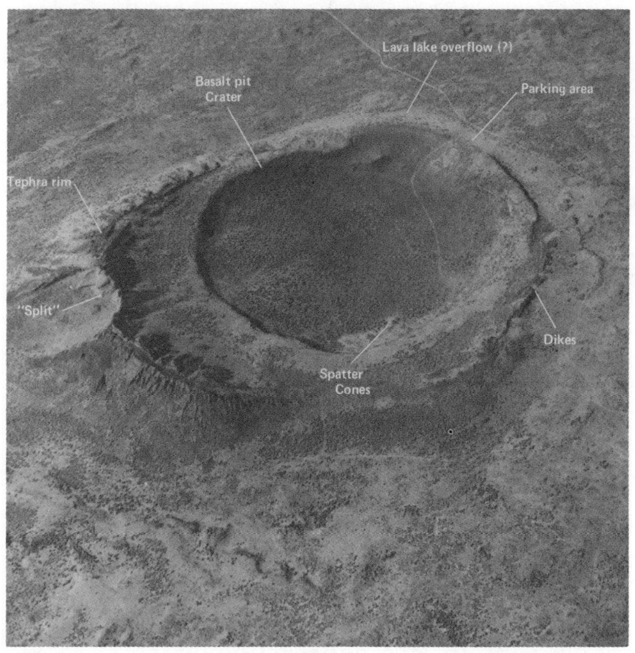

SPLIT BUTTE: Oblique aerial view southwestward across Split Butte (NASA-Ames photograph).

deposits. Beds range in thickness from 2-10 cm in the upper zone and from a few tens of centimeters to over a meter in the lower zone. Accidental basaltic fragments are abundant and range from granules to 1.25-m-sized blocks. In many places blocks are visible in outcrop, with bedding sags in the ash layers below. Lithic fragments are most abundant in the basal tephra layers. The tephra is composed of dense clasts and scoriaceous grains of partially palagonitized sideromelane and a matrix of very finely divided volcanic ash.

A central lava lake was retained by the tephra ring, and basaltic lavas are in disconformable contact with the tephra. Minor lake overflow occurred on the low southwest section of the ring. The lava lake margin is preserved as a narrow circular shelf of basalt, but the central portion has subsided to form a pit crater 20 m deep and 420 m across. Two low mounds of spatter occur along the pit crater scarp. The spatter consists of highly oxidized scoria and may represent a degassing outlet for post-subsidence liquids.

Four en echelon basalt dikes intruded the tephra on the northwest section of the ring. The nearly vertical dikes are 20-30 cm thick and have baked tephra along the contacts. The dikes trend tangentially to the ring and can be traced for over 250 m.

From field observations and laboratory studies of the ash, Split Butte appears to be a tuff ring or maar resulting from the interaction of basaltic magma and groundwater, followed by an effusive phase of eruption. The ash consists of clear, light brown sideromelane that is locally palagonitized, indicating rapid quenching in a water-rich environment. Clasts typically are blocky and angular and have few vesicles, which are small and spherical. Both the clast shape and arrested vesicle growth are evidence of quenching by water. Accretionary lapilli are abundant in the tephra and typically consist of a glassy or lithic nucleus coated with an irregular or incomplete layer of fine ash. Though accretionary lapilli may form by other means, their presence indicates a moist ash cloud during their deposition.

A change in eruption style from explosive to effusive led to the emplacement of the central lava lake. Cessation of the interaction may have resulted from the consumption of groundwater supplies, the removal of access of water to the vent, or an increased eruption rate of magma. Although the lava lake was relatively quiet, as evidenced by the lack of spatter, motion within the lake was sufficient to erode most of the inward-dipping tephra. Slumping of ash into the lava lake caused the disconformable contact visible between the lake basalts and thick tephra deposits in the east. Conformable contacts are prevalent along the low tephra deposits in the west.

How to get there: *Split Butte lies 40 km west-northwest of American Falls, Idaho; road distance from American Falls is 71 km. Travel across the Snake River from American Falls on Idaho State Highway 39 toward Aberdeen for 9.5 km; turn north on North Pleasant Valley Road to Crystal Ice Cave. Continue past Ice Cave on Bureau of Land Management road toward Minidoka. At a distance of 94 km from American Falls, turn left.*

References

Womer, M. B., Greeley, R., and King, J. S., 1980, The geology of Split Butte – A maar of the south-central Snake River Plain, Idaho: *Bull. Volc. 43*, 435-471.

Womer, M. G., Greeley, R., and King, J. S., 1982, Phreatic eruptions of the eastern Snake River Plain of Idaho: *in* Bonnichsen, B., and Breckenridge, R. M. (eds.), *Cenozoic Geology of Idaho*, Idaho Bureau Mines Geology Bull. 26, 453-464.

Ronald Greeley

WAPI
Idaho

Type: *Monogenetic low-shield volcano*
Lat/Long: *42.7°N, 113.2°W*
Elevation: *1,635 m*
Eruptive History: *2.270 yr BP*
Composition: *Basalt*

The Wapi lava field is one of several Holocene to Pleistocene volcanic fields on the Snake River Plain. In many respects, it is typical of the older fields of low shields that make up the present surface of the plain. It covers a large (300-km^2) area that is elongate in the north–south direction and has three prominent lobes extending east, west, and northwest from the main mass of the field.

The slope of the field is typical of the low shields of the Snake River Plain; over distances of 10-20 km, slopes are typically <1°. This flat slope is a consequence of the very fluid pahoehoe lavas and relatively high rates of effusion that typify the Snake River Plain. The only area of the field having a steeper slope is in the vicinity of **Pillar Butte**, the summit region of the field, where the slopes range from 5° to 7°.

The field is composed of numerous compound flow units. Multiple flow units along the south and west sides of the field suggest that the flows are rather thin (15-25 m). In the center of the field at Pillar Butte, the total thickness may exceed 100 m. Near the margins of the field, the flow units are larger, tend to have greater local relief (~10 m), and are characterized by large pressure ridges, pressure plateaus, and collapse depressions. Most flows have a filamented pahoehoe texture, with relatively minor portions broken into slabby pahoehoe or aa lavas.

Pressure ridges are common and can be divided into two types: longitudinal ridges with axes parallel to flow direction (termed flow ridges) and transverse ridges with axes across the direction of flow. Most of the medial cracks in the pressure ridges are empty, although some are filled with lava squeezeups. Pressure ridges range in length from a nearly continuous 10 km to ~8 m and range in height from 1 to 15 m. Some of the ridges have a rounded domical form, but most are relatively flatsided with sides dipping 30° to 40°. The best-formed and largest pressure ridges occur on the eastern side of the field. The medial cracks of the major pressure ridges lack flowouts and generally are 6 to 10 m deep, indicating that cracking occurred late in the history of the flow.

Collapse depressions are abundant in some areas of the Wapi Field. They range from 1 to 35 m across and are from 2 to 10 m deep. The largest and best-formed collapse depressions are on the eastern side of the field. The long axes of the collapse depressions are generally parallel to the flow direction. Collapse depressions commonly contain lava squeezeups which were extruded subsequent to collapse.

The steep profile of Pillar Butte is at least partly attributed to the relatively low-volume flows that did not travel very far from the central vent, and the higher proportion of viscous aa flows. The butte is a prominent mass of agglutinate and layered flows, possibly injected with dikes; the mass rises ~18 m above the general summit region on the south side. This prominent structure served as a landmark for early travelers along the Oregon Trail, and still serves as a general reference point in the south-central Snake River Plain.

The Pillar Butte summit region contains at least 11 eruptive centers, identified by pit craters and former lava lakes. Flows from these centers are typically short and often contain small lava tubes and channels. Many of the tubes and channels were used repeatedly by subsequent flows, sometimes forming roofs over previous channels, or draining into older lava tubes through skylights.

How to get there: *Wapi lava field lies 40 km west-northwest of American Falls, Idaho. Travel across the Snake River from American Falls on Idaho State Highway 39 toward Aberdeen for 9.5 km; turn north on North Pleasant Valley Road toward Crystal Ice Cave. At a distance of 38 km from American Falls, turn left (west) on a jeep trail to the edge of the lava field.*

References

Greeley, R., 1982, The style of basaltic volcanism in the eastern Snake River Plain, Idaho: *in* Bonnichsen, B., and Breckenridge, R. M. (eds.), *Cenozoic Geology of Idaho*, Idaho Bureau Mines Geology Bull. 26, 407-421.

Champion, D. E., and Greeley, R., 1977, Geology of the Wapi Lava Field, Snake River Plain, Idaho: *in* Greeley, R., and King, J. S. (eds.), *Volcanism of the Eastern Snake River Plain, Idaho*, NASA, 133-151.

Ronald Greeley

VOLCANOES OF NEVADA

Sheldon–Antelope, 256
Buffalo Valley, 256
Steamboat Springs, 257
Lunar Crater, 258
Reveille Range, 260
Clayton Valley, 261
Timber Mountain, 261

SHELDON–ANTELOPE
Nevada

Type: *Monogenetic volcanic field*
Lat/Long: *41.8°N, 119.1°W*
Eruptive History: *Quaternary*
Composition: *Basalt*

Three or four flows of Quaternary age basaltic lava straddle the Nevada–Oregon border in an area relatively remote from other young volcanism. The flows (20-35 km long) issued from small shield volcanoes. Only one isotopic age of 1.2 Ma is available for the flows and little geologic mapping has been published.
How to get there: *The lava flows are in the Charles Sheldon National Antelope Range, which is crossed by Highways 140, 34A, and 8A.*

Reference

Luedke, R. G., and Smith, R. L., 1981, Map showing distribution, composition, and age of Late Cenozoic volcanic centers in California and Nevada: *USGS Map I-1091C.*

Charles A. Wood

BUFFALO VALLEY
Nevada

Type: *Monogenetic volcanic field*
Lat/Long: *40.35°N, 117.3°W*
Length: *15 km*
Width: *4-5 km*
Elevation: *1,420-1,750 m*
Eruptive History: *3.05 Ma to 0.92 Ma*
Composition: *Olivine basalt*

The Buffalo Valley volcanic field is located along the eastern margin of Buffalo Valley just north of the Fish Creek Mountains caldera (~24 Ma). The field is comprised of 14 vents and associated flows which form a northeast-trending zone, ~5 km wide and 15 km long, along the northwest flank and piedmont of the Fish Creek Mountains. Both cones and flows are relatively small. Most of the vents are surmounted by breached cinder cones of highly variable size and shape. Several of these cones occur as contiguous pairs or triplets with north to northeast

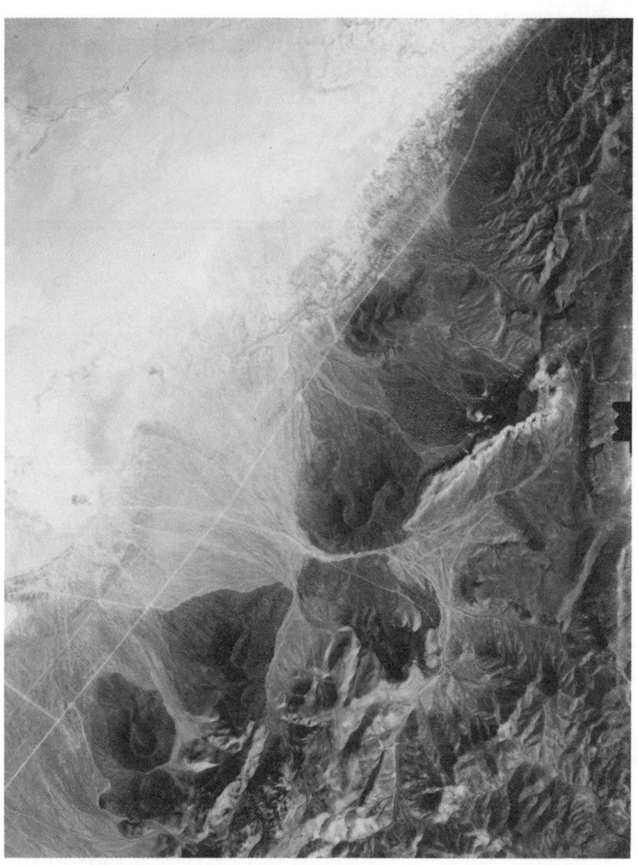

BUFFALO VALLEY: Vertical aerial view of complex cinder cones and minor lava flows. USGS NHAP aerial photograph HAP 80-161-43.

alignments that generally parallel the overall trend of the field. Cone heights range from ~50 to 100 m and cone diameters from ~150 to 500 m. Flow areas are each <0.5 km^2, and the combined area of all the cones and flows is ~10 km^2.

K-Ar dates from four flows within the field indicate that most of the cones and flows are early Pleistocene, although one flow yielded a late Pliocene age. Cone slopes are slightly to moderately dissected by uniformly spaced, first- and second-order gullies, and flow surfaces have been profoundly modified by both alluvial and aeolian deposition. The presence of these cones and flows indicates relatively little net erosion or deposition on the northwest flank and piedmont of the Fish Creek Mountains during the past 1.4 to 3 Ma.

How to get there: *The Buffalo Valley volcanic field is situated along the southeast margin of Buffalo Valley in north-central Nevada. The field is located ~235 km east-northeast of Reno, Nevada, and ~5 km southwest of Battle Mountain, Nevada. The field is accessible from Nevada State Highway 305 via a graded dirt road that runs along the northwest margin of the field.*

References

Morton, J. L., Silberman, M. L., Bonham, H. F., Jr., Garside, L. J., and Noble, D. C., 1977, K-Ar ages of volcanic rocks, plutonic rocks, and ore deposits in Nevada and eastern California - determinations run under the USGS–NBMG cooperative program: *Isochron West 20*, 19-29.

Wollenberg, H., Bowman, H., and Asaro, F., 1977, Geochemical studies at four northern Nevada hot spring areas: *Lawrence Berkeley Lab. Rpt. LBL-6808*, 69 pp.

John C. Dohrenwend

STEAMBOAT SPRINGS
Nevada

Type: *Rhyolite dome volcanic field*
Lat/Long: *39.37°N, 119.72°W*
Elevation: *1,415 m*
Eruptive History: *2.53 to 1.14 Ma*
Composition: *Rhyolite*

A small volcanic field of domes and flows occurs at the south end of Truckee Meadows from ~20 km SSW of Reno, Nevada, to ~12 km south-southeast of Reno, aligned along a northeasterly trend. The western dome is 3 km south-southwest of Steamboat Hot Springs, a questionable dome underlies the hot springs, and 4 other domes are northeast of the springs. The southwest dome and one of the northeast domes are the largest, being 1 km in diameter and nearly 150 m in maximum relief.

Two domes have radiometric ages of 1.21 and 1.14 Ma, but 2 obsidian whole-rock ages for one of the domes are ~3.0 Ma; the anomaly is unresolved. The age of the thermal springs is possibly as much as 3.0 Ma. Certainly by 1.1 Ma activity had started, and it continues to the present time. The active sinter deposits include gold, silver, mercury, antimony, arsenic, thallium, and boron; small quantities are commercial in grade.

Southwest of most of the domes is a ~3-km-long basaltic andesite unit with an age of 2.53 Ma. The vent of this flow is on the long axis of

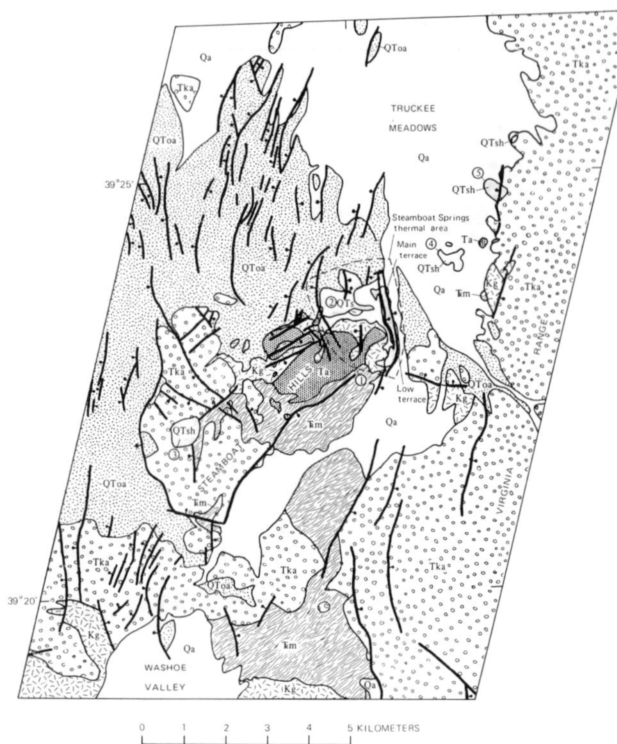

STEAMBOAT SPRINGS: Geologic map of the Steamboat Springs region from Silberman et al. (1979).

the Steamboat Hills, 2 km southwest of the hot springs and is underlain by the highest borehole temperature found in the area: ~229°C.

Bimodal basalt and rhyolite seem to be typical of volcanism that encroached on the Sierran front from the east and northeast during the past 3-5 Ma, as parts of a common trend that resulted in the famous Comstock Lode a few kilometers farther east 10 to 12 Ma ago. We do not know how or why young volcanic rocks are so minuscule at the surface, but very extensive andesite and rhyolite tuffs are abundant in the Comstock (Virginia City) region and perhaps were essential in heating and preparing the ground for continuing thermal activity and young mineralization.

How to get there: *Steamboat Springs is 16 km south of Reno, Nevada, and the thermal areas are easily accessible by car or bus from US Highway 395 (eastern side) or Nevada Highway 431 (north edge).*

References

Thompson, G. A., and White, D. E., 1964, Regional geology of the Steamboat Springs area, Washoe County, Nev.: *USGS Prof. Pap. 458A.*

Silberman, M. L., White, D. E., Keith, T. E. C., and Dockter, R. D., 1979, Duration of hydrothermal activity at Steamboat Springs, Nevada, from ages of spatially associated volcanic rocks: *USGS Prof. Pap. 458D.*

Donald E. White

LUNAR CRATER
Nevada

Type: *Monogenetic volcanic field*
Lat/Long: *38.25°N, 116.05°W*
Elevation: *1,585 to 2,255 m*
Length: *40 km*
Width: *10 km*
Eruptive History: *4.2 Ma to ~0.015 Ma*
Composition: *Subalkaline basalt to basanite*

The Lunar Crater volcanic field, an apparent middle to late Pliocene and Pleistocene continuation of the Reveille Range volcanic field immediately to the southwest, is superposed across the 25-m.y.-old Lunar Lake caldera, a crudely circular topographic basin on the crest of the Pancake Range. The field contains ~95 late Pliocene and Pleistocene vents and at least 35 associated lava flows contained within a northeast-trending zone, up to 10 km wide and ~40 km long, that extends obliquely across the flanks and crest of the range. Vents include cinder cones, elongate fissures, and at least two maars. **Lunar Crater**, a nearly circular maar ~130 m deep and 1,050 m wide, is the most distinctive feature of the field. A second maar, ~550 m wide and 65 m deep, occurs at the south end of a northeast-trending chain of coalesced cinder cones. Several other northeast-trending alignments of closely spaced and coalesced cinder cones characterize the field. Lava flows range up to 1.9 km wide and 6.1 km long with thicknesses from <3 m to as much as 25 m. Progressive degradation of the cones and flows is very similar to that displayed by other basaltic volcanic fields in the southwest Basin and Range (including the Cima, Crater Flat, and Coso fields). Many of the flows in the northeast and central parts of the field are veneered with varying thicknesses of air-fall tephra. In other areas, all but the youngest flows are mantled with extensive deposits of aeolian silt and fine sand.

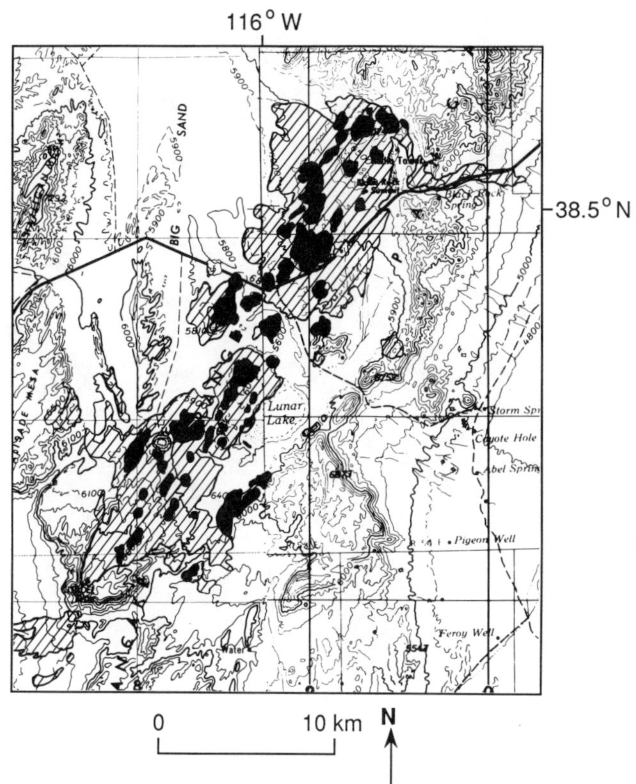

LUNAR CRATER: Modified USGS topographic map with the lava flows indicated by diagonal lines and the vents by solid black.

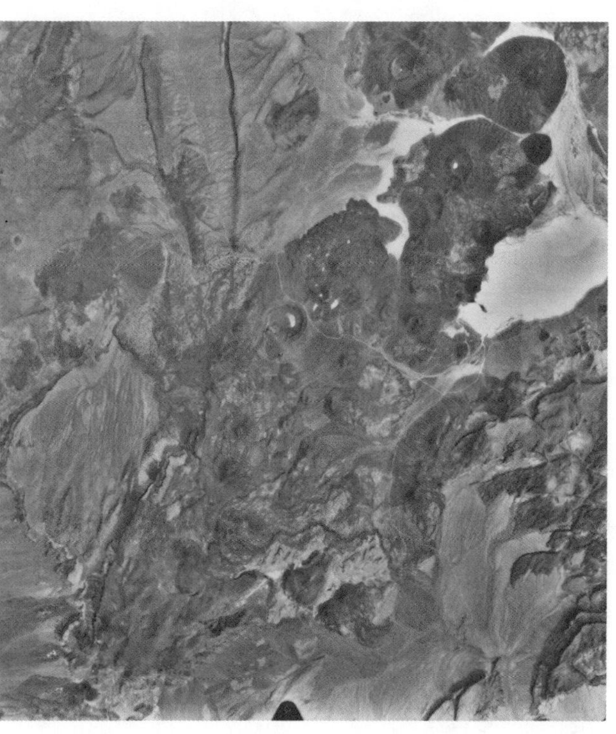

LUNAR CRATER: Central and southern parts of the field, showing the dry Lunar Lake (top right), Lunar Crater (center), and cones and flows. National High Altitude Photography Program vertical aerial view.

Chemical composition of the Lunar Crater lavas ranges from Ne to Hy normative with varying degrees of incompatible element enrichment. Also, along any section perpendicular to the northeast trend of the field, activity has spanned at least 2 to 3 m.y., with the youngest rocks being the most Ne normative. These compositional variations suggest fractional crystallization, differences in primary melts, and a time-progressive shift in the locus of melting and/or in the mantle sources.

How to get there: *The Lunar Crater volcanic field is in the central Great Basin ~105 km east-northeast of Tonopah, Nevada, and 140 km southwest of Ely, Nevada. US Highway 6 runs through the center of the Lunar Crater field and most areas of the field are readily accessible via graded dirt roads.*

LUNAR CRATER: Oblique aerial view of Lunar Crater, a nearly circular maar ~130 m deep.

References

Foland, K. A., Kargel, J. S., Lum, C. L., and Bergman, S. C., 1987, Time-spatial-compositional relationships among alkali basalts in the vicinity of the Lunar Crater, south central Nevada: *Geol. Soc. Am. Abst. with Program 19*, 666.

Scott, D. H., and Trask, N. J., Geology of the Lunar Crater Volcanic Field, Nye County, Nevada: *USGS Prof. Pap. 599-I*, 22 pp.

John C. Dohrenwend

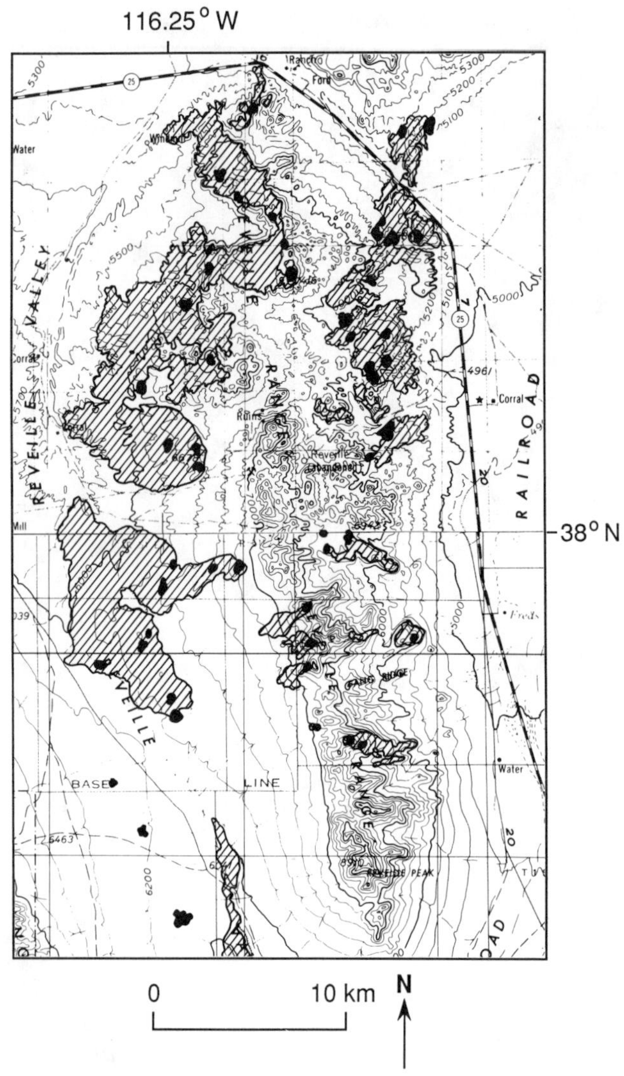

REVEILLE RANGE: Modified USGS topographic map with the lava flows indicated by diagonal lines and the vents by solid black.

REVEILLE RANGE
Nevada

Type: *Monogenetic volcanic field*
Lat:/Long: *38.1°N, 116.2°W*
Elevation: *1,535 to 2,265 m*
Length: *45 km*
Width: *15-18 km*
Eruptive History: *Two principal periods of activity: 6.2 to 5.3 Ma; 4.2 to 3.8 Ma*
Composition: *Predominantly alkali-olivine basalt and hawaiite*

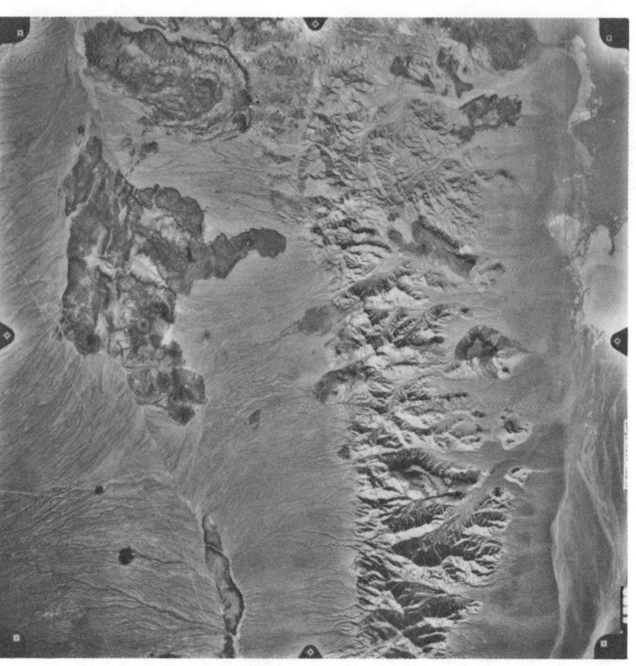

REVEILLE RANGE: Vertical aerial view of the southern part of the Reveille Range. Latest Miocene and early Pliocene basaltic lava flows (dark areas) are scattered across the crest and flanks of the range and cover large areas of the adjoining piedmonts. USGS U-2 aerial photograph 2640.

The Reveille Range volcanic field contains ~50 vents and associated lava flows within a north–south-trending zone, ~18 km wide and 45 km long, that is superposed across most of the length and width of the range. Vents and flows are widely scattered within this zone and cover a combined area of only ~140 km^2 (~17% of the total area of the zone). Flows range up to nearly 3 km in width and 7 km in length. Both vents and flows are highly degraded. The loose tephra carapaces of most cones have been largely eroded to reveal inner frameworks of dikes, sills, and agglutinate layers. Flow surfaces are almost completely devoid of original surface morphology and commonly are mantled by 1-4-m-thick blankets of aeolian silt and fine sand. Surfaces along flow margins are commonly stripped to the dense rock typical of flow interiors, and flow surfaces display a pronounced upward convexity transverse to the flow direction.

Because of their limited age range and wide topographic distribution across the crest, flanks, and adjoining piedmonts of the Reveille Range, these lava flows constitute an ideal datum for documenting latest Tertiary and Quaternary erosion within a typical upland area of the Basin and Range Province. Average dissection rates have ranged between 3.0 and 1.5 cm/10^3 yr in crestal and upper flank areas and between ~1.5 and 0.1 cm/10^3 yr in lower flank and upper piedmont areas. Middle piedmont areas have remained essentially unchanged since latest Miocene time.

How to get there: *The Reveille Range volcanic field is located in the central Great Basin ~105 km east-northeast of Tonopah, Nevada, and 140 km southwest of Ely, Nevada. Nevada State Highway 25 runs along the north and west sides of the range, and the northern and western parts are readily accessible via graded dirt roads.*

References

Dohrenwend, J. C., Turrin, B. D., and Diggles, M. F., 1985, Topographic distribution of dated basaltic lava flows in the Reveille Range, Nye County, Nevada: Implications for late Cenozoic erosion of upland areas in the Great Basin: *Geol. Soc. Am. Abst. with Program 17*, 352.

Ekren, E. B., Rogers, C. L., and Dixon, G. L., 1973, Geologic and Bouguer gravity map of the Reveille Quadrangle, Nye County, Nevada: *USGS Misc. Geol. Invest. Map I-806*, scale 1:48,000.

John C. Dohrenwend

CLAYTON VALLEY
Nevada

Type: *Cinder cone*
Lat/Long: *37.8°N, 117.65°W*
Height: *85 m*
Diameter: *715 m*
Elevation: *1,405 to 1,490 m*
Eruptive History: *0.39 Ma*
Composition: *Alkali-olivine basalt*

The Clayton Valley cinder cone is a solitary cone and flow at the northern end of Clayton Valley on the northeast piedmont of the Silver Peak Range. The closest major late Tertiary or Quaternary volcanic center, the 6.1-4.8-Ma **Silver Peak caldera**, is situated on the crest of the Silver Peak Range ~25 km to the southwest. However, the Clayton Valley cone does not appear to be genetically related to this caldera. The cone, ~85 m high and 715 m in diameter, is deeply breached on its east (downslope) side. The associated flow, originally at least 2 km long and 1 km wide, has been tentatively dated (on the basis of a single K-Ar analysis) at 0.39 Ma. The crest of this late middle Pleistocene cone is sharp, and its outer slopes have been superficially dissected by uniformly spaced, shallow gullies. The associated flow has been eroded by late Quaternary alluvial processes and partly buried by late Quaternary alluvium.

CLAYTON VALLEY: Oblique aerial view of the breached eastern flank of the Clayton Valley cinder cone. Erosion of this 0.39-Ma cone has been limited largely to shallow gullying of its outer flanks.

How to get there: *Clayton Valley is in western Nevada, ~45 km southwest of Tonopah, Nevada. The Clayton Valley cone lies just east of Nevada State Highway 265 at a point ~27 km south of the junction of this highway with US 95.*

Reference: None

John C. Dohrenwend

TIMBER MOUNTAIN
Nevada

Type: *Polygenetic volcanic centers*
Lat/Long: *37.1°N, 116.5°W*
Elevation: *1,065 to 1,675 m*
Eruptive History: *Three episodes:*
12 to 8.5 Ma: Waning stage of Timber Mountain caldera cycle
9.0 to 6.5 Ma
3.7 to <0.01 Ma
Composition: *Hawaiite to alkali basalt*

Late Cenozoic volcanic centers of the Timber Mountain volcanic field include spatially isolated small volume basaltic scoria cones and associated lava flows. Volcanic activity in this field switched from predominantly silicic to predominantly basaltic at ~10 Ma. The volume of basalt eruptions declined drastically at ~8 Ma, but small eruptions continued through 6.5 Ma. Following a gap in activity, basaltic eruptions resumed at 3.7 Ma, with a progressive decline in the volume through the Holocene.

Basaltic centers associated with the oldest episode of activity were localized within and along ring-fracture zones of calderas of the Timber Mountain volcanic field. In association with the decline in eruption volume of basalt, volcanic activity was restricted to the northeast part of the Timber Mountain area. Since 3.7 Ma, all basaltic centers have been in the southwest part of the volcanic field.

There are three major occurrences of post-4-Ma basaltic activity in the Timber Mountain field: (1) **Crater Flat**: A series of deeply dissected 3.7-Ma basalt scoria cones and lava flows are present in eastern Crater Flat. Four more basaltic centers (1.2 Ma) are aligned along a north-northeast trending arc in central Crater Flat. The youngest center in this area is the basalt of **Lathrop Wells** (0.1 to 0.01 Ma), at the south end of Crater Flat. (2) **Buckboard Mesa**: The basalt of Buckboard Mesa (2.8 Ma) erupted in the northeast segment of the moat zone of Timber Mountain caldera. (3) **Sleeping Butte**: Two other Quaternary basalt centers occur at Sleeping Butte (0.3 to 0.01 Ma), 30 km north of Beatty, Nevada, on the south flank of the Black Mountain caldera complex.

A polygenetic eruptive history is a unique and characteristic feature of the three youngest, and possibly all, Quaternary centers of the Timber Mountain volcanic field. These centers (cinder cones and lava flows) formed during brief episodes of eruptive activity that were separated by long periods of inactivity. The duration of activity at individual centers spanned several tens of thousands to more than one hundred thousand years. Preliminary data suggest that polygenetic centers show progressive changes in the volume and style of eruptions. Initial eruptions tend to be larger volume with a low ratio of scoria to lava. Successive eruptions are smaller with a higher scoria/lava ratio. The last activity at the two youngest centers were small-volume (10^5 m^3) eruptions that deposited a scoria-fall mantle on the slopes of pre-existing scoria cones.

How to get there: *The Buckboard and Sleeping Butte volcanic centers are within the controlled boundaries of the Nevada Test Site or the Nellis Bombing and Gunnery Range. These sites are not accessible to the public. The southern center of the Sleeping Butte group is visible from Highway 95, ~16 km north of Beatty. The Crater Flat volcanic field is ~150 km northwest of Las Vegas, Nevada. Lathrop Wells volcanic field is accessible via a short dirt road off Highway 95, located 12 km northwest of Lathrop Wells, Nevada. Crater Flat is reached via a maintained dirt road near Steve's Pass which intersects Highway 95 ~20 km northwest of Lathrop Wells.*

References

Vaniman, D. T., Crowe, B. M., and Gladney, E. S., 1982, Petrology of hawaiite lavas from Crater Flat, Nevada: *Cont. Min. Petr. 80*, 341-357.

Crowe, B. M., Self, S., Vaniman, D., Amos, R., and Perry, F., 1983, Aspects of potential magmatic disruption of a high-level radioactive waste repository in southern Nevada: *J. Geol. 91*, 259-276.

Bruce M. Crowe

VOLCANOES OF WYOMING

Yellowstone Plateau, 263
Leucite Hills, 266

YELLOWSTONE PLATEAU
Wyoming, Idaho

Type: *Ash-flow caldera field*
Lat/Long: *44.58°N, 110.53°W*
Elevation: *2,400 m*
Volcanic History:
2.0, 1.3, 0.6 Ma: caldera-forming eruptions
0.15, 0.11, 0.07 Ma: rhyolite lavas
Composition: *Rhyolite and basalt*

The Yellowstone Plateau spans the continental divide between the Northern and Middle Rocky Mountains, at an average elevation of ~2,400 m. The plateau lies at the center of one of the Earth's largest volcanic fields, entirely postdating 2.5 Ma. The major eruptions of the volcanic field were exceedingly voluminous, but their products are only surficial expressions of the emplacement of a batholithic volume of rhyolitic magma to high crustal levels in several episodes. The total volume of magma erupted from the Yellowstone Plateau volcanic field since 2.5 Ma probably approaches 6,000 km^3.

This great magmatic volume and the enormous calderas produced by the largest pyroclastic eruptions are associated with a surprisingly subtle morphology. The Yellowstone caldera, the youngest of three nested and overlapping calderas, is filled by younger rhyolitic lavas, and is readily recognizable in only one or two sectors. The two older, nested calderas, however, form part of a conspicuous circular basin at the west edge of the volcanic field, called Island Park, which is enclosed along its eastern margin by a younger constructional lava platform at the west edge of the Yellowstone Plateau.

The Yellowstone Plateau volcanic field erupted a bimodal assemblage of basalt and rhyolite in three cycles of activity. Each cycle began with eruptions of both basalt and rhyolite; with time, the largest volume of rhyolite vented as lavas from developing ring-fracture systems. The climax of each cycle was marked by extremely rapid and voluminous eruptions of rhyolitic magma as ash flows from the ring-fracture system – hundreds to thousands of km^3 being ejected in a few hours or days – and by collapse of the source area to form a large caldera. Post-collapse volcanism in each caldera has tended to fill it with

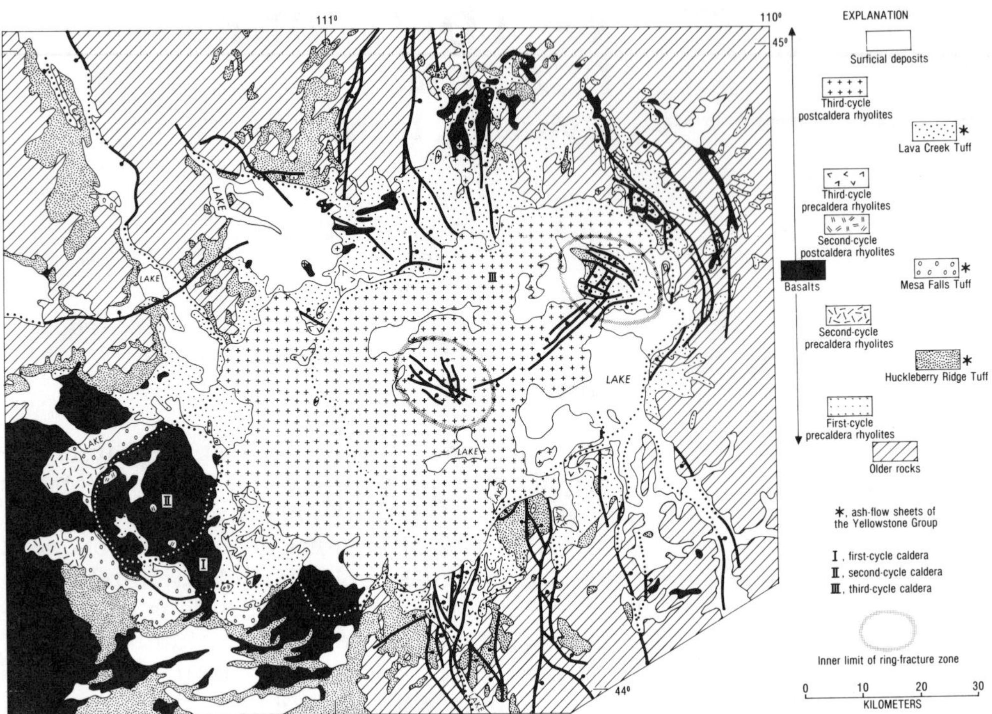

YELLOWSTONE: Simplified geologic map, showing 0.6-Ma Yellowstone caldera, resurgent centers, and rhyolitic plain. Map from Christiansen (1984).

rhyolitic lavas. Throughout each cycle of mainly rhyolitic volcanism, both basaltic and some rhyolitic lavas continued to erupt on the margins of the volcanic field, but no basalts erupted within the major active rhyolitic source areas.

The ash flows erupted at the climax of each cycle form the three largely welded cooling units of the Yellowstone Group, providing a framework for the stratigraphy of the volcanic field. The 2,500-km^3 Huckleberry Ridge Tuff erupted at 2.0 Ma, the 280-km^3 Mesa Falls Tuff at 1.3 Ma, and the 1,000-km^3 Lava Creek Tuff at 0.6 Ma. Each of these great eruptions produced fallout ash deposits over large parts of the western United States, leaving recognizable remnants as far east as the Mississippi River. The first and third cycles were sustained by enormous bodies of rhyolitic magma that accumulated to batholithic size, the highest parts of each intruding and deforming its roof to form compound ring-fracture zones. When a major eruption began from one of these high-level portions of the batholithic chamber, the violent degassing triggered contemporaneous or successive eruptions from the adjacent or overlapping ring-fracture zones, producing composite ash-flow sheets and compound calderas that embrace the cluster of ring-fracture zones.

The compound caldera that formed during the climactic first-cycle Huckleberry Ridge eruption – largest of the three – spanned at least 80 km from **Island Park** (at the margin of the basalt-covered Sŋake River Plain, west of Yellowstone National Park), past the northern Teton Range and Jackson Hole on the south, to the center of the Yellowstone Plateau. The second-cycle **Henrys Fork** caldera is the smallest of the three, ~20 km; both it and the surface outcrop of the Mesa Falls Tuff are restricted to the Island Park area. The third cycle began with the eruption of a series of voluminous rhyolitic lavas from all sectors of a growing fracture system that embraced two adjacent ring-fracture zones. The compound third-cycle Yellowstone caldera, related to the Lava Creek Tuff eruption, is 70 × 40 km across in the center of the Yellowstone Plateau. The caldera is resurgent, with an early post-collapse dome uplifted within each of its two segments, followed by emplacement of early post-resurgence rhyolitic lavas from the enclosing ring-fracture zones.

Renewed magmatic activity has produced voluminous lavas in the Yellowstone caldera since ~150 ka, perhaps even indicating a fourth

YELLOWSTONE: Landsat MSS (80-m resolution) view of Yellowstone region. Yellowstone Lake (dark, right of center here, but white on accompanying map) and others are most conspicuous features on this image. Resurgent domes are hardly noticeable. Island Park caldera (II on map) indicated by arrow.

YELLOWSTONE: Aerial view to the west along the north rim of the Yellowstone caldera, looking toward the Madison River Canyon, which breaches the caldera in the distance. In the middle ground is a downdropped or slumped block of the caldera rim between an outer rim and the large rhyolitic lava flows that fill the caldera. Photograph by L. J. P. Muffler.

volcanic cycle. Following emplacement of a large rhyolitic lava flow in the western ring-fracture zone, renewed uplift of the resurgent dome occurred, reflecting insurgence of magma into the caldera system. Since that time, voluminous rhyolitic lavas (several individual flows exceeding 50 km^3) have filled the central part of the caldera and overflowed its western rim. These lavas were emplaced in three major episodes at ~150 ka, 110 ka, and 70 ka, each time erupting from both the western and eastern sides of the western ring-fracture zone to form the Madison and Central plateaus, respectively. The aggregate volume of these lavas is ~1,000 km^3. Deformation, probably related to continued magmatic activity beneath the Yellowstone caldera, continues with caldera-wide uplift and subsidence at rates as high as 2 cm/yr.

Virtually no eruptive products of the Yellowstone volcanic system have compositions between ~52% and 68% SiO_2; a few intermediate compositions occur in mixed-lava complexes on the margins of the volcanic field, but they clearly represent shallow commingling of basaltic and rhyolitic magmas. Probably >90% of the total volume of erupted products has SiO_2 greater than 76%. Virtually all rhyolite is characterized by abundant phenocrysts of quartz and sanidine, and most has abundant sodic plagioclase phenocrysts as well. Only a few post-collapse rhyolitic lavas have more plagioclase than sanidine phenocrysts; a small number of flows even lack sanidine. Mafic minerals are dominantly ferroaugite and subordinate fayalite. Some units contain hornblende, but biotite is exceedingly rare as a phenocryst. The basalt of the Yellowstone Plateau volcanic field is olivine-tholeiite; some of the most voluminous younger basalt erupted through the older rhyolitic centers of Island Park.

The Yellowstone caldera region hosts the world's largest known hydrothermal system, highlighted by numerous geysers. This hydrothermal system accounts for an average heat flow from the caldera area 40 times greater than the global average. Although the latest eruptions were ~70,000 yr ago, the immense hydrothermal system and a variety of geophysical characteristics indicate that magma still underlies the Yellowstone caldera at a shallow depth. A large negative gravity anomaly, low magnetic intensity, high electrical conductivity, shallow swarm seismicity, and large delays and high attenuation of seismic waves are all consistent with this inference.

Since the middle Miocene, the northern basin-range region, including the part of the Rocky Mountains around Yellowstone, has been broken by extensional normal faults. South of the Yellowstone Plateau, these faults trend

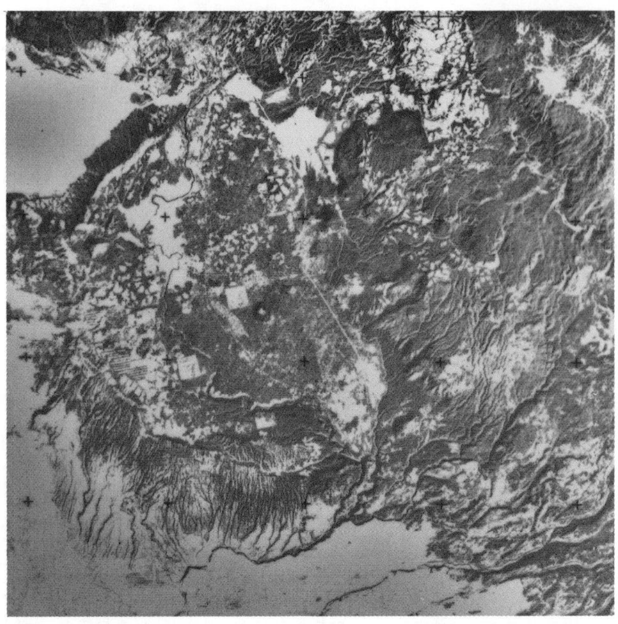

YELLOWSTONE: Landsat RBV (30-m resolution) vertical view of Island Park caldera with its sharp rim on the west; the eastern side has been covered by post-caldera rhyolites.

mainly northward; north of the plateau, they trend northwest to westward. Their displacements form fault-block ranges and basins that break across many of the older structures, especially south of the plateau; to the northwest, however, the basin-range faults reactivate some of the older faults. Vents for the intracaldera rhyolites of the Madison and Central plateaus were localized and controlled by basin-range faults that extend into the Yellowstone caldera from its margins. Throughout the region, late Cenozoic volcanism is characterized by abundant basalt or by bimodal associations of rhyolite and basalt. The Yellowstone Plateau volcanic field is the active part of a mainly rhyolitic system, consistently associated with basalt and with basin-range extension, that has propagated northeastward at ~4 cm/yr since the middle Miocene. The track of this system of propagating volcanism, or "hot spot," is expressed physiographically as the subsided and basalt-covered floor of the eastern Snake River Plain.

The Yellowstone Plateau volcanic field provides an outstanding example of the effects of a voluminous long-lived magmatic system within a continental plate. The abundance of rhyolites, the complexity of evolution of exceedingly large magmatic systems, and intimate relations between magmatic and tectonic processes are all well exemplified.

How to get there: *Yellowstone is the oldest national park in the world and one of the leading tourist attractions in North America. US Highways 16, 212 and 289 go through the Park. Many hydrothermal and volcanic features are well marked.*

References

Christiansen, R. L., 1984, Yellowstone magmatic evolution: Its bearing on understanding large-volume explosive volcanism: in *Explosive Volcanism: Inception, Evolution, and Hazards*, Washington, Nat. Acad. Sci., 84- 95.

Dzurisin, D., and Yamashita, K. M., 1987, Vertical surface displacements at Yellowstone caldera, Wyoming, 1976-1986: *J. Geophys. Res. 92*, 13,753-13,766.

R. L. Christiansen

LEUCITE HILLS
Wyoming

Type: *Monogenetic volcanic field*
Lat/Long: *41.8°N, 108.9°W*
Relief: *76 m*
Field Size: *~35 × 65 km*
Volcanic History: *1.1 Ma*
Composition: *Potassium-rich, mafic alkaline rocks*

Some volcanic rocks have such intriguing chemical compositions that the eruptive systems that emplaced them and the resulting landforms are soon forgotten, and only the mineralogy and isotopic anomalies elicit comment. The Leucite Hills of southern Wyoming are probably the premier example in North America of this myopic obsession. Since 1903 every paper describing the Leucite Hills has concentrated on the unique potassium-rich rocks – wyomingite, orendite, and madupite – and their enclosed ul-tramafic nodules; the general geology of the volcanic field is known only from studies 90 years old!

Only 22 exposures of volcanic material occur in the Leucite Hills, a collection of isolated lava flows, dikes, necks, and cinder cones scattered over an area of ~2,000 km^2. The vents and the exposures of the field itself have a general northwest–southeast alignment, but the reason for this apparent tectonic control is not evident. The volcanic field cuts through sedimentary formations of late Cretaceous–Tertiary age, suggesting (perhaps with the Dotsero volcanics of north-central Colorado) an incipient northward extension of the Rio Grande rift valley.

How to get there: *The Leucite Hills are accessible from Superior, Wyoming, which is ~10 km north of US Highway 30, east of Rock Springs, Wyoming.*

References

Cross, W., 1897, Igneous rocks of the Leucite Hills and Pilot Butte, Wyoming: *Am. J. Sci. 4*, 115-141.

Volkner, R., et al., 1984, Md and Sr isotopes in ultrapotassic volcanic rocks from the Leucite Hills, Wyoming: *Cont. Min. Petr. 87*, 359.

Charles A. Wood

VOLCANOES OF COLORADO

Dotsero, 268

DOTSERO
Colorado

Type: *Cinder cones and flows*
Lat/Long: *39.7°N, 107.0°W*
Elevation: *1,733 m*
Eruptive History: *2 Ma to ~4,000 yr BP*
Composition: *Nepheline-normative alkali basalt*

One of the youngest eruptions in the continental US produced an explosion crater, a lahar, and a 3-km-long lava flow in north-central Colorado near the town of Dotsero. The eruption is dated at 4,150 yr BP, based upon a ^{14}C date from wood under scoria. The Dotsero crater is 700 m wide and 400 m deep. This activity was the last (thus far) of a small resurgence of basaltic volcanism in this area that began ~1.9 Ma, following a 7-m.y. hiatus of volcanism. The earliest of the recent activity produced lava flows (**Triangle Peak**) 16 km north-northwest of Aspen, Colorado. At 0.64 Ma two small cinder cones and a 6.5-km-long basalt flow erupted ~70 km north of Aspen near the town of **McCoy**. An additional undated but young cinder cone (**Willow Peak**) and flow formed near the confluence of the Eagle and Colorado rivers.

These eruptions occurred along rivers or on their gravel floodplains. At least one eruption vented along a known fault. Rapid canyon cutting since 1.5 Ma suggests that renewed tectonic activity might have provided conduits for magma to leak to the surface. Whether the tectonic deformations permitting the eruptions are related to northward extension of the Rio Grande rift, an edge-effect from the Colorado Plateau, or some other cause is unknown.

How to get there: *Dotsero is on US Interstate Highway 70 heading west from Denver. McCoy is on a branch highway (131) going north from I-70; the nearby town of Volcano is presumably the location of the 0.64-Ma cinder cone.*

References

Larson, E. E., Ozima, M., and Bradley, W. C., 1975, Late Cenozoic basic volcanism in northwestern Colorado and its implications concerning tectonism and the origin of the Colorado River system: *in* Curtis, B. F. (ed.), Cenozoic History of the Southern Rocky Mountains, *Geol. Soc. Am. Mem. 144*, 155-178.

Leat, P.T., et al., 1989, Quaternary volcanism in northwestern Colorado: Implications for the roles of asthenosphere and lithosphere in the genesis of continental basalts: *J. Volcan. Geotherm. Res. 37*, 291-310.

Charles A. Wood

VOLCANOES OF UTAH

Honeycomb Hills, 269
Fumarole Butte and Smelter Knolls, 270
Black Rock Desert, 271
Twin Peaks, 273
Mineral Mountains and Cove Fort, 273
Kolob and Loa, 275

HONEYCOMB HILLS
Utah

Type: *Silicic dome*
Lat/Long: *39.71°N, 113.58°W*
Elevation: *1,733 m*
Height: *210 m*
Diameter: *900 m*
Eruptive History: *4.7 Ma*
Composition: *Rhyolite*

The small rhyolite dome of the Honeycomb Hills is of interest because it displays chemical and mineralogical features characteristic of a pegmatite magma. Twelve meters of pyroclastic deposits are overlain by passively emplaced domal lavas. Erosion has exposed the internal structure of the dome, including the central conduit with its concentric, vertical flow banding and an increase in size and abundance of phenocrysts. The lavas contain phenocrysts of plagioclase, sanidine, quartz, fluorbiotite, and fluortopaz. Accessory phases, commonly found in rare element pegmatites, include fluorite, zircon, thorite, fluocerite, as well as the niobates columbite, fergusonite, samarskite, and ishikawaite. The lavas are characterized by extreme enrichments in Li (340 ppm), Be (270), B (240), Rb (1960), Cs (80), and U (30). Initial magmatic concentrations of fluorine were high, with 2.3% F in vitrophyre, and 2-3% F in melt inclusions in quartz. Eruption of the silicic (70-75%), cool (640°C), and phenocryst-rich (15-45%) lavas was aided by high contents of fluorine, which reduced the solidus and viscosity.

How to get there: *The Honeycomb Hills are in westernmost central Utah, ~180 km southwest of Salt Lake City. Proceed west on the paved road past the Intermountain Power Project that intersects US Highway 6, ~17 km north of Delta, Utah. After passing Topaz Mountain in the Thomas Range, turn left on the graded gravel road to Callao, Utah, and proceed over Sand Pass at the southern end of the Fish Springs Range. The Honeycomb Hills are on the north side of the road ~20 km east of the community of Callao. There are no services once you leave US 6 – have a full tank of gas and carry water.*

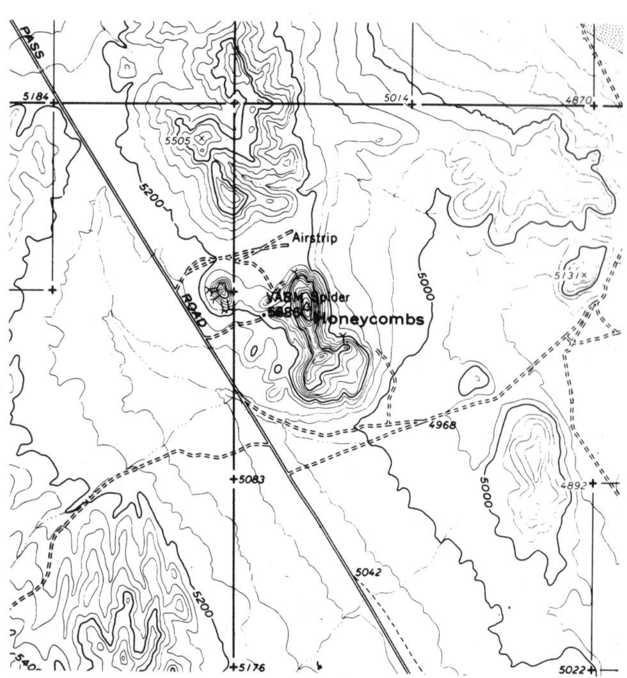

HONEYCOMB HILLS: USGS topographic map. Contour interval 40 ft.

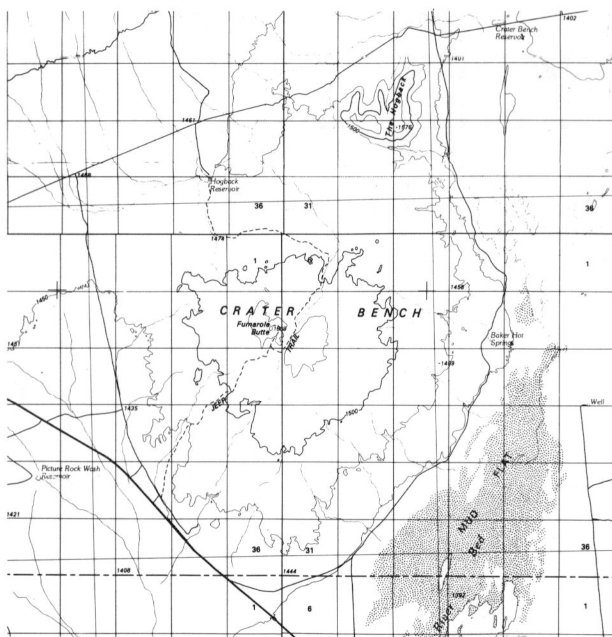

FUMAROLE BUTTE: USGS topographic map. Contour interval 50 m.

References

Christensen, E. H., Sheridan, M. F., and Burt, D. M., 1986, The geology and geochemistry of Cenozoic topaz rhyolites from the western United States: *Geol. Soc. Am. Spec. Pap. 205*, 82 pp.

Congdon, R. D., and Nash, W. P., 1988, High fluorine rhyolite: an eruptive pegmatite magma at the Honeycomb Hills, Utah: *Geology* (in press).

William P. Nash

FUMAROLE BUTTE AND SMELTER KNOLLS
Utah

Type: *Shield and dome*
Lat/Long: *39.5°N, 112.8°W*
Elevation: *Fumarole Butte (F.B.), 1,609 m; Smelter Knolls (S.K.), 1,555 m*
Height: *F.B., 207 m; S.K., 138 m*
Diameter: *F.B., 12 km; S.K., 5 km*
Eruptive History:

6.2 Ma	*F.B. (rhyolite)*
6.1 Ma	*S.K. (basaltic andesite)*
6.0 Ma	*F.B. (tholeiitic basalt)*
3.4 Ma	*S.K. (rhyolite)*
0.9 Ma	*F.B. (basaltic andesite)*
0.31 Ma	*S.K. (tholeiitic basalt)*

Composition: *Basalt, basaltic andesite, rhyolite*

Fumarole Butte is a Quaternary basaltic andesite shield volcano. It overlies minor outcrops of Tertiary basalt and rhyolite erupted 6.1 Ma, contemporaneously with rhyolite of the Keg and Thomas Ranges 10-30 km to the north and west. Fumarole Butte has normal magnetic polarity indicating eruption during the Jaramillo event. The volcanic neck which provides the name Fumarole Butte is at the center of the volcano and rises ~30 m above the gentle slope of the shield. The volcano was inundated briefly by Lake Bonneville, and there are scattered remnants of lacustrine deposits. Benches developed at the Provo level (1,463 m). Crater Springs (also known as Baker Hot Springs and Abraham Hot Springs), on the eastern margin, produced thermal water (87-90°C) at an estimated discharge rate of ~17 l/s in the summer of 1967.

The Smelter Knolls consist of several coalescing rhyolite domes and flows with an estimated total erupted volume of 2.2 km^3. No pyroclastic deposits are exposed, although flow

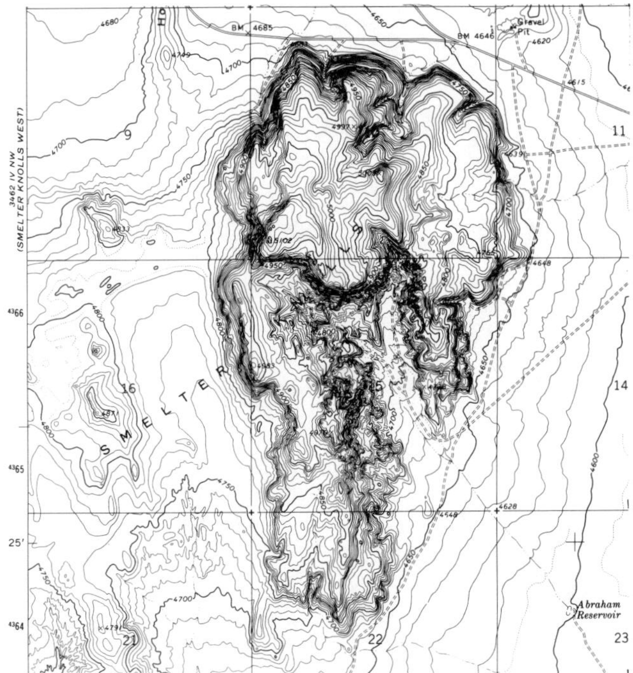

SMELTER KNOLLS: USGS topographic map. Contour interval 10 ft.

breccias and basal vitrophyres are present. The lava is topaz rhyolite with high concentrations of Li (145 ppm), Be (14), Rb (530), Cs (24), Th (58), and U (15).

The two main bodies of the Smelter Knolls are separated by a north–south-trending normal fault, consistent with the orientation of normal faults in Fumarole Butte 15 km to the north. Nearby mafic volcanism dated at 6.1 and 0.3 Ma bracket the silicic episode.

How to get there: *Fumarole Butte is in west-central Utah ~30 km north of Delta, Utah. The paved road to the Intermountain Power Project, which intersects US 6 nearly 17 km north of Delta, passes south of the shield. Dirt roads provide access around the shield and to the Hot Springs but are often impassable when wet.*

Smelter Knolls is 24 km west-northwest of Delta, Utah. Access is via paved and graded gravel roads from Delta. Follow signs directing you to the Topaz Internment Camp Site and Memorial. The Smelter Knolls are 6 km to the west.

References

Christensen, E. H., Sheridan, M. F., and Burt, D. M., 1986, The geology and geochemistry of Cenozoic topaz rhyolite from the western United States: Geol. Soc. Am. Spec. Pap. 205, 82 pp.

Peterson, J. B., and Nash, W. P., 1980, Geology and petrology of the Fumarole Butte volcanic complex,Utah: *in* Studies in late-Cenozoic volcanism in west-central Utah. *Utah Geol. Min. Surv. Spec. Studies* 52, 34-58.

William P. Nash

BLACK ROCK DESERT
Utah

Type: *Monogenetic volcanic field*
Lat/Long: *39.0°N, 112.5°W*
Elevation: *1,500 to 1,800 m*
Eruptive History:

1.5 Ma	*Andesite of Beaver Ridge*
0.92 Ma	*Basalt I of Beaver Ridge*
0.67 Ma	*Black Rock volcano*
0.54 Ma	*Basalt II of Beaver Ridge*
0.4 Ma	*Basalt of Deseret*
0.40 Ma	*Rhyolite of White Mountain*
0.13-0.03 Ma	*Pavant lava field*
16,000 yr BP	*Pavant Butte tuff cone*
14,500 yr BP	*Tabernacle Hill tuff cone and flows*
800 yr BP	*Ice Springs cinder cones and flows*

Composition: *Basalt, basaltic andesite, andesite, rhyolite*

The Black Rock Desert volcanic field consists of a rhyolite dome and many mafic cones and flows erupted from about two dozen vents within the past 1.5 Ma. Early flows are covered extensively by sediments from Lake Bonneville. **White** Mountain rhyolite dome (0.1 km^3) is the youngest rhyolite yet dated in Utah. Its pumiceous carapace has been stripped away by Lake Bonneville and deposited as pumice cobbles in a bar southeast of the dome. **Pavant Butte**, a large (275-m-high, 3-km-diam.) tuff cone, was built 16,000 yr BP on top of pahoehoe and aa lavas of the Pavant field, and erupted through a water depth of 85 m in Lake Bonneville. A prominent shoreline terrace is developed on the tuff cone. The then subaerial portion has been entirely converted to palagonite. Eruptions during the waning stages of Lake Bonneville at **Tabernacle Hill** produced two tuff cones and lavas emanat-

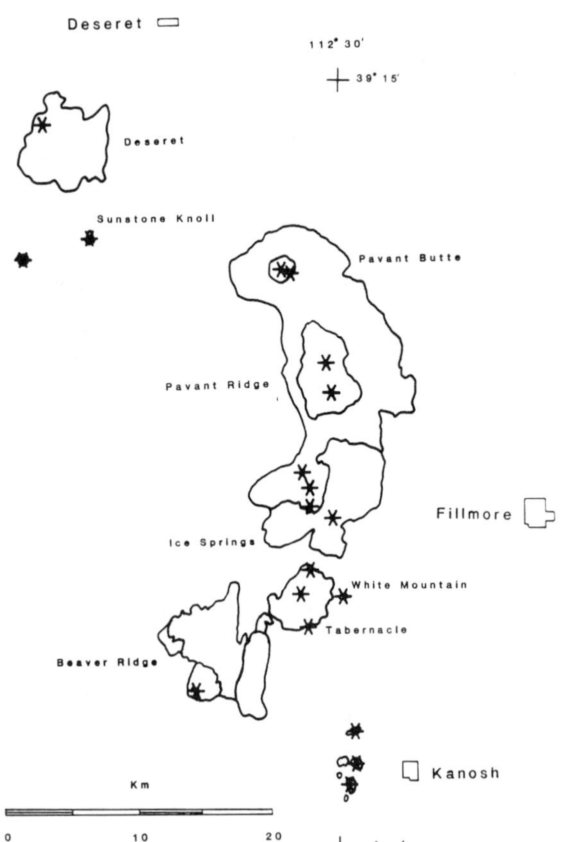

BLACK ROCK DESERT: Sketch map of volcanics east of Fillmore, Utah: Deseret, Pavant, Ice Springs, White Mountain, Tabernacle, and Beaver Ridge.

BLACK ROCK DESERT: Dark, young (800 yr BP) lava flows of Ice Springs volcano.

ing from a central vent which collapsed to form a crater 20 m deep and 0.4 km in diameter. Lava tubes are common, and pit craters dot the surface of the volcano, which is roughly circular and 4 km in diameter. Several features suggest that lavas from the Tabernacle were erupted into Lake Bonneville. Rounded and irregular pillows with glassy rinds and course grained interiors are common on the edge of the flow, and tufa encrustations and wave-rounded cobbles occur as high as 9 m above the base of the flow. Tephra deposits from these two eruptions form distinctive stratigraphic markers in Quaternary Lake deposits of the Sevier and Black Rock Deserts. The most recent activity (~800 yr BP) at Black Rock Desert volcanic field produced the five nested cinder cones and lavas of the **Ice Springs** volcano. Lava flows have a maximum length of 8 km. Both flows and tephra contain abundant xenoliths of partially fused granite. A travertine deposit (1 × 1.5 km) is forming along faults aligned with the structural fabric of the volcanic field.

The location of volcanic vents is strongly controlled by upper crustal Basin and Range structure, with 23 of 22 identified vents falling within the −172 to −187-mgal gravity band that defines the major structural discontinuity bounding the graben of the Black Rock Desert. The mafic lava is hypersthene normative tholeiite, and except for the Pavant lava, has Mg# (MgO/(MgO+FeO)) < 60, indicating that it has undergone some fractionation prior to eruption.

How to get there: *The Black Rock Desert is ~250 km south of Salt Lake City in west-central Utah, immediately west of Interstate 15 between the communities of Kanosh and Holden; several lava flows and cinder cones can be recognized from the freeway. The road west from Fillmore to Flowell provides direct access to the youngest volcanoes. A graded gravel road going south from Flowell leads to the Ice Springs, Tabernac-*

le, and White Mountain vents. North from Flowell, the gravel road to Clear Lake Bird Refuge provides access to Ice Springs and Pavant flows, and Pavant Butte.

References

Hoover, J. D., 1974, Periodic Quaternary volcanism in the Black Rock Desert, Utah: *Brigham Young Univ. Geol. Studies 21*, 3-72.

Oviatt, C. G., and Nash, W. P., 1988, Late Pleistocene basaltic ash and volcanic eruptions in the Bonneville Basin, Utah: *Bull. Geol. Soc. Am.* (in press).

William P. Nash

TWIN PEAKS
Utah

Type: *Bimodal, basalt-rhyolite field*
Lat/Long: *38.7° N, 112.8°W*
Elevation: *1,650 to 2,120 m*
Eruptive History:

2.74 Ma	*Rhyodacite of Coyote Hills*
2.45 Ma	*Rhyolite of Cudahy Mine*
2.6 Ma	*Basalt of Cove Creek*
2.39 Ma	*Rhyolite of North Twin Peak*
2.37 Ma	*Rhyolite of South Twin Peak*
2.22 Ma	*Basalt of Lava Ridge*
2.11Ma	*Basaltic andesite of Burnt Mountain*

Composition: *Basalt, rhyodacite, rhyolite*

Twin Peaks is a late Tertiary, bimodal volcanic field of tholeiite, basaltic andesite, and low- to high-silica rhyolite. Silicic volcanism spanned the period from 2.7 to 2.3 Ma, with two evolutionary cycles beginning with rhyodacite and culminating in high-silica rhyolite. The total volume of silicic eruptives is estimated to be ~12 km^3, much of which is now buried under younger sediments. Silicic volcanics comprise **North Twin Peak** rhyolite dome, **South Twin Peak** rhyolite dome (the largest of the domes, being 400 m high and 1.4 km in diameter), and coalescing domes and flows of the **Coyote Hills**, which define a semi-circular pattern around the 10-km-diameter complex. Mafic lavas are confined primarily to the south. Local subsidence at the end of the first silicic cycle resulted in deposition of 300 m of calcareous lacustrine sediments with interbedded silicic tuffs and mafic lava flows. Resurgent doming in the south produced a doubly plunging anticline with 400 m of relief. Basalt from **Burnt** Mountain erupted on the crest of the resurgent dome and flowed down the eroded flanks of the anticline, indicating that resurgence closely followed cessation of silicic volcanism.

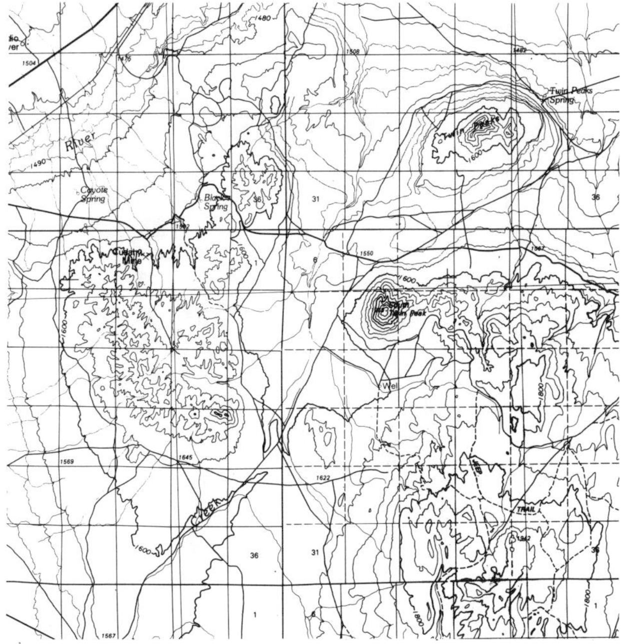

TWIN PEAKS: USGS topographic map. Contour interval 50 m.

How to get there: *Twin Peaks is in west-central Utah ~275 km south of Salt Lake City. Access is via a graded gravel road heading west from the southern Kanosh, Utah, exit on Interstate 15.*

Reference

Crecraft, H. R., Nash, W. P., and Evans, S. H., Jr., 1981, Late Cenozoic volcanism at Twin Peaks, Utah: Geology and petrology. *J. Geophys. Res. 86*, 10,303-10,320.

William P. Nash

MINERAL MOUNTAINS AND COVE FORT
Utah

Type: *Monogenetic volcanic field*

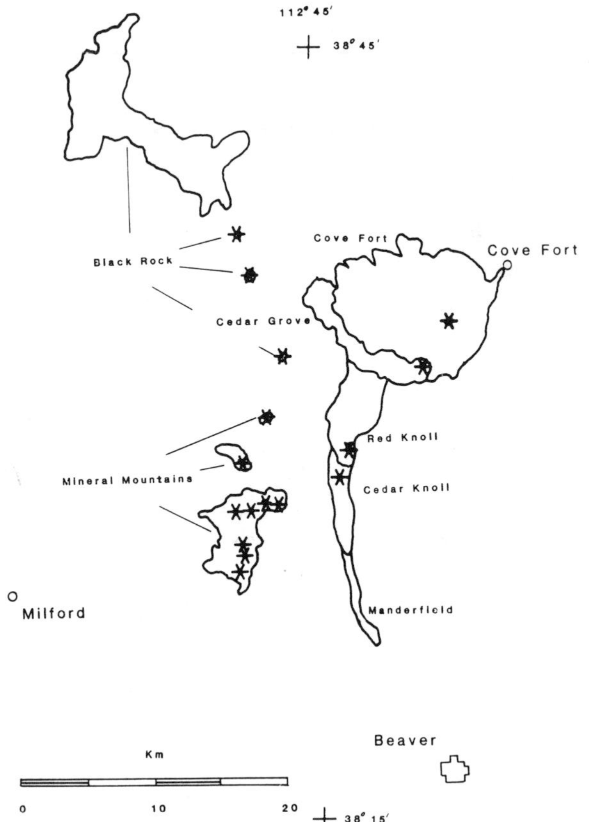

MINERAL MOUNTAINS AND COVE FORT: Sketch map of young volcanics northeast of Milford, Utah: Black Rock, Cove Fort, Red Knoll, Cedar Knoll, Manderfield, and Mineral Mountains.

Lat/Long: *38.5°N, 112.8° W*
Elevation: *1,510 to 2,770 m*
Eruptive History:

1.32 Ma	*Black Rock basalt flows*
1.1 Ma	*Manderfield basalt flow*
1.0 Ma	*Crater Knoll basaltic andesite cinder cone and flows*
0.97 Ma	*Black Rock basalt flows*
0.92 Ma	*Mineral Mountains basalt flows*
0.77 Ma	*Mineral Mountains rhyolite flows*
0.7 -0.5 Ma	*Mineral Mountains rhyolite domes and ash flows*
0.5 Ma	*Cove Fort cinder cone and flows*
0.3 Ma	*Cedar Grove cinder cone and flows*

Composition: *Tholeiite basalt, basaltic andesite, latite, rhyolite*

The Mineral Mountains–Cove Fort volcanic field is a Quaternary bimodal, basalt-rhyolite association with some intermediate composition units. Silicic volcanism began at 0.8 Ma with eruption of two fluid, aphyric, rhyolite flows (3 km long, 80 m thick) along Bailey Ridge and Wildhorse Canyon. Subsequent activity from 0.7 to 0.5 Ma consisted of pyroclastic eruptions and extrusion of at least 11 domes distributed over 10 km along the crest and western flank of the Tertiary Mineral Mountains pluton. Tephra from these eruptions are abundant in the lacustrine deposits of the Beaver Basin. East of the Mineral Mountains are lavas of basalt, basaltic andesite and latite which erupted before the silicic episode of the Mineral Mountains and persisted afterwards. Activity began with outpourings of tholeiitic basalt of the Black Rock field from vents on the eastern margin of the Mineral Mountains, followed by basaltic andesite of the **Manderfield** and **Crater Knoll** fields. Latite lavas were then erupted from **Red Knoll** cinder cone, followed by quartz-bearing basaltic andesite from the topographically dominant Cove Fort cinder cone. The youngest lavas are latite of the **Cedar Grove** field, erupted from a cinder cone on the southwest margin of the Cove Fort field. In these two younger units surface features are readily apparent, including pressure ridges, squeezeups, and pahoehoe textures.

The Roosevelt geothermal field, on the west flank of the Mineral Mountains, is presently producing 25 MW of electrical power. There are Quaternary deposits of siliceous sinter and hydrothermally altered and cemented alluvium, as well as fumarolic activity. The reservoir temperature is 260°C at a depth of 700-1,300 m.

How to get there: *The Mineral Mountains–Cove Fort volcanic field is ~300 km south of Salt Lake City, and ~100 km north of Cedar City, Utah. Access to the silicic volcanoes of the Mineral Mountains is from the west via Milford, Utah. Interstate 15 crosses the Cove Fort flows immediately north of the interchange with Interstate 70. The graded gravel road at the Cove Fort exit provides access to the northern part of the field.*

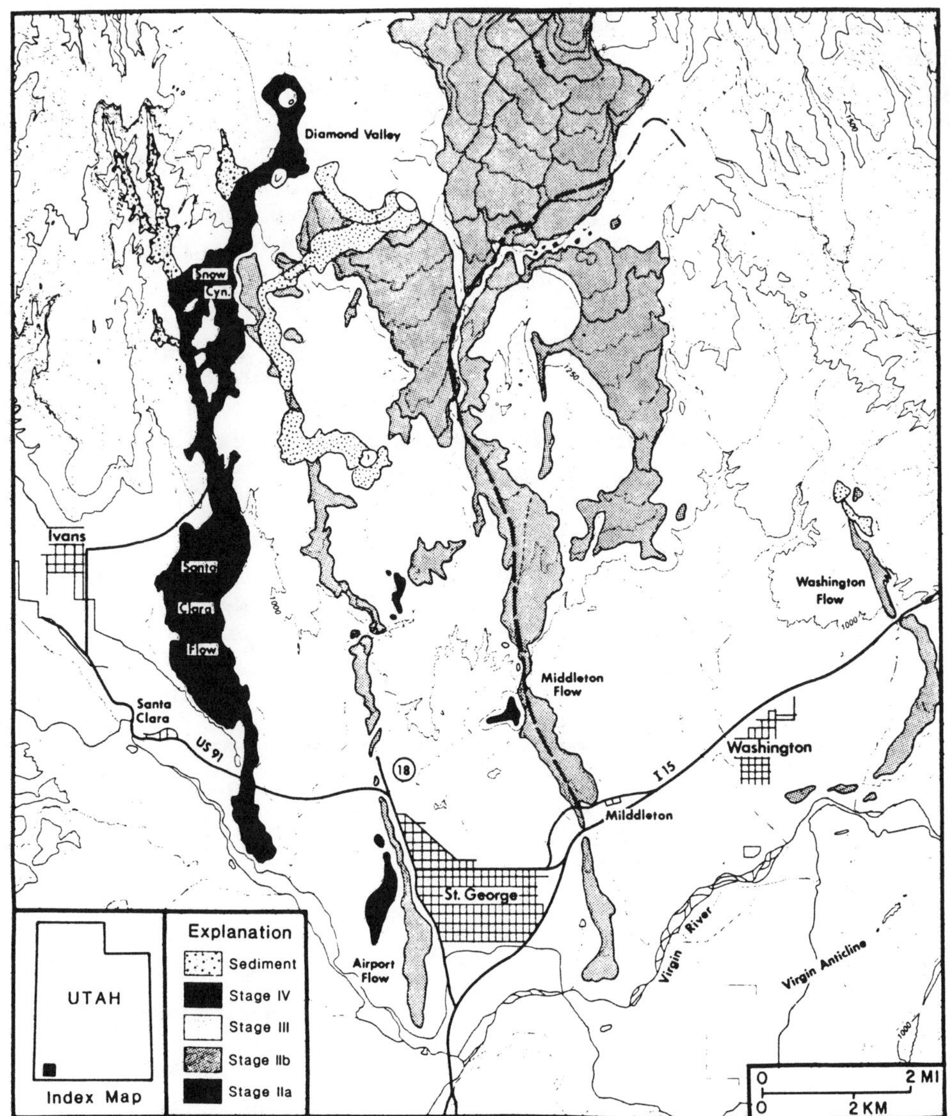

KOLOB AND LOA: Geologic map of lava flows near St. George, Utah (from Hamblin, 1987). Stage IV flows are ~1,000 yr old; Stage III, <0.5 Ma; Stage II, 1-2 Ma.

References

Clark, E. E., 1977, Late Cenozoic volcanic and tectonic activity along the eastern margin of the Great Basin in the proximity of Cove Fort, Utah: *Brigham Young Univ. Geol. Studies 24*, 87-114.

Lipman, P. W., Rowley, P. D., Mehnert, H., Evans, S. H., Nash, W. P., and Brown, F. H., 1977, Pleistocene rhyolite of the Mineral Mountains, Utah: geothermal and archaeological significance: *J. Res., USGS 6*, 133-147.

William P. Nash

KOLOB AND LOA
Utah

Type: *Monogenetic volcanic fields*
Lat/Long: *37.5°N, 112.7° W*
Eruptive History: *6 Ma to 1? ka*
Composition: *Tholeiite basalt, basaltic andesite, latite, rhyolite*

A line of Quaternary lava flows and cinder cones stretches from St. George, Utah, nearly 200 km northeastward to the village of Loa. Additional young vents and flows extend ~50 km north of St. George. Volcanism near St.

George is best known and most spectacular; lava flows erupted from vents in the Pine Mountains flowed downslope into river valleys. Four different episodes of flow emplacement have been recognized, each preserving underlying Mesozoic rock from further erosion. The oldest flows, formed 3-6 Ma, are up to 300 m above their surroundings, and younger flows occur at ~120 m (1-2 Ma), on the present drainage (0.5 Ma), and fill stream valleys (a few thousand years). The most recent flows came from two cinder cones in Diamond Valley, 16 km north of St. George. The cones, ~400 wide and 60 m high, are the sources for the **Santa Clara** flow which traveled 16 km to the south. Geomorphological features to be seen along the flow include inverted valleys, lava dammed lakes, displaced drainages, and 120-m-high lava cascades.

Volcanism along the northeast trend to Loa is poorly known. The occurrence of andesite near Panguitch, Utah, is consist with other volcanic fields in Utah, which seem to have a higher ratio of evolved to basic rocks than is common other places in the southwest.

How to get there: *Many of the volcanic features can be reached easily by automobile from St. George, Utah. The Santa Clara flow can be viewed along State Highway 18, west of St. George in and around Snow Canyon State Park. Other Quaternary vents are distributed along the highway between St. George, Hurricane, and Virgin,and north along US 91 and I-15. Access to volcanoes in the northeast part of the field is along State Highway 14 from Cedar City to US 89 via Navajo Lake. The Loa volcanic field is accessible via Highway 72, which connects Loa and Escalante, Utah.*

Reference

Hamblin, W. K., 1987, Late Cenozoic volcanism in the St. George basin, Utah: *Geol. Soc. Am. Cent. Field Guide-Rocky Mt. Sect.*, 291-294.

Charles A. Wood

VOLCANOES OF ARIZONA

Uinkaret, 277
San Francisco, 278
Sunset Crater, 280
Mormon, 282
Hopi Buttes, 283
Springerville, 285
San Carlos, 286
Sentinel Plain, 287
Geronimo, 287

UINKARET
Arizona

Type: *Monogenetic cones and flows*
Lat/Long: *36.5°N, 113.1°W*
Elevation: *1,555 m*
Eruptive History: *1.2 Ma to 12,500 yr BP*
Composition: *Basanite, alkali olivine basalt, hawaiite, tholeiitic basalt*

Late Cenozoic volcanism extends across a broad region from southwestern Utah to the north rim of the Grand Canyon in western Arizona. In the Grand Canyon region, lavas are strongly alkalic, whereas transitional varieties and tholeiitic basalt, together with minor andesite, occur further north in Utah. The lavas lie astride the Basin and Range–Colorado Plateau transition zone, which is characterized by major Late Cenozoic normal faulting.

The relative ages of lavas can be readily established from morphological relations, with the oldest capping mesas or buttes, and the most recent occupying present drainage valleys. Many of these young flows have no soils developed on them and have well preserved flow features and associated cinder cones. The youngest measured age is 12,500 yr BP for a young flow at the Grand Canyon.

The Uinkaret volcanic field at the north rim of the Grand Canyon in the Grand Canyon National Monument is especially noteworthy. The lavas are alkalic and commonly contain peridotite inclusions. **Vulcan's Throne**, a Quaternary cinder cone on the rim of the Canyon, is cut by recent fault movement on the Toroweap fault. Late Cenozoic lava flows have repeatedly flowed down Toroweap Valley and several adjacent valleys into the Grand Canyon, at times forming large lava dams. Flows are exposed on the walls of the Grand Canyon, often interbedded with fluvial and lacustrine sediments, up to 600 m above present river level. Several of the dams are estimated to have been at least 200 m high. The most recent flows in the Grand Canyon have cascaded over the rim of the Esplanade to the river 1,000 m below. Within the river itself is a volcanic neck, **Vulcan's Forge,** 25 m in diameter and rising 15 m above the river.

How to get there: *Volcanic features between St. George, Utah, and Grand Canyon National Mon-*

ument can be reached by an unpaved, scenic route over Mount Trumbull. It is not well suited for passenger cars. A more direct route to the Monument is to proceed east from St. George to Hurricane, then southeast on Utah 59, which becomes Arizona 389. At 8.5 km past Pipe Springs National Monument, turn south on the gravel road to Grand Canyon National Monument (~90 km). There is a campground on the rim of the inner gorge of the Grand Canyon, and a rail descends to the Colorado River. Once you leave paved roads, either at St. George or Arizona 389, there are no services. The round trip from St. George to the Grand Canyon and return via Arizona 389 is ~470 km. A "must" is the excellent guidebook by Hamblin and Best (1970), which includes a detailed road log complete with 23 air photos covering the whole field trip route.

References

Best, M. G., and Brimhall, W. H., 1974, Late Cenozoic alkalic basaltic magmas in the western Colorado Plateaus and the Basin and Range transition zone, USA, and their bearing on mantle dynamics: *Geol. Soc. Am. Bull. 81*,1677-1690.

Hamblin, W. K., and Best, M. G., 1970, The western Grand Canyon district: *Guidebook to the Geology of Utah, 23*, Utah Geological Society, 156 pp.

William Nash

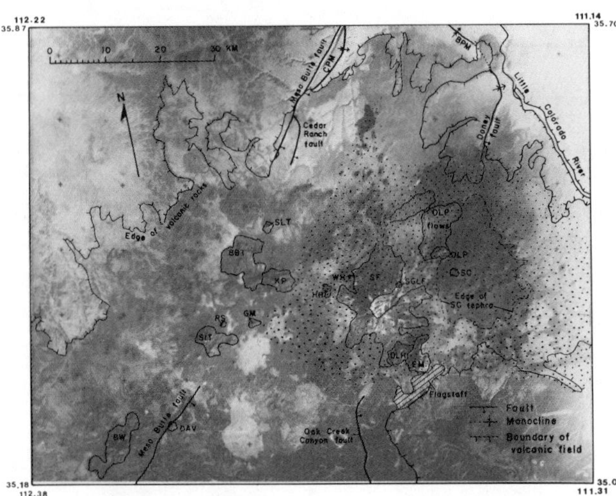

SAN FRANCISCO: Major features of San Francisco volcanic field. Major western silicic centers: Bill Williams Mountain (BW), Sitgreaves Mountain (SIT), and Kendrick Peak (KP), with the adjacent trachyte of Bull Basin Mesa (BBT). Major eastern intermediate to silicic centers: San Francisco Mountain (SF) and O'Leary Peak (OLP). Additional dacite domes or dome complexes: Davenport Hill (DAV), Dry Lake Hills (DLH), and Elder Mountain (EM). Additional rhyolite domes: RS Hill (RS), Government Mountain (GM), Slate Mountain (SLT), Hochderffer Hills (HH), White Horse Hills (WH), and Sugarloaf (SGLF). Sunset Crater cinder cone (SC) and related tephra blanket. Area dominated by vents of Brunhes age (stippling). Coconino (CPM) and Black Point (BPM) monoclines. NASA Landsat RBV image E-31041-17154-D.

SAN FRANCISCO
Arizona

Type: *Polygenetic volcanic field with basaltic cones and silicic stratovolcano*
Lat/Long: *35.4°N, 111.75°W*
Elevation: *3,850 m*
Eruptive History: *Episodic eruption from ~ 6 Ma to <1 ka; generally northeast to east progression*
Composition: *Basalt to rhyolite*

The San Francisco volcanic field is one of several dominantly basaltic volcanic fields of late Cenozoic age near the south margin of the Colorado Plateau. The volcanic field, which is predominantly of Pliocene and Pleistocene age, is just north of a broad transition zone between the Colorado Plateau and Basin and Range provinces. K-Ar ages indicate a general northeastward progression of volcanism during the past 15 Ma across the transition zone from central Arizona onto the plateau.

The San Francisco field lavas, ranging in composition from basalt to rhyolite, erupted through Precambrian basement and the overlying kilometer-thick cover of nearly horizontal sedimentary rocks, mostly of Paleozoic age. The volume of volcanic rocks is ~500 km^3; they cover an area of ~5,000 km^2.

Major structural features of the San Francisco field include the high-angle Mesa Butte and Oak Creek Canyon fault systems and the Black Point and Coconino Point monoclines. Some, if not all, of these structures probably had their origins during the Precambrian and have been subsequently reactivated. Late Cenozoic faulting continued until at least 0.5 Ma, but deformation of the lavas has been minimal. Lavas of pre-Brunhes age are locally broken by high-angle faults, generally of small displace-

SAN FRANCISCO: San Francisco Mountain, an eroded stratovolcano. Humphreys Peak, the highest point in Arizona, at an elevation of 3,850 m above sea level, is just left of center. A basalt cinder cone is in the middle ground. View is eastward.

ment. With few exceptions the lavas of Brunhes age are not faulted.

Most of the San Francisco field is covered by basalt flows ranging in age from ~6 Ma to <1 ka. Flows erupted from ~600 individual vents, most of which are marked by cinder cones. The basalt is predominantly alkali-olivine basalt and associated plagioclase-rich basalt of hawaiitic composition. Included among the basaltic rocks is basaltic andesite, which has the composition of SiO_2- and K_2O-enriched basalt and commonly contains sieved plagioclase, augite, and olivine and may contain hypersthene amphibole or quartz. Much basaltic andesite was erupted simultaneously from the same vents as alkali-olivine basalt and may be best interpreted as contaminated basalt.

Five conspicuous centers at which eruptions of intermediate to silicic lavas were concentrated occur within the volcanic field. Those in the western part of the field – from southwest to northeast, Bill Williams Mountain, Sitgreaves Mountain, and Kendrick Peak – coincide approximately with the Mesa Butte fault zone. Several smaller silicic domes occur along the same trend. K-Ar ages summarized from the western centers and the nearby domes suggest northeastward progression of eruptive activity:

Volcanic center	Period of activity (Ma)
Bill Williams Mountain	4.2-2.8
Sitgreaves Mountain	2.8-1.9
RS Hill	2.9
Government Mountain	2.1
Kendrick Peak	2.7-1.4
Slate Mountain	1.9-1.5

SAN FRANCISCO: Part of Sunset Crater, Bonito lava flow (produced in Sunset Crater eruption), and O'Leary Peak. Black cinders, deposited during an early phase of the Sunset eruption, mantle pre-Sunset cones and are in turn overlain by the Bonito flow. View is northward.

The western centers consist of closely spaced silicic domes and some viscous flows. **Bill Williams** Mountain, the oldest of the western centers, is composed primarily of andesite and dacite. **Sitgreaves** Mountain consists almost entirely of rhyolite domes. **Kendrick Peak** consists dominantly of dacite and rhyolite domes and flows that were intruded and partly buried by andesite. Trachyte and benmoreite that erupted from the west flank of Kendrick Peak, and scattered basaltic vents that overlie rocks of the western silicic centers, are interpreted as younger and not genetically related to the western-center magma systems.

Two additional intermediate to silicic centers, San Francisco Mountain and **O'Leary Peak** in the eastern part of the volcanic field, may have been localized by the Oak Creek Canyon and Doney fault zones, and are mostly younger than the western centers. The upper part of San Francisco Mountain is a truncated stratovolcano built primarily of porphyritic andesitic and dacitic flows and pyroclastic deposits erupted from ~1.0 to 0.4 Ma. These lavas overlie still older dacite and rhyolite, however, that are as old as 2.8 Ma.

Except for a small volume of plagioclase-rich basalt, basaltic lavas are conspicuously absent within San Francisco Mountain, even though basalt of the same general age was erupted abundantly to the west, north, and east. This virtual absence of basalt may reflect the presence of an evolved magma chamber beneath the stratovolcano that blocked ascent of basaltic magma to the surface.

Trachyte of Bull Basin Mesa, erupted just west of Kendrick Peak, is related by age and composition to a group of benmoreite domes, cinder cones, and flows characterized by high Na_2O contents (~5.0-6.5%) and young ages (1.6-0.33 Ma). The trachyte and these benmoreites occur northwest, west, and south of San Francisco Mountain. Their locations roughly define an arc about the western part of the stratovolcano, and benmoreite xenoliths in a basaltic cinder cone extend the arc north of San Francisco Mountain. Possibly the trachyte and benmoreites mark the western limit of the San Francisco Mountain magma chamber.

The **Sugarloaf** rhyolite dome erupted on the northeast flank of San Francisco Mountain ~0.22 Ma. At about the same time, the O'Leary Peak volcanic center, which includes a pair of dacite-porphyry domes, several dacite flows, an andesite flow, and small rhyolite domes, erupted ~9 km to the northeast. One of the dacite-porphyry domes has been dated at 0.24 Ma and a dacite flow at 0.17 Ma. The O'Leary Peak volcanic center and Sugarloaf are the youngest major silicic eruptives in the San Francisco volcanic field.

A deep linear valley transecting the northeast quadrant of San Francisco Mountain was formed between 0.4 Ma, when the youngest unit of the composite cone was erupted, and 0.22 Ma, when Sugarloaf dome was emplaced as the valley's northeast end. After the Sugarloaf eruption, glaciation further sculptured the valley. Although erosion was important in forming the valley, it seems likely that the erosion was localized by collapse of the northeast quadrant of the volcano. The valley, the aligned vent systems of O'Leary Peak, Sugarloaf, the San Francisco Mountain stratovolcano, and a well-defined, linear aeromagnetic low are all colinear and thus may have formed under the influence of common structural control.

K-Ar dating and paleomagnetic measurements show that eruptive activity at intermediate and silicic volcanic centers was broadly coeval with nearby basaltic volcanism. Furthermore, the predominant age of the surface volcanic rocks increases westward and southwestward. Thus, Brunhes-age lavas, comprising much of San Francisco Mountain, all of the O'Leary Peak center, and >200 basalt cinder cones with their related flows, occur only in the eastern half of the volcanic field. The western half of the volcanic field is underlain by Matuyama and pre-Matuyama lavas. Most are of Matuyama age, but pre-Matuyama lavas dominate in the far southwestern parts at Bill Williams Mountain and among the nearby, southwesternmost basalt cones and flows.

The youngest (<0.2 Ma) basaltic lavas occur north and east of San Francisco Mountain. The most recent eruption, which began ~1065 A.D., produced the Sunset Crater cinder cone, three lava flows, and an extensive air-fall tephra sheet.

How to get there: *The San Francisco volcanic field is in north-central Arizona; Flagstaff is at its southern border. US Interstate 40 passes along the southern edge of the volcanic field, and US Highways 180 and 89, and Arizona Highway 64 transect the volcanic field. Secondary roads, mostly unpaved, provide good access to almost all the San Francisco field.*

References

Wolfe, E. W., Ulrich, G. E., Moore, R. B., Newhall, C. G., Holm, R. F., and Bailey, N. G., 1987, Geologic maps of the San Francisco volcanic field, north-central Arizona: *USGS Misc. Field Studies Map MF-1956-1960.*

Tanaka, K. L., Shoemaker, E. M., Ulrich, G. E., and Wolfe, E. W.,1986, Migration of volcanism in the San Francisco volcanic field, Arizona: *Geol. Soc. Am. Bull. 97*, 129-141.

Edward W. Wolfe

SUNSET CRATER
Arizona

Type: *Scoria cone*
Lat/Long: *35.37°N, 111.50°W*
Elevation: *2,447 m*
Height: *300 m*
Volcano field diameter: *60 km*
Eruptive History: *Formed 1064?-1068 A.D.*
Composition: *Alkali olivine basalt*

SUNSET CRATER: Profile view provided by Peter Mouginis-Mark.

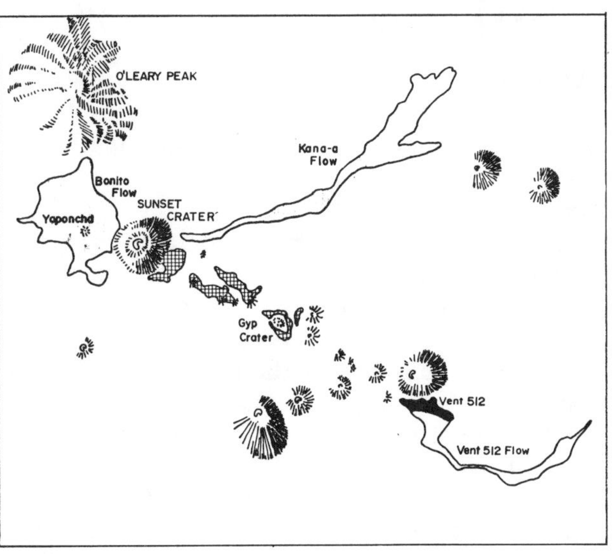

SUNSET CRATER: Map of Sunset Crater area (Pillies, 1979).

Sunset Crater is one of the youngest scoria cones in the contiguous United States. The cone is named for the topmost cap of oxidized, red spatter which makes it appear bathed in the light of the sunset. In the 1920s, H. S. Colton saved the cone from severe damage by averting the attempt of a Hollywood movie company to blow it up in order to simulate an eruption. This led to the establishment of the National Monument at Sunset Crater.

The Sunset eruption products are a classic example of monogenetic strombolian volcanism. The eruption began with the opening of a 15-km-long fissure, accompanied by curtain of fire activity and the growth of a small lava flow at the southeast end. Strombolian fountaining then localized near the northwest end and Sunset scoria cone grew, with the simultaneous deposition of a widespread scoria fall layer. At the same time the 11-km-long Kana-a lava flow issued from the cone. This was followed by further cone building and the production of the Bonito lava flow. This flow may have come from the base of the cone, as it has rafted portions of the cone on its surface. The last phase of the eruption featured low-level fountaining that repaired the the cone and deposited the cap of scoria and spatter, which oxidized bright red due to retained heat.

The Sunset event had a severe effect on the Sinagua Indians living in the area at the time, leading to a temporary exodus from the region. Anthropologists accept an unpublished model of the duration of the eruption based on archeomagnetic data. This proposes that the eruption lasted for 100-200 yr, with various phases occurring spasmodically over this period. Such a long, drawn-out eruption is unprecedented. Volcanologic studies of strombolian eruptions, including Sunset, suggest that a duration of a few years is more realistic. The age of the eruption is known from an archaeological site in the vicinity of the crater.

Unusual features of the Sunset eruption products are the comparatively large volume for a strombolian event (~3 km^3 equally shared between the scoria cone, scoria fall, and lava flows) and the wide, sub-plinian dispersal of the fall units, implying a high mass discharge rate of magma. Partly because of these factors, Sunset Crater falls at the upper end of the size range of scoria cones found on Earth.

How to get there: *From US Route 66 in Flagstaff take US Route 89 north toward Cameron. After ~15 km look for the right turn for Sunset Crater National Monument.*

References

Pillies, P. J., Jr., 1979, Sunset Crater and the Sinagua: A new interpretation: *in* Sheets, P. D., and Grayson, D. K. (eds.), *Volcanic Activity and Human Ecology*, Academic, New York, 459-485.

Holm, R. F., 1987, Significance of agglutinate mounds on lava flows associated with monogenetic cones: An example at Sunset Crater, Arizona: *Geol. Soc. Am. Bull. 99*, 319-324.

Stephen Self

MORMON
Arizona

Type: *Mono- and polygenetic volcano field*
Lat/Long: *35.0°N, 111.5°W*
Elevation: *1,500 to 2,530 m*
Eruptive History: *Continuous activity from 13.6 to 3.1 Ma*
Composition: *Basanite to rhyolite*

The Mormon volcanic field is one of a group of late Tertiary to Quaternary predominantly basalt fields that rim the Colorado Plateau in the southwestern United States. Located south of Flagstaff, Arizona, and north of the Verde and East Verde Rivers, the field straddles the Mogollon Rim, the boundary between the Colorado Plateau and the Transition Zone to the Basin and Range province. The Mormon volcanic field covers an area of >2,500 km^2 and contains over 250 basalt cinder cones, spatter cones, and lava cones, several shield volcanoes, and numerous large volume sheet flows. Silicic centers composed of andesite to rhyolite flows, domes, and pyroclastic deposits are scattered through the field. These volcanic deposits were erupted upon a basement of sedimentary rocks that include all of the Grand Canyon sequence. The underlying structure of this basement with its steeply dipping faults exerted a strong control on the character of the eruptive units. Tephra and lava cones are elongate or form aligned groups with northwest trends. Numerous basaltic feeder dikes that have northwest trends can also be observed in eroded areas. The volcanic style of the field changed as it evolved. Early erupted large-volume basalt sheet lavas were covered by small-volume shield volcanoes and scoria cones with associated lava flows. Intermediate and silicic magmas were emplaced late during the scoria cone activity. Erosion has subdued the profile of most cones and buried the tops of lava flows beneath mantles of soil; the area is heavily forested with ponderosa pine.

Almost all of the larger volcanic structures are composed of intermediate to silicic rocks. From the southern extent of the field to the northern limit, these centers/mountains are Hackberry, Apache Maid, Round, Table, and Mormon. Hackberry and Mormon mountains are the largest centers; **Hackberry** Mountain

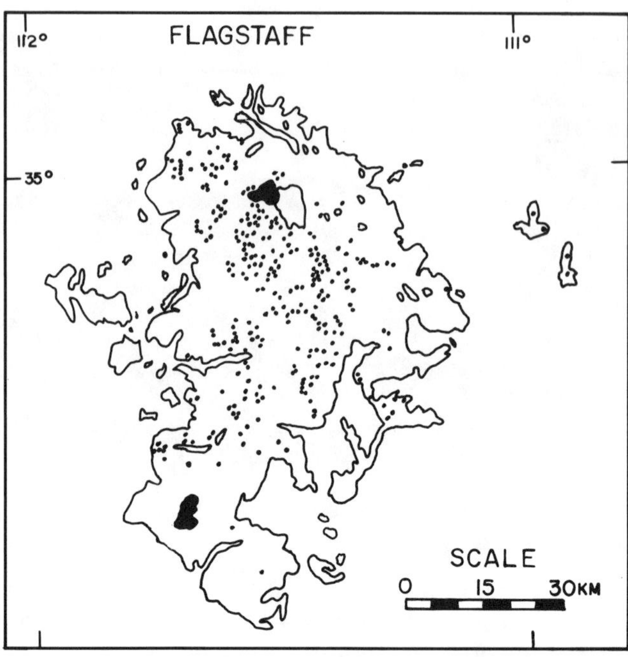

MORMON: Mormon Volcanic field with basaltic cinder cones shown as small black dots and major silicic center as irregularly shaped black areas. Mormon Mountain is in the northern part of the field and Hackberry Mountain is in the southern part. Apache Maid and Stoneman Lake are approximately halfway between these silicic centers.

(late Miocene) contains pyroclastic andesite to rhyolite ash flow deposits. Mormon Mountain (late Pliocene) is a low (~800-m relief) asymmetrical shield volcano composed of andesite and dacite flows, some with well-preserved blocky flow tops and platy terminations, and a rhyolite dome. **Apache Maid**, an asymmetrical scoria cone 300 m high and 1.6 × 2.1 km in diameter, clearly exhibits a three-phase, polygenetic volcanic history. Stage 1 is characterized by basalt pyroclastic eruptions that built the main cone. Stage 2 was dominated by spatter and numerous summit eruptions of lava. Stage 3 produced flank eruptions of mixed basalt and andesite lavas. **Table Mountain** is a relatively thick (45-60 m) andesite lava flow with a crumbly breccia top, and **Round Mountain** is an andesite dome. Another interesting feature, **Stoneman Lake**, is circular in plan and 90 m deep with steep-sided walls of basalt. It is interpreted as a collapse feature above a breccia pipe.

Rock types in the Mormon volcanic field vary from olivine nephelinite to rhyolite; no silica gap is observed. The field is dominated by silicic alkalic basalt and hawaiite. Basaltic lavas contain a variety of xenoliths, ranging from ultramafic cumulates (dominantly pyroxenites) to lower crustal granulites that are partially melted.

How to get there: *The Mormon volcanic field is ~30 km south of Flagstaff, Arizona, and 150 km north of Phoenix, Arizona, and is almost entirely contained within the Coconino National Forest. From Flagstaff the road to Lake Mary continues on, cutting through the center of the field; numerous, well-graded Forest Service roads provide access to the field.*

References

Gust, D. A., and Arculus, R. J., 1986, Petrogenesis of alkalic and calcalkalic volcanic rocks of Mormon Mountain volcanic field, Arizona: *Contr. Min. Petr. 94*, 416-426.

Holm, R. F., Nealey, L. D., Conway, F. M., and Ulrich, G. E., 1989, A. First Day Field Trip - Mormon Volcanic Field, Excursion 5A, Field Guide to Miocene-to-Holocene Volcanism and Tectonism of the Southern Colorado Plateau, Arizona. *IAVCEI (1989) Field Guide.*

David Gust

HOPI BUTTES
Arizona

Type: *Monogenetic volcano field*
Lat/Long: *35.45 °N, 110.1° W*
Elevation: *1,650-2,080 m*
Volcano Field Diameter: *60 km*
Eruptive History: *8.5 to 4.2 Ma*
Composition: *Monchiquite and limburgite tuff*

The dark brown Hopi Buttes dominate the surrounding red-, white-, and tan-colored landscape of the Painted Desert in north-central Arizona by commonly rising to heights of 180 m. Individual buttes are underlain by a diatreme, or by a complex of diatremes, whereas most of the buttes are capped by monchiquite flows. Several sediment-filled maars form inconspicuous low hills, while others may be buried beneath the alluvium. The diatremes and maars erupted into the late Miocene–early Pliocene Hopi Lake.

The volcanic rocks of the diatremes and

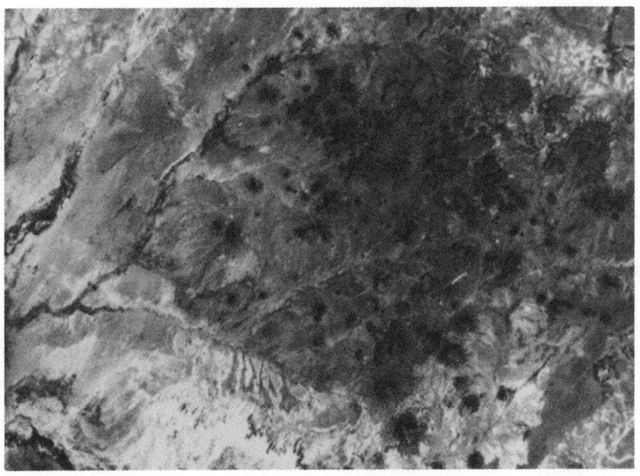

HOPI BUTTES: Landsat image of the volcanic field. The dark-colored buttes, each a diatreme or complex of diatremes, dominate the surrounding Painted Desert. The buttes are restricted to a circular area ~60 km wide.

maars are monchiquite and limburgite tuffs; both contain augite (most abundant), olivine, and biotite phenocrysts. These rocks are distinguished from normal alkali basalt of the southern Colorado Plateau by (1) the absence or paucity of orthopyroxene, hornblende, and plagioclase, and the presence of zeolites and feldspathoids, and (2) their extreme silica undersaturation and high H_2O, TiO_2, and P_2O_5 contents. Volcanic rocks that form tuffaceous deposits (limburgite tuffs) are both air fall and water-laid in origin. The major difference between the monchiquite and limburgite tuffs is textural; that is, they contain similar mineralogy and chemistry, but the monchiquite is coarser grained, forms flows, and has augite microlites in the groundmass, in contrast to the glassy matrix of the limburgite tuffs. The latter are commonly composed primarily of brown glass, and calcite amygdules and open-space fillings surrounded by black opaque iron oxides. Many of the diatremes contain maar-lake sediments, which include fine-grained clastic and carbonate (travertine) rocks, limburgite tuff and tuff breccia, agglomerate, blocks of sedimentary rocks derived from the vent walls (most notably the Wingate Sandstone), and monchiquite dikes, necks, and flows. The violent eruptions of these diatremes may have resulted from the exsolution of magmatic gas as magma ascended through the crust. Alternatively, they may have been due to the interac-

HOPI BUTTES: Morale Claim maar adjacent to the southwest side of the monchiquite flow-capped Red Clay mesa. The whitish-yellow lacustrine (travertine) sediments contrast sharply with the black monchiquite flow.

tion of magma with water-saturated rocks beneath Hopi Lake.

The age of the Hopi Buttes volcanic rocks appears to straddle the Miocene–Pliocene boundary. Nineteen K-Ar age determinations on flows, dikes, and tuff beds by M. Shafiquallah and P. E. Damon established that most volcanism and maar crater formation occurred between 8.5 and 6 Ma, with minor igneous activity extending until about 4.2 Ma.

Many trace elements, most notably the incompatible ones, are unusually abundant in the monchiquites compared to normal basaltic volcanism: Ag, As, Ba, Be, Ce, Dy, Eu, F, Gd, Hg, La, Nd, Pb, Rb, Se, Sm, Sn, Sr, Ta, Tb, Th, Tn, U, V, Zn, and Zr. Uranium was apparently sufficiently enriched in the magma and the residual hydrothermal fluids to form uraniferous travertine deposits. Twenty-three of the maars have travertine beds with uranium concentrations >100 ppm, and the **Morale Claim** diatreme contained sufficient uranium (as high as 5,000 ppm) to have supported minor uranium mining during the 1950s. The absence of an Eu anomaly suggests that the magma did not reside for any appreciable time at depths <50 km. Because the monchiquite is probably mantle-derived, with little crustal influence, yet contains incompatible elements such as U along with high CO_2 and H_2O (commonly believed to be concentrated by crustal contamination), the magmas may have

HOPI BUTTES: USGS topographic map.

been derived from a mantle inhomogeneity that lay beneath the continental shield.

How to get there: *Take I-40 east from Flagstaff, Arizona, and then follow either Route 87 north from Winslow (east side of the volcanic field) or Route 77 north from Holbrook (west side).*

References

Shoemaker, E. M., Roach, C. H., and Byers, F. M., Jr., 1962, Diatremes and uranium deposits in the Hopi Buttes, Arizona: *in Petrological Studies – A Volume in Honor of A. F. Buddington*: New York, Geol. Soc. Am., 327-355.

Wenrich, K. J., 1989, Hopi Buttes Volcanic Field, Excursion 8A: *in* Chapin, C. E., and Zidek, Jiri (eds.), *Field Excursions to Volcanic Terrains in the Western United States*, IAVCEI, Santa Fe, NM, 41 pp.

SPRINGERVILLE
Arizona

Type: *Monogenetic volcanic field*
Lat/Long: *34.25°N, 109.58°W*

Elevation: *2,200 to 3,100 m*
Relief: *900 m*
Volcanic Field Size: *50 × 80 km, 2,600 km^2*
Volcanic Field Volume: *~300 km^3*
Eruptive History: *2.5 to 0.3 Ma*
Composition: *47% alkali basalt, 28% hawaiite, 24% tholeiite, 2% mugearite and benmorite*

The Springerville volcanic field is noted for its size and volume (about half that of the San Francisco volcanic field, 300 km to the northwest), and its "classic" cinder cone field morphology. More than 380 vents and lava flows are clustered together in a ranch lands setting of grasslands, ponderosa pines, and aspens. The field was unmapped and geologically largely unknown until 1979. Like other Colorado Plateau marginal volcanic fields, it is characterized by a variety of basaltic rocks of transitional to alkalic affinities.

A large stratovolcano, **White Mountains Baldy**, lies immediately south of the field, and in many respects resembles the large stratovolcanoes associated with the cinder cone fields of both the San Francisco and Mount Taylor volcanic fields. However, White Mountains Baldy is morphologically degraded and much older (~8 Ma), whereas the volcanic field is morphologically youthful and <2.5 Ma. Therefore, on the basis of the absence of a clear age or volcanological connection between the two volcanic features, the White Mountains Baldy volcano is not further considered.

The cinder cones that make up most of the vents in the field are eroded only slightly; several have unbreached summit craters, but most have channeled flanks. A few lava flows also exhibit youthful and rough surface textures, consistent with their measured ages of only a few hundred thousand years. Unusual vents in the Springerville field include maars (5, including the youngest, **Cerro Hueco**), young cinder cones (**Twin Knolls**, estimated by K-Ar dating as 0.3 Ma), and viscous domes (**Wolf** Mountain).

Based on detailed correlation of vent locations and independently determined relative and absolute ages of corresponding lava flows, the locus of volcanism within the field migrated eastward at ~2.9 cm/yr. This is apparent in the distribution of morphologically youthful vents: the youngest are on the eastern side near Springerville, and the older units are generally on the western side of the field near Show Low. The greatest concentration of vents and the greatest diversity of lava types and vent structures occur near the center of the field. The estimated eruption recurrence interval from 2.0 to 0.3 Ma is 4,600 yr. The average effusion rate is ~1.5×10^{-4} km^3/yr.

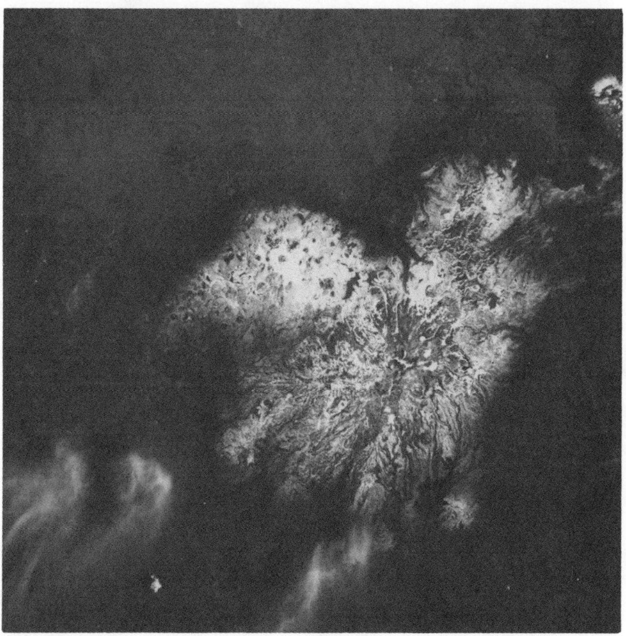

SPRINGERVILLE: Space Shuttle photograph of the Springerville volcanic field. North to the upper left; the topographic map covers only the northern half of the snow-covered field.

Linear fissure-type vents are not common, and in this respect the field differs from the Mount Taylor–Mesa Chivato field, 200 km to the northeast. Many vents are closely linked to associated Quaternary tectonic deformations, and some flows locally show evidence of slight tilting and folding. Structure and elevation drop down along at least two regional northwesterly-trending arcuate topographic steps or scarps which have a variety of types of deformation (folds, strike-slip faulting, pull-apart basins, and linear graben). These structural patterns may result from the location of the field at the intersection of the Rio Grande rift structural trend (northeast) and the northwest–southeast Transition Zone between the Basin and Range–Colorado Plateau.

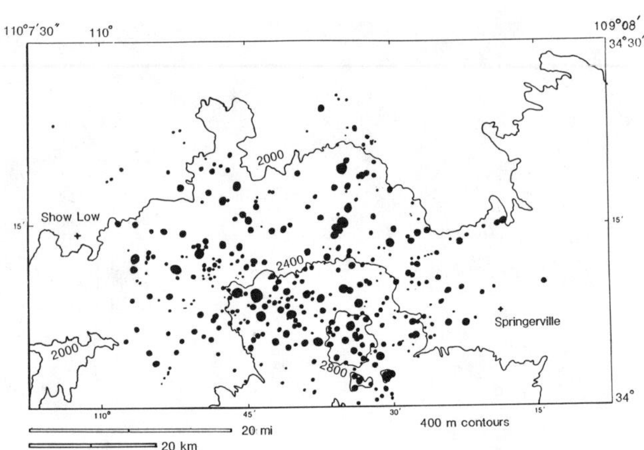

SPRINGERVILLE: Regional topographic map showing the location, shape, and size (in black) of vent deposits (mainly pyroclastic cones).

SPRINGERVILLE: View west from the highest point in the Springerville volcanic field (Greens Peak) toward clustered cinder cones of the central and higher elevation parts of the field.

How to get there: *The Springerville volcanic field lies between the towns of Show Low and Springerville, Arizona, along Highways 60 and 260; it is 75 km south of the Petrified Forest National Park. The topographic summit of the field is accessible from either highway by following signs to "Greens Peak" Lookout.*

References

Condit, C. D., Aubele, J. C., Crumpler, L. S., and Elston, W. E., 1989, Patterns of volcanism along the the southern margin of the Colorado Plateau: the Springerville field: *J. Geophys. Res. 94*, 7975-7986.

Crumpler, L. S., 1989, Lithologic map of the central part of the Springerville volcanic field, east-central Arizona: *USGS Misc. Invest. map series*, in prep.

L.S. Crumpler, Jayne C. Aubele, and C.D. Condit

SAN CARLOS

Arizona

Type: *Monogenetic volcanic field*
Lat/Long: *33.25°N, 110.25°W*
Elevation: *1,000 m*
Eruptive History: *4.2(?) to 1.0 Ma*
Composition: *Basanite to hawaiite*

The San Carlos volcanic field is the smallest of the late Neogene fields in Arizona and is the southernmost member of the Jemez Lineament volcanoes. Cone and lava flow remnants capping deeply dissected Gila River gravels cover <50 km^2 total area. The volcanic field has long been known for its abundance of peridotite and kindred nodules in the basalt of Peridot Mesa, the largest remnant. Many of the lherzolite nodules contain gem-quality olivine megacrysts >2 cm in size.

The **Peridot Mesa** vent is the only one of four vents having any volcanological distinction. It is a diatreme/tuff ring with a rather extensive lava flow (containing the nodules) which may be "rootless," having originated in a fountain over the pre-basalt tuff ring.

How to get there: *The field is ~40 km east of Globe, Arizona, and is entirely within the San Carlos Apache reservation; it is not possible casually to visit either the mines or the other volcanoes. Available samples have been collected only by people who "know somebody" in the tribe.*

References

Shafiqullah, M., et al., 1980, K–Ar geochronology and geologic history of southwestern Arizona and adjacent areas: *Ariz. Geol. Digest 12*, 201–260.

Wohletz, K. H., 1978, The eruptive mechanism of Peridot Mesa vent, San Carlos, Arizona: *Ariz. Bur. Geol. Min. Tech. Spec. Pub. 2*, 167-173.

Daniel J. Lynch

SENTINEL PLAIN: Typical low-aspect-ratio lava cone. None of the flows in this field retain any primary flow-top structures.

SENTINEL PLAIN
Arizona

Type: *Monogenetic volcanic field*
Lat/Long: *32.8°N, 113.2°W*
Elevation: *160 m*
Eruptive History: *3.3 to 1.3 (?) Ma*
Composition: *Alkali basalt*

The Sentinel Plain volcanic field comprises 600 km^2 of thin (<2 m) basalt flows erupted onto a low-relief surface of aggraded Gila River gravel west of Gila Bend. Four additional isolated vents with 10-20 km^2 of similarly thin basalt lie along a line extending 50 km to the northeast across the S-shaped bend in the Gila River. All but one of the vents in this field are low-aspect-ratio lava cones. Vent morphology indicates basalt effusion without appreciable pyroclastic activity.

The extensive main part of the field contains only 12 identifiable eruptive centers. Cone contours constitute elongate ovals oriented northwest–southeast, and pairs of cones are aligned in this direction. Cone flanks and flow surfaces are entirely of erosional-aggradational origin with significant accumulation of soil carbonate. Contacts between flows from different vents are obscure. Flow bases exposed along the river in the northern part of the field are remarkably rubble-free.

The **Arlington** cone, northeasternmost of the separate cones, is morphologically identical to the others and appears to be monogenetic. However, 6 K-Ar ages determined in site studies for the nearby Palo Verde nuclear power generating station fall between 3.3 and 1.3 Ma, a range that encompasses all the other ages in the field. Although the K-Ar ages indicate the field was active for 2 m.y., the physiography suggests penecontemporaneous fissure eruption of flood-like basalt in the main part of the field.

How to get there: *The Sentinel Plain volcanic field is ~125 km southwest of Phoenix, Arizona. Interstate Highway 8, from Gila Bend to Yuma, passes through the northern part of the field, but most of it is in the Luke Air Force Base Bombing Range; it is unsafe to leave public roads.*

SENTINEL PLAIN: Landsat view; arrows point to the outlying cones along the "Gila Bend," A is the Arlington cone. The main part of the field is extensively covered with soil carbonate and blown sand.

Reference

Shafiqullah, M., et al., 1980, K-Ar geochronology and geologic history of southwestern Arizona and adjacent areas: *Ariz. Geol. Digest 12*, 201–260.

Daniel J. Lynch

GERONIMO
Arizona

Type: *Monogenetic volcanic field*
Lat/Long: *32.5°N, 109.25°W*
Elevation: *1,300 m*
Eruptive History: *3.2 to 0.3 Ma*
Composition: *Basanite to hawaiite*

The Geronimo (or **San Bernardino**) volcanic field covers an area of 850 km^2 in the northern end of the San Bernardino valley in southeast-

GERONIMO: ***Paramore*** *maar crater and tuff ring, looking north. This maar, ~800 m long, is younger than the surrounding cones. The tuff contains numerous ultramafic nodules. Photograph by P. Kresan.*

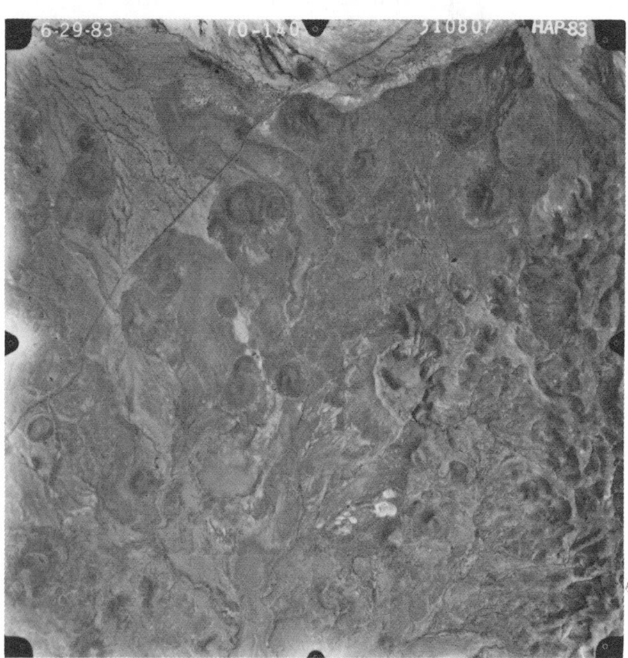

GERONIMO: Eastern portion, including Paramore crater. National High Altitude Photography Program vertical view.

ern Arizona. One hundred forty eruptive centers have been identified; five are tephra-ring maar craters, and the rest are pyroclastic cones in various stages of erosion. The cones are composed of agglutinate or a heterogeneous mixture of pyroclast types; the majority appear to have

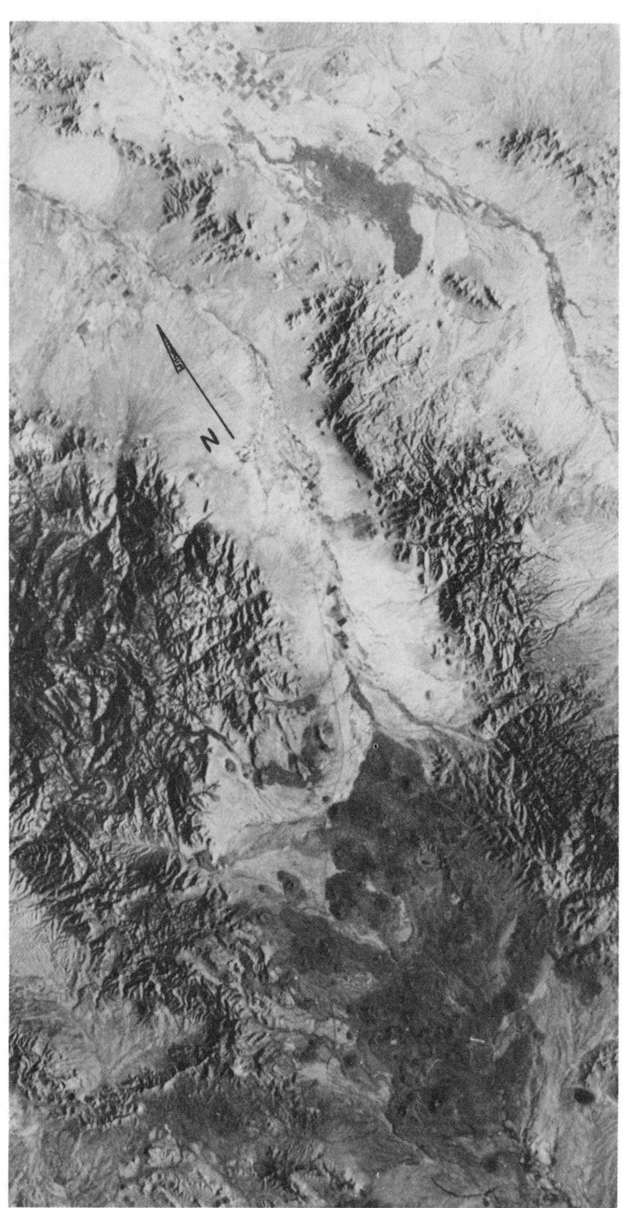

GERONIMO: The Geronimo field from a Skylab photograph. The dark-toned basalt is easily seen. The Animas flow (top) lacks a pyroclastic cone at its source.

been breached during construction. Lava flow surfaces are eroded flat or are filled with aeolian material. Eroded flow margins constitute prominent landforms in the mountains on both sides of the valley.

The San Bernardino valley is an asymmetric graben, deepest on the west. Crustal extension is still active in the area as shown by a century-old, 4-m-high fault scarp that extends 60 km from a point south of the field. Some lava flows have been offset by post-3.2-Ma normal faulting in the adjacent mountains. Despite the evidence for east–west extension beneath the field, there is no clear stress-field control of vent location or distribution.

Lavas and tephras contain abundant nodules of spinel lherzolite, wherlite, websterite, clinopyroxenite, harzburgite, amphibole peridotite, and two-pyroxene granulite. Megacrysts of plagioclase, aluminous augite, kaersutite hornblende, and magnetite are common. Strontium and neodymium isotope studies of the nodules, megacrysts, and lavas conclude that the mantle source of the basalts produced magma in a billion-year-earlier partial melting episode, and that this source was enriched in light rare-earth elements prior to producing the modern magmas.

How to get there: *The Geronimo volcanic field is in the southeast corner of Arizona, ~60 km east of Bisbee. Highway 80, from Douglas to the New Mexico border, cuts through the field, which is largely on private ranch land.*

References

Kempton, P. D., Dungan, M. A., and Menzies, M. A., 1982, Petrology and geochemistry of ultramafic xenoliths from the Geronimo volcanic field: *Terra Cognita 2*, 222.

Lynch, D. J., 1978, The San Bernardino volcanic field of southeastern Arizona, in Callender, J. F., Wilt, J. C., and Clemons, R. E. (eds.), *Land of Cochise*, New Mexico Geol. Soc. 29th Field Conf. Guidebook, 261-268.

Daniel J. Lynch

VOLCANOES OF NEW MEXICO

Taos, 290
Raton–Clayton, 292
Ocate, 293
Jemez, 295
Cerros del Rio, 297
San Felipe, 299
Taylor, 300
Albuquerque, 302
Cat Hills, 303
Zuni–Bandera, 303
Lucero, 305
Red Hill, 306
Carrizozo, 308
Jornada del Muerto, 309
Kilbourne Hole, 310
Potrillo, 311

TAOS
New Mexico

Type: *Monogenetic volcanic field*
Lat/Long: *36.83°N, 105.83°W*
Elevation: *2,135 m (main Taos Plateau) to 3,087 m (Ute Mountain)*
Eruptive History:
10.0 Ma: Quartz latite
5-4.6 Ma: Main basaltic activity
1.8 Ma: Youngest volcanism
Composition: *Basalt to rhyolite*

The Taos Plateau volcanic field is the most voluminous concentration of lavas along the axis of the Rio Grande rift, covering ~7,000 km^2 and ranging from basalt to rhyolite. Despite their location in a well developed rift setting, the lavas were erupted from monogenetic central volcanoes, including at least 35 shields and cones clustered within a 30×50-km region of the central Taos Plateau. The larger volcanoes form an imperfect concentric pattern ~40 km in diameter: tholeiitic shields are located centrally within the volcanic field, andesite volcanoes at intermediate positions, and dacite farthest out. Two small silicic lava domes are also near the field's center. Individual volcanoes tend to be petrologically uniform. Basaltic rocks are volumetrically dominant, and volumes of rock types decrease as the silica content increases.

The most widespread and voluminous basalt (~200 km^3) is distinctive coarse-grained, diktytaxitic flows of high-alumina tholeiite (**Servilleta** Basalt), erupted from broad, low shields. The upper surface of the Servilleta flows forms the Taos Plateau. The Servilleta Basalt is traceable at least 150 km north–south along the axis of the rift, and has a minimum thickness of 200 m where exposed along the Rio Grande gorge. Servilleta flows extend in the subsurface at least as far north as Alamosa, Colorado. Much smaller volumes of silicic alkalic basalt and xenocrystic basaltic andesite occur in local flows erupted from cinder cones. Shield volcanoes of olivine andesite (26 km^3) and nearly aphyric, two-pyroxene (31 km^3) dacite are as much as 10 km in diameter and 1 km high. Andesite and dacite were erupted mainly later in the evolution of the volcanic field, but overlapped with basaltic volcanism.

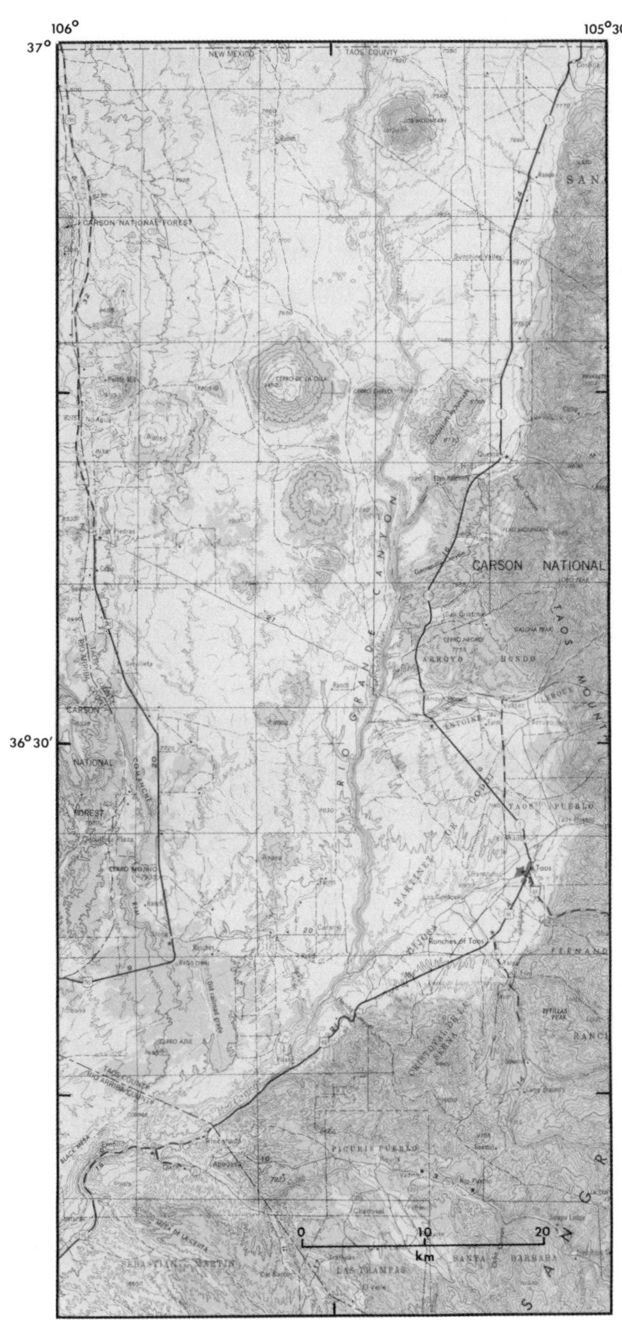

TAOS: USGS topographic map.

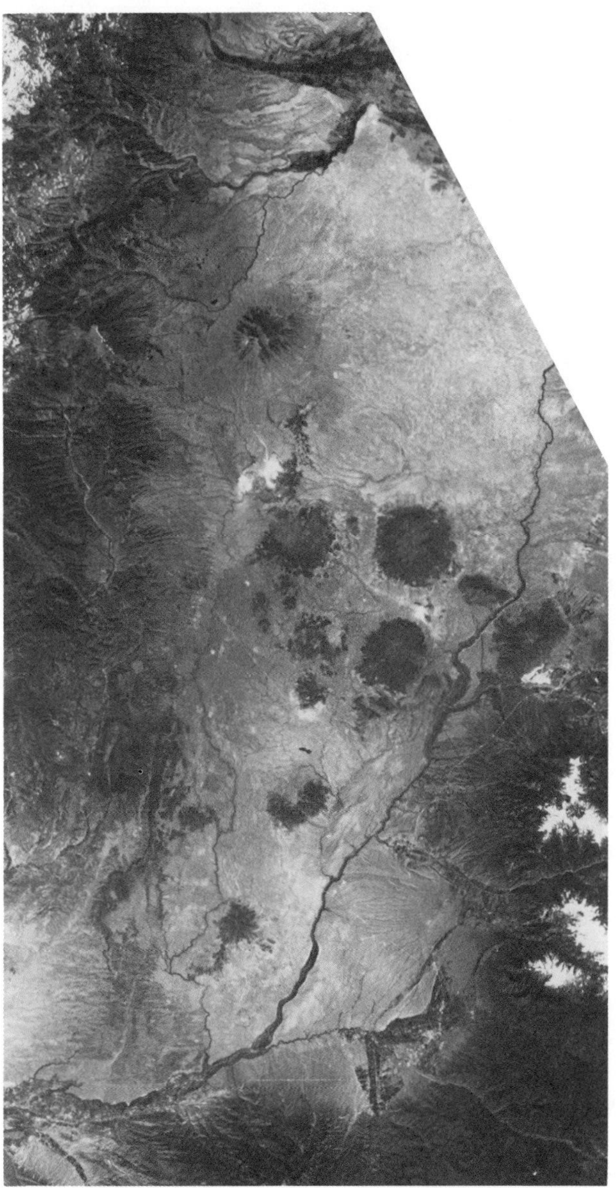

TAOS: Skylab photograph of Taos Plateau volcanics.

The Taos Plateau volcanic rocks resulted from complex interaction between mantle-derived basaltic magmas and lower continental crust. Andesite and dacite evolved along complex differentiation paths by fractional crystallization, assimilation of lower crust, mafic recharge, and magma mixing. Most of the voluminous tholeiitic basalt is hybrid, produced by 10-35% admixture of andesitic and dacitic magmas, similar to those erupted concurrently with the Servilleta Basalt. The broadly concentric compositional pattern of vents, along with geochemical modeling and petrographic evidence, suggests a consanguineous origin for most of the volcanic rocks. Basalt, derived by partial melting of continental lithospheric mantle in response to rifting events, provided the heat necessary for prolonged assimilation of crust, resulting in intermediate composition magmas. The increase in SiO_2 content of lavas with distance from the basaltic vents reflects a decreasing rate of replenishment away from the main mafic conduit system.

How to get there: *The Taos Plateau volcanic field fills the northern end of the Rio Grande rift valley, north-northwest of Taos, New Mexico. Highways 285 and 3 skirt the west and east sides of the field, respectively. Secondary roads lead to many of the cones.*

References

Lipman, P. W., and Mehnert, H. H., 1979, The Taos Plateau volcanic field, northern Rio Grande rift, New Mexico: *in* Riecker, R. E. (ed.), *Rio Grande Rift: Tectonics and Magmatism*, Am. Geophys. Un. Spec. Pub., 289-311.

McMillan, N. J., and Dungan, M. A., 1988, Open system magmatic evolution of the Taos Plateau volcanic field, northern New Mexico: 3. Petrology and geochemistry of andesite and dacite: *J. Petrol.* 29, 527-557.

Peter W. Lipman and W. Scott Baldridge

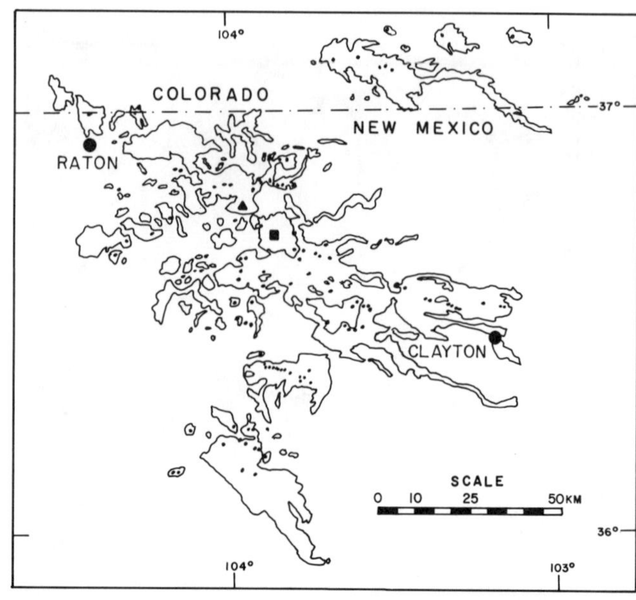

RATON-CLAYTON: Map of the Raton–Clayton volcanic field located in northeastern New Mexico. Basaltic cinder cones are indicated by small black dots. Location of Capulin Mountain is indicated by the black triangle, and Sierra Grande, an andesitic stratovolcano, by the black square.

RATON-CLAYTON
New Mexico

Type: *Monogenetic volcano field*
Lat/Long: *36.5°N, 104.3°W*
Elevation: *1,550 to 2,700 m*
Eruptive History: *Three episodes:*
8.2 Ma
7.2 to 2.2 Ma
1.9 Ma to ~2.3 ka
Composition: *Dacite to olivine melitite nephelinite*

The Raton–Clayton volcanic field, in the extreme northeastern corner of New Mexico, is Pliocene to Holocene in age. Approximately 125 basaltic to nephelinitic cinder cones, ranging in age from >1.0 Ma to 2,300 yr BP, are distributed throughout the field, with a concentration of feldspathoidal compositions near the town of Des Moines. Many cones have associated lava flows. Perhaps the most impressive cone is the youngest, which is fortunately protected as **Capulin** National Monument. The rim of this steep-sided cinder cone is ~1.7 km in circumference, and stands 305 m high, with a crater depth of 125 m. A variety of andesitic and dacitic volcanic necks and domes also occurs throughout the field; **Sierra Grande**, a large stratovolcano 15 km in diameter and 600 m in height, is composed of numerous flows of a distinctive and homogeneous andesite.

Volcanism in the Raton–Clayton field began with the eruption of alkali basal, now preserved as the relatively high Raton, Barella, and Clayton mesas. Eruptions were separated by long periods of inactivity, indicated by the inverted topography of the lava flows. Andesitic and dacitic volcanism was contemporaneous with alkali basalt activity and restricted to well-defined, usually small centers. Mafic, feldspathoidal volcanism began at ~2 Ma, perhaps continuing through the eruption of the youngest unit, the Capulin Basalt. The occurrence of such feldspathoidal lavas, ranging in composition from olivine melitite nephelinite to basanite, is noteworthy; exceptionally few rocks of similar composition have been described in the Cenozoic volcanics of New Mexico. These lavas also contain a wide variety of xenoliths including peridotites as well as crustal rock types.

RATON-CLAYTON: National High Altitude Photography Program vertical aerial view of Capulin cinder cone and associated lava flows (top left) and Sierra Grande stratovolcano (bottom right).

How to get there: *The Raton–Clayton volcanic field is in northeastern New Mexico and southeastern Colorado. Raton is ~125 km east of Taos, New Mexico. US Highway 64 cuts directly through the center of the field and connects the towns of Raton and Clayton.*

References

Muehlburger, W. R., Baldwin, B., and Foster, R. W., 1967, High Plains, northeastern New Mexico: *New Mex. Bur. Mines Min. Res. Scenic Trips of Geological Past 7.*

Phelps, D. W., Gust, D. A., and Wooden, J. L., 1983, Petrogenesis of the mafic feldspathoidal lavas of the Raton–Clayton volcanic field, New Mexico: *Cont. Min. Petr. 84,* 182-190.

David Gust

OCATE
New Mexico

Type: *Monogenetic volcanic field*
Lat/Long: *36.12°N, 104.75°W*
Elevation: *<1,800 to >3,000 m*
Eruptive History: *8.3-0.8 Ma, most volume erupted from ~5-2 Ma*
Composition: *Basalt, andesite and dacite*

The Ocate (or **Mora**) volcanic field consists of numerous basaltic to dacitic flows and ~50 cinder cones exposed over a broad region of the southern Rocky Mountains (Sangre de Cristo Range) and western Great Plains, on the uplifted eastern flank of the Rio Grande rift. The total volume of volcanic rocks presently exposed is ~90 km^3. The physiographic expressions of flows reflect their relative ages. The oldest flows cap the highest mesas, up to 600 m above the modern drainage, and successively younger lava flows cap correspondingly lower mesas. The youngest, 10-30 m above the present drainage, were erupted onto surfaces reflecting the modern stream network. Geomorphic surfaces preserved beneath flows 5.7 m.y. in age and older are warped and locally displaced across reactivated Laramide fault zones. This tectonic deformation is broadly correlative with renewed rifting and with uplift of adjacent ranges of the southern Rocky Mountains.

Volcanic rocks are divided into five compositional groups: alkalic olivine basalt, transitional olivine basalt, olivine andesite, xenocrystic basaltic andesite, and dacite. The variety of intermediate compositions resulted from mixing of basalt with silicic magma (evidenced by inclusions of rhyolitic glass in quartz and plagioclase xenocrysts), formed by fusion of crustal rocks, in combination with fractional crystallization and wall rock assimilation. Alkali olivine basalt eruption was at a maximum early and, to a lesser extent, late in the history of the Ocate field, whereas transitional basalt and intermediate composition rocks were most voluminous during the period of greatest volcanic activity at 5-2 Ma. The temporal pattern of volcanism is consistent with a maximum in upper mantle heat flux from 5-2 Ma, resulting in a higher degree of partial melting at shallower mantle depths and thus producing the less alkaline, transitional basalt. The increase in volume of mafic magmas introduced into the crust during this period resulted in a greater degree of crustal melting, thereby producing the associated intermediate rocks.

How to get there: *The Ocate volcanic field is on the east side of the Rio Grande rift in northern New Mexico. Volcanic cones and flows partially surround theTurkey Mountains feature, which*

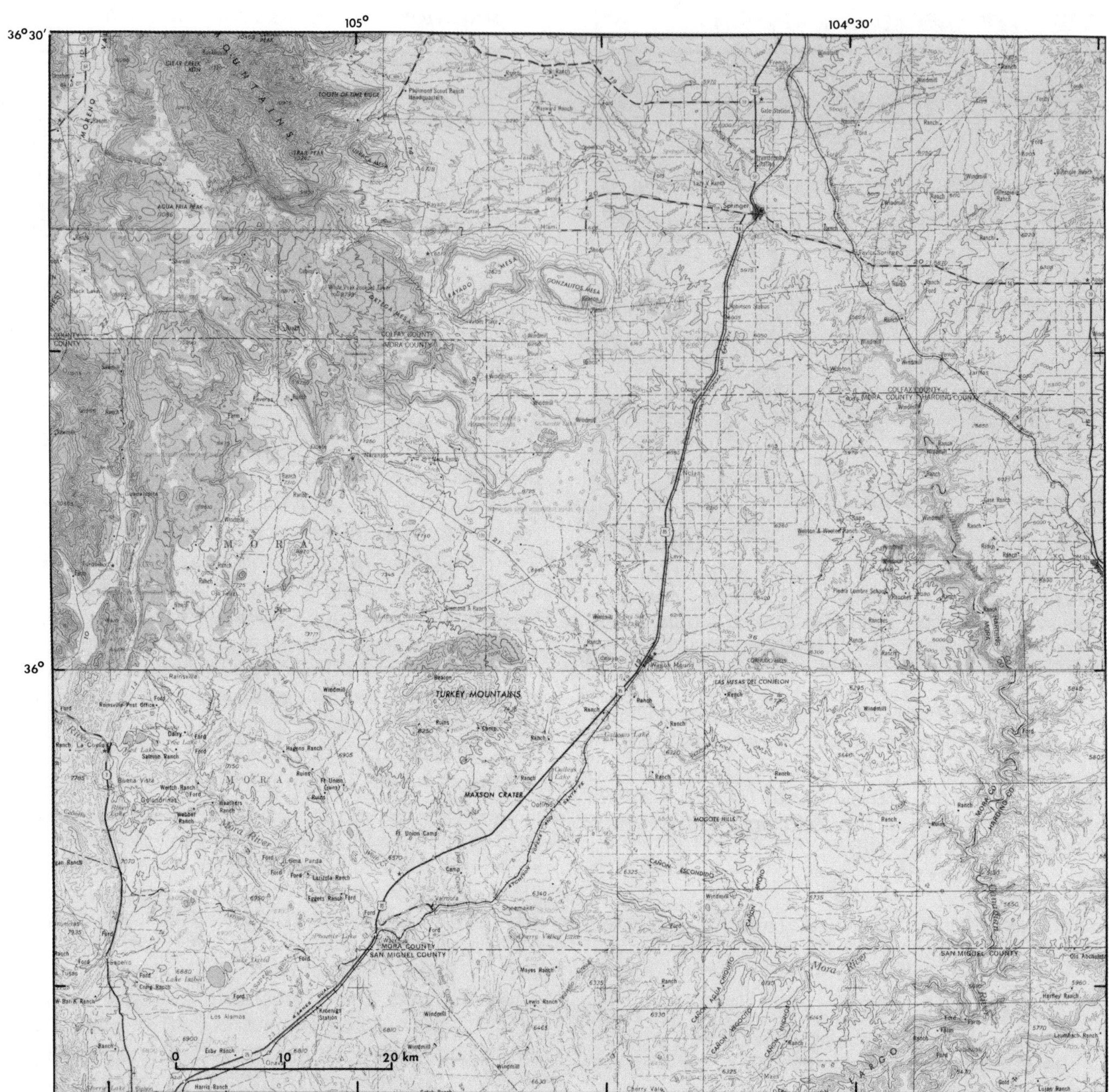

OCATE: USGS topographic map.

looks like a resurgent caldera, but is actually a laccolith. Highways 3 and 85 (US 25), north of Las Vegas, New Mexico, enclose most of the Ocate field, and secondary Highways 120 and 121 cross the field.

References

O'Neill, J. M., and Mehnert, H. H., 1980, Late Cenozoic physiographic evolution of the Ocate volcanic field, north-central New Mexico: *USGS Open-file Rpt. 80-928.*

Nielson, R.L., and Dungan, M.A., 1985, The petrology and geochemistry of the Ocate volcanic field, north-central New Mexico: *Geol. Soc. Am. Bull. 96*, 296-312.

W. Scott Baldridge

JEMEZ
New Mexico

Type: *Ash-flow volcanic complex*
Lat/Long: *35.88°N, 106.53°W*
Elevation: *3,525 m to 1,830 m*
Volcanic Field Size: *85 × 50 km*
Caldera Diameter: *22 km*
Eruptive History:
14–10 Ma: Tholeiite basalt and high-silica rhyolite
10–7 Ma: Andesite and high-silica rhyolite
7–2 Ma: Basalt, andesite, dacite
2 Ma to present: Rhyolite, high-silica rhyolite, minor basalt
No historic activity
Composition: *Tholeiitic basalt to high-silica rhyolite*

The Jemez volcanic field is a long-lived, large-volume (~2,000 km^3), active continental volcanic center, with the famous 1-Ma **Valles** caldera (22-km diameter) as its centerpiece. The field is best known for the voluminous Bandelier Tuff, two major ignimbrites, each 200-400 km^3 dense rock equivalent volume, aged 1.45 and 1.12 Ma. Caldera formation may have begun as long ago as 2.5 Ma, but the caldera depression seen today is mainly due to collapses of the volcanic pile in response to eruption of the older Bandelier Tuff.

The volcanic field is located at the intersection of two structures. North-trending faults of the western side of the Rio Grande rift pass through the volcanic center and are crossed by the northeast-trending Jemez Lineament. Some as yet undetermined deep-seated process at this structural intersection has caused this region to have a long magmatic history.

Activity in the Jemez Mountains began with eruption of Rio Grande rift-related olivine tholeiitic basalt and scattered alkalic basalt in the south of the area, accompanied by high-silica rhyolite lava domes. From 13 to ~10 Ma basalt and rhyolite, with voluminous andesite and dacite, formed composite volcanoes and lava dome complexes with attendant pyroclastic and epiclastic aprons, whose eroded remnants now make up the southern part of the field. From 10 to ~7 Ma a prodigious amount of andesite, with some basalt, rhyodacite, and high-silica rhyolite, was erupted in the southern to west-central part of the field to form a cluster of (or maybe one) large composite volcanoes. This Paliza Canyon-age volcanism produced just under half the volume of volcanics presently making up the field. Considerable volumes of material of this age are found in volcaniclastic fans on the southern edge of the Jemez Mountains.

From 7 to 5 Ma there was a slight but distinct lull in magma output and a hiatus in basalt volcanism, which has otherwise persisted at a low level throughout the history of the field. After ~5 Ma activity stepped up again, this time with the locus of volcanism being in the central to northeast part of the field. A large (~400 km^3), dominantly dacitic lava dome complex, the **Tschicoma** center, was built by eruptions that persisted until ~2 Ma. Part of this center now forms the highest peak in the Jemez Mountains, Chicoma Mountain (3,525 m). Basaltic volcanism began again at ~4 Ma and has continued into the Pleistocene. Three basaltic lava fields were formed, the **El Alto** on the northeast edge of the Jemez Mountains, the Cerros del Rio on the eastern flanks in the Rio Grande gorge, and the **Santa Ana Mesa** field in the south. As the dacitic dome complex grew a volcanogenic alluvial fan developed, extending into the Rio Grande rift valley. Conglomerates, lacustrine deposits, and mudflows of the Puye fan occur in prominent cliffs on the east side of the Jemez Mountains.

The buildup toward the "climax" of volcanic activity perhaps began as long ago as 2.5 to 3 Ma with the eruption of small, high-silica rhyolite ignimbrites and pyroclastic fall deposits from vents approximately in the middle of the cluster of volcanoes forming the field. These have similar compositions to the much more voluminous magmas erupted in the Bandelier events after 2 Ma, to which they are related in a complex way. Also, at about this time a crater or caldera formed on the Tschicoma dome complex. This depression used to be known as the **Toledo** caldera, but owing to current uncertainty regarding its origin, it is now called the Toledo embayment.

The first very large explosive high-silica rhyolite eruption occurred at 1.45 Ma, and began with a plinian outburst from a vent (or

JEMEZ: Landsat Return Beam Vidicon image E-31055-16530-C.

vents) in the middle of the field. The plinian phase was followed by an immense outpouring of pyroclastic flows, and the resulting pumice fall deposit and ignimbrite is known as the Otowi Member (or Lower) Bandelier Tuff. Pyroclastic flows reached some 40 to 50 km from the center of the the field. Some of the flows ponded in the caldera depression that began to form before the end of the eruption. Cliffs of outflow ignimbrite up to 150 m thick dominate the canyon scenery in much of the Jemez Mountains. Air-fall ash related to this eruption would have covered considerable parts of New Mexico, Texas, and neighboring states. A volume on the order of 300 km^3 of magma was erupted. Caldera formation swallowed up considerable parts of the Paliza Canyon andesitic and the Tschicoma dacitic centers. The present caldera is largely the product of this eruption. A period of resurgence of the caldera floor may have occurred after the eruption; this was followed by formation of several plinian deposits and lava domes erupted both inside the caldera and in the Toledo embayment.

At 1.12 Ma another similar large eruption occurred, forming the Tshirege Member (or Upper) Bandelier Tuff (200-250 km^3 of magma). The early plinian deposit was this time dispersed by winds blowing toward the northwest; it must have covered considerable parts of Utah and Colorado with ash fall. The ignimbrite was laid down over a similar area (>2,500 km^2) to the Lower Bandelier ignimbrite and can be seen also in most cliffs around the Jemez Mountains. Caldera collapse was more or less confined within the earlier depression but deepened it, especially on the eastern side. A trapdoor-type collapse mechanism hinged on the western side has been proposed for both caldera-forming events. A period of resurgence lasting ~100,000 yr pushed up the caldera fill (ignimbrites, lavas, and sediments) to form a spectacular resurgent block or dome. Around this a series of rhyolite lava domes formed at ~100,000-yr intervals up to 150 ka. These are a magnificent example of post-caldera ring-dome volcanism and were erupted in almost perfect counter-clockwise order. The latest eruption, at ~150 ka, produced a plinian pumice fall deposit, a welded ignimbrite, and an 8-km-long flow of rhyolite obsidian. These units, the El Cajete pumice, the Battleship Rock ignimbrite, and the Banco Bonito lava, can be well seen in the main road that follows Canon de San Diego and turns east across the caldera. The composition of this group of units, and some other, late, post-caldera products, is a low-silica rhyolite of very complex, mixed magma origin.

The Valles caldera must still be considered an active volcano. Rhyolite volcanism usually has a slow tempo; thus eruptions occur on a time scale of hundreds of thousands of years. The western, and particularly southwestern, part of the caldera appears to be most active and has hot springs, fumaroles, and acid sulfate springs in several places. These are surface manifestations of a large, long-lived hydrothermal system that was emplaced about 1 Ma, just after the second caldera formed.

Two special features of the Jemez Mountains volcanics deserve special mention: **Redondo** Peak (36.88°N, 103.56°W) is the resurgent block or dome. At 3,431 m, the second highest point in the Jemez Mountains, the flat, domelike top of Redondo Peak stands 840 m above the caldera floor. Surrounding it, the moat is full of lava domes, or forms expansive, high altitude grasslands (the Valles Grande). There must be few places in the world where a major caldera and resurgent block can be so easily and clearly viewed. The resurgent center is composed mainly of densely welded Upper Bandelier ignimbrite and is crossed by a major set of faults, the middle ones defining the apical graben. These faults trend along the Jemez Lineament.

Soda Dam (35.79°N, 106.70°W) is a unique travertine deposit. Canon de San Diego, which cuts down from the caldera following the Jemez fault zone (again a surface expression of the Jemez Lineament), contains a number of hot springs. Water from the hydrothermal system under the caldera leaks down against the faults and at one location comes to the surface against a tiny horst of Precambrian basement granitoids. Here a series of travertine deposits has grown over the past 1 m.y., the latest a damlike structure ~100 m long from ~5 ka. Unfortunately the dam was breached by roadworks in the 1950s; no longer fed by its parent hot spring, it is therefore in a state of slight decay due largely to human damage. The travertine precipitate is from dissolution of Mississippian limestones that underlie the caldera and fault zone.

How to get there: *The Jemez Mountains are 20 km west-northwest of Santa Fe, New Mexico. The town of Los Alamos is built on the Bandelier ignimbrite plateau on the eastern flanks. New Mexico Highway (Route) 4, paved all the way, loops through the mountains and traverses most of the volcanic highlights. Join Route 4 at San Ysidro on State Highway 44 from Bernalillo to Farmington. It can be followed via Los Alamos to join Highway 64 (the Taos–Santa Fe road) at Projoaque.*

References

Bailey, R. A., and Smith, R. L., 1978, Guide to Jemez Mountains and Espanola Basin: *New Mex. Bur. Mines Min. Res. Circ. 163*, 184-196.

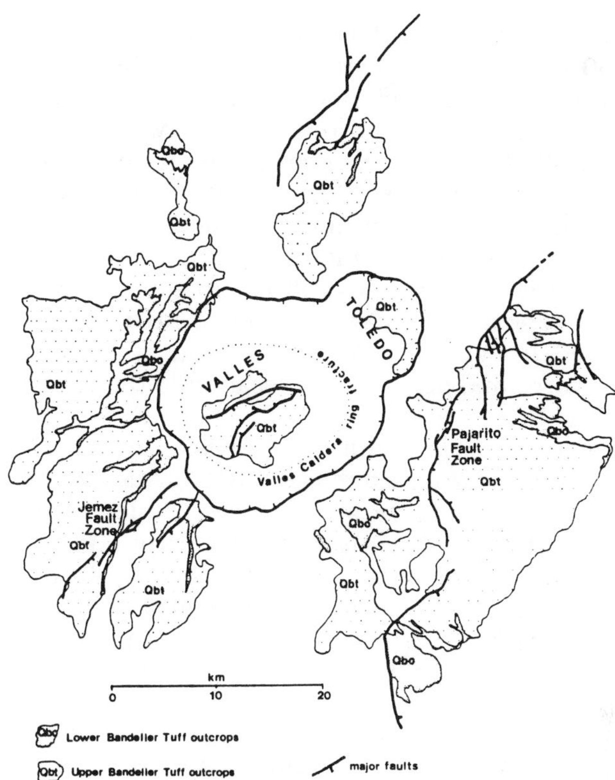

JEMEZ: Sketch map of Jemez volcanic complex showing Bandelier Tuff from Valles and Toledo calderas. Map from W. M. Kite, unpublished M.S. thesis, Arizona State University, 1984.

Self, S., Goff, F., Gardner, J. N., Wright, J. V., and Kite, W. M., 1986, Explosive rhyolite volcanism in the Jemez Mountains: Vent locations, caldera development, and relation to regional structures: *J. Geophys. Res. 91*, 1779-1799.

Stephen Self

CERROS DEL RIO
New Mexico

Type: *Monogenetic volcanic field*
Lat/Long: *35.67°N, 106.!7°W*
Elevation: *2,285 m*
Height: *450 to 664 m*
Volcanic Field Size: *19 km north–south × 8 km east–west*
Eruptive History: *2.9 to 2.2 Ma*
Composition: *Hawaiite, minor basanite, and basaltic andesite*

The Cerros del Rio volcanic field consists of ~60 vents, predominantly cinder-spatter cones with associated flows. These are primarily hawaiite in composition but include two basaltic andesite flows. Andesite eruptions have formed five highly visible "flow-domes," resembling dissected mesas, up to 60 m thick, thinly foliated in the interior, and glassy and massive toward the top. Dissected phreatomagmatic vents and deposits of unknown source area are exposed in White Rock Canyon and subsidiary canyons, and occur in one location in the center of the field.

Activity began with the extrusion of hawaiite, which forms extensive flows on the south and north margins of the area. Cinder cone-forming and phreatomagmatic eruptions occurred contemporaneously, particularly in the vicinity of White Rock Canyon, which could be interpreted as evidence for an early throughgoing stream in the area of the modern Rio Grande river. This activity was followed by eruptions of andesite flow-domes. Throughout these eruptions, less voluminous extrusions of hawaiite occurred along a zone of vents roughly trending northwest–southeast in the center of the field. Following the andesite activity, hawaiite eruptions continued along the central zone. The field probably extended north of the present White Rock Canyon, but any vents in this area were covered by material from Valles caldera in the Jemez Mountains.

The Cerros del Rio volcanic field is notable because of its location within the Rio Grande rift at a point where the rift is laterally offset 30 km to the east. The unusual combination of alkalic basalt and andesite may be due to the location of the field on the rift offset. The northwest-trending line of vents that dominantly erupted hawaiite across the field's center may reflect a flexure line between the rift's subsiding basins. This assemblage of volcanic lithologies more closely resembles volcanic fields of the Basin and Range province than rocks of Colorado Plateau marginal volcanic fields or of other rifts.

The Cerros del Rio volcanic field is also notable because it records the complex interaction between the Rio Grande river and volcanism. The Rio Grande has cut the present White Rock Canyon within the past 1 m.y., but the long history of aggradation of fluvial sediments, down-

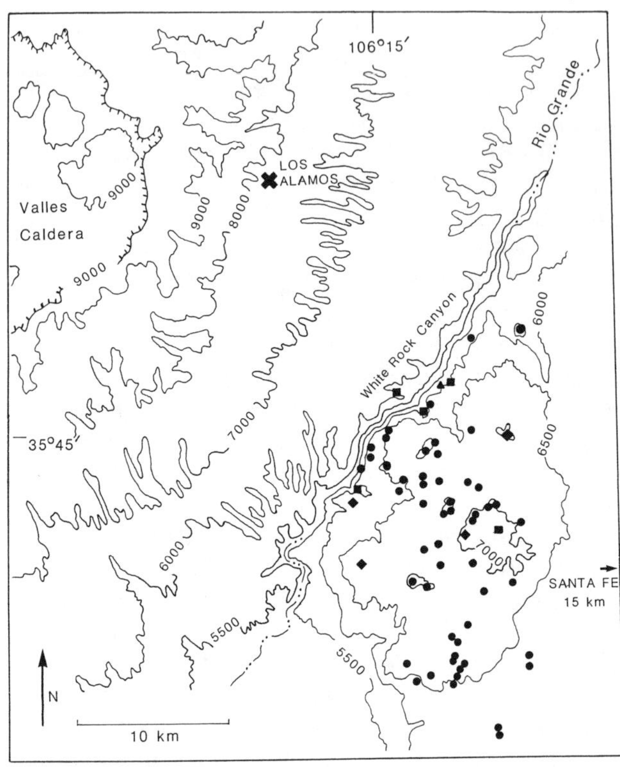

CERROS DEL RIO: Regional topography and vent distribution of the Cerros del Rio volcanic field. Circles = cinder-spatter cones; squares = phreatomagmatic vents or deposits; diamonds = silicic flow-domes.

CERROS DEL RIO: Montoso maar (V-shaped shadowed canyon) in Cerros del Rio volcanic field, looking across White Rock Canyon from Bandelier National Monument. Maar rim is ~1 km wide; Montoso Peak, an andesite vent, is visible in the background.

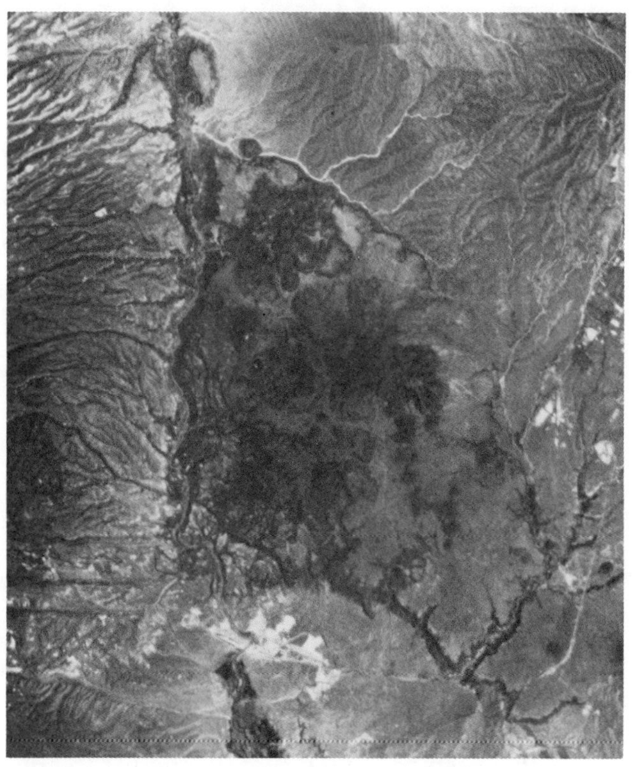

CERROS DEL RIO: Skylab photograph.

cutting by the river, and in-filling with volcanic deposits, as well as tectonic activity associated with margins of the Rio Grande rift resulted in a complex vertically stacked volcanic field, rather than the more typical areally extensive field. Where the river has dissected the field, a thick sequence of intercalated maar deposits, lake beds, river gravels, basaltic intrusions, and lava flows is exposed, with some repetition of lava flow outcrop due to toreva block faulting in White Rock Canyon.

One of the more unusual vents in this field is the **Montoso** Maar, which is exposed by a side canyon, revealing a complete traverse section from the rim to the internal intrusions. Phreatomagmatic deposits, from buried or eroded source vents, outcrop in other side canyons to White Rock and exhibit stratified layering and cross-bedding due to base surge.

Although the Cerros del Rio field is close to Santa Fe it is not frequently visited, except by local woodcutters seeking Pinon trees. A small population of wild horses, protected by federal law, lives in the field.

How to get there: *Entrance to the field is from the southeast by a dirt road off of Agua Fria Road or off of Airport Road, both in Santa Fe. Follow US Forest Service signs to Caja del Rio Grant. The easiest access into White Rock Canyon is by Buckman Road, Sante Fe, or by trail from Bandelier National Monument. White Rock Canyon is infamous for rattlesnakes.*

References

Aubele, J. C., 1979, The Cerros del Rio volcanic field, New Mexico: *New Mexico Geol. Soc. Guidebook, 30th Field Conf., Santa Fe Country,* 243-252.

Baldridge, W. S., 1978, Petrology and petrogenesis of Plio-Pleistocene basaltic rocks from the central Rio Grande rift, New Mexico and their relation to rift structure: *in* Neumann, E. R., and Ramberg, I. B. (eds.), *Petrology and geochemistry of Continental Rifts,* Reidel, Dordrecht, 71-78.

Jayne C. Aubele

SAN FELIPE
New Mexico

Type: *Monogenetic volcano field*
Lat/Long: *35.50°N, 106.50°W*
Elevation: *1,585 to 1,950 m*
Area: *100 km^2*
Eruptive History: *2.6 to 2.5 Ma*
Composition: *Alkali-rich tholeiite*

The San Felipe volcanic field has the largest volume of eruptive material in the Albuquerque basin portion of the Rio Grande rift. The field lies north of the confluence of of the Jemez and Rio Grande rivers, high atop the Santa Ana Mesa. Many faults, with throws up to 100 m and mostly oriented north–south, cut the field. An explosive eruption initiated the activity, as indicated by a 6-m-thick laminated basalt tuff under the earliest flows. Three main lava flows erupted from two parallel north–south fissures ~6.5 km apart; subsequently, 3 satellite centers with at least 2 more flows erupted through the early flows to the southeast, southwest, and west. On the main ridge along the east fissure, a low shieldlike mound formed: San Felipe Peak. This ridge also has 26 cinder centers; 40 cinder centers occur along the west alignment. These cinder centers, presumably relict cin-der cones, are typically circular patches, low mounds, and in some cases due to erosion, slight depres-

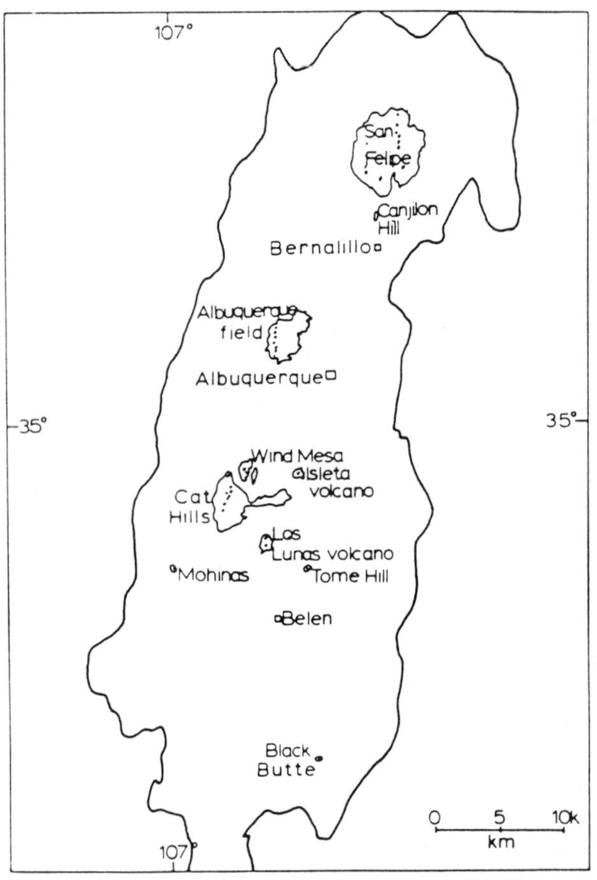

SAN FELIPE: Sketch map of San Felipe, Albuquerque, Cat Hills, and other minor volcanics in the Albuquerque basin. Map from Kelly and Kudo, 1978.

sions. Many of the centers have been intruded by plugs, dikes, and cone sheets. South of the main field lies the **Canjilon Hill** maar/tuff ring complex, which has four subcenters of tephra, flows, sills, dikes, and cone sheets. One crater or collapse basin is filled with lava (lava lake?). San Felipe and Canjilon Hill fields have basalt with alkali contents >4%, and range from hypersthene to quartz normative. Phenocrysts of olivine, augite, and plagioclase are common.

How to get there: *Take I-25 north of Algodones to the San Felipe Pueblo exit. Go through the pueblo onto the mesa top. Permission to enter must be obtained from the pueblo governor.*

Reference

Kelly, V. C., and Kudo, A. M., 1978, Volcanoes and related basalts of Albuquerque Basin, New Mexico: NM Bureau Mines Min. Res. *Circular 156*, 30 pp.

A.M. Kudo

TAYLOR
New Mexico

Type: *Stratovolcano and volcanic field*
Lat/Long: *35.25°N, 107.58°W*
Elevation: *2,100 to 3,445 m*
Height: *1,345 m (Mount Taylor)*
Volcano Diameter: *20 km*
Volcanic Field Size: *~30 × 60 km*
Eruptive History: *Mount Taylor: 3.3 to 1.5 Ma; Volcanic field: 3.5 to 1.5 Ma*
Composition: *Mount Taylor: alkali basalt, hawaiite, mugearite, porphyritic trachyte, rhyolite; Volcanic field: alkali basalt, basanite, hawaiite, mugearite, aphyric trachyte, and rhyolite*

The Mount Taylor volcano and surrounding field of basaltic vents and trachytic domes is one of several Plio-Pleistocene volcanic fields on the margin of the Colorado Plateau. Volcanic features of the area figured prominently in early exploration reports of the volcanic geology of the western United States, and yet the detailed lithology and volcanic features have been mapped only recently.

Mount Taylor is a large continental-type stratovolcano, which consists of a variety of generally more alkalic petrologic rock types than the stratovolcanoes of the Cascades. The volcano is essentially a concentration of extrusive domes of silicic porphyritic lavas and late pyroclastic materials with trachytic affinities. Although a few cinder cones and basaltic lava flows occur on its flanks, including some basalt flows bearing ultramafic nodules, most of the volume of the main cone consists of thick porphyritic flows, pyroclastic materials, and mudflows. Pyroclastics and numerous radial dikes are exposed in the large erosional amphitheater at the center of the volcano, and mudflows and thick alluvial deposits are interbedded with late basalt flows on the lower flanks. The central amphitheater may have been partially excavated by late explosive activity like that of Mount St. Helens, or, because the interior of the volcano consists of more easily eroded pyroclastic units, by later erosion. The flanks of the main volcano are not deeply eroded, and although a number of relatively deep radial canyons are cut into the flanks, the outline of in-

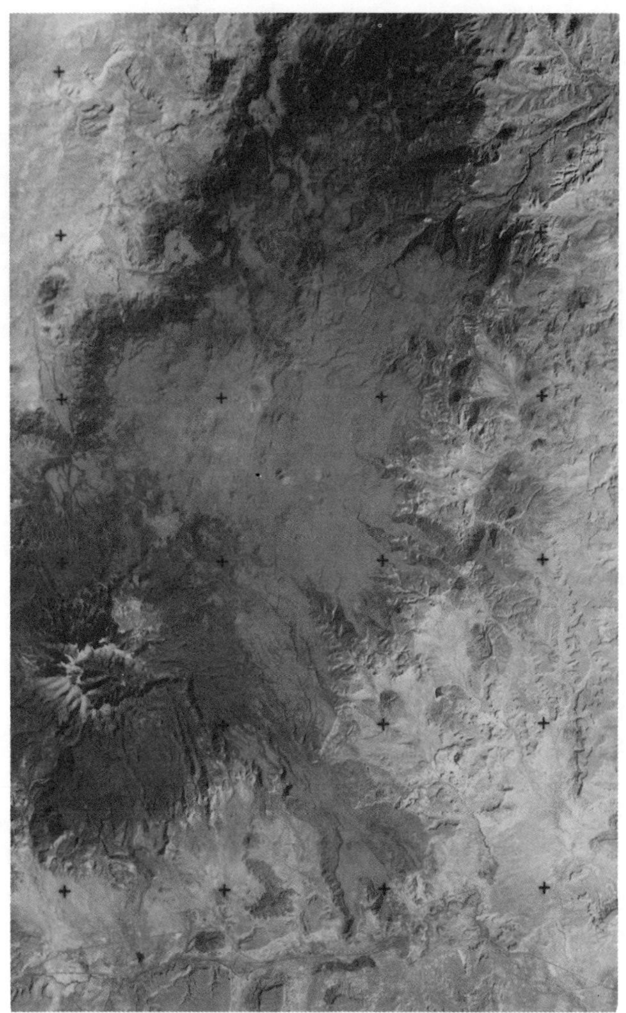

TAYLOR: Landsat Return Beam Vidicon image of Mount Taylor (bottom) and the Mesa Chivato plateau extending to the north.

dividual viscous flow units, basalt flows, and mudflows can still be easily mapped.

The surrounding volcanic field is of interest because of the large number of maar-type volcanic craters, unusually prominent fissure-type eruption patterns, trachyte domes, excellent exposures of volcanic necks, half-sectioned cinder cones, and a rare and complete (for North America) alkali-basalt through trachyte petrologic suite. Altogether there are >250 vents in addition to those responsible for the Mount Taylor volcano. Most of these features occur in the large, mostly basalt-capped mesa or plateau, **Mesa Chivato**, that extends to the northeast from the base of the Mount Taylor volcano. The present volume of Mount Taylor and the

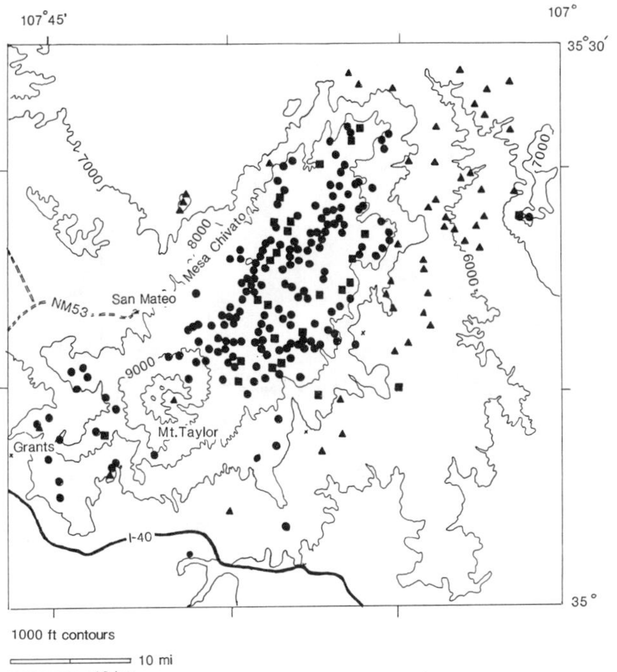

TAYLOR: Topographic map showing vent distribution for Mount Taylor and Mesa Chivato. Circles = cinder cones; squares = maars; triangles = volcanic necks. Contour interval 1,000 ft.

Mesa Chivato volcanics is ~220 km^3, with an estimated original volume of ~300 km^3.

The cinder cones and associated vents are not deeply eroded, although most are ~2 m.y. old. Slopes range from 15 to 25°, a consequence of erosive removal of loose cinders near the crater rims, the exposure of inward-dipping pyroclastic beds there, and deposition of the removed cinders as broad debris fans near the bases of the cones. Maars and large collapse depressions are especially common; some collapse features are >1 km in diameter and encompass entire cinder cones.

Both maars and collapse depressions, as well as cinder cones, volcanic necks, and viscous domes, are all frequently aligned along distinct northeast-oriented fissure-like trends. Strong northeasterly fissure patterns are evidence that approximately east–west regional extension occurred at the time of volcanism. This faulting may be related possibly to continued extension along the margins of the Rio Grande rift, which lies only a few tens of kilometers to the east. Vertical faults paralleling the fissure trends displace a few flows and cinder cones, providing

TAYLOR: View from the northeast end of the Mount Taylor field toward three of the northernmost volcanic necks: ***Cerro Santa Clara*** *(foreground),* ***Cabezon*** *(rear), and* ***Cerro Guadaloupe*** *(right).*

evidence for continued extension following the last eruptions in the field.

How to get there: *To reach Mount Taylor take I-40 west from Albuquerque to Grants, then New Mexico 547 (Lobo Canyon road) north, following Forest Service signs to La Mosca Peak. To reach the Mesa Chivato volcanic field take I-40 to Milan, then north on New Mexico 53 to San Mateo. Follow the San Mateo bypass and Forest Service access road onto mesa.*

References

Lipman, P. W., Pallister, J. S., and Sargent, K. A., 1978, Geologic map of the Mount Taylor quadrangle, Valencia County, New Mexico: *USGS Map GQ-1523.*

Crumpler, L. S., 1982, Volcanism in the Mount Taylor region: *NM Geol. Soc. Guidebook 33rd Field Conf., Albuquerque County II*, 291-298.

L. S. Crumpler

ALBUQUERQUE
New Mexico

Type: *Monogenetic volcano field*
Lat/Long: *35.12°N, 106.75°W*
Elevation: *1,750 to 1,800 m*
Eruptive History: *0.17 to 0.07 Ma*
Composition: *Olivine tholeiite*

The Albuquerque volcanics lie ~11 km west-northwest of downtown Albuquerque. Two

ALBUQUERQUE: View toward north of small spatter cones built along the Albuquerque fissure line. Photograph by L. Crumpler and J. Aubele.

north–south alignments, slightly offset, suggest eruption along fissures oriented N2°E in the southern part and N3°E in the northern part of the volcanic field. The three earliest flows cover an area of ~60 km^2 and are eroded on the eastern edge, forming what is locally called the "Volcano Cliffs," which have a large concentration of petroglyph art. Later flows are more restricted and localized. Most of the cones have cinders, but considerable amounts of lava and spatter are found even on the summits. The highest cone (**Vulcan** or **J Cone**) is in the southern alignment ~55 m above the highest lava flows. The base of this cone has a thick unit of cinders, but toward the top a lava dome formed which has been split subsequently into north–south halves by an explosion crater which has only remnant east and west rims. Semiradial dikes and dribble flows cut the dome.

Slight differentiation has been detected in the basalt. The older flows have olivine and plagioclase phenocrysts in a matrix of plagioclase, augite, opaques, and glass. Pigeonite occurs in the groundmass in the youngest rocks. All flows are hypersthene normative.

How to get there: *Take I-40 west from the I-25/I-40 interchange in Albuquerque, to Coors Blvd. north to Atrisco north-northwest (or to Montano and west) to Petroglyph Park and onto the escarpment. Go south to the model airplane park and take road westerly to the largest cone.*

References

Kelly, V. C., and Kudo, A. M., 1978, Volcanoes and related basalts of Albuquerque Basin, New Mexico: NM Bureau Mines Min. Res. *Circular 156*, 30 pp.

Baldridge, W. S., 1979, Petrology and petrogenesis of

Plio-Pleistocene basaltic rocks from the central Rio Grande rift, New Mexico, and their relation to rift structure: *in* Riecker, R .E. (ed.) *Rio Grande rift: Tectonism and magmatism*, Am. Geophys. Un. Spec. Pub., 323-353.

A. M. Kudo

CAT HILLS

New Mexico

Type: *Monogenetic volcanic field*
Lat/Long: *34.83°N, 106.83°W*
Elevation: *1,540 to 1,750 m*
Eruptive History: *0.14 Ma*
Composition: *High-alkali olivine basalt*

These volcanics are among the youngest in the Albuquerque basin exclusive of the activity within the Valles caldera to the north. Four major flows issued from a fissure zone trending slightly east of north. The first two flows extended the farthest and now cover >65 km^2; the later flows built up the areas closest to the fissures. Twenty-three cinder cones with summit craters formed along the alignment of the fissure zone. Many of the cones have closed rims and silted playas on the crater floors. The largest cinder cone is >20 m high, almost 150 m in diameter with a crater ~45 m wide. Well-shaped bombs are rare, most fragments being irregular and averaging 7 to 10 cm in diameter.

The basalt contains phenocrysts of olivine and plagioclase in an intergranular groundmass of augite, plagioclase, olivine, and opaques. The modal olivine to plagioclase ratio decreases from early to late flows. Chemically, the basalt has silica contents between 48.5 and 49.7% and high alkali contents of 4 to 5%. The basalt trends from hypersthene normative to slightly nepheline normative with age.

Additional smaller and older volcanic centers lie north and east of Cat Hills. **Wind Mesa** is a small, undated shield volcano, cut by north–south faults, north of Cat Hills. Silica values range from 47 to 52%. To the northeast is the 2.78-Ma **Isleta** volcano, which is a maar with related alkalic basalt lava flows. Southeast of Cat Hills is a pile of andesite flows (**Los Lunas** volcano) totaling ~60-90 m in thickness, and dated at 1.3-1.1 Ma.

CAT HILLS: View from east along I-25 of the Cat Hills cinder cones and lava flows. Photograph by L. Crumpler.

How to get there: *Take I-25 south of Albuquerque to the Isleta Pueblo turnoff; proceed west on a dirt road to the field. Permission to enter and travel in the Cat Hills volcanic field must be obtained from the governor of the pueblo.*

References

Kelly, V. C., and Kudo, A. M., 1978, Volcanoes and related basalts of Albuquerque Basin, New Mexico: NM Bureau Mines Min. Res. *Circular 156*, 30 pp.
Baldridge, W. S., Perry, F. V., and Kudo, A. M., 1982, Petrology and geochemistry of the Cat Hills volcanic field, central Rio Grande rift, New Mexico: *Geol. Soc. Am. Bull. 93*, 635-643.

A. M. Kudo

ZUNI–BANDERA

New Mexico

Type: *Monogenetic volcano field*
Lat/Long: *34.82°N, 108.13°W*
Elevation: *1,880 to 2,700 m*
Eruptive History: *Sporadic activity from 4 Ma to ~0.001 Ma*
Composition: *Tholeiitic and alkalic basalt*

The Zuni–Bandera lava field in western New Mexico forms part of the Jemez volcanic lineament. The field occupies a large valley south of Grants, New Mexico, and is bounded to the west by the Zuni uplift. Stretching 90 km long and 1 to 35 km across, it covers an area of 2,460 km^2 with a composite thickness of 20 to 60 m of lava. Between 62 and 123 km^3 of lava was erupted from 74 vents that tend to be aligned along faults and fissures. Oriented N38°E, the pri-

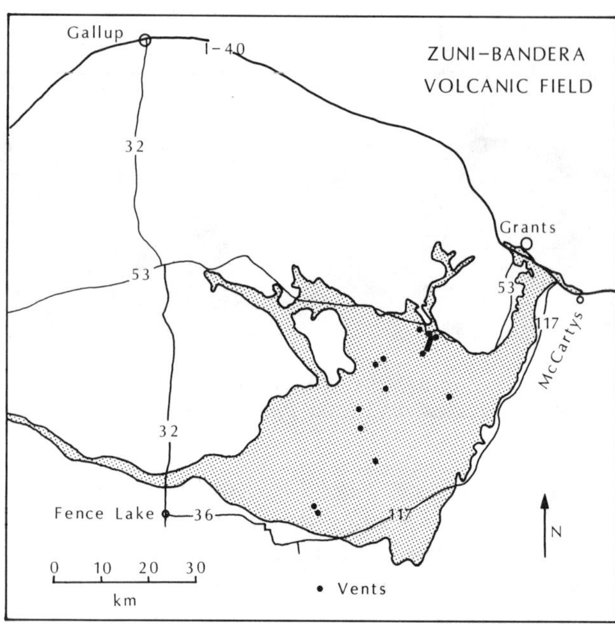

ZUNI-BANDERA: Sketch map.

ZUNI-BANDERA: Relatively small central cinder cone source for the McCartys lava flow in the Zuni–Bandera volcanic field. The lumpiness of the cone and the jumbled lava in the foreground are non-primary features, resulting from aerial dive-bombing target practice during World War II. Unexploded 100-lb bombs are still present. Photograph by L. Crumpler and J. Aubele.

mary chain of vents is oblique to the Jemez lineament but parallels the eastern edge of a positive gravity anomaly. Another segment of vents is located along north–south normal faults that cut the Zuni uplift. Various names have been applied to the field including **Malpais**, **McCartys**, **Zuni**, and **Bandera**.

The Zuni–Bandera volcanic field provides excellent examples of a variety of vent types and lava flow morphologies. Eruptive centers, representing complex histories of both effusive and pyroclastic activity, are marked by cinder cones, spatter ramparts and cones, small shields, maars, and collapse pits. Lava flows exhibit pahoehoe, aa, and block surface textures and may be extremely long, as exhibited by the 90-km-long Fence Lake flow to the west and the 60-km-long **McCartys** flow to the east. Lava tubes up to 28.6 km long form some of the most striking morphological features within the field. These tubes are restricted to pahoehoe flows but are not limited to a specific lava composition. Other surface morphologies indicative of tube-fed lava characterize some of the flows and include pressure ridges, tumuli, linear squeezeups, grooved lava, and collapse pits.

Although overlapping relationships indicate that much of the Zuni–Bandera field was emplaced penecontemporaneously, the degree of weathering, thickness of aeolian mantles, and some stratigraphic relationships suggest that activity may have shifted northward with time. The most recent eruptions produced the McCartys flow along the eastern margin and the late-stage Bandera eruption to the west. Early activity was contemporaneous with the emplacement of Mount Taylor to the northeast and the Springerville volcanic field to the southwest.

The Zuni–Bandera field is of significant interest because both tholeiitic and alkalic basalt erupted within a relatively small area over a short time span. No distinct patterns of compositional variations with time, spatial distribution, or vent type are observed. Most lavas within the field are aphanitic, with >85% finely crystalline groundmass and small phenocrysts of olivine, pyroxene, and plagioclase. Some bombs from Bandera Crater are cored by ultramafic nodules and country rock.

Ice has accumulated within several lava tubes in which subsurface temperatures are maintained at a constant -0.5°C by stagnant cold air in the summer. An example of this interesting occurrence of ice caves is located at Bandera Crater and is open to the public.

How to get there: *The Zuni–Bandera field is located 112 km west of Albuquerque, New Mexico, and ~390 km northwest of El Paso, Texas. Ac-*

cess is provided by I-40 across the northern part of the field, by State Highway 53 along the northwest margin, and by State Highway 117 along the eastern and southern margins. Much of the area is privately owned, however, so the best access is at rest areas on I-40 and at Bandera Crater, which can be reached by a short drive from State Highway 53.

References

Ander, M. E., Heiken, G., Eichelberger, J., Laughlin, A. W., and Huestis, S., 1981, Geologic and geophysical investigations of the Zuni volcanic field, New Mexico: *Los Alamos Nat. Lab. Rpt. LA-8827-MS*, Los Alamos, New Mexico, 39 pp.

Laughlin, A. W., Brookins, D. G., and Causey, J. D., 1972, Late Cenozoic basalts from the Bandera lava field, Valencia County, New Mexico: *Geol. Soc. Am. Bull. 83*, 1543-1552.

EileenTheilig

LUCERO
New Mexico

Type: *Monogenetic volcano field*
Lat/Long: *34.42°N, 107.20°W*
Elevation: *1,573-2,395 m*
Eruption History: *3 periods of activity:*
8.3 to 6.2 Ma
4.3 to 3.3 Ma
1.1 to ~0 Ma
Composition: *Basanite, alkali olivine basalt, tholeiite, tholeiitic basalt*

Eruptive rocks of the Lucero area (**Puertocito**) comprise a group of vents and flows occurring as isolated lava-capped mesas and buttes, scattered over an area >2,000 km^2 on the southeastern margin of the Colorado Plateau. The total volume of volcanic rocks presently exposed is 7.3 km^3. The oldest flows occur mainly along the crest of the Sierra Lucero, a westward dipping faulted monocline flanking the Rio Grande rift valley. Flows of intermediate age, cropping out at lower elevations, were erupted mainly along north-northeast-trending fracture zones onto surfaces that reflect a drainage network substantially different from that of the present. Lavas of the youngest cycle were erupted on or close to the modern land surface at the lowest elevations. Several of the youngest vents were mainly or entirely pyroclastic: The youngest eruption in the field produced a tuff ring not more than a few tens of thousands of years old. **Cerro Verde**, a prominent shield cone 280 m high, was the source of a 42-km-long lava flow that followed Rio Grande river tributaries.

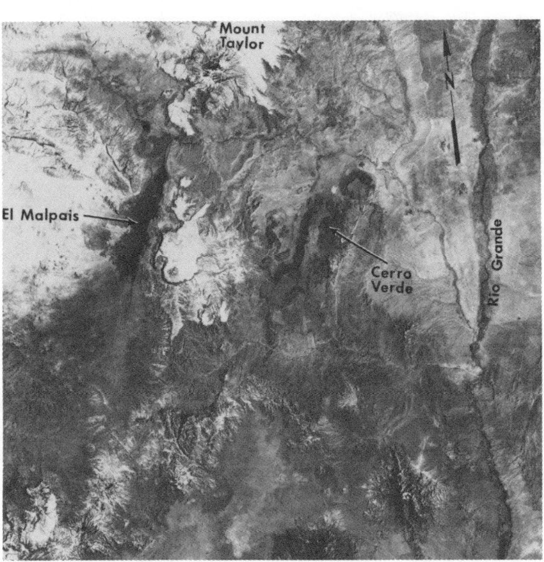

LUCERO: Landsat Thematic Mapper image of a portion of the southeastern Colorado Plateau and central Rio Grande rift in New Mexico. Basalt-capped mesas of the Lucero volcanic area are visible in the center of the picture. Albuquerque is in the upper right; the Rio Grande river flows from north to south near the right edge. Mount Taylor is in the upper left center of the image. South and slightly west of Mount Taylor is El Malpais (part of the Bandera flow), comprised of Quaternary basalt lava flows.

Compositions erupted from each vent comprise distinctly different, although partially overlapping, basaltic populations. They define a trend (from oldest to youngest) toward relatively larger volumes of tholeiitic basalt and an increasingly bimodal distribution with respect to alkalies versus silica. Lavas of the oldest cycle are predominantly basanite and alkali-olivine basalt (SiO_2 = 45-48.7%). The intermediate-cycle lavas are both alkaline-olivine basalt and related, more silicic types, and tholeiitic basalt. The alkalic rocks are, on the average, more evolved than those of the oldest cycle. Lavas of the youngest cycle tend, with the exception of the flow from Cerro Verde, to be strongly alkaline and silica-undersaturated (olivine basanite to mugearite). The Cerro Verde flow, by

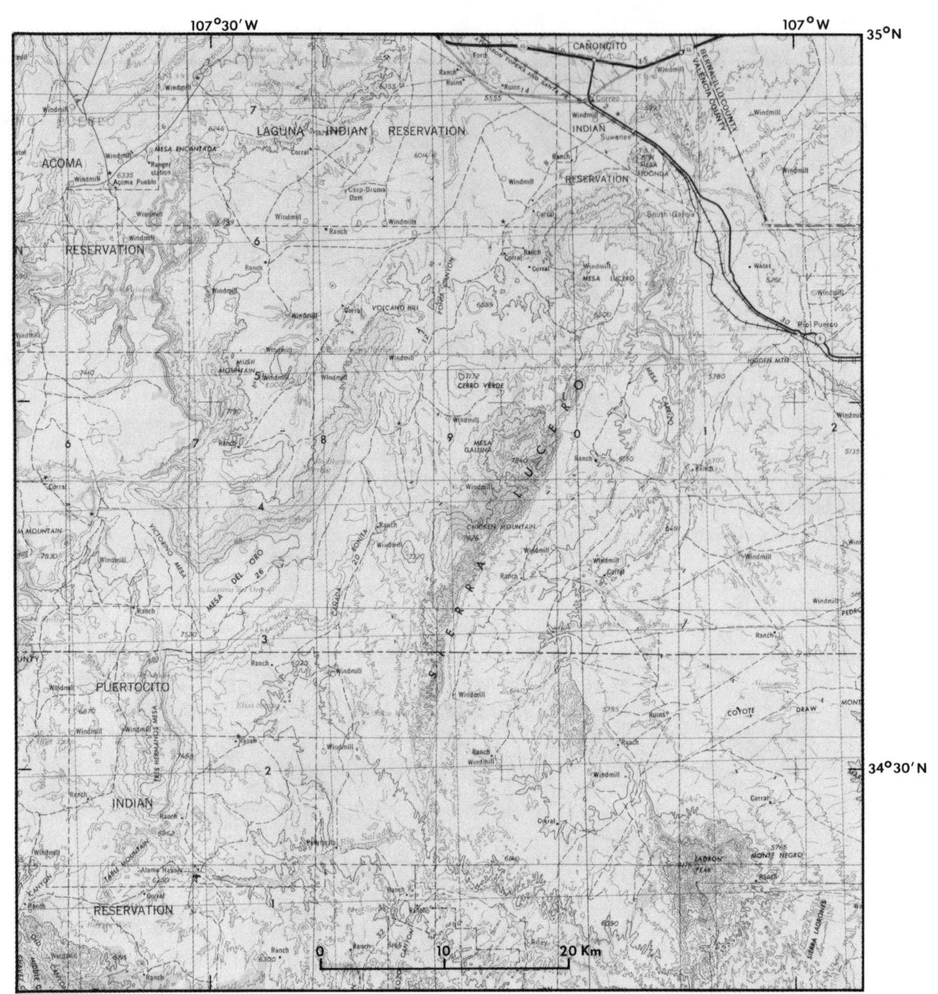

LUCERO: Topographic map of the Lucero volcanic area.

contrast, is strongly subalkaline (SiO_2 = 49.5-51.6%; Na_2O+K_2O = 3.3-4.0%).

How to get there: *State Highway 5, joining Las Lunas to US Highway 66, cuts through the northern portion of the Lucero volcanic field.*

References

Wright, H. E., Jr., 1946, Tertiary and Quaternary geology of the lower Rio Puerco area, New Mexico: *Geol. Soc. Am. Bull.* 57, 383-456.

Baldridge, W. S., Perry, F. V., and Shafiqullah, M., 1987, Late Cenozoic volcanism of the southeastern Colorado Plateau: I. Volcanic geology of the Lucero area, New Mexico: *Geol. Soc. Am. Bull.* 99, 463-470.

W. Scott Baldridge

RED HILL
New Mexico

Type: *Monogenetic cinder cone field*
Lat/Long: *34.25°N, 108.83°W*
Elevation: *2,200 to 2,300 m*
Relief: *100 m*
Eruptive History: *0.9 to 0.023 Ma*
Composition: *Unknown, probably alkali basalt*

Very little is known about the Red Hill (also called **Quemado**) volcanic field. It lies only 24 km east of the large Springerville, Arizona, volcanic field and is morphologically and (based on reconnaissance data) petrologically similar to the basaltic volcanism on the eastern side of the Springerville field. The fact that many vents appear morphologically young, as are

RED HILL: Panorama across the center of Red Hill maar as seen from the east rim.

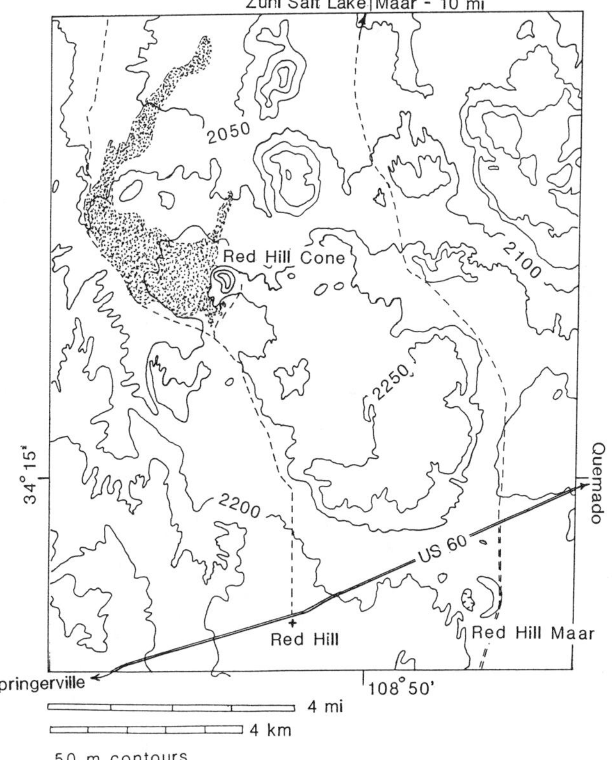

RED HILL: Topographic map of locations of two of the younger vents in the Red Hill field. Stippled area shows extent of lava flow associated with the Red Hill cinder cone. This area is only 16 km south of Zuni Salt Lake maar.

vents on the eastern edge of the Springerville field, strengthens the impression that the two areas of volcanism are related. The field is relatively dispersed, with the younger vents distributed between older basalt-capped mesas of unknown, but possibly late Cenozoic age.

The Red Hill cone is a youthful vent consisting of reddish cinders, and is morphologically slightly more weathered than basaltic cones dated at 0.3 Ma in the Springerville field, but slightly less weathered than cones dated at 0.8 Ma. This implies a tentative age assignment of ~0.5 Ma for Red Hill. The associated lava flow is olivine basalt with preserved pressure ridges on its surface, but a moderately thick soil cover is locally present.

Red Hill maar is interesting as an example of a moderately youthful maar associated with faulting and fissure-type eruptive activity. Youthful fault scarps cut underlying basalt to the west and are clearly visible as linear scarps on topographic maps. The maar and a smaller elongate vent to the south are aligned parallel to the fault scarps, implying that, unlike the Springerville field, regional tensile stresses exerted significant control on vent shapes.

Like the **Zuni Salt Lake** maar to the north, a small cinder cone occurs on the floor of the Red Hill maar, and is offset from the center of the crater. Cinders interbedded with maar deposits suggest that the eruption started with pyroclastic cinder and ash activity, evolved to more explosive maar-type eruptions, and later returned to the eruption of ash and cinders. The late ash blankets most of the crater rim-forming maar deposits, and ash was apparently dispersed by strong southwesterly winds to the northeast where ash covers the countryside for 5 to 10 km along US Highway 60.

How to get there: *Red Hill lies along US Highway 60, ~39 km east of Springerville, Arizona, and 37 km west of Quemado, New Mexico. The Red Hill cinder cone and lava flow can be reached by traveling 9 km north on a road intersecting Highway 60 at the villa of Red Hill. The Red Hill maar lies ~1 km south along a road that intersects Highway 60 ~5.5 km east of Red Hill.*

Reference

Bikerman, M., 1972, New K-Ar ages on volcanic rocks from Catron and Grant Counties, NM: *Isochron West* 3, 9-12.

L. S. Crumpler and Jayne C. Aubele

CARRIZOZO
New Mexico

Type: *Monogenetic volcano field*
Lat/Long: *33.77°N, 106.00°W*
Elevation: *1,565 to 1,730 m*
Eruptive History: *Two undated periods of activity during the Holocene*
Composition: *Subalkaline basalt*

The Carrizozo volcanic field, comprised of the Broken Back and Carrizozo lava flows, is an isolated Quaternary basaltic lava field in south-central New Mexico. The older Broken Back flow extends 16 km southward and 11 km eastward into the upper Tularosa Valley. Lava originated from two cinder cones in the vicinity of **Broken Back Crater** in the northwest part of the flow. These cones are 60 to 75 m high and have been partly mined. The younger Carrizozo flow erupted from three cinder cones 11.5 km east of Broken Back Crater. The most recent cone is the 26-m-high **Little Black Peak**. Extending down the Tularosa Valley 68 km southwestward from this vent, is one of the longest basaltic flows emplaced by a tube-fed mechanism. The flow ranges in width from 0.8 to >8 km, and is 50 m thick 3 km south of State Highway 380. Approximately 4.2 km^3 of lava were erupted to cover 328 km^3. The surface of the flow is predomi-

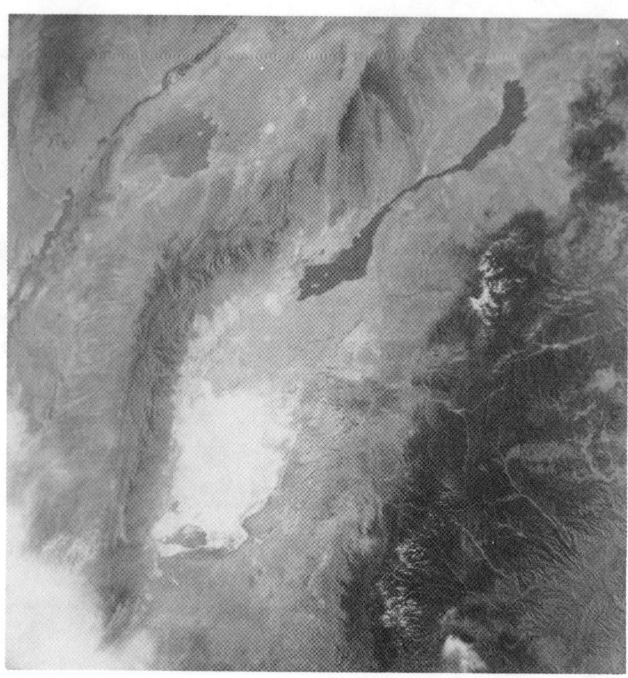

CARRIZOZO: Space Shuttle photograph (STS-3-10-614) of Carrizozo (snake-like on upper right) and Jornada del Muerto volcanic fields (left) and White Sands.

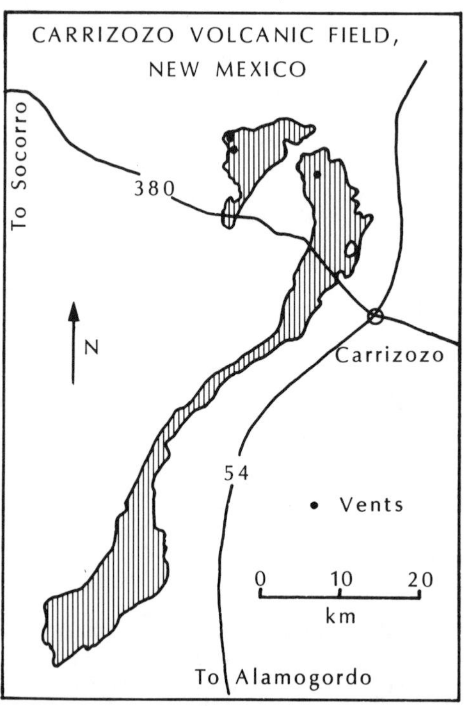

CARRIZOZO: Sketch map of flows.

nantly pahoehoe with patches of aa and is characterized by pressure ridges, tumuli, collapse pits, and linear squeezeups. Two chemically distinct eruptive units have been identified within a large sinkhole, but are difficult to identify from surface morphology.

The exact age of the Carrizozo field is undetermined, but based on the small degree of modification, it is estimated to be recent. Because of the fresh surficial flow features, lack of erosional dissection, and intact conditions of the latest cone, the Carrizozo flow may be 1,000-1,500 yr old. The Broken Back flow is almost overlapped by the Carrizozo lava field and is more heavily modified by alluvium and soil development indicating a greater age.

In thin section the Carrizozo basalt consists of sparse to abundant olivine phenocrysts in a fine-grained matrix. The matrix in the interior of the flows is crystalline and consists of andesine to labradorite laths, augite, and olivine but becomes glassy near the surface.

How to get there: *The Carrizozo lava field is west of Carrizozo and north of White Sands, New Mexico. It is ~100 km east of Socorro, New Mexico, and 225 km north of El Paso, Texas. Access is viaUS Highway 380, which crosses the flow 7.4 km south of Little Black Peak. The Valley of Fires State Park is located within the Carrizozo flow. Southward, the flow enters White Sands Missile Range and is inaccessible.*

References

Weber, R. H., 1964, Geology of the Carrizozo Quadrangle, New Mexico: *in* Ash, S. R., and Davis, L. V. (eds.), *Guidebook of the Ruidoso Country, Fifteenth Field Conference,* NM Geol. Soc.

Renault, J., 1970, Major-element variations in the Potrillo, Carrizozo, and McCartys basalt fields, NM Bur. Mines Min. Res. *Circular 113,* 22 pp.

Eileen Theilig

JORNADA DEL MUERTO
New Mexico

Type: *Large lava flow from a single vent; shield?*
Lat/Long: *33.53°N, 106.87°W*

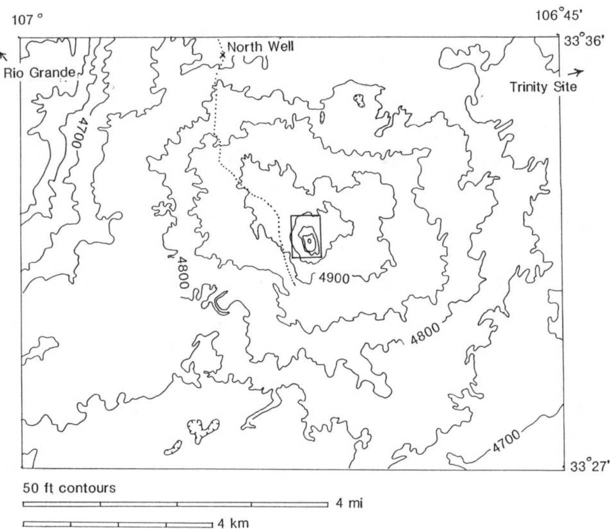

JORNADA DEL MUERTO: Topographic map of overall shieldlike shape of Jornada del Muerto. Square indicates vent region.

JORNADA DEL MUERTO: Vent region of shield seen from ~1.5 km to the southwest. From this distance the vent appears as a broad oval rampart with a relatively flat summit.

Elevation: *1,430 to 1,555 m*
Height: *Volcanic field: 122 m;*
Central vent area above lava flows: 30 m;
Cinder cone in central vent: 24 m
Diameter: *Volcanic field: 16 × 24 km;*
central vent: ~550 m;
base/crater cinder cone: 244/76 m
Eruptive History: *0.76 Ma*
Composition: *Alkali basalt*

The Jornada del Muerto field lies a few kilometers east of a prominent westward bend in the north–south run of the Rio Grande river through New Mexico, and on the western edge of the White Sands Missile Range. The name translates from the medieval Spanish as "Route of the Dead Men," in reference to the frequent outcome of those who tried to save time on the trail from Mexico City to Santa Fe by making a straight cutoff of the river bend across the "malpais" or volcanic "badland." The area is barren, devoid of water, remote, and even today, difficult to reach. As a consequence few geologists have visited the central crater and little is known about the details of the eruptive history, petrology, and volcanology.

The Jornada del Muerto field is of interest because it is similar in volume (~13 km^3) to the slightly younger, but equally spectacular McCartys (Zuni–Bandera) and Carrizozo flows. Each of these is a voluminous and young single-vent lava field in which the central vent fed multiple flow units with total volumes of a few cubic kilometers. The Jornada vent is structurally unusual, however, and differs from the relatively simple cone structures forming the vents of the other two large flows. Viewed from a distance, the profile is of a flat-topped squat cone, or rampartlike shape, resembling the raised rim of Meteor Crater, Arizona. The interior of the rampart forming the vent region is not depressed deeply, but instead contains a small cinder cone resting on a larger and downwardly concave perched platform. The platform is made of multiple sheets of lava, is relatively smooth, and resembles the surface of a solidified lava lake. The platform is isolated from the outwardly sloping ramparts by a narrow circular graben. The trough may have originated by extension about the margins of the lava lake during cooling and sagging of the interior. The cinder cone formed over the north end of the platform, has an unbreached summit pit, and is in the early stages of radial drainage formation on its flanks.

Large lava tubes are prominent south of the central vent area, parts of which have collapsed leaving long sinuous troughs. These tubes were mined for bat guano early in the century, and a few ruins associated with the operation remain.

A few additional late cinder cones, maars, and lava flows occur mostly south of Jornada. Localities include east of the river in the Elephant Butte Reservoir, near Eagle and Truth or Consequences, and to the northwest of Jornada at Black Butte near San Marcial. All these appear to be basalt a bit older (~2-3 Ma) than that at Jornada. Because of their proximity to the central channel of the Rio Grande river, these volcanic areas are locally either dissected or interbedded with alluvial floodplain materials.

How to get there: *A field vehicle with high ground clearance and four-wheel drive is desirable. Take a ranch road south from Highway 801 at a point 8-16 km east of San Antonio and I-25. Roads and gate locks change, but the objective is to work south toward Val Verde and the northern edge of the flow at North Well. From North Well a trail traverses the surface of the flow to the Bat Caves just >2 km south of the central vent.*

References

Aubele, J. C., Crumpler, L. S., Loeber, K. N., and Kudo, A. M., 1976, Maar and tuff rings of New Mexico: *NM Geol. Soc. Spec. Pub. 6*, 109-114.

Bachman, G. O., and Mehnert, H. H., 1978, New K-Ar dates and the late Pliocene to Holocene geomorphic history of the central Rio Grande region, New Mexico: *Geol. Soc. Am. Bull. 89*, 283-292.

L. S. Crumpler and Jayne C. Aubele

KILBOURNE HOLE
New Mexico

Type: *Maar*
Lat/Long: *31.97°N, 106.97°W*
Elevation: *1,292 m*
Height: *0-45 m*
Volcano Diameter: *2.4 × 3.4 km*
Crater Diameter/Depth: *2.3 × 3.3 km/135 m*
Eruptive History: *80,000 yr BP*
Composition: *Basalt mixed with sand and silt*

Kilbourne Hole, the best known of the Potrillo maar volcanoes, sits astride the north–south-trending Fitzgerald fault, surrounded by the late Cenozoic **Afton** basalt flow. The maar was formed by steam explosions due to the heating of water-saturated sand and silt strata by rising basaltic magma. From the bottom of the crater to the top of the rim the following units are exposed: (1) Santa Fe Group sediments, (2) olivine basalt (Afton basalt), (3) bedded hyalo-

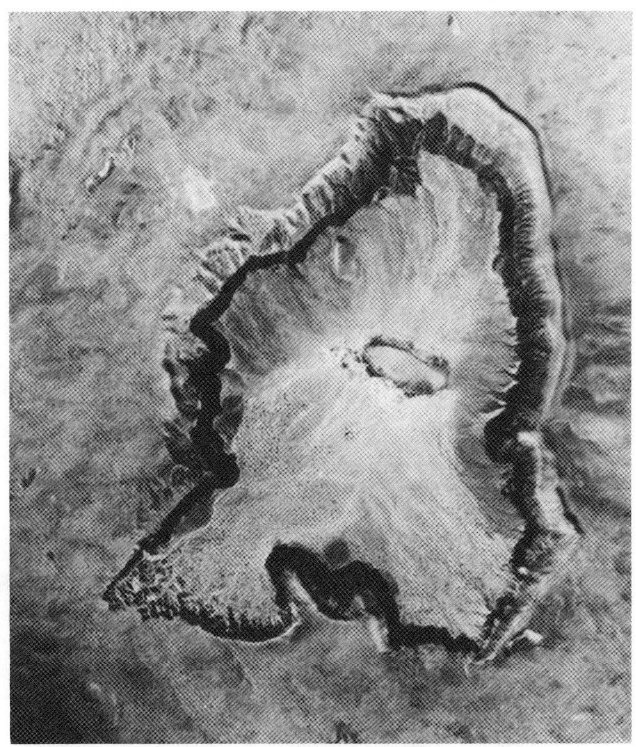

KILBOURNE HOLE: Vertical aerial photograph.

clastic tuffs (base surge and air fall) and vent breccia, and (4) Holocene wind-blown sand.

Four other explosion craters occur in the Potrillo basalt field. **Hunt's Hole** is a shallower and smaller (1,385 m wide) version of Kilbourne Hole. Further south on the Mexican border is **Potrillo** Maar, a 4,920 × 3,385-m elliptical crater with several cinder cones and basalt flows on its floor. **Malpais** is a circular tuff ring ~1,230 m in diameter, 25 km west of the Potrillo Maar in the West Potrillo Mountains. The ring is made of well-bedded ash, lapilli tuff, and lapilli breccia. Post-explosion activity constructed a cinder cone and basalt dikes and flows. **Riley** is a circular tuff cone, 925 m wide, ~12 km north of Malpais. Riley's rim is made of poorly bedded hydroclastic lapilli tuff and tuff breccia, and the floor of the crater contains a basalt flow.

How to get there: *The explosion craters are 55 to 75 km west and west-northwest of El Paso, Texas. Access to Kilbourne, Hunt's, and Potrillo is easy; Malpais is moderate, and Riley difficult.*

References

Cordell, L., 1975, Combined geophysical studies at Kilbourne Hole maar, New Mexico: *N. Mex. Geol. Soc. Guidebook, 26th Field Conf.*, 273-281.

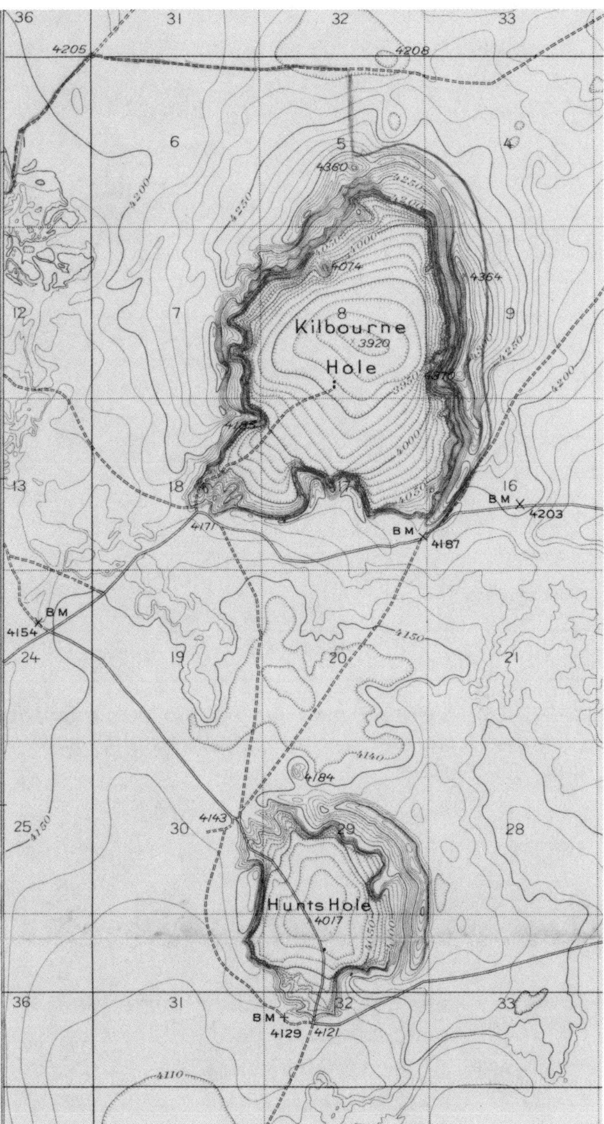

KILBOURNE HOLE: USGS topographic map of Kilbourne and Hunt's Hole. Contour interval 10 feet.

Seager, W.R., 1987, Caldera-like collapse at Kilbourne Hole maar, New Mexico: *New Mexico Geology 9*, 69-73.

Jerry Hoffer

POTRILLO
New Mexico

Type: *Monogenetic cone field*
Lat/Long: *31.9°N, 107.2°W*
Elevation: *1,695 m*

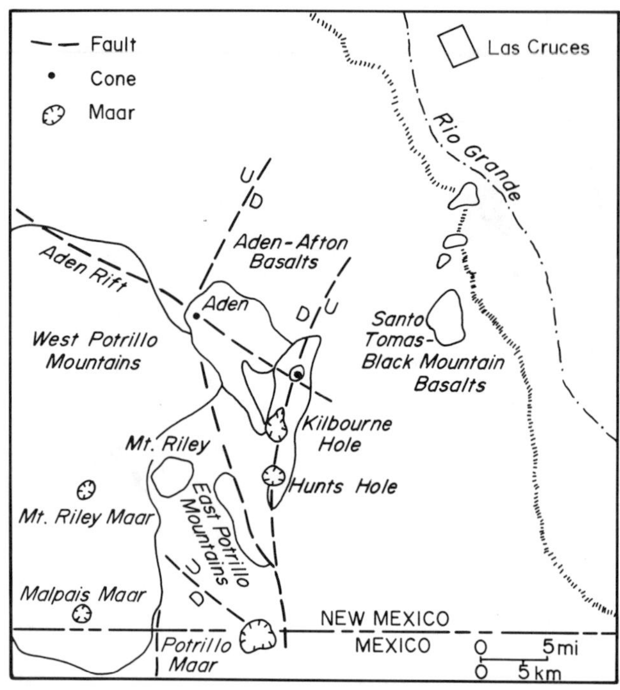

POTRILLO: Location map for Potrillo volcanics and Kilbourne Hole region. Up (U) and down (D) sides of faults marked.

POTRILLO: Aden crater from the northeast. Aden is composed of a base built up of gently sloping basalt flows and capped by a steep rim of spatter and driblet.

Height: *200 m*
Volcano Field Size *50 × 35 km*
Eruptive History: *17 radiometric dates from 2.65 to 0.15 Ma; No historic eruptions*
Composition: *Alkaline olivine basalt*

The Potrillo volcanic field is a typical monogenetic volcanic field lying along the west margin of the Rio Grande rift in southern New Mexico. The Potrillo field can be divided into three volcanic regions: The West Potrillo field consists of >150 cinder cones, 2 maar volcanoes, and associated flows, all covering ~1,250 km^2.

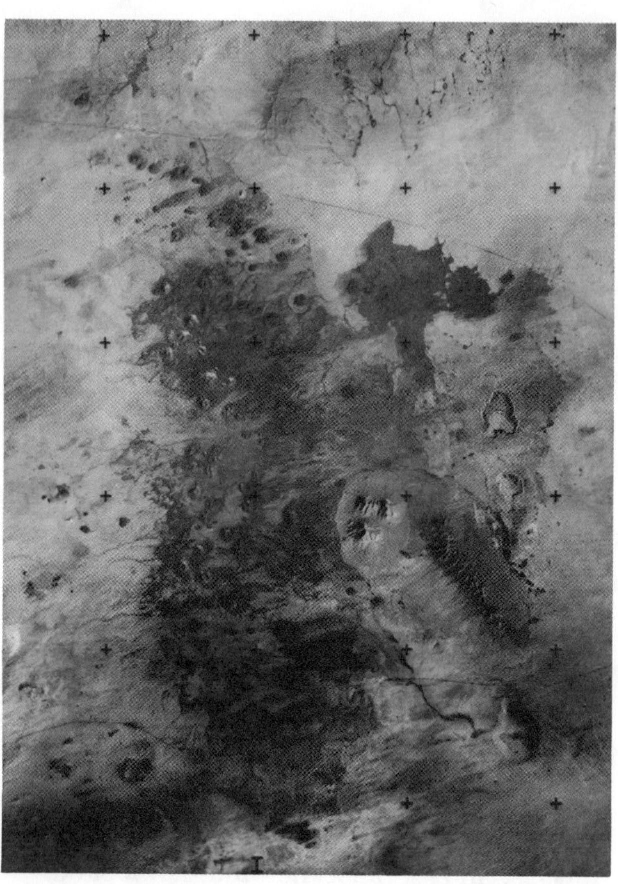

POTRILLO: Landsat Return Beam Vidicon view of entire Potrillo and Aden–Afton volcanic fields. In this 30-m-resolution scene Kilbourne Hole, Potrillo Maar, and most of the features indicated on the location map are visible.

The Aden–Afton field (~230 km^2) includes predominantly young flows, 3 cinder cones, 3 maar volcanoes (see Kilbourne Hole), and Aden Crater, a small shield cone. The **Black Mountain–Santo Tomas** basalt (40 km^2) consists of 4 eruptive centers in a north–south line near the Rio Grande River. The West Potrillo field is apparently the oldest part of the entire Potrillo volcanic field, but radiometric ages up to 2.65 Ma have been obtained from Santo Tomas.

The most unusual feature in the volcanic field is **Aden Crater**, a beautifully preserved small shield volcano, formed in 5 stages: (1) extrusion of thin lava flows to build a shield; (2) construction of a spatter rampart around a lava lake dated at 0.53 Ma; (3) collapse of the western portion of the solidified lava lake; (4) construction of small spatter mounds on the flanks and within the cone; and (5) persisting fumaroles.

How to get there: *Drive 68 km northwest of El Paso, Texas, or 38 km southwest of Las Cruces, New Mexico. Access is easy.*

References

Hoffer, J. M., 1976, Geology of the Potrillo Basalt Field, southcentral New Mexico: New Mexico Bureau of Mines, *Circular 149*, 30 pp.

Kahn, P. A., 1987, Geology of Aden Crater, Dona Ana County, New Mexico: M.S. Thesis, Univ. Texas, El Paso, 90 pp.

Jerry Hoffer

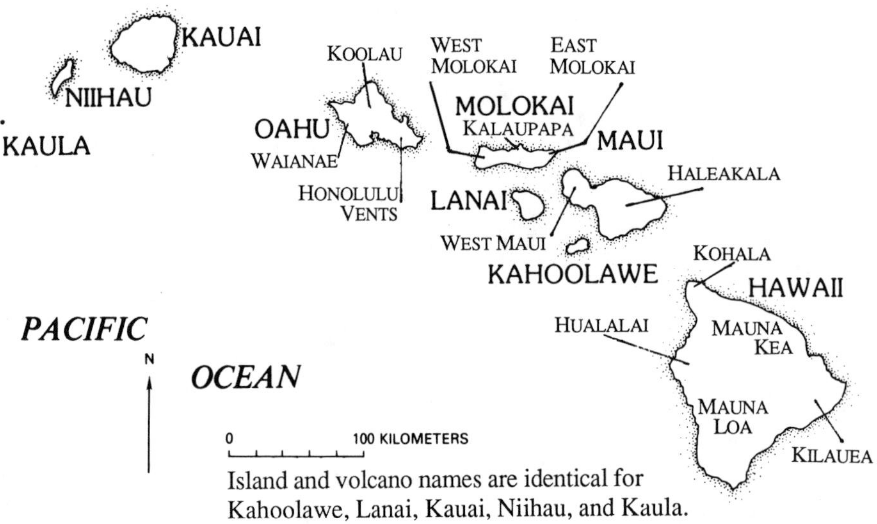

HAWAII FIGURE 1: Major volcanic centers of Hawaii (modified from USGS Prof. Pap. 1350).

VOLCANO TECTONICS OF ALASKA OF THE HAWAIIAN ISLANDS

Charles A. Wood

Hawaii is the only state in the USA built entirely of volcanic materials. Each of its islands is made up of one or more massive shield volcanoes rising from the ocean floor or from the flanks of its neighbors. The Hawaiian Islands are the southernmost end of a largely submerged chain of similar shield volcanoes that extends 3,400 km to the northeast and then, bending toward the north, another 2,300 km as the Emperor Seamounts, reaching to Kamchatka. About 100 separate volcanoes, with a total volume of ~1 million km^3 (Bargar and Jackson, 1974), make the Hawaiian–Emperor chain the most massive point-source volcanic outpouring on Earth.

Hot Spot Traces

Nearly 150 years ago, James Dana, the first geologist to visit Hawaii, recognized the volcanic nature of the Hawaiian Islands, as well as a northwest to southeast progression from old age to youthful eruptive activity (Appleman, 1987). From that time until the 1960s most volcanological research concerned monitoring the ongoing eruptions at Kilauea and Mauna Loa. In 1963, the Canadian geophysicist J. T. Wilson proposed that the Hawaiian Islands and other linear volcanic island chains in the Pacific formed over deep, stationary magma source regions (later called hot spots) and were carried away as the ocean floor moved to the northwest. Clague and Dalrymple (1987) provide a history of development of this idea and demonstrate that considerable evidence supports it. Perhaps the most convincing is that the ages of the volcanoes decrease from 75-80 Ma for the Meiji Seamount about to be subducted beneath Kamchatka, to 40 Ma for the Hawaiian–Emperor bend at ~32°N, to 0.4-0 Ma for Kilauea volcano and the yet-to-emerge Loihi seamount south of the big island of Hawaii.

Detailed radiometric dating and petrologic studies provide excellent characterization of the Hawaiian–Emperor chain volcanoes (summarized in Clague and Dalrymple, 1987), yet the basic question of what caused the hot spot remains unanswered. Clague and Dalrymple point out defects in all proposed mechanisms, but conclude by endorsing Morgan's (1971) hypothesis that plumes of hot, geochemically primitive material convectively rise through the mantle and interact with the lithosphere beneath Hawaii to yield the observed volumes and compositions of lavas. However the hot spot originates, its great volume output and tremendous duration of operation are unique on Earth. Interestingly, however, the largest volcanoes on Mars and Venus each have, within a factor of 2, the same volume as the entire Hawaiian–Emperor chain. Is this a coincidence or do the processes of magma generation and ascent to the surface have similar limitations on three planets with presumably different geologic histories?

Magma Generation and Storage

During the 1959-60 eruption of Kilauea, earthquakes were recorded from a maximum depth of 60 km beneath Hawaii. Decker (1987) uses this datum, plus petrologic inferences, to suggest that magma is generated in a zone from 60 to 170 km beneath Hawaii. More refined seismic studies additionally imply that a broad, hot region exists at ~350 km depth south of Hawaii (a hot spot plume?), feeding heat and perhaps material upward to the magma generation zone. Decker (1987) reviews evidence for believing that garnet peridotite in the mantle is melted, with the melt gradually aggregating and rising toward the surface by buoyancy effects (its density is lower than that of unmelted rocks) and/or fracturing. The process is efficient enough to provide a steady flow of 0.1 km^3/yr of magma to Kilauea and probably as much to Mauna Loa (Swanson, 1972).

The existence of calderas – typically 3-5 km wide on young shields, but twice that size on most older ones – on Hawaiian shields is prima facie evidence for near-surface magma storage. Geophysical support for this includes an aseismic region under Kilauea's caldera that defines a volume of at least partially melted material from 3 to 6 km deep (Klein et al., 1987). Rather than being a liquid sphere of magma, as is often assumed in seismic and tilt modeling, the Kilauea reservoir is apparently an interconnected plexus of magma conduits (Ryan et al., 1981).

Walker (1987) has proposed a new theory of origin for Hawaiian calderas inspired by his mapping of downsagged caldera fill and a vast number of dikes in and around the erosion-exposed Koolau caldera. Caldera formation occurs by subsidence due to the excess loading of summit regions by a large number of intruded dikes. Subsidence is facilitated by the existence of a thermally weakened lithosphere above the Hawaiian hot spot.

In contrast to this new proposal, detailed sonar mapping of Hawaii's submerged flanks provides evidence supporting the older explanation that calderas form due to collapse as a magma chamber partially empties during major eruptions (Holcomb et al., 1988). The sonar reveals large, apparently young lava flows along the deeply submerged extensions of the rift zones of Kilauea, Mauna Loa, and Loihi. Thus, the lack of correlation between subaerial lava flow volumes and caldera collapse volumes (e.g., Walker, 1988) may simply reflect the lack of knowledge of the submarine component of an eruption. As both Walker's theory and the newly buttressed older one are based on detailed observations, it is likely that both mechanisms apply; the real debate will probably be over the relative importance of each through time at individual volcanoes and from volcano to volcano.

Evolutionary Stages of Hawaiian Volcanoes

Based upon his detailed mapping of many of the volcanoes, Stearns (1940) proposed that Hawaiian shields evolved through a series of stages. His recognition of a progression of morphology and rock composition still provides the framework for understanding Hawaiian shield volcano evolution. Peterson and Moore (1987) have proposed a revised series of 7 stages, but the more schematic 4-stage history of Langenheim and Clague (1987) is outlined here. The first, *preshield stage* was forced upon geologists by the recent exploration of the submarine Loihi volcano growing off the south coast of Hawaii. Loihi is a young shield volcano, exactly where the hot spot theory projects the next addition to the Hawaiian–Emperor chain. Two characteristics of Loihi do not fit previous theories, however: its alkalic basalt composition and its summit caldera. Thus, the preshield stage, based on a single volcano (perhaps even anomalous for Hawaii?), assumes that the earliest stage of submarine activity builds an alkalic shield with summit calderas and rift zones (e.g., Fournari et al., 1988).

The *shield stage* is a 0.5-1-m.y. period when vast amounts of tholeiitic basalt erupt effusively to enlarge greatly the alkalic shield volcano. Langenheim and Clague (1987) suggest that near the end of this shield stage a caldera forms and may be filled with alkalic basalt as well as tholeiite; the recent evidence from Loihi shows that calderas can form in the preshield stage. A capping of alkalic basalt and differentiated lavas form during the *postshield stage.* These flows are less fluid and more likely to be associated with formation of cinder cones than were the shield stage eruptions. After a substantial hiatus, the *rejuvenation stage* completes the volcano's magmatic lifetime with eruptions producing small cinder cones and short flows of silica-poor lava (alkalic basalt, basanite, nephelinite). Subsidence and erosion eventually cause the volcanic carcass to disappear below sea level, where very low rates of erosion preserve it until the plate tectonic conveyor belt reaches Kamchatka and the volcano is destroyed by subduction.

Loihi is the only example of the preshield stage, but dredged samples of tholeiite suggest that Loihi has entered the shield stage. Kilauea and Mauna Loa are both at the shield stage, and all three of the other volcanoes on the island of Hawaii are capped by postshield alkalic rocks. Most of the volcanoes on the older Hawaiian Islands are in the rejuvenation stage, although Lanai became dormant ~1 Ma while still in the shield stage. Koolau skipped the postshield phase, and West Molokai has not yet been rejuvenated (Langenheim and Clague, 1987).

Eruptive stages of Hawaiian volcanoes

Volcano	Preshield	Shield	Postshield	Rejuven.
Kaula	?	x	x	x
Niihau	?	x	x	x
Kauai	?	x	v[a]	x
Waianae	?	x	x	x
Koolau	?	x		x
W. Molokai		?	x	x
E. Molokai	?	x	x	x
Lanai	?	x		
Kahoolawe	?	x	x	x
West Maui	?	x	x	x
East Maui	?	x	x	x
Kohala	?	x	x	
Hualalai	?	x	x	
Mauna Kea	?	x	x	
Mauna Loa	?	x		
Kilauea	?	x		
Loihi	x	x?		

[a]Though Kauai has more than one caldera, only one passed through the postshield alkalic stage; hence the v entry (half an x) in the table.
Source: Simplified (and Kaula added) from Langenheim and Clague (1987).

Rift and Caldera Eruptions

Most eruptions on Hawaiian shields occur at only two locations: within calderas and along rifts. The caldera location represents lavas that have come from the magma chamber due to increased hydrostatic pressure resulting from the input of new batches of magma (Blake, 1981). Kilauea's Halemaumau crater had persistent activity for 101 yr, ending in 1924 (Peterson and Moore, 1987). Such lava-lake activity is a rare type of volcanism, present at only 3 other volcanoes in the world.

Each Hawaiian shield volcano has 2 or 3 rift zones which appear to form in response to regional stress fields or imposed gravitational stress from neighboring, buttressing shields (Fiske and Jackson, 1972). Erosion exposes massive dike complexes at rifts on older Hawaiian shields (Walker, 1987), and recently active rifts are marked by spatter and small cinder cones, collapse pits, and fissures. Rift lavas also come from summit magma chambers, driven out by radial pressures. Decker (1987) reports that seismic evidence suggests that rift structures extend from a few kilometers beneath the surface to the bottom of the volcanic edifice.

References

Hawaii is probably the best studied volcanic province on Earth, although modern geologic studies of the volcanoes – rather than of current eruptive phenomena – are relatively recent. The very best introduction to the huge literature on Hawaii, and the most important contribution to it, is the 1987 *Volcanism in Hawaii* (USGS Prof. Pap. 1350) edited by Decker, Wright, and Stauffer (abbreviated *VIH* throughout this chapter's references). This very heavy, two-volume bible is the starting point for all future research on Hawaiian volcanism. For work in the field rather than the library, no volcanologist who goes to Hawaii should be without one or both of the excellent single-volume books by Stearns (1985) and Macdonald et al. (1983).

Appleman, D. E., 1987, James D. Dana and the origins of Hawaiian volcanology: The US Exploring Expedition in Hawaii: *VIH*, 1607-1618.

Bargar, K. E., and Jackson, E. D., 1974, Calculated volumes of individual shield volcanoes along the Hawaiian–Emperor Chain: *USGS J. Res. 2*, 545-550.

Blake, 1981,Volcanism and the dynamics of open magma chambers: *Nature 289*, 783-785.

Clague, D. A., and Dalrymple, G. B., 1987, The Hawaiian–Emperor volcanic chain, Part I. Geologic evolution: *VIH*, 5- 54.

Decker, R. W., 1987, Dynamics of Hawaiian volcanoes: An overview: *VIH*, 997-1018.

Fiske, R. S., and Jackson, E. D.,1972, Orientation and growth of Hawaiian volcanic rifts – the effect of regional structure and gravitational stresses: *R. Soc. Lond., Proc. 329A*, 299-326.

Fournari, D. J., Garcia, M. O., Tyce, R. C., and Gallo, D. G., 1988, Morphology and structure of Loihi seamount based on Seabeam sonar mapping: *J. Geophys. Res. 93*, 15,227-38.

Holcomb, R. T., Moore, J. G., Lipman, P. W., and Belderson, R. H., Voluminous submarine lavas flows from Hawaiian volcanoes: *Geology 16*, 400-404.

Klein, F. W., Koyanagi, R. Y., Nakata, J. S., and Tanigawa, W. R., 1987, The seismicity of Kilauea's magma system: *VIH*, 1019-1186.

Langenheim,V. A. M., and Clague, D. A., 1987, The Hawaiian–Emperor volcanic chain, Part II. Stratigraphic framework of volcanic rocks of the Hawaiian Islands: *VIH*, 55-84.

Macdonald, G. A., Abbott, A. T., and Cox, D. C., 1983, *Volcanoes in the Sea* (2nd ed.): Univ. of Hawaii Press, Honolulu, 517 pp.

Morgan, W. J., 1971, Convection plumes in the lower mantle: *Nature 230*, 42-43.

Peterson, D. W., and Moore, R. B., 1987, Geologic history and evolution of geologic concepts, Island of Hawaii: *VIH*, 149-189.

Ryan, M. P., Koyanagi, R. Y., and Fiske, R. S., 1981, Modelling the three-dimensional structure of macroscopic magma transport systems: application to Kilauea Volcano, Hawaii: *J. Geophys. Res. 86*, 7111-7129.

Stearns, H. T., 1940, Geology and ground-water resources of the Islands of Lanai and Kohoolawe, Hawaii: *Hawaii Div. Hydrogr. Bull. 6*, 177 pp.

Stearns, H. T., 1985, *Geology of the State of Hawaii* (2nd ed.): Pacific Books, Palo Alto, 335 pp.

Swanson, D. A., 1972, Magma supply rate at Kilauea volcano, 1952-1971: *Science 175*, 169-170.

Walker, G. P. L., 1987, The dike complex of Koolau Volcano, Oahu: Internal structure of a Hawaiian rift zone: *VIH*, 961-993.

VOLCANOES OF HAWAII

Kaula, 319
Niihau, 320
Kauai, 321
Waianae, 323
Koolau, 324
Honolulu Vents, 325
West Molokai, 327
Wailau, 328
Kalaupapa and Mokuhooniki, 330
Lanai, 331
Kahoolawe, 332
West Maui, 333
Haleakala, 335
Kohala, 336
Hualalai, 337
Mauna Kea, 339
Mauna Loa, 342
Kilauea, 344

KAULA: Low oblique aerial view.

KAULA
Hawaii

Type: *Tuff cone atop shield volcano*
Lat/Long: *21.65°N, 160.55°W*
Elevation: *160 m*
Relief: *160 m at sea level, 4.4 km below sea level*
Volcano Diameter: *~1 km a.s.l., ~30 km b.s.l.*
Crater Diameter: *0.7 km*
Eruptive History:
Main shield: 5-4 Ma
Post-erosional lavas: ?-1.8 Ma
No historic activity
Composition: *Nephelinitic tuff including tholeiite, phonolite, nephelinite blocks*

Kaula is a flat-topped shield volcano completely submerged below sea level except for about half of a tuff cone which rises 160 m above sea level. The cone is made of nephelinitic tuff with accidental blocks of spinel peridotite, garnet pyroxenite, and lavas (tholeiite, alkalic basalt, nephelinite, and phonolite). The tuff cone sits on the southeast side of a wave-cut platform that is a separate shield volcano from its neighbor, Niihau volcano.

How to get there: *Kaula Island is a target for aerial bombing by the US military. Obtain permission before trying to visit. Many unexploded 500-lb bombs are present on the island. The tiny island also has a bird population of >50,000 (with the bombing they must have unusual population fluctuations). Mantle xenoliths are abundant and fresh on Kaula.*

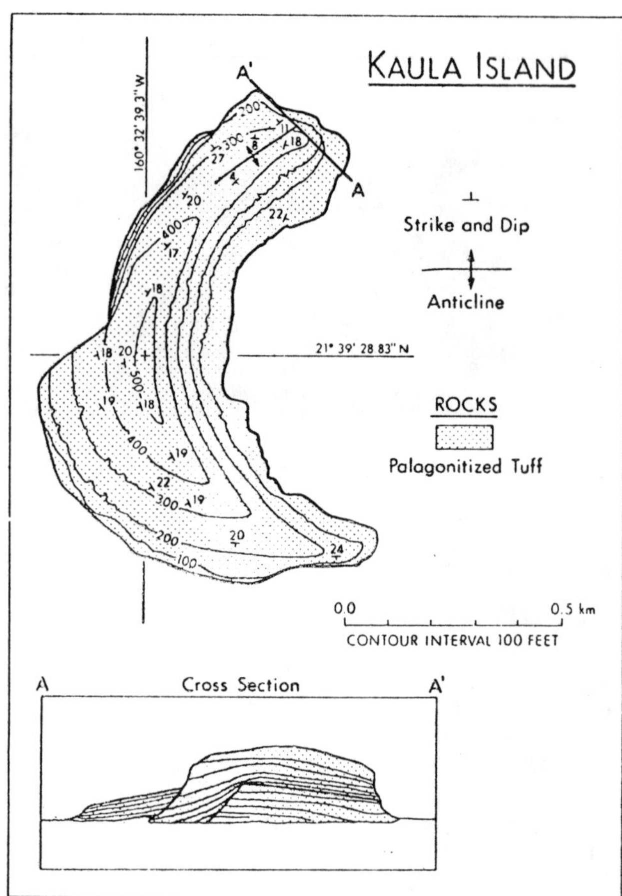

KAULA: Topographic and geologic (one unit) map of Kaula from Garcia et al. (1986).

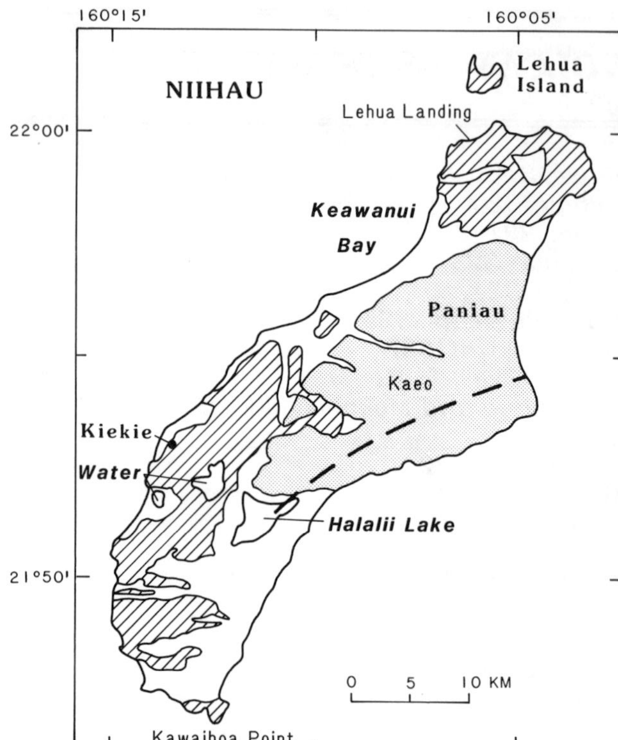

NIIHAU: Simplified geologic map. Sedimentary units: white; Kiekie Basalt: lined; Paniau Basalt: light stipple; rift zone: dashed line.

References

Garcia, M. O., Frey, F. A., and Grooms, D. G., 1986, Petrology of volcanic rocks from Kaula Island, Hawaii: *Cont. Min. Petr. 94*, 461-471.

Palmer, H. S., 1936, Geology of Lehua and Kaula Islands: *B.P. Bishop Mus. Occas. Pap. 12*, 3-36.

Michael Garcia

NIIHAU
Hawaii

Type: *Shield volcano*
Lat/Long: *21.9°N, 160.1°W*
Elevation: *391 m*
Eruptive History:
Tholeiitic shield stage: ~5.5-5.0 Ma
Alkalic postshield stage: 4.9 Ma
Alkalic rejuvenated stage: 2.5-0.4 Ma

Composition
Tholeiitic shield stage: tholeiitic basalt and olivine tholeiitic basalt
Alkalic postshield stage: alkalic basalt
Alkalic rejuvenated stage: alkalic basalt

The island of Niihau consists of a deeply eroded shield volcano mantled by lava of the alkalic rejuvenated stage on the north, west, and south sides. The Paniau Basalt forms the central highland area and consists of tholeiitic basalt and olivine tholeiitic basalt of the shield stage and remnants of a single alkalic postshield stage vent at Kaeo. Several dikes exposed near the eastern coastline also consist of alkalic basalt and presumably fed alkalic postshield stage that have been removed by erosion.

The exposed shield-stage lava was erupted from a southwest rift system; the summit of the shield (and former caldera?) were presumably northeast of the present island. The shield-stage lava flows are cut by numerous dikes, ranging in thickness from ~5 cm to 5.2 m, exposed in the sea cliffs on the east and southeast sides of the island. The eastern side of the island was re-

NIIHAU: Oblique air photograph of Lehua Island, a breached tuff cone of the rejuvenated-stage Kiekie Basalt.

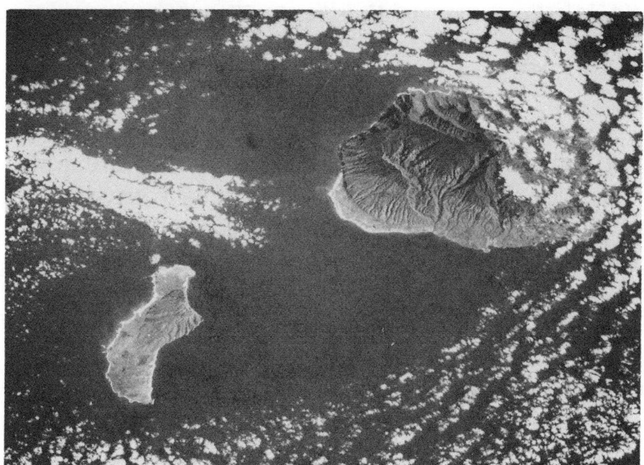

NIIHAU: Space Shuttle photograph of Niihau (left) and western Kauai. Photograph STS-61C-51-103.

moved mainly by downfaulting that resulted in a giant submarine landslide to the south-southeast. A period of volcanic quiescence after eruption of the alkalic postshield lavas lasted about 2.4 Ma and was followed by eruption of the rejuvenated-stage alkalic lava of the Kiekie Basalt, which consists entirely of alkalic basalt. **Lehua** Island off the north shore is a breached tuff cone of the Kiekie Basalt, as is the cone at the island's south point. Six other Kiekie vent edifices are low lava shields; the remaining identifiable structure is a deeply weathered cinder cone.

Niihau Island is perched atop a much larger submarine shield. Numerous cones and flows on the seafloor to the north and west were perhaps erupted at the same time as the Kiekie Basalt.

The most noteworthy feature on Niihau is the magnificent dike swarm, exposed on the east coast, which marks the location of the southwest rift zone.

How to get there: *Niihau Island is privately owned, and access is restricted. No commercial airlines or boats service the island; some helicopter tours fly around it Much of the geology can be viewed on a clear day from Kauai; particularly good vantage points occur along State Highway 550 on the way to Waimea Canyon.*

References

Stearns, H. T., 1947, Geology and ground-water resources of the island of Niihau, Hawaii: *Hawaii Div. Hydrogr. Bull. 12*, 38 pp.

Macdonald, G. A., Abbott, A. T., and Peterson, F. L., 1983, *Volcanoes in the sea*: Honolulu, Univ. Hawaii Press, 517 pp.

David A. Clague

KAUAI
Hawaii

Type: *Shield volcano*
Lat/Long: *22.1°N, 159.5°W*
Elevation: *1,598 m*
Eruptive History:
Tholeiitic shield stage: 5.2-4.0 Ma
Alkalic postshield stage: 3.92 Ma
Alkalic rejuvenated stage: 3.65-0.52 Ma
Composition:
Tholeiitic shield stage: tholeiitic basalt and picritic tholeiitic basalt
Alkalic postshield stage: mugearite and hawaiite
Alkalic rejuvenated stage: alkalic basalt, basanite, nephelinite, and melilitite

Kauai, the westernmost of the principal Hawaiian Islands, is a single shield volcano, ~40 km wide at sea level, that rises ~6,700 m above the ocean floor. The volume of the shield is ~29.3×10^3 km^3, making it the third largest in the Hawaiian–Emperor volcanic chain. The volcano has been deeply incised by stream erosion that has created spectacular canyons. The northwest coast is a sheer sea cliff rising up to 500 m above sea level. Beneath the sea, the submarine slopes around Kauai are dotted with small cones and flows, particularly to the west-northwest and southeast. These cones and flows are probably related to the subaerial Koloa volcanics. Large submarine landslides have formed on the flanks of Kauai; three particularly large

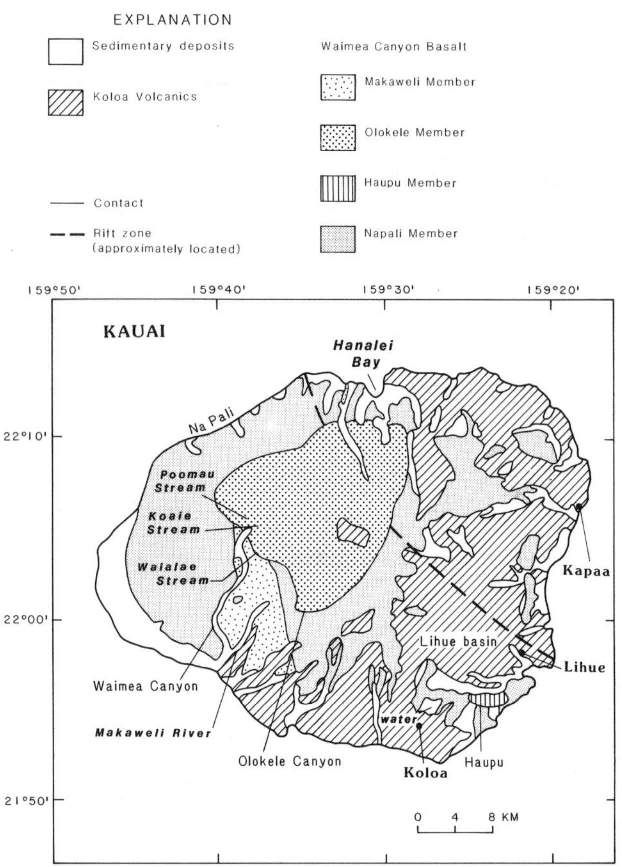

KAUAI: Simplified geologic map. The rift zone indicated (dashed line) is diffuse compared to those on other Hawaiian volcanoes.

KAUAI: East wall of Kalalau Valley. The knife-edge ridges result from erosion and chemical weathering of the lavas.

ones moved toward the north, northeast, and south. The landslide to the south may be related to formation of the Makaweli graben. Several small elongate ridges below sea level may be extensions of rift zones but none is well developed, confirming that Kauai is unique among Hawaiian volcanoes in having no well-developed rift zones. Dikes radiate from the summit caldera in all directions, although they are more abundant toward the northeast and west-southwest.

The lava flows have been divided into the shield-stage lava of the Waimea Canyon Basalt and the rejuvenated-stage vents and flows of the Koloa Volcanics. The Waimea Canyon Basalt is subdivided into the following units, from older to younger: (1) the Napali Member, which constitutes the main shield of the volcano; (2) the Olokele Member, which fills the enormous (16 × 19 km) summit caldera; (3) the Haupu Member, which fills a small elongate flank caldera (~2 × 4 km) in the southeastern part of the island; and (4) the Makaweli Member, which fills a 6.5-km-wide graben extending southward from the summit. The lava flows of the Waimea Canyon Basalt consist mostly of tholeiitic basalt and picritic tholeiitic basalt except for rare flows of mugearite and hawaiite that occur in the uppermost parts of the Olokele and Makaweli Members. These rare flows constitute the alkalic postshield stage on Kauai.

The Koloa volcanics consist of alkalic rejuvenated-stage vents and flows of alkalic basalt, basanite, nephelinite, and melilitite. The flows erupted from >40 vents distributed primarily around the eastern and southern parts of the island. Some of these vents are crudely aligned along rifts oriented N60°E. A few vents are inferred from the locations of small gabbroic intrusions. Koloa lavas cover about half the surface of the eastern half of the island and reach a maximum exposed thickness of 650 m in Hanalei Valley. The Koloa volcanics erupted between 3.65 and 0.52 Ma, although all but two dated flows are between 2.01 and 0.52 Ma. Activity peaked ~1.2 Ma. All dated flows older than 1.7 Ma occur in the west-northwest half of the island, and all flows younger than 1.5 Ma occur in the east-southeast half. The various lithologies have no spatial or chronologic pattern. The flows of the Koloa volcanics are near-primary magmas and commonly contain xenoliths of dunite, harzburgite, and lherzolite.

The most noteworthy features of the volcano are the enormous size of the caldera, the presence of a flank caldera, and the presence of the

KAUAI: Photomosaic taken from Waimea Canyon Lookout showing the major volcanic members of the Waimea Canyon Basalt. Right edge is looking south-east and left edge is looking north-northeast.

graben south of the summit caldera. The relations of the shield, caldera-filling, and graben-filling lavas can be clearly seen from state Highway 550 on the west rim of Waimea Canyon. The many flows of the Koloa volcanics may be sampled easily along road cuts and sea cliffs, but the field relations are generally obscured by the thick soils developed on most of these flows. Some of the spatter and cinder cones in the southern part of the island are surprisingly well preserved considering their ages of ~1.2 Ma.

How to get there: *Kauai airport near Lihue has frequent air connections to Honolulu on several air carriers. Rental cars are available at the airport but should be reserved in advance. Numerous tour companies run guided tours of the island that include trips to Waimea Canyon, the most scenic and geologically the most interesting part of the island.*

References

Clague, D. A., and Bohrson, W., 1987, Waimea Canyon, Kauai, Hawaii: *Geol. Soc. Am. Centennial Field Guide, Cord. Sect.*, 1-4.

Clague, D. A., and Dalrymple, G. B., 1988, Age and petrology of alkalic postshield and rejuvenated stage lava from Kauai, Hawaii: *Cont. Min. Petr. 99*, 202-218.

David A. Clague

WAIANAE

Oahu, Hawaii

Type: *Shield volcano*
Lat/Long: *21.42°N, 158.17°W*
Elevation: *1.2 km*
Relief: *1.2 km above sea level; ~5.8 km above ocean floor*
Volcano Diameter: *35 ×15 km above sea level*
Crater Diameter: *~8 km*
Eruptive History:
Main shield: 3.8-2.95 Ma
Mauna Kuwale rhyodacite flow: 3.2 Ma
Alkalic cap activity: ~2.95-2.4 Ma
Post-erosional activity: Pleistocene?
No historic activity
Composition:
Tholeiitic basalt, icelandite, rhyodacite;
Alkalic basalt, hawaiite, mugearite

Waianae is the older of the two volcanoes comprising the island of Oahu. It is made of lavas erupted during three of the identified stages of Hawaiian volcanism, including shield, alkalic cap, and rejuvenation (post-erosional) stages; the pre-shield stage is apparently not exposed. Products of the shield and rejuvenation stages are included in the Waianae volcanics, which were primarily erupted from two principal rift zones striking ~N60°W and S15°E from the center of the volcano, near present-day Kolekole Pass. A possible third, poorly developed rift zone has an orientation of ~N65°E. A well-developed caldera ~6-9 km in diameter was likely present throughout the shield-building history. Early shield-building lavas (Lualualei Member of the Waianae volcanics) are tholeiitic basalt. Later shield-building lavas (Kamaileunu Member) include transitional and alkalic lithologies. The Kamaileunu Member includes the most differentiated tholeiitic lithology known from the Hawaiian Islands, the Mauna Kuwale rhyodacite flow. This ~100-m-thick flow was erupted inside the Waianae caldera and is a glassy, hornblende-biotite-hypersthene rhyodacite, which overlies ~60 m of glassy icelandite. Waianae is the only Hawaiian volcano in

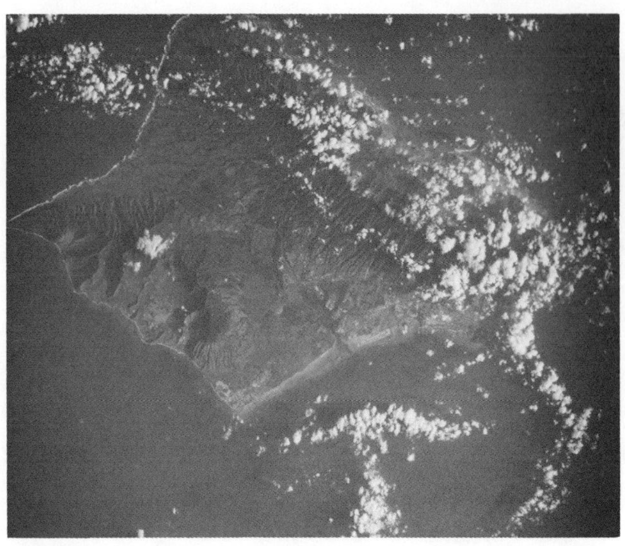

WAIANAE: Space Shuttle photograph of the island of Oahu with Waianae (left) and Koolau (right, under clouds) shields. Photograph STS-61C-34-024.

WAIANAE: Leeward (west) slopes of Waianae volcano, Oahu, with Makaha Valley (left) and Waiamae Valley (right) in the foreground. The highest peak is Mount Kaala, the highest point in Oahu. Photograph by G. Macdonald.

which such silicic variants have been recognized.

The alkalic cap of the Waianae volcano is represented by hawaiite with rarer alkali basalt and mugearite of the Palehua Member. These lavas occur in the higher stratigraphic and topographic elevations of the volcano. Post-erosional volcanism is sparsely represented by the Kolekole volcanics, a series of cinder cones and associated mafic flows containing a variety of crustally derived xenoliths. The Kolekole volcanics have not been radiometrically dated; they may be overlain by Koolau volcanics, which range in age from ~2.7 to 1.8 Ma.

How to get there: *From Honolulu airport take H-1 west past Pearl Harbor and Waipahu. H-1 turns into Farrington Highway near Makakilo, at the south end of the Waianae Range. Farrington Highway runs along most of the leeward, west coast of Waianae volcano.*

References

Sinton, J. M. (1986), Revision of stratigraphic nomenclature of Waianae Volcano, Oahu, Hawaii. *USGS Bull. 1775-A*, A9-A15.

Stearns, H. T. (1946), Geology of the Hawaiian Islands. *Hawaii Div. Hydrogr. Bull. 8*, 106 pp.

John M. Sinton

KOOLAU
Oahu, Hawaii

Type: *Lava shield*
Lat/Long: *21.40°N, 157.80°W*
Elevation: *0.96 km; 5.5 km above ocean floor*
Eruptive History: *2.7 to 1.8 Ma*
Composition: *Tholeiitic basalts*

Koolau is a large basaltic lava-shield volcano dated at ~1.8 to 2.7 Ma that constitutes the Koolau Range on Oahu. It rises to a present summit elevation of 960 m above sea level, or ~5,500 m above the deep ocean floor. The estimated volume is 21,000 km^3 (about half that of Mauna Loa). Koolau is strongly elongate on a northwest–southeast axis, and the subaerial length is 57 km. Many planezes – cuestalike remnants of the original volcano surface – occur around the south and west foot of the volcano, sloping seaward at 4° to 13°. Deep valleys cut the higher parts of the range, and erosional embayments scallop the northeast part and penetrate the axis of the volcano; these embayments have been attributed to mass-wasting by great landslides.

So far as is known the Koolau lava shield is composed exclusively of tholeiitic basalt; the volcano lacks the alkalic-basaltic capping that characterizes many other Hawaiian volcanoes (e.g., Waianae). Sections are composed almost

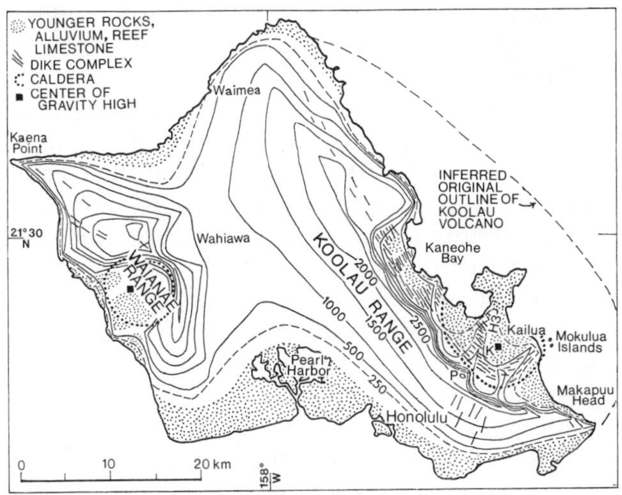

KOOLAU: Sketch map of Oahu showing generalized contours for the eroded Koolau and Waianae volcanoes; elevations in feet above sea level. A major portion of the Koolau edifice is missing, attributed to great slides probably totaling >500 km^3. Fine and accessible exposures occur at Makapuu Head, the Pali lookout (P), and Waimea. Fine exposures of the axial dike complex occur in the Mokulua Islands, Kapaa Quarry (K), and H3 road cuts (H3).

exclusively of lava flows, pahoehoe and aa being more or less equally represented; pyroclastic or sedimentary interbeds are very scarce. The lava flows average 4 to 5 m thick, and the pahoehoe flows are invariably subdivided into flow units that average 0.6 m thick. The best and most accessible exposures are at Makapuu Head – a magnificent example of an infilled lava tube – in coastal cliffs 0.5 km south of the Point, the Pali Lookout, and Waimea.

The main rift zone is marked by an intense dike complex exposed at an erosion depth of ~1 km along the long axis of Koolau volcano. The dike intensity varies from 40% to 90% (400 to 1,200 dikes/km across-strike) and drops off to 2% (20 dikes/km) in <1 km at the margins. Where widest, the complex consists of an estimated 7,400 subparallel dikes averaging 0.73 m wide. The finest exposures of dikes are on the Mokulua Islands, in the Kapaa Quarries, and in the road cuts along H-3 Freeway. A minor southwest rift zone trending toward Diamond Head also occurs. The dikes most commonly have 70°-80° dips.

A caldera occupies the low-lying ground centered on Kailua and is a 13-km-wide area of predominantly centripetal dips and thick massive lava flows. The rocks in parts of this area show propylitic alteration. A 320-mgal positive Bouguer gravity anomaly centered here is interpreted to mark the intrusive core, and seismic refraction profiling indicates that rocks with a high P-wave velocity (7.7 km/s) occur at a shallow (1-km) depth. A noteworthy feature is the extreme diminution in dike intensity to 2% (10 dikes/km) in the center of the caldera. This is attributed to very rapid subsidence in the center that rapidly carried dikes down in the subsiding lava prism to below outcrop level.

How to get there: *The Koolau dike swarm is easily viewed from Highway 61 (Pali Highway), just past the junction of Highway 61 and 83, where it is exposed by a road cut. The proposed Koolau caldera may also be seen from Highway 61 as you head north toward Kailua. Look to the left just after the drive-in theater. The oval-shaped caldera floor is now a swampy area and is easily distinguished from the surrounding topography.*

References

Roden, M. F., Frey, F. A., and Clague, D. A., 1984, Geochemistry of tholeiitic and alkalic lavas from the Koolau Range, Oahu, Hawaii: Implications for Hawaiian volcanism: *Earth Planet. Sci. Lett.69*, 141-158.

Walker, G. P. L., 1987, The dike complex of the Koolau volcano, Oahu: Internal structure of a Hawaiian rift zone: *VIH*, 961-993.

George P. L. Walker

HONOLULU VENTS
Oahu, Hawaii

Type: *Monogenetic volcanic field*
Lat/Long: *21.35°N, 157.80°W*
Elevation: *0 to 530 m*
Eruptive History: *1 Ma to ~0.01 Ma*
Composition: *Alkali basalt, basanite, nephelinite, melilite nephelinite*

The field of 30 to 40 scattered vents that comprise the Honolulu Group or Series is superimposed on the eastern part of the Koolau tholeiitic lava shield of eastern Oahu. It includes

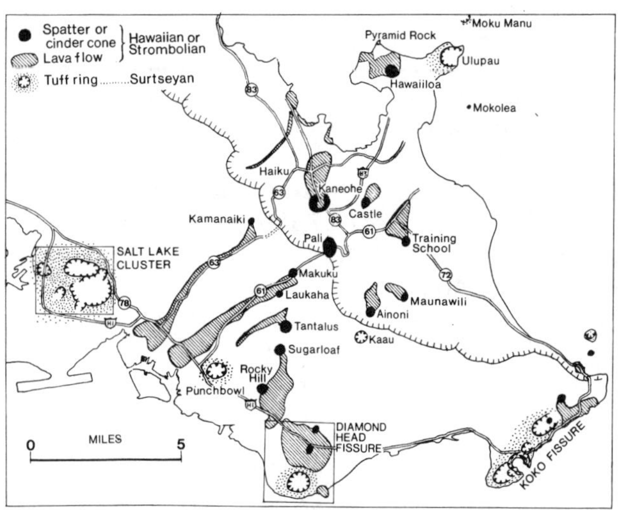

HONOLULU GROUP VENTS: Sketch map of eastern Oahu showing monogenetic volcanoes of the Honolulu Group (after Stearns, 1939, Hawaii Div. Hydrography, Bull. 2, 75 p.). Surtseyan tuff rings occur near the coast, whereas eruptions farther inland produced mainly cinder/spatter cones and lava flows.

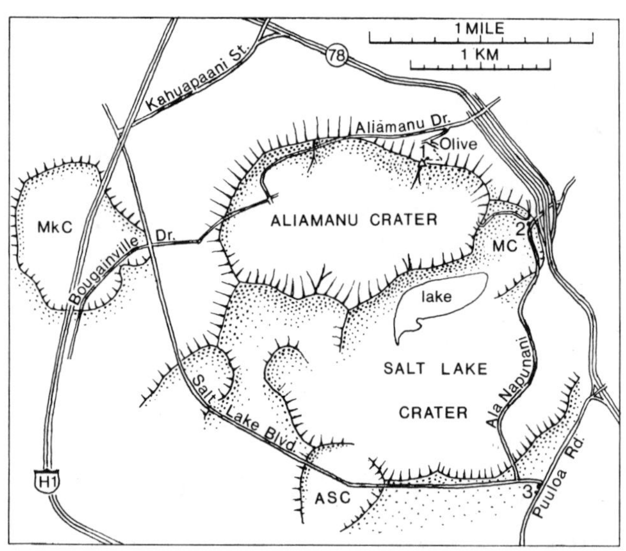

HONOLULU GROUP VENTS: Craters of the Salt Lake cluster. 1 is a good viewpoint; 1-3 are good exposures of the tuffs; and 3 shows the base of the tuffs and good base-surge deposits. MkC: ***Makalapa*** *crater; ASC:* ***Aliamanu*** *School crater.*

some notable scenic landmarks (e.g., Diamond Head and Hanauma Bay), some petrologically significant lavas (e.g., the melilite nephelinite flow of Moiliili with its coarse segregation veins), some well-known occurrences of mantle-derived pyroxenite, garnet pyroxenite, and lherzolite (in the tuffs of Salt Lake and Aliamanu craters), and some fine examples of dune-bedded base-surge deposits (e.g., in the tuffs of Salt Lake and Koko fissure).

The tholeiitic activity of the Koolau shield ceased ~1.8 Ma, and an interval of ~1 m.y. elapsed before the first of the Honolulu Group volcanics was erupted. The Honolulu Group magmas created new pathways to the surface, apparently unrelated to earlier structures. For example, the 9.5-km-long Koko fissure trends at right angles to the main Koolau rift zone, suggesting that the plumbing systems of the Koolau volcano had by then cooled down.

Honolulu Group vents situated in or near the sea gave rise to eruptions of surtseyan style and constructed tuff rings having diameters of 0.3 to 1.8 km. The highest tuff ring is **Koko** Crater; persistent trade winds caused deposition of its ash predominantly on the southwest side, where a cone 368 m high accumulated. Next highest are **Diamond Head** (232 m) and **Ulupau Head** (208 m). The widest craters are **Salt Lake** (1.8 km) and Diamond Head (1.2 km). All of the surtseyan tuff rings consist predominantly of fine-grained pyroclasts, generally palagonized and with zeolitic pore fillings. Some beds (e.g., on Koko Crater) are rich in accretionary lapilli, including some (generally <1 cm in size) that originated when raindrops fell through ash clouds and accreted ash particles, and in others (up to >5 cm in size) having a core of scoria, that grew by accretion as they rolled downslope. Some tuffs (e.g., at **Hanauma Bay**) are rich in fragments of reef limestone cored out by the explosions. The seaward side of the Koko fissure shows conspicuous intraformational slump structures and water-cut gullies.

Honolulu Group vents situated inland gave rise to eruptions of strombolian or Hawaiian styles; they formed cinder and spatter cones and lava flows. Noteworthy among the lava flows is that of **Moiliili**, which ponded on the floor of Manoa Valley, and the low-angle pahoehoe shield (75 m high × 2.5 km across) of **Kaimuki** on the Diamond Head fissure. The north rim of Kaimuki crater is notable for the great number of thin lava flow units (repeated overflows from a lava pond?) that compose it. A dike that occurs on the shore at the western side of Black Point is assumed to have fed the Black Point flow. The lava flows from the Pali and Hawaiiloa vents are rich in ultramafic inclusions.

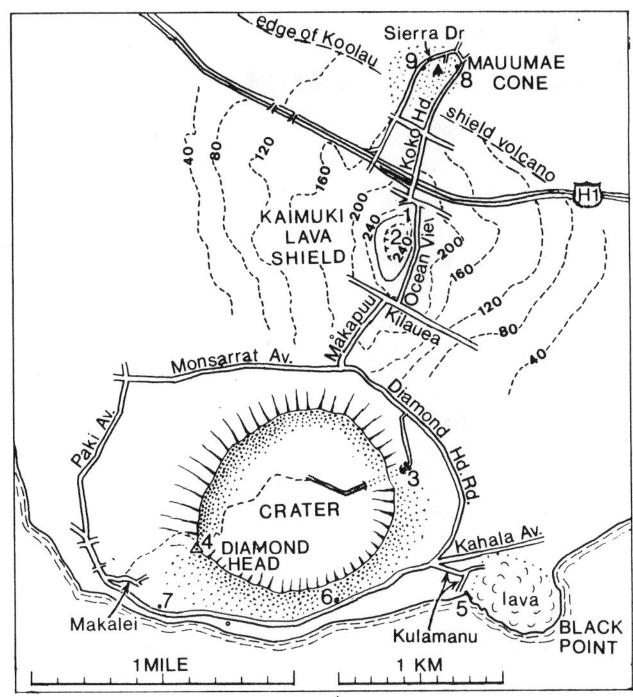

HONOLULU GROUP VENTS: Volcanoes of the Diamond Head fissure. Diamond Head is a surtseyan tuff ring. Kaimuki is a pahoehoe lava shield, shown by contours at 40-ft intervals, and capped by a crater rimmed by very thin lava flows. ***Mauumae*** *is a cinder/spatter cone with lava flows. 1-9 are good outcrops. 1: very thin lava flows; 4: good viewpoint, elevation 760 ft, best reached by trail from inside crater; 5: dike intruding reef limestone.*

It is possible that the latest eruption of the Honolulu Group, that which gave rise to the Moiliili lava and the extensive fire-fountain scoria fall of **Tantalus** that blankets the ridges west of Manoa Valley, was <10,000 yr ago; the monogenetic field must be regarded as active and likely to erupt again.

How to get there: *Honolulu Group vents are, as their name implies, in and near Honolulu, and virtually all are accessible by roads. Some, such as Hanauma Bay, an ocean-breached tuff cone, are beautiful state parks; one* ***(Punchbowl)*** *is a World War II cemetery, and others are covered by buildings (Tantalus) or are being mined for aggregates (Salt Lake Crater).*

References

Lanphere, M. A., and Dalrymple, G. B., 1980, Age and strontium isotopic composition of the Honolulu Volcanic Series, Oahu, Hawaii: *Am. J. Sci. 280-A*, 736-751.

Winchell, A., 1947, Honolulu Series, Oahu, Hawaii: *Geol. Soc. Am. Bull. 58*, 1-48.

George P. L. Walker

WEST MOLOKAI
Hawaii

Type: *Shield volcano*
Lat/Long: *21.13°N, 157.20°W*
Elevation: *421 m*
Eruptive History:
Tholeiitic shield stage ended at ~1.89 Ma
Alkalic postshield stage at 1.76 Ma
Composition: *Tholeiitic basalt to hawaiite*

West Molokai volcano (also named Mauna Loa) makes up the western half of the island of Molokai. The volcano is a broad, slightly dissected, shield of tholeiitic basaltic lava that was highly fluid and formed thin-bedded (0.6-m-thick) aa and pahoehoe flows. Most of these flows are sparsely porphyritic with either olivine, plagioclase, or orthopyroxene phenocrysts, but picritic tholeiitic basalt also occurs. Many of the tholeiitic basalt flows are characterized by anomalously high rare-earth abundances. The shield-stage flows dip 2° to 10° away from the principal southwest and subordinate northwest rift zones, which join near the summit. Some dikes are exposed where coastal erosion has cut into the rift zones.There is no clear evidence of a summit caldera, but the eastern flank of the volcano consists of a series of fault blocks exposing flows that have been extensively altered by hydrothermal fluids.

About 16 small spatter and cinder cones are preserved on the surface of West Molokai, particularly in the northwest quadrant of the volcano; those that have been sampled range in composition from hawaiite to basalt transitional between alkalic basalt and hawaiite. These cones, and numerous other deeply weathered small hills that are probably also cones, constitute the alkalic postshield stage. One of the youngest of these cones, **Waiele**, erupted a flow underlain by up to 1.2 m of red soil. No ultramafic xenoliths have been observed in lavas from West Molokai. The lavas of West Molokai volcano are overlain to the east by the lavas

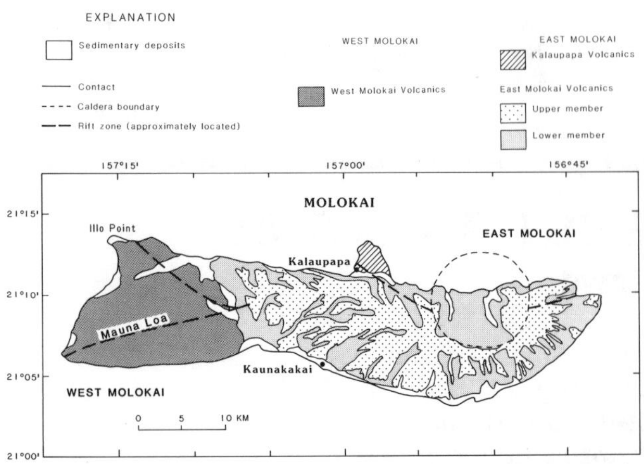

WEST MOLOKAI: Simplified geologic map of Molokai showing the distribution of the West Molokai (or Mauna Loa) and East Molokai volcanics and the orientation of the rift zones. A number of small vents and associated alkalic basalt and hawaiite flows of the postshield alkalic stage occur in northwestern West Molokai. The rest of the exposed lavas are tholeiitic basalt of the shield stage (dark pattern). Various units of East Molokai are indicated by the other patterns; dashed lines are rift zones; circle shows the location and size of the East Molokai caldera.

of the East Molokai volcano. Calcareous beach sand has been blown inland by the tradewinds to form a dune field along the north coast.

Beneath the sea, West Molokai extends west-southwest as Penguin Bank, which has a flat surface at a depth of 55 m below sea level. Penguin Bank may be a subsided portion of West Molokai volcano or a separate shield volcano.

The most noteworthy characteristic of West Molokai volcano is the apparent lack of a caldera complex; it is the only Hawaiian volcano with no clear evidence of a summit caldera.

How to get there: *Molokai has frequent air connections from Honolulu. Rental cars are available at the airport. Take Highway 46 to the west; the eastern flank of West Molokai volcano is only ~2 km from the airport. A 4-wheel drive is needed for many of the unpaved roads.*

References

Stearns, H. T., and Macdonald, G. A., 1947, Geology and ground-water resources of the island of Molokai, Hawaii: *Hawaii Div. Hydrogr. Bull. 11*, 113 pp.

Clague, D. A., 1987, Petrology of West Molokai Volcano: *Geol. Soc. Am. Abst. with Program 19*, 366.

David A. Clague

WAILAU (EAST MOLOKAI)

Hawaii

Type: *Shield volcano*
Lat/Long: *21.17°N, 156.83°W*
Shield dimensions:
Length: 150 km
Maximum width: 70 km
Elevation: 1,210 m; 6,800 m above ocean floor
Caldera dimensions:
Width: 12 km
Depth: filled, but formerly >1,000 m
Eruptive history:
Growth of main shield ~2-1.5 Ma
Postshield eruptions ~1.5-1.3 Ma
Growth of Kalaupapa shield ~0.6-0.3 Ma
Composition: *Basalt to trachyte*

Wailau is a shield volcano forming the eastern two-thirds of Molokai. It overlaps the West Molokai volcano and is overlapped by the Lanai, West Maui, and Haleakala volcanoes. Only a small part of Wailau is exposed on Molokai; the shield grew largely beneath the sea and has subsided >1.5 km since it ceased growing. In addition to having most of its south flank covered by younger volcanoes, much of its north flank has been removed by a giant landslide. Present evidence indicates that this landslide split the upper part of the volcano, carrying away the northern half of its filled caldera.

The oldest rocks exposed in the subaerial part of the shield, on the west side of Haupu Bay, are south-dipping tholeiitic basalt flows in a sequence ~200 m thick, having reversed magnetic polarity. Above this is a normal-polarity sequence 200-300 m thick that probably represents the Olduvai epoch of 1.88-1.66 Ma. This sequence is in turn overlain by 300 m of interlayered alkalic and tholeiitic basalt having reversed polarity and a K-Ar age >0.5 Ma. The oscillatory changes in composition in this upper sequence may reflect cyclical changes in eruption frequency caused by repeated caldera collapses. Along the central coast between Haupu Bay and Papalaua Valley is a thick section of caldera-filling lava flows of normal polarity. The edge of the caldera, with caldera-filling flows ponded above talus along the caldera wall,

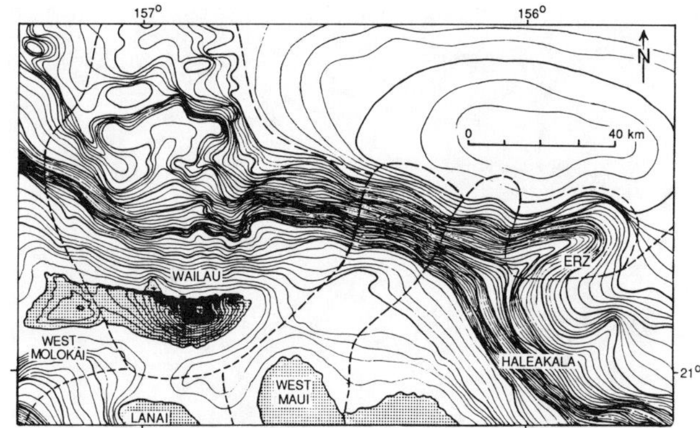

WAILAU: Topographic map, with subaerial portion shaded. Irregular topography north of Molokai is proximal part of giant landslide. Dashed lines indicate approximate contacts between Wailau and its neighbors. Submarine ridge labeled ERZ, 50-100 km east of Molokai, is thought to have grown along the distal east rift zone of Wailau; current evidence suggests that the central stretch of this ridge is overlain by lava flows from West Maui and Haleakala. Contour interval 100 m; contours not shown on islands of Maui and Lanai.

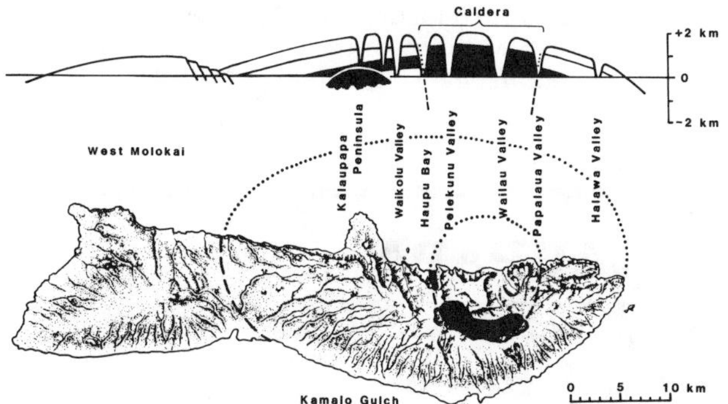

WAILAU: Physiographic map of Molokai and profile of its north coast. Black areas of map are caldera and separate pit crater as inferred by Stearns and Macdonald (1947); dashed and dotted circle is caldera as inferred by Holcomb (1985). Dotted semi-ellipse indicates portion of present subaerial shield thought to have collapsed. Black and white areas of profile indicate lava sequences having normal and reversed magnetic polarity, respectively. East flank is less exposed above sea level than west flank, probably owing to southeastward tilting of Wailau under weight of younger volcanoes.

is exposed at the head of Haupu Bay and along the southwest wall of Pelekunu Valley. Much of the lava and breccia near the caldera margin is highly altered. The caldera fill and upper flanks of the volcano are capped by postshield flows of mugearite with lesser amounts of hawaiite and trachyte of reversed polarity; some of these flows have been dated at 1.49-1.35 Ma. Rejuvenation-stage eruptions of alkalic basalt and basanite at ~0.4 Ma built a satellitic lava shield that rises 123 m above sea level to form the Kalaupapa peninsula.

Wailau is of interest primarily due to the superb views of its interior. Not only is a good cross-section well exposed by the giant landslide, but deep stream dissection has opened many other views into the caldera, rift zones, and upper flanks of the volcano. Many dikes are exposed in the caldera fill and along the northwest rift zone. Despite the erosion, the original form of the shield's southern flank is still well preserved.

How to get there: *Molokai is served several times daily by flights from other islands and once daily by ferries from Honolulu and Maui. Superb views can be had from helicopters and other aircraft. There is no bus service, but rental cars are available. Paved roads provide easy access to most of the southwest flank and lower south and east flanks; higher south and east flanks are accessible via jeep and foot trails. The Kalaupapa Peninsula and a few other parts of the north coast can be reached by steep foot trails, but much is accessible only by boat or helicopter; permission is required. Much of the north coast is thickly vegetated; Kamalo Gulch cuts deeply into the less vegetated south flank but does not penetrate the caldera.*

WAILAU: Oblique aerial view of Molokai, looking west. Dissected caldera is in middle distance. Photograph by US Army Air Force, ca. 1946.

References

Holcomb, R. T., 1985, The caldera of East Molokai Volcano, Hawaii: *Nat. Geogr. Soc. Res. Rpt. 21*, 81-87.

Stearns, H.T., and Macdonald, G.A., 1947, Geology and ground-water resources of the Island of Molokai, Hawaii: *Hawaii Div. Hydrog. Bull. 11*, 113 pp.

Robin Holcomb

KALAUPAPA AND MOKUHOONIKI
Molokai, Hawaii

Type: *Monogenetic volcanoes*
Lat/Long: *21.20°N, 156.97°W (Kalaupapa)*
21.13°N, 156.72°W (Mokuhooniki/Kanaha)
Elevation: *123 m (Kalaupapa)*
62 m (Mokuhooniki/Kanaha)
Eruptive History: *Kalaupapa: 1.24 to 0.34 Ma; Mokuhooniki/Kanaha not dated*
Composition: *Basanite to basalt, transitional between alkali basalt and tholeiitic basalt*

Two rejuvenation-stage volcanoes occur on the island of Molokai. One is the pahoehoe lava shield of the Kalaupapa peninsula that projects out 4 km from the foot of the great cliffline midway along the north coast. The other is the eroded tuff cone forming the small islands (total area of 0.06 km^2) of Mokuhooniki and Kanaha, off the eastern end of Molokai.

The crater of Kalaupapa, called **Kauhako**, has a rim diameter of 0.5 × 0.65 km and rises 123 m above sea level at Puu Uao. The vent is marked by a funnel-like pit on the crater floor containing a pond ~50 m across. The water level in this pond is little above sea level and the pond has been plumbed to a depth of 255 m. The shield has slopes from 6° near the center decreasing to 1-2° near the coast. These low-angle slopes extend offshore for 0.5 to 1 km, but beyond the 50-m isobath the slope steepens to 25 or 30°. It is inferred that this slope change occurs at the passage zone from subaerial lava to flow-front lava-delta deposits, and marks sea level at the time of the Kalaupapa eruption.

A broad discontinuous lava channel extends north from Kauhako Crater, and several lava tubes are also seen. The walls of the crater and channel are built of >200 thin lava-flow units representing repeated overflows from a lava pond. A pyroclastic cone at the foot of the cliff 1.4 km southwest of Kauhako Crater appears to belong to the same eruption and is the vent from which most of the degassing occurred.

Kalaupapa is regarded as a monogenetic volcano, the product of one eruption. K-Ar ages vary from 1.24 to 0.34 Ma, and on morphological grounds the younger age is more plausible.

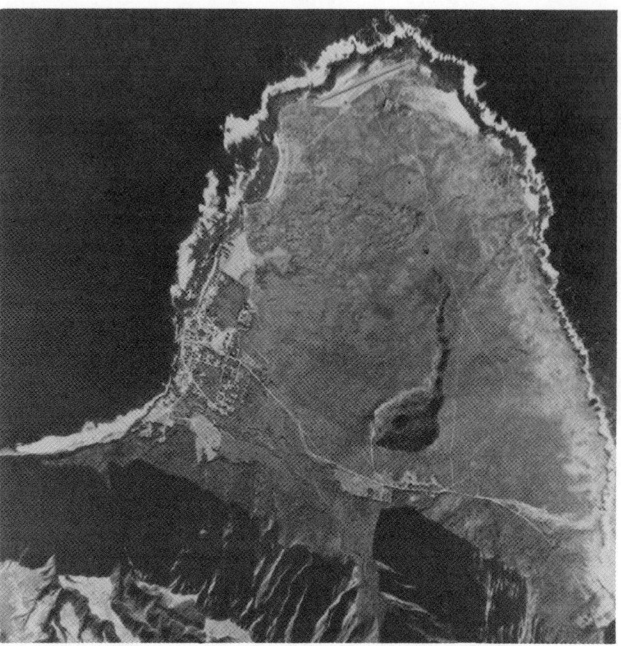

KALAUPAPA: Vertical aerial photograph of lava shield and lava channel/tube. Photograph by R. M. Towill Corp., Honolulu; provided by C. Coombs.

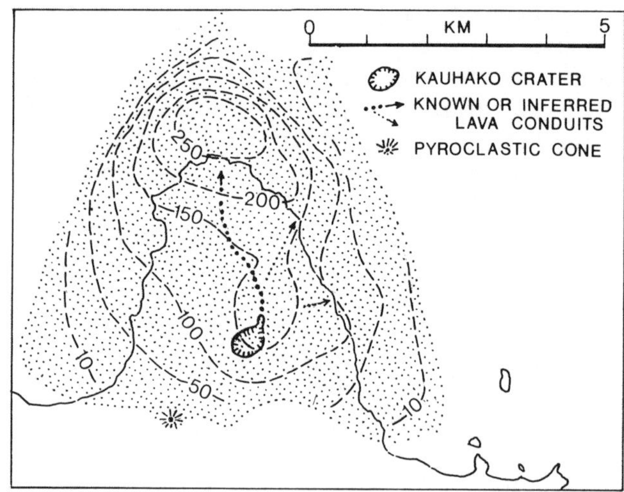

KALAUPAPA: Isopach map of the Kalaupapa volcano lava flows; thickness in meters (after Walker and Coombs, in press). Stippled area is approximate inferred extent of the volcano, much of which is submerged.

It has an estimated volume of 3 km^3 (88% below sea level), and is hence the largest known monogenetic volcano in the Hawaiian Islands. The lava contains phenocrysts of olivine, augite, and plagioclase. Sr-isotopic and trace element ratios are strikingly similar to the rejuvenation-stage Honolulu Group on Oahu, indicating a similar mantle source.

How to get there: *The peninsula of Kalaupapa is served by Air Molokai and Aloha Island Air, two small commuter airlines, from either Honolulu via Molokai's main Hoolehua airport, or by hiking or riding a mule down a winding 5-km trail along the sheer mountain wall (pali) bounding the southern side of the peninsula. Hiking or riding time is ~4 hr round-trip. All incoming visitors must be escorted while visiting the peninsula. Tour packages are available from Honolulu that include round-trip transportation, a half-day visit, and an escort. Lunch is also included as no restaurant or public store is available to guests of the settlement.*

References

Clague, D. A., et al., 1982, Age and petrology of the Kalaupapa basalt, Molokai, Hawaii: *Pacific Sci. 36*, 411-420.

Walker, G. P. L., and Coombs, C., in prep., Volcanology of Kalaupapa volcano, Kalaupapa National Historical Park, Molokai: Hawaii's largest monogenetic volcano.

George P. L. Walker

LANAI
Hawaii

Type: *Shield volcano*
Lat/Long: *20.83°N, 158.92°W*
Shield Dimensions:
Length: 60 km
Maximum width: 50 km
Elevation: 1,030 m; 5.5 km above ocean floor
Caldera Dimensions:
Palawai Basin: 5 km wide × >250 m deep
Miki Basin: 2 km wide × >30 m deep
Eruptive History:
Growth of northeast flank ~1.8-1.4 Ma
Growth of southwest flank ~1.4-1.2 Ma
Composition: *Tholeiitic basalt*

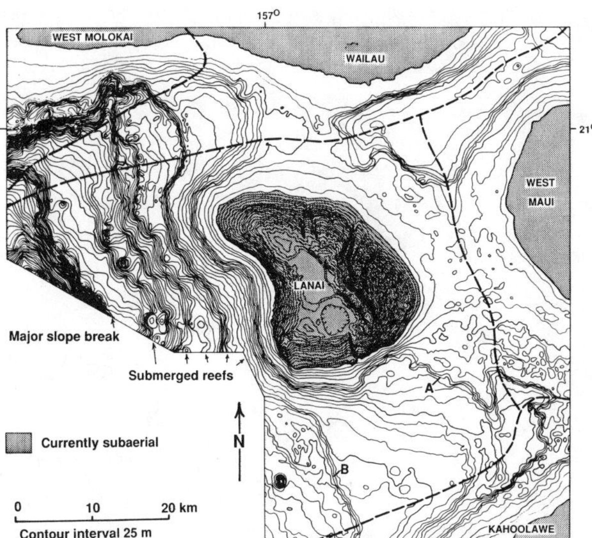

LANAI: Topographic map showing upper part of Lanai volcano, with currently subaerial portion shaded and contacts with neighboring volcanoes shown approximately by dashed lines. Major slope break inferred to represent shoreline of subaerial volcanic shield at end of shield-building stage; submerged terraces inferred to represent successive intervals of Pleistocene reef growth as shield subsided. Straight reef segments A and B inferred to overlie landslide structures active during and after shield growth, respectively. Contour interval 25 m (not shown on islands of Molokai, Maui, and Kahoolawe).

Lanai volcano forms the southwest flank of the Hawaiian Ridge between the northern and southern strands of the Molokai Fracture Zone. It overlaps Wailau volcano to the north and is overlapped by West Maui and Kahoolawe volcanoes to the east and southeast. Lanai has subsided since its growth ceased; the major slope break that developed at sea level at the end of shield growth now lies at a depth of ~1,250 m and is 20 km west of the present island.

The island consists of a semicircular southern segment and an elongate northern segment separated by a northwest-trending zone of normal faults, downthrown to the southwest. The crest of the northern segment rises 600 m above the southwestern plateau. The windward northeast flank is incised by several canyons; Maunalei Gulch is the deepest at 300 m. Lavas of the southwestern plateau lap against scarps of the axial fault zone; the plateau evidently grew along the fault scarps following displacement. The plateau is indented by the Palawai and Miki basins and is deeply mantled by a residual red

LANAI: Space Shuttle photograph of Lanai (center), Molokai (top), West Maui (top right), and Kahoolawe (bottom right). Photograph STS-61C-51-026.

soil planted with pineapple. Exposed lavas have reversed magnetic polarity and K-Ar ages of 1.5-1.2 Ma; they are almost entirely tholeiitic basalt, but trachyte has been found recently on the southeast flank. Widespread boulder gravel consisting of basalt and reef debris occurs on the south flank as high as 375 m. This gravel may have been deposited by a train of giant sea waves originating ~50 km southwest of the island. Uranium-series dating indicates an age of ~105 ka for coral fragments in the deposit.

The submarine (but formerly subaerial) flank above the major slope break is terraced by Pleistocene reefs that record fluctuations in sea level during subsidence of the volcano. Presently available bathymetry shows five distinct terraces west of the island. The two shallowest reefs south and southeast of the island have linear segments that are aligned with subaerial fault scarps and probably grew on submarine continuations of those scarps. A giant landslide on the seafloor southwest of Lanai appears to originate from the volcano.

The unbuttressed southwest flank of Lanai evidently underwent repeated episodes of collapse as the shield grew. The earliest collapse that is presently inferred from the zone of axial faulting must have occurred at ~1.4 Ma, prior to growth of the southern plateau. Another collapse, which occurred after shield growth ceased at ~1.2 Ma but before the growth of a reef at ~0.4 Ma, is indicated by the Kaholo Pali fault scarp along the southwest shore of the island. A later collapse may have generated the gravel deposit at some time after 0.1 Ma.

LANAI: View of Kaholo Pali, ~300 m high, looking north-northwest from Palaoa Point on southwest tip of island. Cliff inferred to be headwall of giant landslide that occurred after end of shield building but before growth of youngest submerged reefs.

How to get there: *Lanai airport is served several times daily by flights from other islands. Paved roads lead to the south, west, and northeast coasts, and most localities can be reached by jeep or by foot. Rental cars and jeeps are available. Most of the island is privately owned; inquiries should be sent to the Koele Company, the principal landowner, prior to arrival. Maunalei Gulch is closed in order to avoid contamination of the island's water supply.*

References

Stearns, H. T., 1940, Geology and groundwater resources of the islands of Lanai and Kahoolawe, Hawaii: *Hawaii Div. Hydrogr. Bull. 6*, 176 pp.

Moore, G. W., and Moore, J. G., 1988, Large-scale bedforms in boulder gravel produced by giant waves in Hawaii: *in* Clifton, H. E. (ed.), *Geol. Soc. Am. Spec. Pap. 229, Sedimentologic Consequences of Convulsive Geologic Events*, 101-110.

Robin Holcomb

KAHOOLAWE
Hawaii

Type: *Shield volcano*
Lat/Long: *20.55°N, 156.6°W*

KAHOOLAWE: Space Shuttle photograph of Kahoolawe (center), and southern portion of Maui, including the southwest rift zone of East Maui volcano. Photograph STS-61C-51-027.

Elevation: *456 m*
Relief: *456 m above sea level, 4.2 km above seafloor*
Volcano Diameter: *18 × 11 km above sea level, 55 × 40 km at seafloor*
Crater Diameter: *Interpreted caldera 4.4 km*
Eruptive History:
Main shield: 1-2 Ma
Possible younger, "post-erosional" eruption of unknown age
No historic activity
Composition: *Tholeiitic basalt and hawaiite*

Kahoolawe volcano is one of the smallest shield volcanoes in the Hawaiian Islands. Wave erosion has exposed a caldera. The pre-caldera lavas dip gently (3-10°) away from the caldera and are thin (0.5-2 m thick). The caldera-filling lavas are horizontal and thick (3-10 m). Post-caldera lavas cap both units. Pre-caldera and caldera-filling lavas are tholeiitic. The post-caldera lavas range from tholeiitic to hawaiitic. Some of the post-caldera lavas are hydrothermally altered and contain extreme enrichments in Y and some REE.

How to get there: *This island is a US military bombing target. Do not try to visit the island without permission from the US military! By boat, approach from the east side, where you will see the best view of a dissected Hawaiian volcano.*

References

Stearns, H. T. (1940) Geology and groundwater resources of the islands of Lanai and Kahoolawe, Hawaii: *Hawaii Div. Hydrogr. Bull. 6*, 173 pp.
Garcia, M. O., Leeman, W., Garlach, D. G., and West, H. (in press) Petrology, isotope geochemistry and age of lavas from Kahoolawe volcano, Hawaii: *Cont. Min. Petr.*

Michael Garcia

WEST MAUI
Maui, Hawaii

Type: *Shield volcano*
Lat/Long: *20.92°N, 156.58°W*
Elevation: *1.78 km above sea level*
Volcano Diameter: *18 × 29 km above sea level*
Crater Diameter: *5 km*
Eruptive History:
Main shield: 1.97-1.29 Ma
Alkalic cap stage: 1.2 Ma
Post-erosional activity: Pleistocene?
No historic activity
Composition: *Tholeiitic basalt, alkali basalt, benmoreite, trachyte, rare hawaiite*

West Maui is the older and smaller of the two volcanoes comprising the island of Maui. It contains lavas erupted during three of the volcanic stages characteristic of Hawaiian volcanism. The Wailuku Basalt forms the shield stage and mainly consists of thin-bedded pahoehoe lava flows, comprising more than 90% of the subaerial shield. Aa flows and discontinuous beds of vitric tuff are dispersed throughout the exposed section, but become more plentiful in the late stages of shield activity. Lowermost Wailuku Basalts are tholeiitic, and an ~50-m-thick Upper Member has larger and more abundant phenocrysts, more clinopyroxene phenocrysts, and an increasing abundance of vitric tuffs and alkalic basalts and hawaiites. A central caldera has been identified, roughly coincident with the Iao Valley. Wailuku Basalts were erupted from this caldera and from two rift zones extending

WEST MAUI: From the sea, looking to the northeast. The large valley in the center is Olowalu Valley, and to the right is Ukumehame Valley. Low hills on the north (left) slopes of Olowalu Valley are Honolua trachytes. The south rift zone of West Maui is on the right skyline. Photograph by G. Macdonald.

northwest and southeast from the caldera region. The alkalic cap of West Maui volcano, the Honolua volcanics, in many places unconformably overlies the Wailuku Basalt. The time break between latest Wailuku and Honolua volcanism can be no more than 100,000-150,000 yr. Honolua volcanics mainly are massive aa or block lava flows and endogenous domes of highly differentiated alkalic benmoreites and trachytes. Compositions between hawaiite and benmoreite (i.e., mugearite) have not been found at West Maui; thus a distinct compositional gap is present on this volcano. Although minor trachytes occur elsewhere in Hawaii, most notably on Hualalai and Kohala volcanoes, the Honolua volcanics of West Maui are the most differentiated formation known in the Hawaiian Islands. Analysis of the outcrop pattern indicates that the Honolua volcanics could have been formed by ~40-60 eruptions over a period of ~300,000 yr, indicating a decrease in eruption frequency of 1-2 orders of magnitude over modern Hawaiian shields. Although most Honolua eruptions were from a poorly defined zone trending northeast–southwest, this new alignment probably formed in response to the growth of neighboring East Maui (Haleakala) volcano. The rejuvenation stage of West Maui, the Lahaina volcanics, consists of the products of only four small eruptions; all are basanites. The distribution of Lahaina vents bears no systematic relation to the rift zones from which Wailuku and Honolua eruptions were concentrated. The age of Lahaina volcanic activity is not well known, but Lahaina volcanics locally overlie Pleistocene deposits. Lahaina volcanics are the last eruptive products of West Maui volcano.

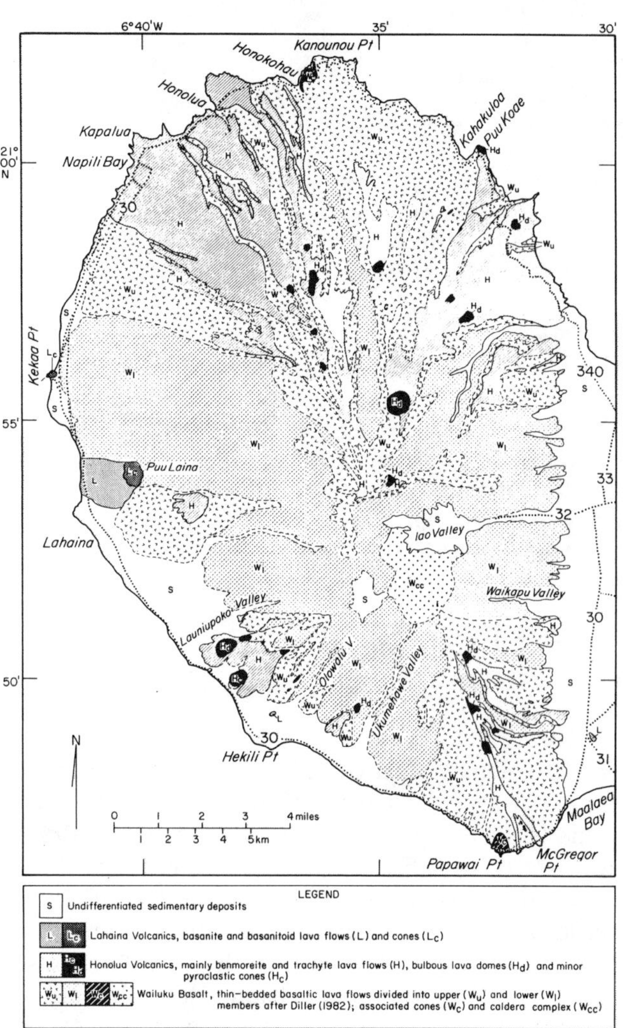

WEST MAUI: Geologic map.

How to get there: *From Kahului Airport, Maui, take Route 38 west toward Kahului. The west slopes of West Maui volcano and Iao Valley are accessible by following Route 36 to Wailuku and on to Iao Valley. The best exposures are obtained by following Route 38 until it merges with Route 30 which encircles most of West Maui.*

References

Sinton, J. M., Diller, D. E., and Chen, C.-Y., 1987, Geology and petrological evolution of West Maui Volcano: *in* Sinton, J. M. (ed.), *Field Trip Guide to Maui,* Geol. Soc. Am. 83rd Ann. Cord. Sect. Mtg., Hilo, Hawaii, 13-30.

Stearns, H. T., and Macdonald, G. A., 1942, Geology and ground water resources of the island of Maui: *Hawaii Div. Hydrogr. Bull.* 7, 344 pp.

John Sinton

HALEAKALA: Space Shuttle view of the island of Maui with West Maui shield (left) and Haleakala shield (right). Photograph STS-61C-49-049.

HALEAKALA
East Maui, Hawaii

Type: *Shield volcano*
Lat/Long: *20.75°N, 156.25°W*
Elevation: *3.05 km above sea level*
Volcano Diameter: *53 × 42 km above sea level*
Crater Diameter: *12 × 4 km*
Eruptive History:
Main shield: >0.83 Ma
Alkalic cap stage: 0.91-0.35 Ma
Holua dike: 0.49 Ma
Post-erosional activity: 0-? Ma
Historic activity: 1790 A.D. ± 3 yrs
Composition: *Tholeiitic basalt, ankaramite, alkali basalt, basanitoid, trachyte, mugearite, and hawaiite*

Haleakala is the easternmost of two shield volcanoes forming the island of Maui, and is the third largest of all Hawaiian shield volcanoes, having a total volume of ~29,300 km^3. Lavas from three of the four known stages of Hawaiian volcanism are exposed at Haleakala. Volcanic activity was concentrated primarily along the southwest and east rifts and also along the subordinate northwest rift. The most impressive feature is Haleakala Crater, a large erosional caldera at the summit of the volcano created by the coalescence of the heads of two large valleys due to extensive erosion.

HALEAKALA: View across the summit caldera with youthful cinder cones and lava flows. Photograph by C. A. Wood.

The subaerially exposed section of shield-building stage lavas (Honomanu Formation) consists of thin-bedded, highly vesicular, pahoehoe flows of intercalated tholeiitic and alkalic basalt. The vast majority of subaerially exposed lavas were erupted during the alkalic cap stage and are divided into two formations. The older Kumuiliahi Formation is composed of thin-bedded, pahoehoe flows of alkali basalt and hawaiite, and is exposed only within Haleakala Crater. The younger Kula Formation forms the bulk of the exposed alkalic cap; eruptive products consist primarily of thick-bedded aa and pahoehoe flows of alkalic basalt, hawaiite, and mugearite. Indicative of the partly explosive nature of eruptive activity are numerous, interbedded ash deposits exposed in the walls of

KOHALA: Profile view seen from Mauna Kea; view to the east-northeast. Photograph by J. D. Griggs, USGS.

Haleakala Crater, and numerous cinder cones covering the flanks of the volcano. Unusual rock occurrences in Haleakala Crater include a single dike of trachyte exposed in the northwest wall, an amphibole-bearing mugearite flow on the upper west wall, and several bosses of micro-diorite along the northern and western walls. After ~250,000 yr of inactivity, post-erosional basanitoids, alkalic basalts, and hawaiites of the Hana Formation were erupted, primarily along the principal rift zones and at the summit of the volcano where lava flows and cinder cones (up to 180 m high) completely cover the floor of Haleakala Crater. The only known historical eruption at Haleakala occurred near La Perouse Bay and was witnessed by Hawaiians.

How to get there: *Haleakala is east of Kahului Airport, which sits atop Haleakala lavas, and can be reached via Haleakala Highway (37). The north and east flanks of the volcano are crossed by the Hana Highway (360), the west flank by the Kula Highway (37), and the south flank by the unpaved Piilani Highway (31).*

References

Macdonald, G. A., 1978, Geologic map of the crater section of Haleakala National Park, Maui, Hawaii: *USGS Misc. Invest. Ser. Map I-1088.*

Chen, C.-Y., and Frey, F. A., 1985, Trace element and isotope geochemistry of lavas from Haleakala Volcano, East Maui: Implications for the origin of Hawaiian basalts: *J. Geophys. Res. 90*, 8743-8768.

Howard B. West

KOHALA
Hawaii

Type: *Shield volcano*
Lat/Long: *20.12°N, 155.70°W*
Elevation: *1.7 km*
Relief: *1.7 km above sea level; 6.7 km above seafloor*
Volcano Diameter: *22 × 36 km above sea level; 75 ×100 km at seafloor*
Crater Diameter: *None*
Eruptive History:
Main shield construction: 0.46-0.30 Ma
Alkalic cap stage: 0.26-0.06 Ma
No historic activity
Composition: *Tholeiitic basalt, mugearite, and hawaiite*

Kohala is the oldest and northernmost of the five shield volcanoes forming the island of Hawaii. Subsidence during and since shield formation has been ~1 km. Deep valleys on the windward (east) side of the volcano expose >1-km sections of tholeiite lavas. The volcano was built along a northwest–southeast rift zone, now marked by cinder cones and domes. The shield is capped by thick (3-20-m) blocky lava flows of hawaiite to trachyte which erupted from a graben along the crest of the volcano. Arcuate faults suggest that these alkalic rocks may bury a summit caldera. Eruptions during the alkali-cap (or postshield) stage occurred on average every 1,250-1,500 yr. The eruption rate was 4×10^4 m^3/yr, compared to 30×10^6 m^3/yr for Kilauea and Mauna Loa volcanoes.

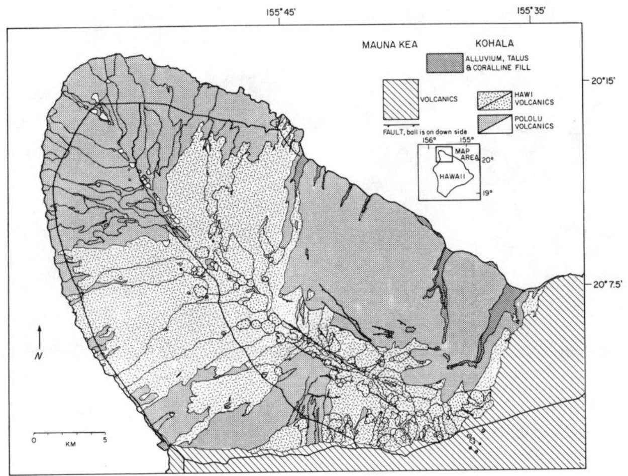

KOHALA: Schematic geologic map showing shield-building Pololu volcanics and postshield Hawi volcanics.

KOHALA: Space Shuttle view (STS61C-51-31) with Mauna Loa volcano to the lower right.

How to get there: *Fly to Hilo or Kona airports on the island of Hawaii. Drive north from either airport to Kohala. Two paved roads cross the mountain. On the southeast side, Waipo Valley shows an excellent cross-section of the flank of the volcano. The upper road across Kohala allows you to see the alkali cap lavas and their vents, including a trachyte dome at a scenic overlook which also provides distant views of three other shields: Mauna Kea, Mauna Loa, and Hualalai.*

References

Feigenson, M. D., Hoffmann, A. W., and Spera, F. J., 1983, Case studies on the origin of basalt II. The transition from tholeiitic to alkalic volcanism on Kohala volcano, Hawaii: *Cont. Min. Petr. 84*, 390-405.

Spengler, S. R., and Garcia, M. O., 1988, Geochemistry of Hawi lavas, Kohala volcano, Hawaii: *Cont. Min. Petr.* (in press).

Michael Garcia

HUALALAI

Hawaii

Type: *Shield volcano*
Lat/Long: *19.7°N, 155.85°W*
Elevation: *2.5 km*
Eruptive History:
Tholeiitic shield stage: complete ~120 ka
Trachyte of alkalic postshield stage: 105 ka
Alkalic basalt of postshield stage: ~100 ka to 1801 A.D.
Composition:
Tholeiitic shield stage: tholeiitic basalt and picritic tholeiitic basalt
Alkalic postshield stage: alkalic basalt, trachyte, alkalic basalt transitional to hawaiite, and ankaramite

Hualalai is the westernmost shield volcano on the island of Hawaii. The volcano has a poorly developed north rift zone, a moderately developed south rift zone, and a well-developed northwest rift zone. The rift zones are 2-5 km wide and are loci of numerous spatter and cinder cone eruptions. A few vents on the volcano are marked by broad lava shields built during long-lived eruptions. The surficial lavas (Hualalai volcanics) belong to the alkalic postshield stage and include alkalic basalt, ankaramite, basalt transitional to hawaiite, and a single trachyte cone (Puu Waawaa) and flow extending north; ~95% of the surface area is Holocene, and ~55% is <3 ka. Tholeiitic basalt of the shield stage has been dredged and recovered by submersible from the submarine extension of the northwest rift zone and has been recovered below alkalic lavas in several water wells. Tra-

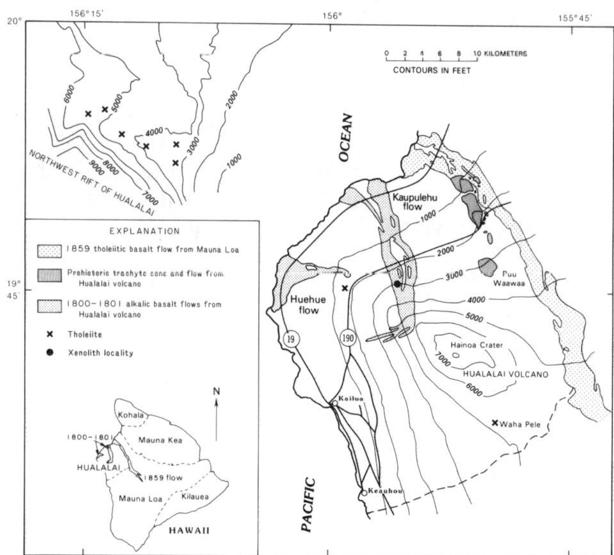

HUALALAI: Simplified geologic map. Crosses indicate locations where tholeiitic basalt of the shield stage has been dredged offshore along the northwest rift, occurs as xenoliths in a tuff (Waha Pele), or has been penetrated in a drillhole near the vent of the Huehue flow.

chyte, in addition to the flow and cone at Puu Waawaa, also occurs near the bottom of one water well, as blocks in the tuff at Waha Pele, and as small xenoliths and cinders at several other basaltic vents.

The **Puu Waawaa** trachyte cone is a large pumice and cinder cone with steep sides; the cone is nearly 1.6 km in diameter. The slopes of the cone are fluted, mostly by erosion, though the channels may have been created initially by glowing avalanches sweeping down the slopes. The surface deposits are a mixture of trachyte obsidian, banded obsidian, and pumice. The eruption was Plinian, with tall eruption columns that repeatedly collapsed. The pumice from this eruption is widely distributed in wells on the north side of the mountain. The trachyte flow that issued from the base of the cone is >275 m thick and extends 9.7 km northward from the cone. The trachyte has been partially covered by later flows of alkalic basalt from Hualalai and tholeiitic basalt flows from Mauna Loa.

The most recent eruptions of Hualalai occurred along the northwest rift in 1800-1801, when 5 vents erupted alkalic basalt having a total volume of ~300×10^6 m^3. Two of these vents issued large flows that reached the ocean.

HUALALAI: Oblique air photograph of the summit of Hualalai looking west. Photograph by R. Moore.

The upper, apparently earlier, of these two main vents, produced the enormous Kaupulehu aa flow that entered the ocean as two discrete flow lobes. One lobe destroyed a Hawaiian village that lay in its path. The lower vent is flanked by a spatter rampart trending down the slope of the volcano. The Huehue flow erupted from this vent is almost all tube-fed pahoehoe and formed a large lava delta where it entered the sea.

Many of the alkalic basalt flows and vent deposits contain common to abundant xenoliths. Many lithologies are represented including dunite, wehrlite, websterite, various types of gabbro, anorthosite, and syenite; lherzolite is absent. The 1800 Kaupulehu flow is famous for a locality at 915-m elevation, adjacent to a telephone relay station, ~3.2 km above Highway 190, where xenoliths occur as cobble deposits. These xenoliths were entrained in the host magma, carried quickly to the surface and erupted, carried downslope inside a lava-tube system, and apparently deposited as xenolith beds when the tube system became blocked. The xenoliths have diameters as large as 35 cm. Some of the gabbroic xenoliths are cumulates from ocean crustal layer 3, whereas others are cumulates of Hualalai tholeiitic basalt that probably accumulated in a magma storage reservoir at the base of the ocean crust. Still others are cumulates of Hualalai alkalic basalt.

The submarine northwest rift zone of Hualalai extends offshore nearly 70 km as a fairly well-defined ridge. In shallow water, the rift has been mantled by two coral terraces at depths of ~150 and 400 m. A third deeper and older terrace is present to within a few kilometers of the rift at a depth of ~750 m. The two deeper terraces are

HUALALAI: Profile view seen from the east. Photograph by J. D. Griggs, USGS.

draped in some places by flows of tholeiitic basalt, indicating that tholeiitic eruptions continued past the age (~128 ka) of the 400-m terrace. No alkalic basalt has been recovered offshore along the rift zone; the rift zones apparently became inactive in their distal parts as the rate of magma supply decreased near the end of the tholeiitic stage.

How to get there: *The Kona airport is located on the 1801 flow of Hualalai; this airport has frequent flights on several air carriers from Honolulu, Hilo, and Maui. Rental cars are available at the airport and at several locations in Kailua. Numerous tour companies run guided tours of the Kona coast, but none of these tours focuses on the geology. The xenolith beds are located on private property; permission to enter must be obtained from Hualalai Ranch or the Lands Division of the King Kamehameha Schools, Bishop Estate, located in Keahou. The Puu Waawaa trachyte cone is also on private land; permission to enter can be obtained from West Hawaii Concrete Company, which operates a gravel quarry at the base of the cone.*

References

Clague, D. A., 1987, Hawaiian xenolith populations, magma supply rates, and development of magma chambers: *Bull. Volc. 49*, 577-587.

Moore, R. B., Clague, D. A., Rubin, M., and Bohrson, W. A., 1987, Hualalai volcano, Hawaii: A preliminary summary of geologic, petrologic, and geophysical data: *VIH*, 571-585.

David A. Clague

MAUNA KEA
Hawaii

Type: *Shield volcano*
Lat/Long: *19.82°N, 155.47°W*
Elevation: *-5,000 to 4,205 m*
Eruptive History:
Submarine eruptions began ~0.8 Ma
Basaltic postshield volcanism began by at least 0.3 Ma
Hawaiitic postshield activity began 60-70 ka
Youngest known eruptions occurred 4-5 ka
Composition: *Basalt, hawaiite to benmoreite*

Mauna Kea is the second largest, in subaerial surface area, of the five shield volcanoes that comprise the Island of Hawaii. Its summit, the highest point in the Hawaiian chain, towers more than 9 km above the adjacent seafloor. Volcanic loading has caused pronounced downwarping of the oceanic crust beneath Hawaii; consequently, the most deeply buried lavas on Mauna Kea lie nearly 15 km beneath the volcano's summit.

Eruptive activity of Mauna Kea was contemporaneous in part with growth of the neighboring volcanoes Kohala, Hualalai, and Mauna Loa. Thus, Mauna Kea lavas interfinger with lavas of these nearby volcanoes. However, surface lavas of Mauna Kea are largely younger than those of Kohala and overlap northward and northwestward onto Kohala. Conversely, surface lavas of Mauna Loa are largely younger than those of Mauna Kea, which they overlap northward.

Mauna Kea's estimated volume is ~33,000 km^3. Approximately 95% of this volume is inferred to consist of shield-stage lavas which are

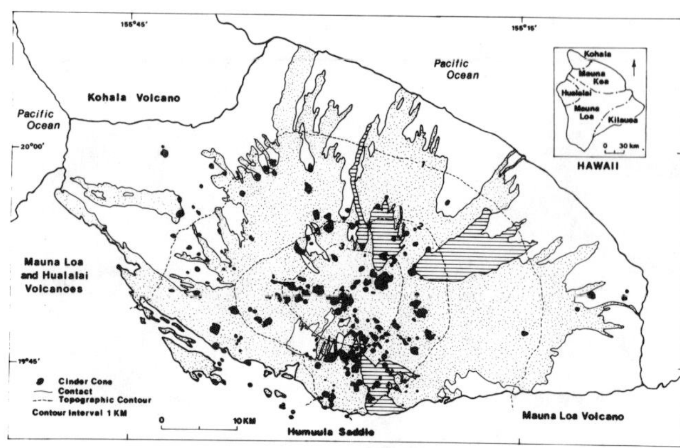

MAUNA KEA: Simplified geologic map. Horizontal-line pattern: younger (postglacial) Laupahoehoe lava flows; dot pattern: older Laupahoehoe lava flows; unpatterned: Hamakua lava flows.

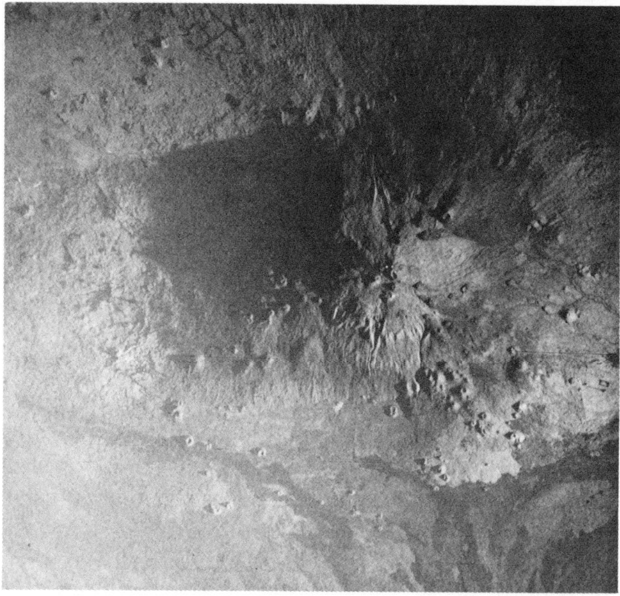

MAUNA KEA: Space Shuttle photograph of Mauna Kea and near-summit cinder cones. Photograph STS-61C-51-032.

exposed only below sea level; postshield lava now caps the entire subaerial surface of the volcano. Rare dredge samples and inference from the currently active shield volcanoes, Kilauea and Mauna Loa, suggest the Mauna Kea shield-stage lava is predominantly tholeiitic basalt erupted at a rate on the order of 0.1 km^3/yr.

Postshield lavas occur as two lithologically and compositionally distinct formations. The older formation, the Hamakua volcanics, consists essentially of basalt. The younger formation, the Laupahoehoe volcanics, contains no basalt; it consists instead of felsic hawaiite, mugearite, and some benmoreite. The Hamakua volcanics comprise a postshield basaltic cap, and the Laupahoehoe volcanics comprise a still younger hawaiitic cap.

The Hamakua volcanics form most of the lower slopes of Mauna Kea; Hamakua basalt also crops out in kipukas on the upper slopes. Although pahoehoe flows occur locally, most Hamakua flows are dense, thick (>3 m) aa flows with rubble at their tops and bases. Flows range up to more than 20 km in length and up to ~2 km wide. Cinder cones formed at all vents that are exposed on the lower slopes; some of the upper-slope lavas erupted from fissure vents. Most of the Hamakua cones are simple crescentic structures in which part of the wall was rafted away as lava welled out of the conduit. The cones range in height from ~25 to 140 m, and in diameter from 120 to >900 m. Their shapes have been degraded by erosion and mass wasting.

The estimated volume of Hamakua basalt is 850 km^3. The lavas range in age from ~300 to 700 ka, giving an average lava-supply rate of ~0.004 km^3/yr. This is between 1 and 2 orders of magnitude less than the rate of supply inferred for the shield stage.

The Laupahoehoe volcanics are nearly continuous over the upper slopes of Mauna Kea; few flows extend to sea level. Most Laupahoehoe eruptions formed prominent cinder cones and thick lava flows composed predominantly of aa. The cinder cones range in diameter from <100 m to >1 km. Maximum height is ~200 m. The concentration of vents and lava flows on the upper slopes cause the rugged local relief that makes the upper slopes and summit of Mauna Kea stand in such marked contrast to the much smoother slopes of the Mauna Loa shield.

Laupahoehoe lava flows are mostly aa. However, relatively dense, massive pahoehoe occurs in the proximal areas of some flows. Flow thickness is commonly ~5-10 m; aa 25 m or more in thickness occurs locally. Gully walls expose thick, dense, massive flow interiors overlying and capped by aa rubble. In some outcrops the massive facies and aa-breccia facies are complexly interlayered within single flows. The longest flows extend 15-20 km from their

MAUNA KEA: South flank. Road in foreground is on dark, Holocene, Mauna Loa basalt flows that overlap grass-covered, late Pleistocene, hawaiite/mugearite flows of Mauna Kea in Humuula Saddle. Distinct cinder cones all belong to the Laupahoehoe volcanics. Makanaka end moraine forms light-colored, gullied bench high on the slope. Horizontal distance from saddle road to summit is ~15 km; elevation difference is 2 km. View is northward. (Hawaiian Volcano Observatory photograph 88.7.11EW120A#1.)

MAUNA KEA: Laupahoehoe cinder cones and lava flows on the upper west flank of Mauna Kea. Flows with prominent ridge-and-trough fabric emanate from the lower flanks of the bases of the cones. Some flows breached their source cones. Base of nearest cinder cone, in left foreground, is ~600 m across. North flank of Mauna Loa is at upper right. View is southeastward. (Hawaiian Volcano Observatory photograph 88.7.11EW135A#5.)

sources. Flow widths generally range from ~0.5 to 7 km, but some elongate flow lobes as narrow as 50 m extend hundreds of meters down steep slopes.

On the steeper slopes, networks of discontinuous, locally anastomosing, longitudinal ridges and troughs give the Laupahoehoe flow surfaces a braided or corrugated appearance. Some of the ridges are levees that bounded lava-channel segments. Others are narrow, steep-sided flow lobes; some of these preserve narrow lava channels along their axes. On more gentle slopes, particularly in the terminal parts of flows, the Laupahoehoe aa tended to advance as broad sheets. There, channels tend to be broader and levees less well defined than on the steeper slopes, and hummocks and transverse pressure ridges abound.

Many Laupahoehoe eruptions apparently produced eruption columns rich in ash and lapilli. Fallout formed local air-fall deposits, which were reworked by wind, water, and mass wasting to produce ash-rich fluvial, colluvial, and aeolian units that discontinuously mantle much of the surface of Mauna Kea.

The volume of Laupahoehoe flows and pyroclastic deposits is ~25 km^3. These lavas were erupted from ~65 to 4 ka, giving an average lava-supply rate of ~0.0004 km^3/yr – an order of magnitude less than the average supply rate for the underlying Hamakua volcanics. Younger (postglacial) Laupahoehoe lavas issued from 17 separate vents on the upper slopes of Mauna Kea. Field relations and radiocarbon ages indicate that at least 11 of these vents, located on the north, northeast, and south flanks, were active between ~4 and 7 ka.

Deposits of three distinct glacial episodes are found on Mauna Kea, which is the only Hawaiian volcano known to have been glaciated. The two older glacial units, which consist mostly of diamicton interpreted as till, are intercalated with basalts of the Hamakua volcanics. The older, known as the Pohakuloa Glacial Member of the Hamakua volcanics, is bracketed by basalt flows dated at ~0.15 and 0.115 Ma. The younger of these two, called the Waihu Glacial Member of the Hamakua volcanics, is bracketed by basalt flows dated at ~85 and 70 ka. The Pohakuloa drift is found mainly in a few canyon-wall exposures. The Waihu drift forms a line of kipukas at ~3-km elevation on the upper southwest flank of Mauna Kea.

The youngest glacial deposits, intercalated with hawaiitic Laupahoehoe lavas, comprise the Makanaka Glacial Member of the Laupahoehoe volcanics. The unit is represented by an annulus of till, largely between elevations of ~3.3 to 3.6 km, and local, fringing deposits of outwash gravel. Lava-flow surfaces within the annulus of Makanaka till were deeply scoured by glacial ice, widely exposing massive flow interiors. Many of these outcrops are striated and weakly polished. K-Ar dating indicates that the Makanaka glaciation was under way by ~40 ka. ***How to get there:*** *Paved roads circle Mauna Kea, and a steep, partly paved and partly graded, gravel road from Humuula Saddle to the summit provides good access to the south flank. Ranch roads and unimproved roads, some very rough, provide additional access.*

References

Porter, S. C., 1979, Hawaiian glacial ages: *Quat. Res. 12*, 161-187.

Wolfe, E. W., Wise, W. S., and Dalrymple, G. B., *in prep.* The geology and petrology of Mauna Kea Volcano, Hawaii: A study of postshield volcanism.

Edward W. Wolfe and William S. Wise

MAUNA LOA: Space Shuttle image of Mauna Loa and its southwest and east-northeast rift zones. Mauna Kea in upper right. Photograph STS-51-I-37-062.

MAUNA LOA
Hawaii

Type: *Shield volcano*
Lat/Long: *19.48°N, 155.60°W*
Elevation: *4,169 m above sea level, ~9 km above seafloor*
Eruptive History: *Birth: ~1 Ma (?)*
Oldest exposed lavas: ~0.2 Ma
1750? to 1984: 39 known eruptions
Composition: *Tholeiitic basalt*

Mauna Loa is considered the world's largest active volcano, with an estimated volume of ~40,000 km^3, based on the assumption that the volcano extends from the surface to the top of the isostatically depressed oceanic crust. However, because of adjacent volcanoes interfingering with Mauna Loa, and because older volcanic edifices may be buried at depth, the volcano's actual volume is probably less. The area of the volcano above sea level is >5,000 km^2; another 4,000 km^2 of Hawaii's submarine flanks are covered by Mauna Loa lavas. Mauna Loa's lavas are well-exposed in desolate, arid areas >2,000 m elevation, but are covered with dense rain forest in wet areas on the lower flanks. Older surface lavas are weathered and covered by ash and soil in these areas.

MAUNA LOA: Pit craters near the summit of Mauna Loa, and the broad Mokua'aweoweo caldera on the near horizon. Mauna Kea in background. View from south. USGS photograph by P. Lipman.

The Mauna Loa shield is composed entirely of relatively thin flows of aa and pahoehoe, averaging 3-4 m in thickness. Individual flows are typically <1 km wide, but some are >50 km long. Pyroclastic rocks are insignificant in volume, consisting of minor ejecta and vitric ash

MAUNA LOA: Broad profile from the lower slopes of Mauna Kea. Photograph by C. A. Wood.

near eruptive vents and very small areas of phreatic explosion debris near the summit. Mauna Loa's eruptive products are entirely of tholeiitic basalt composition, with major element chemical variation principally controlled by olivine content. Olivine and plagioclase are the only megascopically resolvable minerals in lava flows, but many flows are aphyric. Xenoliths are uncommon and where found consist of cognate microgabbros.

Mauna Loa has the most detailed prehistoric chronology of any volcano on Earth, based on >300 ^{14}C dates on charcoal recovered from beneath flows. About 500 different Holocene lava flows and flow remnants have been mapped, although many more are buried by younger flows. The areal distribution of dated flows shows that ~40% of the Mauna Loa edifice is covered every 1,000 yr. Historical coverage rates are higher than this, however, and 22 flows have erupted outside the summit caldera since the first identifiable flow, that of 1843. About 4.2 km^3 of lavas have been erupted since 1843, in individual eruptions of <0.001 to >0.5 km^3 volume. The latest eruption occurred in 1984, when 50 km^2 of Mauna Loa was covered by ~0.22 km^3 of lava.

Mauna Loa's summit is indented by a 3 × 5-km caldera up to 180 m deep – **Mokua'aweoweo**. Extensive net infilling of the caldera has occurred during the past two centuries. Mokua'aweoweo is elongated northeast–southwest and consists of several coalescing collapse craters, mostly formed in prehistoric time. Rift zones extend from Mokua'aweoweo ~50 km to the east-northeast, and 100 km to the southwest (partly below sea level). These rift zones are loci of abundant subparallel, linear eruptive vents, and are marked by hundreds of spatter cones and linear spatter ramparts. Eruptions occur either along these rift zones, within Mokua'aweoweo, or from radial fissures on the northwest flank. Feeder dikes for Mauna Loa lavas are nowhere exposed except on the walls of Mokua'aweoweo and summit pit craters.

Mauna Loa slowly inflates prior to eruptions, and quickly deflates as lava is erupted at the surface. Geodetic and seismic measurements show that this deformation is caused by volumetric changes of a magma storage area located at 3-5-km depth directly below Mokua'aweoweo.

Inactive normal step-faults, largely buried by younger flows, cut the lower southeast flank of the volcano. Recent studies show that the steep western flank of Mauna Loa reflects the generation of enormous pre-Holocene landslides from that unbuttressed flank. These fed multiple submarine debris avalanches which blanket vast areas of the seafloor west and northeast of Hawaii.

How to get there: *Mauna Loa is encircled by a paved road along its lower flanks. Access to the higher slopes is provided via the Humuula Saddle road, which crosses between Mauna Loa and Mauna Kea volcanoes at 1,800 m. From the Humuula Saddle the Mauna Loa Observatory Road reaches 3,400 m. A trail leads from there to the summit, which is usually snow-covered in win-*

ter months. Resthouses are maintained by the National Park Service on the edge of Moku-a'aweoweo and on the northeast rift zone at 3,000-m elevation, facilitating backcountry camping.

References

Lockwood, J. P., and Lipman, P. W., 1987, Holocene eruptive history of Mauna Loa volcano, Hawaii: *VIH*, 509-535.

Lockwood, J. P., Lipman, P. W., Petersen, L. D., and Warshauer, F. R., 1988, Generalized ages of surface lava flows of Mauna Loa volcano, Hawaii: *USGS Map I-1908*.

Jack P. Lockwood

KILAUEA

Hawaii

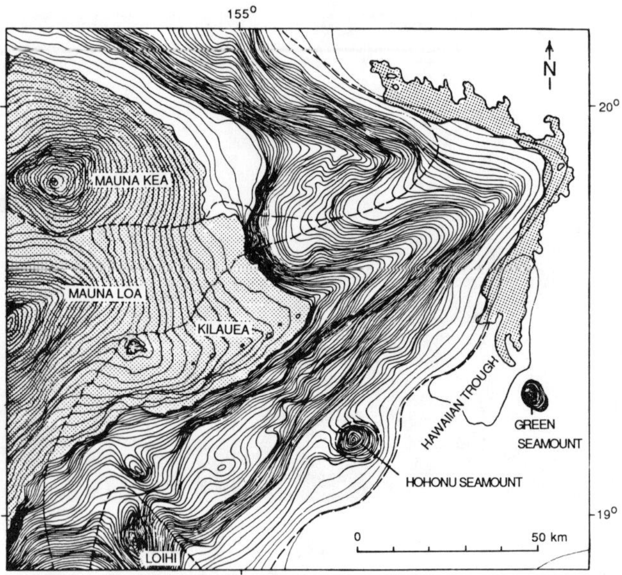

KILAUEA: Topography of Kilauea, which is separated by dashed lines from neighboring volcanoes. Shading at left indicates subaerial parts of Kilauea, Mauna Loa, and Mauna Kea; remainder of this area is submarine. Shading at right indicates extensive sheet flows that have spread from the distal end of Kilauea's east rift zone into the Hawaiian Trough. Contour interval 100 m.

Type: *Shield volcano*
Lat/Long: *19.42°N, 155.28°W*
Shield Dimensions *:*
Length: 180 km
Maximum width: 75 km
Elevation: 1,250 m, 6,750 m above seafloor
Caldera Dimensions:
Modern caldera: 3.5 by 5 km
Powers caldera: 6 by 8 km
Depth: variable, has exceeded 270 m
Eruptive History:
200-2 ka: Growth and caldera collapse
2-0.2 ka: Collapse and filling of Powers caldera
A.D. 1790: Last major collapse of modern caldera
1790-1960: Summit eruptions dominant
1960-Present: Flank eruptions dominant
Composition: *Tholeiitic basalt*

Kilauea lies at the southeast end of the Hawaiian chain. It is the youngest major shield now protruding above sea level, but much of its bulk is under water.

The subaerial part of Kilauea covers ~1,500 km^2. Its modern caldera contains an annulus of arcuate blocks stepped downward toward an inner sink, which is partly filled by a lava shield indented by **Halemaumau** crater. The modern caldera is thought to be nested within the buried rim of an older caldera. Two rift zones radiate from the summit. The subaerial east rift zone is 4-6 km wide; it has an upper segment dominated by pit craters and satellitic lava shields, and a lower segment marked by closely spaced fissures, faults, and grabens having net seaward subsidence. The southwest rift zone is <4 km wide and displays obvious seaward subsidence. The Koae fault system extends east-northeast between the rift zones, with faults downthrown mainly toward the caldera. The south flank is dominated by normal faults (Hilina system) having seaward displacements that commonly exceed 100 m. The north flank is free of surface structures but is draped across an inactive distal segment of Mauna Loa's east rift zone and shows some signs of a buried fault inland from Kilauea's steep northeast submarine slope, where large landslides may occur. In all parts of Kilauea except the north flank structures have been recently active and display repeated offsets overlapped by lava flows and tephra.

The submarine Puna Ridge, extending east-northeast of Hawaii, has grown along the east rift zone of Kilauea. The convex upper flanks of

KILAUEA: Oblique aerial view across Kilauea caldera toward Mauna Kea. Halemaumau Crater, left center, is ~1 km wide. Flow of 1954 extends east of Halemaumau; several other flows have been erupted within the caldera since this photograph was made by the US Navy in 1954.

the Puna Ridge consist largely of pillowed lava flows, whereas the concave lower flanks consist of mixed lava and debris. Extending along the base of the ridge are elongate sheet flows erupted from the distal part of the rift zone and having individual volumes of a few cubic kilometers. The submarine south flank offshore from the subaerial part of Kilauea consists largely of giant landslide blocks mixed with pyroclastic debris produced along the shoreline.

Kilauea probably began to grow at ~0.2 Ma, and the oldest flows of Hilina Basalt now exposed in fault scarps of the subaerial south flank probably date from 50-100 ka. Included within the Hilina Basalt are several tephra layers thought to represent explosive eruptions associated with repeated caldera collapses. Overlying the Hilina Basalt, and separated from it by the Pahala Ash of 10-25 ka, is the Puna Basalt, which veneers nearly all of Kilauea's surface. Ash layers within the Puna Basalt indicate that Kilauea had several explosive eruptions during the Holocene; some probably were associated with caldera collapses. Additional subaerial evidence, however, has been found for only two recent calderas, which may have erased their predecessors. The **Powers** caldera is thought to have developed prior to deposition of the Uwekahuna Ash Member at ~2 ka. That large explosive eruption appears to have been followed by several centuries of caldera filling and eruptive hiatus along upper parts of the rift zones, when the northeast flank of Kilauea became extensively vegetated.

Lava erupted within the Powers caldera began to overflow onto the south flank at ~1.2 ka, roughly coincident with Polynesian settlement. Overflows continued intermittently until ~0.35 ka; by that time the old caldera rim had been entirely covered. Sustained overflows produced some of the longest uncollapsed lava tubes known on Earth. Interspersed with summit activity were eruptions along the rift zones. Frequent flank eruptions along both rift zones during the 18th century produced extensive flow fields and satellitic lava shields. Contemporaneous with or following these flank eruptions was the collapse of the modern caldera, culminating in a large explosive eruption in 1790.

Written records of Kilauea's activity began with arrival of Christian missionaries in 1823. The following century saw nearly continuous eruption in the caldera, punctuated by lava-lake subsidence and eruptions along both rift zones. Continuous summit activity ended in 1924 with deep subsidence and an explosive eruption at Halemaumau. Eruptions occurred intermittently in Halemaumau through 1934 and, following a hiatus, resumed in 1952. Eruptions returned to the east rift zone in 1955, and were frequent on both summit and flank until 1968; but in 1969 summit eruptions became less frequent as sustained activity began along the east rift zone. That activity so far has built lava shields and extensive flow fields and pyroclastic deposits at **Mauna Ulu** (1969-1974) and **Pu'u O'o** (beginning in 1983).

The fluidity of magma in Kilauea's plumbing system makes it possible for events in one part of the volcano to affect other parts far distant. Magma rises at a rate of ~0.11 km^3/yr from the mantle into a reservoir a few kilometers beneath the summit, where it is stored prior to eruption. The magma may be erupted either from the summit, directly above the principal reservoir, or from a rift zone after moving laterally for long distances beneath the surface. For

KILAUEA: Oblique aerial view looking westward toward Puu Kapukapu, a tilted block bounded by scarps of the Hilina fault system, which is thought to be the upper part of a giant landslide on the south flank of Kilauea. Cropping out in these cliffs (300 m high) are the oldest rocks of Kilauea now exposed. Lava flows 500-1000 yr old have spread around and across the block to the shoreline, where they have produced lava deltas that are now subsiding into the sea. USGS photograph by J. D. Griggs.

sustained eruptions, the volcano behaves like an open system, with the volume erupted being balanced by the volume supplied. For brief eruptions, the volcano behaves like a closed system, with rapid eruption of a large volume causing the reservoir to drain and the surface above it to subside. For brief flank eruptions, volume of eruption and collapse is well correlated with elevation of eruption site along the rift zone. Deep submarine eruptions of a few cubic kilometers are thought to cause major caldera collapses as frequently as once every few centuries to a few millennia.

Present evidence indicates that Kilauea undergoes long-term eruptive changes in response to caldera collapses. Large collapses are followed by continuous eruption within the caldera and infrequent eruptions from the rift zones; but as the caldera fills, summit activity wanes and rift eruptions become more frequent and more sustained farther from the summit. These changes are thought to arise from gradual refilling of reservoirs in the summit and rift zones. Filling may be interrupted when the reservoirs are dilated by seaward displacements of the south flank, which are themselves caused in part by repeated intrusions along the rift zones.

Kilauea is of interest especially because of its accessibility and frequent activity, which make it a prime laboratory for study of volcanic processes. Several classic studies were made during the 19th and early 20th centuries, and the Hawaiian Volcano Observatory has operated continuously on the caldera rim since 1912. In 1987 a new building was dedicated and several comprehensive publications were issued on the occasion of the observatory's diamond jubilee.

How to get there: *Principal airports for the Island of Hawaii are at Hilo and Kailua-Kona; sightseeing flights are available via helicopters and fixed-wing aircraft. Kilauea caldera can be reached via a paved highway (and daily bus service) south and west from Hilo (50 km) or south and east from Kailua (150 km). Most of the summit, south flank, and southwest rift zone are located in Hawaii Volcanoes National Park, where a hotel, restaurant, cabins, and camping facilities are available. Paved roads give easy access to much of the south flank and lower part of the subaerial east rift zone, and unpaved roads extend into much of the north flank. Foot trails lead into most areas away from roads on the summit, south flank, and southwest rift zone. A large area of the middle east rift zone and north flank has difficult or hazardous access owing to dense forest, ground cracks, frequent eruptions, and suspicious drug growers. Inquire locally about access to recent eruption* *sites, lava tubes, tree molds, and other features of special interest.*

References

Fiske, R. S., Simkin, T., and Nielsen, E. A., eds., 1987, *The Volcano Letter*. Washington, DC, Smithsonian Institution Press, 530 issues of 1925-1955 republished in 1 volume.

Holcomb, R. T., 1987, Eruptive history and long-term behavior of Kilauea Volcano: *VIH*, 261-350.

Robin Holcomb

Contributor index

Allen, C. C.
Tuya Butte, Canada, 119
Allen, J. E.
Boring Lava, OR, 170
Aubele
Cerros del Rio, NM, 297
Jornada del Muerto, 309
Red Hill, NM, 306
Springerville, AZ, 284
Bacon
Crater Lake, OR, 193
Baldridge
Lucero, NM, 305
Ocate, NM, 293
Taos, NM, 290
Beget
Glacier Peak, WA, 156
Bevier
Chilcotin Basalt, Canada, 136
Rainbow Range, Canada, 131
Brew
Behm Canal and Rudyerd Bay, AK, 95
Duncan Canal, AK, 94
Tlevak Strait and Suemez Island, AK, 95
Brophy
Cold Bay, AK, 50
Kanaga, AK, 24
Charland
Itcha Range, Canada, 134
Chitwood
Devils Garden, OR, 203
Fort Rock Basin, OR, 203
Newberry, OR, 200
Christiansen
Shasta, CA, 214
Yellowstone Plateau, WY, 263
Clague
Hualalai, HI, 337
Kauai, HI, 321
Niihau, HI, 320
West Molokai, HI, 327
Clynne
Lassen, CA, 216
Lassen: other cones, CA, 221
Pre-Lassen centers, CA, 219
Dittmar, 220
Maidu, 220
Snow Mountain, 220
Yana, 220
Condit
Springerville, AZ, 285
Conrey
Jefferson, OR, 177
Crowe
Timber Mountain, NV, 261
Ubehebe, CA, 237
Crumpler
Jornada del Muerto, NM, 309
Red Hill, NM, 306
Springerville, AZ, 285
Taylor, NM, 300
Dohrenwend
Adobe Hills, CA, 230
Black Mountain, CA, 241
Buffalo Valley, NV, 256
Cima, CA, 240
Clayton Valley, NV, 261
Lunar Crater, NV, 258
Reveille Range, NV, 260
Saline Range, CA, 238
Donnelly-Nolan
Clear Lake, CA, 226
Medicine Lake, CA, 212
Duffield
Coso, CA, 239
Foster
Prindle, AK, 108
Fournelle
Isanotski, AK, 49
Roundtop, AK, 49
Shishaldin, AK, 48
Francis
Volcano Mountain, Canada, 118
Garcia
Kahoolawe, HI, 333
Kaula, HI, 319
Kohala, HI, 336
Gillespie
Big Pine, CA, 236
Greeley
Amboy, CA, 243
Craters of the Moon, ID, 248
Snake River Plain, ID, 246
Split Butte, ID, 252
Wapi, ID, 254
Green
Garibaldi Lake, Canada, 143
Grose
Eagle Lake, CA, 222
Gust
Mormon, AZ, 282
Rayton–Clayton, NM, 292
Hamilton
Level Mountain, Canada, 121
Hammond
Indian Heaven, WA, 166
Marble Mountain–Trout Creek Hill Zone, WA, 167
Hart
Diamond Craters, OR, 208
Jordan Craters, OR, 210
Harwood
Tuscan Formation, CA, 244
Hickson
Wells Gray–Clearwater, Canada, 137
Hildreth
Adams, WA, 164
Griggs, AK, 72
Katmai, AK, 71
Mageik, AK, 67
Novarupta, Falling Mountain, and Cerberus, AK, 70
Trident, AK, 68
Hoffer
Kilbourne Hole, NM, 310
Potrillo, NM, 311
Holcomb
Kilauea, HI, 344
Lanai, HI, 331
Wailau (East Molokai), HI, 328
Kay
Bobrof, AK, 24
Bogoslof, AK, 40
Chagulak, AK, 33
Davidof and Khvostof, AK, 20
Great Sitkin, AK, 27
Kasatochi, AK, 28
Kiska, AK, 18
Little Sitkin, AK, 20
Segula, AK, 19
Semisopochnoi, AK, 21
Kienle
Augustine, AK, 79
Buzzard Creek, AK, 85
Ukinrek, AK, 65
King
King's Bowl, ID, 250
Menan Buttes, ID, 251
Kudo
Albuquerque, NM, 302
Cat Hills, NM, 303
San Felipe, NM, 299
Lee-Wong
St. George, AK, 97
St. Paul, AK, 96
Lipman
Taos, NM, 290
Lockwood
Mauna Loa, HI, 342
Lynch
Geronimo, AZ, 287
San Carlos, AZ, 286
Sentinel Plain, AZ, 287
Marsh
Amak, AK, 51
Atka, AK, 29
Buldir, AK, 18
Gareloi, AK, 22
Moffett and Adagdak, AK, 25
Tanaga, AK, 23
Mathews
Garibaldi, Canada, 144
Miller
Aniakchak, AK, 59
Black Peak, AK, 58
Dutton, AK, 51
Emmons and Hague, AK, 52
Fisher, AK, 46
Iliamna, AK, 80
Pavlof and Pavlof Sister, AK, 53
Ugashik and Peulik, AK, 63
Moll-Stalcup
Ingakolugwat Hills, AK, 100
Ingrisarek Mountain, AK, 102
Kochilagok Hill, AK, 101
Kookooligit, AK, 104
Nelson Island, AK, 102
Nushkolik Mountain, AK, 101
St. Michael, AK, 103
Ungulungwak Hill–Ingrichuak Hill, AK, 100
Yukon Delta, AK, 99
Muffler
Salton Buttes, CA, 245

Myers
Amukta, AK, 32
Carlisle, AK, 34
Cleveland and Chuginadak, AK, 35
Herbert, AK, 34
Kagamil, AK, 36
Seguam, AK, 31
Uliaga, AK, 36
Yunaska, AK, 33

Nash
Black Rock Desert, UT, 271
Fumarole Butte and Smelter Knolls, UT, 270
Honeycomb Hills, UT, 269
Mineral Mountains and Cove Fort, UT, 273
Twin Peaks, UT, 273
Uinkaret, AZ, 277
Nye
Espenberg, AK, 106
Gilbert, AK, 44
Imuruk, AK, 106
Koyuk-Buckland, AK, 107
Makushin, AK, 41
Okmok, AK, 38
Recheschnoi, AK, 37
Spurr, AK, 83
Teller, AK, 105
Vsevidof, AK, 37
Wrangell, AK, 88

Pringle
Rainier, WA, 158

Richter
Capital, AK, 89
Drum, AK, 86
Eastern Wrangell, AK, 91
Sanford, AK, 87
Tanada, AK, 90
Riehle
Chiginagak, AK, 61
Edgecumbe, AK, 93
Hayes, AK, 84
Kialagvik, AK, 62
Yantarni, AK, 60

Scott
Bachelor, OR, 185
Self
Jemez, NM, 295
Sunset Crater, AZ, 280
Sherrod
Bald Mountain, OR, 205
High Cascades, OR
Clear Lake to Olallie Butte, 176
Columbia River to Mount Hood, 172
East of Mount Hood to Clear Lake, 175
South of Mount Jefferson to Santiam Pass, 178
South of Three Sisters to Willamette Pass, 187
Willamette Pass to Windigo Pass, 189
Windigo Pass to Diamond Lake, 190
Hood, OR, 173
Sprague River Valley, OR, 196
Thielsen, OR, 191
Western Cascades, OR, 169
Sinton
Waianae, HI, 323
West Maui, HI, 333
Smith
Brown Mountain, OR, 199
McLoughlin, OR, 197
Pelican Butte, OR, 195
Souther
Bridge River Cones, Canada, 139
Cayley, Canada, 142
Edziza, Canada, 124
Hoodoo, Canada, 127
Ilgachuz Range, Canada, 132
Iskut-Unuk River Cones, Canada, 128
Maitland, Canada, 126
Milbanke Sound Cones, Canada, 130
Nazko, Canada, 135
Silverthrone, Canada, 138
The Thumb, Canada, 129
Stasiuk
Meager Mountain, Canada, 141
Swanson, D. A.
Goat Rocks, WA, 160
St. Helens, WA, 161
Swanson , S. E.
Akutan, AK, 43
Denison, Steller, and Kukak, AK, 73
Devils Desk, AK, 75
Douglas, AK, 78
Fourpeaked, AK, 77
Kaguyak, AK, 75
Kejulik, AK, 66
Martin, AK, 67
Pogromni, AK, 44
Snowy, AK, 73
Westdahl, AK, 45

Taylor
Belknap, OR, 182
Broken Top, OR, 183
Sand Mountain, OR, 180
Three Sisters, OR, 184
Washington, OR, 181

Theilig
Carrizozo, NM, 308
Pisgah, CA, 242
Zuni-Bandera, NM, 303

Vallier
Koniuji, AK, 28

Walker, G. P. L.
Honolulu Vents, HI, 325
Kalaupapa and Mokuhooniki, HI, 330
Koolau, HI, 324
Walker, G. W.
Iron Mountain, OR, 207
Silicic Domes of south-central OR, 206
Wenrich
Hopi Buttes, AZ, 283
West
Haleakala, HI, 335
Westgate
Old Crow Tephra, AK, 92
White River Tephra, AK, 92
White
Steamboat Springs, NV, 257
Williams
Long Valley, CA, 234
Wilson
Kupreanof (Stepovak Bay), AK, 55
Wise
Mauna Kea, HI, 339
Wolfe
Mauna Kea, HI, 339
San Francisco, AZ, 278
Wood
Aurora-Bodie, CA, 229
Baker, WA, 155
Dotsero, CO, 268
Inyo Craters, CA, 233
Jackies Butte, OR, 211
Kolob and Loa, UT, 275
Leucite Hills, WY, 266
Mono Craters, CA, 231
Nunivak, AK, 98
Porcupine, AK, 108
Saddle Butte, OR, 209
Sheldon-Antelope, NV, 256
Silver Creek, OR, 207
Simcoe, WA, 165
Sonoma, CA, 228
Sutter Buttes, CA, 225
Togiak, AK, 102

Yount
Dana, AK, 54
Redoubt, AK, 81
Veniaminof, AK, 56

Volcano index

Italics indicate a volcano that does not have its own listing but is mentioned in the discussion of another volcano.

Adagdak, AK, 25
Adams, WA, 164
Aden Crater, NM, 312
Adobe Hills, CA, 230
Afton, NM, 310
Ahkiwiksnuk, AK, 99
Akutan, AK, 43
Albuquerque, NM, 302
Amak, AK, 51
Amboy, CA, 243
Amukta, AK, 32
Anahim Peak, Canada, 132
Ancient Mount Kanaton, AK, 24
Andrew Bay, AK, 25
Aniakchak, AK, 69
Antelope, CA, 223
(Sheldon–)Antelope, NV, 256
Apache Maid, AZ, 282
Arlington, AZ, 287
Armadillo Peak, Canada, 125
Armet Creek, OR, 170
Ash, Canada, 120
Atka, AK, 29
Atwell Peak, Canada, 145
Augustine, AK, 79
Aurora, CA, 229
Aurora–Bodie, CA–NV, 229

Bachelor, OR, 185
Badger Butte, OR, 175
Baker, WA, 155
Bald Mountain, OR, 205, *206*
Ball Point, OR, 175
Bailey, OR, 191
Bandera, NM, 304
(Zuni–)Bandera, NM, 303
Bare, WA, 167
Battle Ax, OR, 169
Bear Wallow Butte, CA, 222
Beaver Butte, OR, 176
Behm Canal, AK, 95
Belknap, OR, 182
Benchmark Butte, OR, 188
Berry, WA, 166
Big Cinder Butte, ID, 249
Big Craters, ID, 250
Big Hole, OR, 206
Big Lava Bed, WA, 166
Big Pine, CA, 236
Big Southern Butte, ID, 248
Bill Williams, AZ, 279
Bird, WA, 166
Black Butte, CA, 215
Black Mountain, CA, 241
Black Mountain (Eagle Lake), CA, 223
Black Mountain, NM, 312
Black Peak, AK, 58
Black Point, CA, 232
Black Rock Desert, UT, 271
Black Buttes, WA, 155
Black Tusk, Canada, 143
Blacks, CA, 222
(The) Blowouts, OR, 203
Blue Box Pass, OR, 175
Binalik, AK, 99
Bobrof, AK, 24
Bobs Hill, OR, 171
(Aurora–)Bodie, CA–NV, 229
Bogoslof, AK, 40
Bogoslof Hill, AK, 97
Bona, AK, 91, 92
Booth Hill, OR, 172
Boring Lava, OR, 170
Borox Lake, CA, 226
Bowden Crater, OR, 211
Bridge River Cones, Canada, 139
Broken Back Crater, NM, 308
Broken Top, OR, 183
Brokeoff, CA, 217
Brown Mountain, OR, 199
Buckboard Mesa, NV, 262
(Koyuk–)Buckland, AK, 107
Buffalo Valley, NV, 256
Bug Butte, OR, 196
Buldir, AK, 18
Bumpass, CA, 218
Burn Butte, OR, 190
Burney, CA, 222
Burnt, UT, 273
Buzzard Creek, AK, 85

Cabezon, NM, 302
Calimus Butte, OR, 196
Camille, AK, 107
Canjilon Hill, NM, 300
Canyon Creek, Canada, 129
Capital, AK, 89
Capulin, NM, 292
Carlisle, AK, 34
Carrizozo, NM, 308
Castle, AK, 91
Castle Rock, AK, 41
Cat Hills, NM, 303
Cayley, Canada, 142
Cedar Grove, UT, 274
Cedar Hill, CA, 229
Cerberus, AK, 70
Cerberus (Semisopochnoi), AK, 21
Cerro Guadaloupe, NM, 302
Cerro Hueco, AZ, 285
Cerro Santa Clara, NM, 302
Cerro Verde, NM, 305
Cerros del Rio, NM, 297
Chagulak, AK, 33
Champs Flat, CA, 224
Chaos Crags, CA, 218
Chaos Jumble, CA, 218
Charlton Butte, OR, 187
Cheching, AK, 102
Chiginagak, AK, 61
Chilcotin Basalt, Canada, 136
China Hat, OR, 206
Christie Hill, CA, 221
Chuginadak, AK, 35
Churchill, AK, 91
Cima, CA, 240
Cinder, Canada, 128
Cinder Butte, CA, 222
Cinder Cone, CA, 218
Cinder Cone, Canada, 143
Cinnamon Butte, OR, 190
(Rayton–)Clayton, NM, 292
Clayton Valley, NV, 261
Clear Lake, CA, 226
Clear Lake Butte, OR, 176
Clear Lake to Olallie Butte, OR, 176
(Wells Gray–)Clearwater, Canada, 137
Cleveland, AK, 35
Clinker Peak, Canada, 144
Cloud Cap, OR, 173
Coffeepot Crater, OR, 210
Cold Bay, AK, 50
Columbia Crest, WA, 159
Columbia River to Mt. Hood, OR, 172
Coso, CA, 239
Cougar, OR, 206
Cove Fort, UT, 273
Cowhorn, OR, 189
Coyote Hills, UT, 273
Crater Butte, OR, 189
Crater Flat, NV, 262
Crater Knoll, UT, 274
Crater Lake, OR, 193
Crater (Mountain), AK, 104
Crater Peak, AK, 84
Crater Ridge, AK, 94
Craters of the Moon, ID, 248
Crazy Hills, WA, 166
Cupit Mary, OR, 187

Dana, AK, 54
Davidof, AK, 20
Davis, OR, 188
Davis Lake, OR, 188
Defiance, OR, 172
Denison, AK, 73
Devastator, Canada, 141
Devils Desk, AK, 75
Devils Garden, OR, 203
Diamond Craters, OR, 208
Diamond Peak, OR, 189, 191
Dittmar, CA, 220
Doe, CA, 221
Dog River Springs, OR, 175
Dotsero, CO, 268
Douglas, AK, 78
Drum, AK, 86
Duncan Canal, AK, 94
Dutton, AK, 51

Eagle Lake, CA, 222
Eagle Peak, CA, 218, 221
East Butte, OR, 206
East Cape, AK, 18
East Crater, WA, 166
East Dome, WA, 162
East Maar, AK, 65
East Molokai, HI, 328
Eastern Wrangell, AK, 91
Eaton Butte, OR, 188
Echo Rock, WA, 158
Edgecumbe, AK, 93
Edziza, Canada, 124
El Alto, NM, 295
Elk, OR, 187
Emigrant Butte, OR, 189
Emmons, AK, 52
Emmons Lake, AK, 52, 92

Espenberg, AK, 106
Eve Cone, Canada, 125

Falling Mountain, AK, 70
Faris, AK, 45
Fisher, AK, 46
Flag Point, OR, 175
Forgotten Crater, OR, 194
Forked Butte, OR, 179
Fort Rock, OR, 204
Fort Rock Basin, OR, 203
Fort Selkirk, Canada, 118
Fourpeaked, AK, 77
Fox, CA, 223
Fuji, OR, 187
Frailey Point, OR, 175
Freaner Peak, CA, 222
Fredericks Butte, OR, 206
Frog Leg Buttes, OR, 175
Frosty Peak, AK, 50
Fuego, OR, 196
Fumarole Butte, UT, 270

Gareloi, AK, 22
Garibaldi Lake, Canada, 143
Garibaldi, Canada, 144
George, CA, 228
Geronimo, AZ, 287
Gifford Peak, WA, 166
Gilbert, AK, 44
Glacier Peak, WA, 156
Glass (Long Valley), CA, 235
Glass (Medicine Lake), CA, 214
Glass Butte, OR, 206
Glass Creek, CA, 233
Goat Rocks, WA, 160
Goat Rocks (St. Helens), WA, 163
Gordon, AK, 91
Gordon Butte, OR, 175
Gosling, AK, 107
Grasshopper Point, OR, 175
Grassy, ID, 249
Great Sitkin, AK, 27
Grewingk, AK, 41
Griggs, AK, 72
Grouse Hill, OR, 194

Hackberry, AZ, 282
Hague, AK, 52
Haleakala, HI, 335
Halemaumau, HI, 344
Hamner Butte, OR, 188
Harter, OR, 169
Hat, CA, 218
Hat Creek Basalt, CA, 222
Hayes, AK, 84
Hayrick Butte, OR, 179
Heavey, CA, 223
Helen, CA, 218
Hells Half Acre, ID, 247
Helmet Peak, Canada, 130
Henrys Fork, WY, 264
Herbert, AK, 34
High Cascades, OR, 172, 175, 176, 178, 187, 189, 190
High Prairie, OR, 170
Highland Butte, OR, 171
Hillman Peak, OR, 193
Hogback, WA, 161
Hogg Rock, OR, 179
Hole-in-the-Ground, OR, 204, 206

Honeycomb Hills, UT, 269
Honolulu Vents, HI, 325
Hood, OR, 173
(East of Mount) Hood to Clear Lake, OR, 175
Hoodoo, Canada, 127
Hootnanny Point, OR, 175
Hopi Buttes, AZ, 283
Horseshoe Island, AK, 71
Hotlum, CA, 216
Howlock, OR, 191
Hualalai, HI, 337
Hunt's Hole, NM, 310

Ice Peak, Canada, 125
Ice Springs, UT, 272
Ilgachuz Range, Canada, 132
Iliamna, AK, 80
Imuruk, AK, 106
Indian Heaven, WA, 166
Inferno Cone, ID, 250
Ingakslugwat Hills, AK, 100
Ingariak Hills, AK, 102
(Ungulungwak Hill-)Ingrichuak Hill, AK, 100
Ingrisarak Mountain, AK, 102
Inskip Hill, CA, 225
Inyo Craters, CA, 233
Irish, OR, 187
Iron Mountain, OR, 207
Iron Mountain (Western Cascades), OR, 169
Isanotski, AK, 49
Iskut Canyon Cone, Canada, 128
Iskut-Unuk River Cones, Canada, 128
Island Park, WY, 264
Isleta, NM, 303
Itcha Range, Canada, 134

J Cone, NM, 302
Jackies Butte, OR, 211
Jarvis, AK, 91
Jefferson, OR, 177
(South of Mount) Jefferson to Santiam Pass, OR, 178
Jemez, NM, 295
Jordan Craters, OR, 210
Jornada del Muerto, NM, 309

Kagamil, AK, 36
Kaguyak, AK, 75
Kahoolawe, HI, 332
Kalaupapa, HI, 330
Kanaga, AK, 24
Kasatochi, AK, 28
Katmai, AK, 71
Kauhako, HI, 330
Kauai, HI, 321
Kaula, HI, 319
Kejulik, AK, 66
Kelsay Point, OR, 190
Kendrick Peak, AZ, 279
Khvostof, AK, 20
Kialagvik, AK, 62
Kilauea, HI, 344
Kilbourne Hole, NM, 310
King Creek, Canada, 129
Kinia, AK, 102
King's Bowl, ID, 250
Kiska, AK, 18
Kitasu Hill, Canada, 130

Kliuchef, AK, 29
Kochilagok Hill, AK, 101
Kohala, HI, 336
Kolob, UT, 275
Konia, AK, 29
Koniuji, AK, 28
Konocti, CA, 226
Kookooligit, AK, 104
Koolau, HI, 324
Korovin, AK, 29
Kostal Cone, Canada, 137
Koyuk-Buckland, AK, 107
Kukak, AK, 73
Kupreanof (Stepovak Bay), AK, 55
Kwolh Butte, OR, 186

Lakeview, OR, 189
Lanai, HI, 331
Larch, OR, 171
Lassen, CA, 216
Lassen: other cones, CA, 221
Lassen Peak, CA, 218
Lathrop Wells, NV, 262
Latour, CA, 221
Lava Butte, OR, 202
Lava Fork, Canada, 129
Lava Peak, AK, 43
Lava Point, AK, 44
Lehua Island, HI, 321
Lemei Rock, WA, 166
Lenz Butte, OR, 172
Leucite Hills, WY, 266
Level Mountain, Canada, 121
Lincoln Plateau, WA, 165
Little Black Peak, NM, 308
Little Brother, OR, 184
Little Hebe Crater, CA, 237
Little Packsaddle, OR, 187
Little Sitkin, AK, 20
Loa, UT, 275
Logan, CA, 222, 223
Lone Butte, WA, 166
Long Butte, OR, 206
Long Prairie, OR, 172
Long Valley, CA, 234
Los Lunas, NM, 303
Lost Jim, AK, 107
Lost Lake Butte, OR, 172
Lucero, NM, 305
Lunar Crater, NV, 258

McCartys, NM, 304
McCool Butte, OR, 188
McCoy, CO, 268
McCulloch, AK, 41
McKay Butte, OR, 206
McLoughlin, OR, 197
Magee, CA, 222
Mageik, AK, 67
Maiden Peak, OR, 187
Maidu, CA, 220, *224*
Maitland, Canada, 126
Maklaks, OR, 188
Makushin, AK, 41
Malpais (Kilbourne Hole), NM, 311
Malpais (Zuni-Bandera), NM, 304
Mammoth, CA, 234
Manderfield, UT, 274
Marble, WA, 167
Marble Mountain-Trout Creek Hill Zone, WA, 167

Martin, AK, 67
Mauna Kea, HI, 339
Mauna Loa, HI, 342
Mauna Ulu, HI, 345
Mauumae, HI, 327
Maxwell Butte, OR, 179
Mazama, OR, 193
Meager Mountain, Canada, 141
Medicine Lake, CA, 212
Menan Buttes, ID, 251
Merriam Cone, OR, 195
Mesa Chivato NM, 301
Metcalf Domes, AK, 41
Milbanke Sound, Canada, 130
Mill Creek, OR, 172
Mineral Mountains, UT, 273
Misery Hill, CA, 215
Mokuhooniki, HI, 330
Moffett, AK, 25
Mokua'aweoweo, HI, 343
Mono Craters, CA, 231
Montoso , NM, 299
Mora, NM, 293
Morale Claim, AZ, 282
Morgan, CA, 221
Mormon, AZ, 282
Morzhovoi, AK, 50
Mud Springs, CA, 229
Mule, OR, 190
Mullet Island, CA, 245

Nanwaksjiak, AK, 99
Nazko, Canada, 135
Nelson Island, AK, 102
Newberry, OR, 200
New Bogoslof, AK, 41
Niihua, HI, 320
North Cinder Peak, OR, 179
North Crater, ID, 249
North Pinhead, OR, 176
North Twin Peak, UT, 273
North Wilson, OR, 176
Northern Coulee, CA, 232
Novarupta, AK, 70
Nunivak, AK, 98
Nushkolik Mountain, AK, 101

Observation Rock, WA, 158
Obsidian Butte, CA, 245
Obsidian Domes, CA, 233
Ocate, NM, 293
Odell Butte, OR, 188
Okmok, AK, 38
Olallie Butte, OR, 177
Old Crow Tephra, AK, 92
O'Leary, AZ, 279
Opal Cone, Canada, 145
Otter, AK, 96

Packsaddle, OR, 187
Pakushin, AK, 42
Panum, CA, 232
Paramore, AZ, 288
Parkdale, OR, 172
Paulina Peak, OR, 201
Pavant Butte, UT, 271
Pavlof, AK, 53
Pavlof Sister, AK, 53
Pelican Butte, OR, 195
Peridot, AZ, 286
Phantom Cone, OR, 194
Pillar Butte, ID, 254
(The) Pinnacle, OR, 173
Pisgah, CA, 242
Pit(t), OR, 197
Pogromni, AK, 44
Pogromni's Sister, AK, 45
Point Kadin, AK, 42
Porcupine, AK, 108
Potrillo, NM, 311
Potrillo Maar, NM, 311
Powers, HI, 345
Pre-Lassen centers, CA, 219
Price, Canada, 143
Prindle, AK, 108
Prospect Peak, CA, 222
Punchbowl, HI, 327
Pu'u O'o, HI, 345
Puu Waawaa, HI, 338
Pyramid, Canada, 138
Pyre Peak, AK, 31

Quartz, OR, 206
Quemado, NM, 306

Ragged Jack, AK, 49
Rainbow Range, Canada, 131
Rainier, WA, 158
Raker Peak, CA, 218
Ranger Butte, OR, 188
Ray, OR, 188
Rayton–Clayton, NM, 292
Reading Peak, CA, 218
Recheschnoi, AK, 37
Red, WA, 166
Red Cinder Butte, OR, 190
Red Cone, OR, 191
Red Hill, NM, 306
Red Island, CA, 245
Red Knoll, UT, 274
Red Lake, CA, 221
Red Rock, CA, 221
Redcloud Cliff, OR, 194
Redondo Peak, NM, 297
Redoubt, AK, 81
Redtop, OR, 189
Reveille Range, NV, 260
Riley, NM, 311
Rock Hill, CA, 245
Rocky Peak, CA, 221
Rodley Butte, OR, 191
Roop, CA, 223
Round, AZ, 282
Round Head, AK, 24
Roundtop, AK, 49
Royce, OR, 188
Rudyerd Bay, AK, 95
Rush Hill, AK, 97

Saddle, OR, 196
Saddle Butte, OR, 209
St. George, AK, 97
St. Helens, WA, 161
St. Michael, AK, 103
St. Paul, AK, 96
Sajaka, AK, 23
Salal Glacier, Canada, 140
Saline Range, CA, 238
Salton Buttes, CA, 245
San Bernardino, AZ, 287
San Carlos, AZ, 286
San Felipe, NM, 299
San Francisco, AZ, 278
Sand Bay, AK, 27
Sand Crater, ID, 247
Sand Mountain, OR, 180
Sandy Glacier, OR, 173
Sanford, AK, 87
Santa Ana Mesa, NM, 295
Santa Clara, UT, 275
Santo Tomas, NM, 312
Sargents Ridge, CA, 215
Sarichef, AK, 29
Sawtooth, OR, 189
Sawtooth, WA, 166
Schreibers Meadow, WA, 155
Scott, OR, 193
Sea Lion Rock, AK, 96
Seguam, AK, 31
Segula, AK, 19
Semisopochnoi, AK, 21
Sentinel Plain, AZ, 287
Servilleta Basalt, NM, 290
Sham Hill, Canada, 140
Sharp Peak, OR, 194
Shasta, CA, 214
Shastina, CA, 215, 216
Sheldon–Antelope, NV, 256
Sheridan, OR, 185
Sherman Crater, WA, 155
Ship Rock, AK, 41
Shishaldin, AK, 48
Sierra Grande, NM, 292
Sifford, CA, 221, 224
Signal Peak, WA, 165
Silicic domes of south-central OR, 206
Silver Creek, OR, 207
Silverthrone, Canada, 138
Simcoe, WA, 165
Sisi Butte, OR, 176
(The) Sisters, AK, 104
Sitgreaves, AZ, 279
Sixbit Point, OR, 187
Sleeping Butte, NV, 262
Ski Heil Peak, CA, 218
Skookum Creek, AK, 91
Smelter Knolls, UT, 270
Snake River Plain, ID, 246
Snippaker, Canada, 129
Snow Mountain, CA, 220
Snow Peak, OR, 169
Snowy, AK, 73
Solo Creek, AK, 91
Sonoma, CA, 228
Sonya Creek, AK, 91
South Cinder Peak, OR, 179
South Inyo Crater, CA, 234
South Pinhead, OR, 176
South Twin Peak, UT, 273
Spectrum Range, Canada, 125
Split Butte, ID, 252
Sprague River Valley, OR, 196
Springerville, AZ, 284
Spurr, AK, 83
Squaw Ridge, OR, 206
Stams, OR, 206
Steamboat Springs, NV, 257
Steel Bay, OR, 194
Steller, AK, 73
Stepovak Bay, AK, 55
Stoneman Lake, AZ, 282
Stuart, AK, 104

Suemez Island, AK, 95
Sugar Bowl, WA, 162
Sugarloaf (Makushin), AK, 42
Sugarloaf (Semisopochnoi), AK, 21
Sugarloaf, AZ, 280
Sugarloaf, CA, 222
Summit Butte, OR, 176
Sunflower Flat, CA, 218
Sunset, ID, 249
Sunset Crater, AZ, 280
Sutter Buttes, CA, 225

Tabernacle Hill, UT, 271
Table, AZ, 282
Table, CA, 222
(The) Table, Canada, 144
Table Rock, OR, 205
Table Top, AK, 42
Tabor, OR, 171
Tahoma Peak, AK, 41
Takawangha, AK, 23
Tanada, AK, 90
Tanaga, AK, 23
Tantalus, HI, 327
Taos, NM, 290
Taylor, NM, 300
Taylor Butte, OR, 187, 196
Tea Table, OR, 206
Tehama, CA, 224
Teller, AK, 105
Tenas Peak, OR, 190
Tern, AK, 102
Thielsen, OR, 191
Thirsty Point, OR, 190
"385," AK, 99
Three Fingered Jack, OR, 179
Three Sisters, OR, 184
(South of) Three Sisters to Willamette Pass, OR, 187
(The) Thumb, Canada, 129
Timber Mountain, NV, 261
Tipsoo Peak, OR, 191
Tlevak Strait, AK, 95
Togiak, AK, 102
Toledo, NM, 295
Tolo, OR, 190
Triangle Peak, CO, 268
Trident, AK, 68
(Marble Mountain-)Trout Creek Hill Zone, WA, 167
Tschicoma, NM, 295
Tuber Hill, Canada, 140
Tumble Buttes, CA, 222
Turpentine Peak, OR, 179
Tuscan Formation, CA, 224
Tuya Butte, Canada, 119
Twin Calderas, AK, 107
Twin Craters, OR, 182
Twin, AK, 99
Twin Knolls, AZ, 285
Twin Peaks, UT, 273
(The) Twins, OR, 187

Ubehebe, CA, 237
Ugashik and Peulik, AK, 63
Uinkaret, AZ, 277
Ukinrek, AK, 65
Uliaga, AK, 36
Ungulungwak Hill-Ingrichuak Hill, AK, 100
(Iskut-)Unuk River Cones, Canada, 128

Valles, NM, 295
Valley of Ten Thousand Smokes, AK, 67, 70
Veeder, CA, 228
Veniaminof, AK, 56
Vent Mountain, 60
Volcano Mountain, Canada, 118
Vsevidof, AK, 37
Vulcan, NM, 302
Vulcan's Forge, AZ, 277
Vulcan's Throne, AZ, 277
Vulcan's Thumb, Canada, 142

Waianae, HI, 323
Waiele, HI, 327
Wailau (East Molokai), HI, 328
Walrus, AK, 96
Wapi, ID, 254
Wart Peak, OR, 205
Washington, OR, 181
Wells Gray-Clearwater, Canada, 137
West Crater, WA, 168
West Maar, AK, 65
West Maui, HI, 333
West Molokai, HI, 327
West Pinhead, OR, 176
West Prospect Peak, CA, 222
Westdahl, AK, 45
Western Cascades, OR, 169
Wests Butte, OR, 176
Whaleback, AK, 48
White, UT, 271
White Mountains Baldy, AZ, 285
White River Tephra, AK, 92
Wide Bay, AK, 42
Willamette Pass, OR, 187
Willamette Pass to Windigo Pass, OR, 189
Williams Crater, OR, 194
Williamson, OR, 187
Willow Peak, CO, 268
Wilson, OR, 176
Wind Mesa, NM, 303
Windigo Pass to Diamond Lake, OR, 190
Wolf, AZ, 285
Wootton's Cone, Canada, 118
Wrangell, AK, 88

Yamsay, OR, 206
Yana, CA, 220, 224
Yantarni, AK, 60
Yapoa Cone, OR, 182
Yellowstone Plateau, WY, 263
Yoran, OR, 189
Yukon Delta, AK, 99
Yunaska, AK, 33

Zanetti, AK, 88
Zuni-Bandera, NM, 303
Zuni Salt Lake, NM, 307